ACKNOWLEDGEMENTS

For a book that began in 2007, we need to thank lots of people who helped along the way. Jana Dean and Wayne Au were involved in advocating for and imagining this book in its early stages. Our "Earth in Crisis" curriculum workgroup in Portland contributed important ideas and feedback along the way. This group never had a formal membership list, but regular participants included Brady Bennon, Ken Gadbow, Mary Grover, Sylvia McGauley, Frank McGowan, Hyung Nam, Julie Treick O'Neill, Zach Post, and Greg Smith.

Thanks, too, to Rethinking Schools editorial associate Adam Sanchez, teacher at Madison High School in Portland, who invited Bill to co-teach a unit on climate change. Some of the curriculum included here was first tested in Adam's classes. Similarly, at Portland's Lincoln High School, where Tim teaches, we worked with Chris Buehler, Julie Treick O'Neill, and Matt Plies on the La Vía Campesina role play and other curriculum included here.

Many of the articles in the book first appeared in *Rethinking Schools* magazine. Rethinking Schools has a rigorous editorial process and articles are improved by the work of this collective, whose membership has included Wayne Au, Terry Burant, Linda Christensen, Kelley Dawson Salas, Grace Cornell Gonzales, Jesse Hagopian, Stan Karp, David Levine, Fred McKissack, Larry Miller, Hyung Nam, Bob Peterson, Adam Sanchez, Jody Sokolower, Melissa Bollow Tempel, Stephanie Walters, Dyan Watson, Kathy Xiong, and Moé Yonamine. Managing editor Jody Sokolower's editorial handiwork improved many of these articles as they made their way first into the pages of *Rethinking Schools*. Bill's colleagues at the Zinn Education Project—Deborah Menkart, Lauren Cooper, Bill Holtzman, Alison Kysia, and Katy Swalwell—offer ongoing wisdom and humor, and help this work connect with new audiences through the Zinn Education Project's growing reach.

Additional Rethinking Schools staff members that breathe life into the project have included Kris Collett, Tegan Dowling, Rachel Kenison, and Mike Trokan. Mike deserves special thanks for worrying about all the details from coordinating the book's budget to being the liaison with editors, art director, printers, indexer, and others we no doubt are unaware of. As the book heads to the printer, Mike is retiring. There is no way to overestimate how central Mike has been to every aspect of Rethinking Schools. He will still help out, but we will miss him dearly as our organizational rudder.

They are too numerous to list here, but everyone whose writing appears in the book graciously allowed us to use their words. We first discovered a number of these articles in *Orion* magazine. Special thanks go to *Orion*'s managing director Madeline Cantwell and managing editor Andrew Blechman for help in contacting *Orion* writers to secure reprint permissions. Similarly, many of the illustrators and photographers whose work we include here regularly contribute to *Rethinking Schools* and work for ridiculously little compensation. Thanks, too, goes to our excellent proofreader Lawrence Sanfilippo, who can spot a typo or misused comma at a hundred yards.

Greg Smith read an early manuscript of the book and provided valuable feedback, including suggestions of specific article excerpts we use in the book. Of course, Greg should not be held responsible for any of our final editorial decisions. Other advice and suggestions came from Sudia Paloma McCaleb, Desiree Hellegers, and Ryan Zinn.

When the two of us needed to get away from the distractions of daily life to focus on the book during the summer of 2013, Susan Katz, of Oregon Physicians for Social Responsibility, kindly lent us her beautiful home in Pacific City, Oregon. Looking out on the Pacific Ocean for a weekend provided a moment-to-moment reminder of the Earth's beauty—something we hope this book will help defend.

Jennifer Morales was the book's first project coordinator. Her feedback was essential as we talked through the big picture we were trying to assemble.

We were thrilled when Catherine Capellaro agreed to step in as production editor in the fall of 2013. Catherine is a former *Rethinking Schools* managing editor, a wonderful writer, and one of the most creative people we know. Catherine guided the project to completion with a sharp editorial and artistic eye, along with kindness and good humor.

Nancy Zucker, our brilliant art director at *Rethinking Schools*, worked late nights on this book while dealing with acts of nature and balancing work on the magazine. We are immensely grateful for Nancy's warmth, patience, and fine visual sensibility.

The artist Ricardo Levins Morales, who we've admired for years, designed the book's cover. We urge everyone to check out Ricardo's extraordinary work.

We are thankful for several individuals and foundations that helped fund this work. Jennifer Ladd, one of Bill's closest friends for decades, put us in touch with Gone Giving, a donor advised fund associated with the New Hampshire Charitable Foundation. Gone Giving provided support at a crucial point in the development of the book. Thanks, too, to the Lannan Foundation for its generous funding at a critical late stage in our work on the book. Other foundations and individuals supporting Rethinking Schools' work include the New Visions Foundation, the Pathfinder Fund, Mary Bills, Kristin Brown, Matt Damon, and Art Peterson.

Finally, Bill writes:

I want to thank my wife and partner in life, Linda Christensen. For 30 years, Linda has shown me what it means to teach for joy and justice. Her vision of teaching and learning finds its way into everything I write. For my part, this book is built on our hikes in Point Reyes, Sedona, the Columbia River Gorge, and in Portland's Forest Park; our conversations about the world and the classroom; and our love for one another, our daughters Anna and Gretchen Hereford, and our grandson Xavier King Hertel. For us, Xavier gives the future a face, a name, and an urgency.

Tim writes:

I'd like to thank Bill for inviting me to work with him on this project. His vision of teaching for justice and a better, more equitable world will always be the standard for which I strive. Working together on this book has given me the education of a lifetime, as well as a good friend.

Most of my work on this book was done while my mom, Karen Swinehart, provided the best childcare one could imagine for my daughters. Thank you, mom, for continuing to share your joy and warmth with the next generation of Swinehart children.

For any environmental ethos that I brought to this book, I owe credit to my wife, Emily Lethenstrom, who has patiently taught me the value of conservation and ecology through her daily actions for the last 15 years. I also want to thank Emily for sharing in the joys and challenges of parenting our daughters, Zadie and Mira. They remind us each day not only of the urgency of living in an age of environmental crisis, but also of the care, beauty, and hope that we must all embody as we forge ahead together toward a better future. ⊕

ABOUT THE EDITORS

Bill Bigelow began teaching high school social studies in Portland, Oregon, in 1978, and taught for almost 30 years. He now works as the curriculum editor for *Rethinking Schools* magazine and is co-director of the Zinn Education Project. He has authored or co-edited numerous books on teaching, including *Strangers in Their Own Country: A Curriculum on South Africa* (Africa World Press), *The Power in Our Hands: A Curriculum on the History of Work and Workers in the United States* (Monthly Review Press), *Rethinking Columbus, Rethinking Our Classrooms, Vols. 1 and 2, Rethinking Globalization: Teaching for Justice in an Unjust World, The Line Between Us: Teaching About the Border and Mexican Immigration*, and *A People's History for the Classroom* (all published by Rethinking Schools). He lives in Portland with his wife, Linda Christensen. ⊕

Tim Swinehart teaches social studies at Lincoln High School, in Portland, Oregon. He also teaches in the masters in teaching program at Lewis and Clark College. He is a frequent contributor to *Rethinking Schools* magazine and a longtime organizer with Portland Area Rethinking Schools and the Northwest Teaching for Social Justice Conference. He began community organizing as a graduate student in 2002 when he and his wife founded the Flagstaff Community Supported Agriculture (CSA) project in Arizona. He lives in Portland with his wife, Emily Lethenstrom, and two daughters, Zadie and Mira. ⊕

CONTENTS

Introduction

Chapter 1:
The Whole Thing Is Connected

3   Introduction

4   Plastics and Poverty
by Van Jones

8   Smarter Than Your Average Planet
by David Suzuki

11   Interconnectedness—The Food Web
by Kate Lyman

13   Reading the World in a Loaf of Bread
by Christian Parenti

15   A Deadly Drought
by Nash Colundalur

17   The World Turned Upside Down
by Leon Rosselson

18   Stealing and Selling Nature
by Tim Swinehart

24   The Commons
by David Rovics

25   Two Views of Nature
by Vandana Shiva

27   Principles of Environmental Justice

29   Teaching ideas

Chapter 2:
Grounding Our Teaching

35   Introduction

36   How My Schooling Taught Me Contempt
for the Earth
by Bill Bigelow

42   A Pedagogy for Ecology
by Ann Pelo

48   Lessons from a Garden Spider
by Kate Lyman

52   "Before Today, I Was Afraid of Trees"
by Doug Larkin

57   Exploring Our Urban Wilderness
by Mark Hansen

61   Looking for Justice at Turkey Creek
by Hardy Thames

67   Students Blow the Whistle on Toxic Oil
Contamination
by Larry Miller and Danah Opland-Dobs

Chapter 3:
Facing Climate Chaos

73   Introduction

74   Farewell, Sweet Ice
by Matthew Gilbert

77   Goodbye, Miami
by Jeff Goodell

79   Teaching the Climate Crisis
by Bill Bigelow

92   Climate Change Mixer
by Bill Bigelow

102   Climate Change Timeline

106   Proof Positive
by Robert Kunzig

109   Ask Yourself These Questions
by George Monbiot

110   Carbon Matters
by Jana Dean

117   Keeling Curve

118   Paradise Lost
by Brady Bennon

124 Retreat of Andean Glaciers Foretells
Global Water Woes
by Carolyn Kormann

127 "Don't Take Our Voices Away"
by Julie Treick O'Neill and Tim Swinehart

143 Climate Change in Kwigillingok
by Lauren G. McClanahan

147 The Thingamabob Game
by Bill Bigelow

154 Remember the Carbon Footprint of War
by Bruce E. Johansen

157 Polar Bears on Mission Street
by Rachel Cloues

161 The Big Talk
by Sandra Steingraber

163 Who's to Blame for the Climate Crisis?
by Bill Bigelow

171 Bali Principles of Climate Justice

174 Teaching Ideas

**Chapter 4:
Burning the Future**

178 "Black Waters"
by Jean Ritchie

179 Introduction

180 The Mystery of the Three Scary Numbers
by Bill Bigelow

191 A Matter of Degrees
by Bill McKibben

194 A Short History of the Three Ages of
Carbon—and the Dangers Ahead
by Michael T. Klare

Coal

200 Coal, Chocolate Chip Cookies, and
Mountaintop Removal
by Bill Bigelow

209 An Insult to the Moon
by Erik Reece

210 "They Can Bury Me in These Hills but I
Ain't Leavin'"
by Jeff Goodell

216 Coal at the Movies
by Bill Bigelow

220 Exporting Coal and Climate Change
by Bill Bigelow

226 This Much Mercury…
by Dashka Slater

Oil

231 Environmental Crime on Trial
by Brady Bennon

238 "We Know What's Goin' On"
by Terry Tempest Williams

241 Dirty Oil and Shovel-Ready Jobs
by Abby Mac Phail

Natural Gas and Fracking

248 Teaching About Fracking
by Julie Treick O'Neill

252 Fracking…Firsthand

255 Fracking Democracy
by Sandra Steingraber

258 Life and Death in the Frack Zone
by Walter Brasch

260 Divesting from Fossil Fuels

261 Teaching Ideas

Chapter 5:
Teaching in a Toxic World

267 Introduction

268 Forget Shorter Showers
by Derrick Jensen

271 Keep America Beautiful?
by Elizabeth Royte

273 Science for the People
by Tony Marks-Block

280 Combating Nail Salon Toxics
by Pauline Bartolone

283 Teaching About Toxins
by Kelley Dawson Salas

288 Reading Chilpancingo
by Linda Christensen

294 A Toxic Legacy on the Mexican Border
by Kevin Sullivan

297 Measuring Water with Justice
by Bob Peterson

303 Transparency of Water
*by Selene Gonzalez-Carillo and
Martha Merson*

309 Facing Cancer
by Amy Lindahl

Teaching in a Nuclear World

314 Outrageous Hope
by Gary Pace

316 "We Want to Stop It Now"
Fukushima's Nuclear Refugees

319 "Kazue, Alive!"
*Hiroshima's Nuclear Refugees
by Kazue Miura*

321 Uranium Mining, Native Resistance, and
the Greener Path
by Winona LaDuke

324 Teaching Ideas

327 Valuable Films on Our Toxic Nuclear
Legacy

Chapter 6:
Food, Farming, and the Earth

331 Introduction

332 Food Secrets
by Michi Thacker

338 Got Milk, Got Patents, Got Profits?
by Tim Swinehart

345 Greening for All
by Marcy Rein and Clifton Ross

348 King Corn
by Tim Swinehart

354 "We Have the Right..."
Youth Food Bill of Rights

355 Hunger on Trial
by Bill Bigelow

361 10 Myths About Hunger
*by Frances Moore Lappé
and Joseph Collins*

366 Food, Farming, and Justice
*by Bill Bigelow, Chris Buehler,
Julie Treick O'Neill, and Tim Swinehart*

386 Teaching Ideas

**389 Afterword:
Resources for the Earth**

397 Index

Rethinking Schools works to credit all artists and photographers appropriately. Please let us know if we have made an error.

INTRODUCTION

It's hard to say where the idea for this book originated. It may have been in 2007 when we looked at *Modern World History*, the new global studies textbook our school district, in Portland, Oregon, purchased. The book began one of its three miserable paragraphs on the climate crisis with the statement "Not all scientists agree with the theory of the greenhouse effect." And it was buried on page 679. This was the best that Portland could offer its high school students? (This widely adopted book, published by Holt McDougal, still anchors the official curriculum for Portland high school students' sole class on today's world.)

Or this book's origins may have been at an excellent teach-in sponsored by the International Forum on Globalization in Washington, D.C., that same year, called "Confronting the Triple Crisis," about climate change, the end of cheap energy, and resource depletion and extinction. A number of the contributors to this book presented at this extraordinary gathering: Vandana Shiva, Frances Moore Lappé, Bill McKibben, Michael Klare, and Jeff Goodell. We came away from that weekend convinced of the enormity of the crisis, but we also understood how each supposedly distinct crisis linked to all the others, and then tied back to the fundamental problem of a global economy driven by the quest for profit. The teach-in was our introduction to Annie Leonard's short film *The Story of Stuff*, which captures many of these connections with humor and common sense.

The decision to launch this book—and how we imagined it—was no doubt heavily influenced by the powerful and interconnected analyses offered by the speakers at this teach-in. But we were dismayed that there was no discussion about what this all meant for K–12 education. How should environmental justice movements partner with the educators who work daily with the millions of young people learning their ecological ABCs—or, perhaps too often, not learning them? Implicitly, the conference suggested that this was knowledge to be shared among adults.

We left inspired and informed, but weighed down by the immense burden of figuring out how to "story" the environmental crisis through curriculum.

Back home in Portland we initiated what we called an "Earth in Crisis" curriculum group, and invited colleagues to discuss and test out teaching ideas with one another. This collective nurtured many of the activities included in this book, and also identified key themes that weave through the book. One of these is that **our curriculum must confront the false dichotomy between the environment and people.** It's a theme that Van Jones addresses directly in his TED Talk on "Plastics and Poverty," included in Chapter One (p. 4). Jones points out that people were rightly concerned about the damage to living systems in the Gulf of Mexico caused by the BP oil spill. But he notes that we often do not seem as concerned when that oil gets to where it is "supposed" to go: for example, to petrochemical plants that dot Cancer Alley between Baton Rouge and New Orleans, where it then poisons the largely poor and African American people who live there. Yes, the "environment" is about polar bears, dolphins, redwood forests, and bees; but it is also about human beings—workers, consumers, families, and community members. We call this book a *people's* curriculum for the Earth because we try to keep the focus on the inextricable link between nature and people.

> Yes, the "environment" is about polar bears, dolphins, redwood forests, and bees; but it is also about human beings—workers, consumers, families, and community members.

And this suggests another theme that emerged in our Earth in Crisis curriculum work in Portland: **Everyone on Earth is affected by the environmental crisis, but we are affected unequally—based on race, class, nationality, or location.** This is mad-

"Is This Book Biased?"

When Rethinking Schools published *Rethinking Globalization: Teaching for Justice in an Unjust World* in 2002, we asked the question: "Is this book biased?" We answered by quoting the great historian and activist Howard Zinn: "In a world where justice is maldistributed there is no such thing as a neutral or representative recapitulation of the facts."

Similarly, we cannot be neutral about the environmental crisis. As articles in this volume point out, to prevent the climate from heating up more than two degrees Celsius (3.6 degrees Fahrenheit), about 80 percent of coal, oil, and gas reserves need to stay in the ground. That's not our opinion. That's the science. (And even this internationally agreed-upon goal dooms species to extinction and island nations to oblivion.) But energy companies have no intention of "stranding" some $20 trillion of assets. They plan to endlessly blast and scrape and dig and drill. The possibility of a habitable Earth is fundamentally incompatible with the business plan of the fossil fuel industry.

So, yes, this book takes sides. As the book's title announces, the curricular materials we provide here are *for* the Earth. We advocate not so much teaching *about* fossil fuels as teaching *against* fossil fuels—and against a system that privatizes decisions on which the future depends.

We are partisan in favor of communities in Kentucky and West Virginia whose health is compromised by the coal industry's mountaintop removal. We are partisan in favor of Indigenous communities around the globe who have so often found themselves victimized by "development." And we are partisan in favor of our own children and grandchildren, who we hope can live in a world that doesn't poison them when they drink the water, breathe the air, or make a living.

In compiling the materials in *A People's Curriculum for the Earth* we did not seek to provide "equal time" to free market proponents whose policy agenda promises more environmental degradation. These perspectives are widely disseminated by the mainstream media—including the textbooks produced by huge corporations. Ours is not a point-counterpoint curriculum that aims to appear "balanced." But nor do we advocate shielding students from perspectives at odds with ours. To the contrary. We believe it's important for students to encounter and wrestle with multiple viewpoints, especially those that advocate business as usual.

We believe there's an important difference between being partisan and being biased. A bias means being unwilling to examine or express one's own premises. And that's exactly what we find in mainstream textbooks: a failure to ask students to consider root causes and to ask critical questions about the ideologies and economic forces that led us to our current predicament.

Partisan teaching takes sides on behalf of the Earth. To paraphrase *Rethinking Globalization*, partisan teaching alerts students to social and environmental injustice, seeks explanations, and encourages activism.

Our collective house is on fire; it's not "biased" to want to put it out. ●

deningly evident with the impact of climate change. Throughout the book we feature stories about individuals, and communities—Matthew Gilbert and the Gwich'in (p. 74), Enele Sopoaga in Tuvalu (p. 96), Anisur Rahman of Antarpara, Bangladesh (p. 98), the Aymara people of Bolivia (p. 137), the Yup'ik teenagers of Kwigillingok, Alaska (p. 143), and too many others to list, whose carbon footprint is virtually nonexistent and yet who are among the first to suffer from its ravages. Similar issues of race and class are at play when it comes to exposure to workplace pollutants ("Combating Nail Salon Toxics," p. 280), lead poisoning of children in urban areas ("Teaching About Toxins," p. 283), or the pollution from mostly foreign-owned manufacturing plants that blankets poor communities around the world with deadly consequences ("Reading Chilpancingo," p. 288).

This is not to say that people are not organizing in response to this toxic trespass, in the expression of ecologist Sandra Steingraber. They are. And some of them are featured in these pages: the Milwaukee students who blew the whistle on oil contamination in their neighborhood (p. 67), Maria Gunnoe's passionate anti-mountaintop removal activism with communities in West Virginia (p. 210), the Indigenous people described in Winona LaDuke's "Uranium Mining, Native Resistance, and the Greener Path" (p. 321)—"resilient in the face of a deep history of genocide and destruction." But there is a fundamental inequality at the heart of the environmental crisis—one that is central to the articles and teaching activities included in this book.

Shorter Showers?

In our Earth in Crisis group, teachers kept returning to our students' responses: They wanted to know what they could do personally. Early in our work, we concluded that **we need to help students recognize the inadequacy of responding to the environmental crisis solely as individuals**. As we mention in the teaching ideas for Chapter Three, "Facing Climate Chaos" (p. 174), there are entire books that urge students to consider their *individual* carbon footprints, suggesting that our personal patterns of consumption are a root cause of global warming. Students are urged to think about the frequency of their baths, their electricity use, the stuff they buy. Yes, of course, we want young people—and everyone—to be mind-

ful of the Earth as we go through our daily lives. And we want students to recognize the power they have—collectively or individually—to make the world a better place. But it's wrong to direct students primarily toward individual solutions to create change.

In his Chapter Five essay, "Forget Shorter Showers," Derrick Jensen confronts this problematic celebration of individual action:

> Consumer culture and the capitalist mindset have taught us to substitute acts of personal consumption (or enlightenment) for organized political resistance. Al Gore's film *An Inconvenient Truth* helped raise consciousness about global warming. But did you notice that all of the solutions presented had to do with personal consumption—changing lightbulbs, inflating tires, driving half as much—and had nothing to do with shifting power away from corporations, or stopping the growth economy that is destroying the planet?

As students' awareness of the environmental crisis grows, this consciousness can be misdirected by social forces that have an interest in how young people respond. The energy industry would much prefer that our students change their lightbulbs, recycle their soda cans, or even install solar panels than organize a demonstration at the state capitol to shut a coal-fired power plant, testify at a public hearing against fracking, or otherwise gum up their fossil fuel machinery.

And there is another way that this celebration of the individual needs to be questioned in a people's curriculum for the Earth. Individual property "rights" have long been seen as synonymous with "liberty." "Liberty! Property!" was a cry of the American Revolution. But there were other more democratic cries as well, like Benjamin Franklin's famous assertion that "Private Property… is a Creature of Society, and is subject to the Calls of that Society, whenever its Necessities shall require it, even to its last Farthing."

> There is a fundamental inequality at the heart of the environmental crisis—one that is central to the articles and teaching activities included in this book.

What happens to the Earth if we respect the "right" of the fossil fuel industry to manage their assets however they please? More and more, the headlines are filled with the answer to that question: superstorms, drought, heat waves, melting glaciers, ocean acidification, species extinction, floods, drowning islands. A curriculum on the climate, and the environmental crisis more broadly, needs to address patterns of ownership and decision-making. **Our curriculum needs to confront the myth that private property is, in fact, private.** The fate of the Earth "belongs" to us all.

Capitalism

Helping students acquire a critical consciousness about the environmental crisis means we need to consistently encourage them to ask "Why?" Why is it that the future of life on Earth has been put at risk? It seems an impossible question to answer unless we **engage students in thinking about the nature of global capitalism.** Throughout the book, we draw students' attention to this broader systemic context within which the environmental crisis is unfolding. Activities like "The Thingamabob Game" (p. 147) and the trial role play "Who's to Blame for the Climate Crisis?" (p. 163) explicitly confront students with the fundamental clash between an economic system that prizes wealth accumulation above all else and people's need for a healthy environment. Capitalism insists that key productive decisions be made on the basis of what will yield the greatest profit. It grants godlike powers to unelected elites whose livelihoods depend not on creating a world of equality and environmental sustainability, but on making the most money. If we're going to help our students not just describe, but explain, the environmental crisis, it is essential that educators name this elephant in our classrooms.

Joy amid Crisis

As this book heads to the printer, the Intergovernmental Panel on Climate Change (IPCC) is about to release what news outlets indicate is its most dire report to date—another in a string of reports, each with more urgent language and frightening scenarios than the one before. The new IPCC report warns that at least three-quarters of known fossil fuel re-

serves must remain in the ground if we are to avoid a 3.6-degree Fahrenheit (two-degree Celsius) rise in global temperatures over preindustrial times (see "The Mystery of the Three Scary Numbers, p. 180 and "A Matter of Degrees," p. 191). The consequence of exceeding these limits would "almost certainly have catastrophic effects, including a mass extinction of plants and animals, huge shortfalls in food production, extreme coastal flooding, and many other problems," according to the *New York Times,* which received a draft of the report.

The news is bad. But **despite the dimensions of the environmental crisis, students can approach this frightening content in ways that are lively and playful**. Not long ago, we participated in a weeklong teach-in for 6th through 8th graders about energy issues at Sunnyside Environmental School, a public school here in Portland. Throughout the week, students heard speakers and participated in activities about everything from mountaintop removal coal mining to catastrophic oil spills to the civilization-threatening consequences of climate change. They also encountered people working on solar and wind power, local food initiatives, and other innovative responses to environmental challenges; but the week definitely offered an adult dose of planetary crisis. Nonetheless, in classrooms we visited during the concluding activist projects that students worked on, these middle schoolers were anything but grim; and their small-group work was electric with idea sharing and laughter. As with adults, we've found that students are able to live with contradiction; students grasp the sadness and injustice at the heart of the environmental crisis while finding joy and humor. For the book, we've selected activities that address key environmental concerns, but these activities do not invite despair. They are engaging, and feature collective work that triggers student playfulness and imagination.

Interconnections

Throughout the final stages of working on this book, we collaborated with Portland teaching colleagues Chris Buehler, Julie Treick O'Neill, and Matt Plies on a role play about La Vía Campesina. Despite the fact that La Vía Campesina may be the largest social movement in the world—with more than 200 million small farmers in its affiliated organizations—it's pretty much impossible to find its work described in today's mainstream textbooks. We conclude *A People's Curriculum for the Earth* with La Vía Campesina efforts because we think that **it highlights the way a deep response to any one crisis—for example, how to feed a world populated by perhaps a billion hungry people—addresses other social and environmental crises.** La Vía Campesina presents a grassroots, agroecological challenge to agribusiness globalized, free market, chemical-drenched, genetically modified prescription for the world's food production. The peasant movement shows that addressing hunger can simultaneously address climate change, inequality, public health, unemployment, forced migration, and much more. These are the kind of interconnections that infuse our curricula with hope—offering students the sense that fundamental change is not only desperately needed but also possible.

Challenging Curricular Apartheid

The teaching we observed at Sunnyside Environmental School showed us what happens when teachers collaborate across disciplines. Unfortunately, in too many schools, the environmental crisis seems to have become a kind of curricular hot potato. No discipline wants to claim the crisis as its own. We get it. We are both high school social studies teachers and we often bump up against our own shaky grasp on scientific concepts, trying to recall details from past biology and chemistry classes. While teaching one climate lesson at Lincoln High School, a student made an assertion about the impact of methane versus carbon dioxide that stumped us both and sent us combing through IPCC reports that evening. We try not to let these moments force us to retreat into the silo that traditionally has been considered social studies. And we've spoken with science teachers who feel that analyzing the social causes and effects of climate change reaches beyond their curricula or of their own knowledge. Similarly, teachers in language arts, mathematics, world languages, business, physical education, or art may wonder "What does this have to do with my class?"

But **in this moment of crisis, it's imperative that we reject artificial barriers between disciplines.** Throughout this book we've featured stories from educators who consciously cross conventional

curricular boundaries—see for example, "Carbon Matters" (p. 110), "Science for the People" (p. 273), "Measuring Water with Justice" (p. 297), and "Facing Cancer" (p. 309). Throughout the curriculum, educators can collaborate to help students become the scientist-activists they need to be. Confronting the toxic injustice that has become one of the defining features of our time requires us immediately to begin constructing a fossil fuel-free world built on principles of ecology and justice, rather than profit and endless growth. No matter which classes we teach, educators need to find ways to help young people develop the analytical tools to understand the causes of the environmental crisis and to exercise their utopian imaginations to consider alternatives.

> The peasant movement shows that addressing hunger can simultaneously address climate change, inequality, public health, unemployment, forced migration, and much more.

Political and Educational Context

In an article in the *Guardian*, Naomi Klein, author of *This Changes Everything: Capitalism vs. the Climate*, laments the "bad timing" of the climate crisis:

> Our problem is that the climate crisis hatched in our laps at a moment in history when political and social conditions were uniquely hostile to a problem of this nature and magnitude—that moment being the tail end of the go-go '80s, the blast-off point for the crusade to spread deregulated capitalism around the world. Climate change is a collective problem demanding collective action the likes of which humanity has never actually accomplished. Yet it entered mainstream consciousness in the midst of an ideological war being waged on the very idea of the collective sphere.

That same war has been waged in the education arena. At the precise moment we need our schools to educate and engage the next generation about the historic global challenges we face, public education is under attack from the same private and corporate interests that have polluted our natural and social environments. Curriculum is being standardized and narrowed to what can be poorly measured by bubble tests. Decisions about what schools should teach and children should learn are being moved away from classrooms and communities to the same politicized bureaucracies and monied interests that are undermining democracy. This too is "bad timing." At a time when we need an urgent national conversation about how schools and curriculum should address the environmental crisis, we're being told that the problems we need to focus on are teacher incompetence, government monopoly, and market competition. The reform agenda reflects the same private interests that are moving to shrink public space—interests that have no desire to raise questions that might encourage students to think critically about the roots of the environmental crisis, or to examine society's unsustainable distribution of wealth and power.

> **For educators, this is the curriculum work of our lives. And, yes, it is a fight, too.**

* * *

This book is not so much "a people's curriculum for the Earth" as it is an invitation to begin to build that curriculum. And it's encouragement to educators to demand the right to effect a curriculum that honestly and deeply addresses the environmental crisis. Some of this work will go on in our classrooms; in meetings with other teachers; in teacher social justice conferences in San Francisco, New York, Milwaukee, Atlanta, Chicago, and Seattle; in our professional organizations; in the pages of *Rethinking Schools* magazine; and at the Zinn Education Project and This Changes Everything websites. And some will go on in our unions, community organizations, and other activist organizations where we fight to teach about crucial issues in the world.

The intertwined social, economic, and environmental crises that confront humanity require us to be audacious. As Naomi Klein writes, this is "the fight of our lives." For educators, this is the curriculum work of our lives. And, yes, it is a fight, too. We need to demand and organize for the right to teach about what really matters, and not be forced to toe the textbook line or obey "rigorous" standards, developed afar, that may or may not help students appreciate and act on this moment in history.

We educators need to imagine, cooperate, create, hope—and at times, defy and resist. And we need to see ourselves as part of a broader movement to build the kind of society that is clean and just and equal and democratic. One that seeks to leave the world better than we found it. ⊕

CHAPTER ONE: The Whole Thing Is Connected

SIMONE SHIN

"It really boils down to this: that all life is interrelated. We are all caught in an inescapable network of mutuality, tied into a single garment of destiny. Whatever affects one destiny, affects all indirectly."

—Martin Luther King Jr.

NASA.GOV

INTRODUCTION
The Whole Thing Is Connected

If we had to distill the articles and teaching activities in *A People's Curriculum for the Earth* down to one key point, it's that "everything is connected."

In his article "Smarter Than Your Average Planet," David Suzuki writes, "Interwoven, interlinked, joined, chained, bonded, coupled—no matter how you describe it, everything in nature seems somehow connected to everything else." Suzuki emphasizes the peril of thinking that there is some *away* where we can pollute without consequence. The hopeful piece of this ecological "an injury to one is an injury to all" is that everyone—everywhere—has an opportunity to make a positive planetary difference.

Van Jones uses the issue of plastic to point out that we may be connected, but we are not connected *equally*. Jones argues we are linked to cycles of production and consumption that are toxic—to nature and also to people. This toxicity falls hardest on poor people and people of color. And he chides environmentalists for often caring more about pollution's effect on nature than on human beings. According to Jones, our society treats *stuff* as disposable, but it also treats *people* as disposable—and both are part of the same process.

Christian Parenti extends the "everything is connected" concept to climate change, charting the chain of events from drought to the price of bread to uprisings in the Middle East. It's just one example. Corporations mining coal in Wyoming or chopping down rain forests in Indonesia similarly contribute to desertification in East Africa and thus to armed conflicts between livestock herders. It's another instance where people are connected, but connected unfairly, as those least responsible for pollution are the ones to suffer the most.

Indian physicist, environmentalist, and philosopher Vandana Shiva contrasts two views of nature—one that denies people's reliance on and connection to the Earth, and the other that affirms them. Shiva critiques the notion that the Earth is dead matter: "The death-of-nature idea allows a war to be unleashed against the Earth." By contrast, she celebrates the Indian Chipko movement: "Women knew that the real value of forests was not the timber from a dead tree, but the springs and streams, food for their cattle, and fuel for their hearths."

And here lies the choice this first chapter highlights: We can continue the "enclosure of the commons," begun so long ago—the privatization and commodification of nature—or we can recognize the fundamental truth that we are all connected and that there is nothing "private" about how we treat the Earth, or each other. ⊕

ERIK RUIN

Plastics and Poverty

Why the whole thing is connected

BY VAN JONES

We've been talking a lot about the horrific impacts of plastic on the planet and on other species, but plastic hurts people too, especially poor people. In the production of plastic, the use of plastic, and the disposal of plastic, the people who have the bull's-eye on their foreheads are poor people. People got very upset when the BP oil spill happened—for very good reason. People thought, "Oh, my God. This is terrible, this oil. It's in the water. It's going to destroy the living systems there. People are going to be hurt. This is a terrible thing, that the oil is going to hurt the people in the Gulf."

What people don't think about is what if the oil had made it safely to shore. What if the oil actually got where it was trying to go? Not only would it have been burned in engines and added to global warming, but there's also a place called Cancer Alley. The reason it's called Cancer Alley is because the petrochemical industry takes that oil and turns it into plastic and, in the process, kills people. It shortens the lives of the people who live there in the Gulf. So oil and petrochemicals are not just a problem when there's a spill; they're a problem when there's not. And what we don't often appreciate is the price that poor people pay for us to have these disposable products.

The other thing that we don't often appreciate is it's not just at the point of production that poor people suffer. Poor people also suffer at the point of use. Those of us who earn a certain income level, we have something called choice. The reason why you want to work hard and have a job and not be poor and broke is so you can have choices, economic choices. We actually get a chance to choose not to use products that have dangerous, poisonous plastic in them. Other people who are poor don't have those choices. So low-income people often are the ones who are buying the products that have those dangerous chemicals in them that their children are using. Those are the people who wind up ingesting a disproportionate amount of this poisonous plastic and using it. And people say, "Well, they should just buy a different product." The problem with being poor is you don't have those choices. You often have to buy the cheapest products. The cheapest products are often the most dangerous.

And if it weren't bad enough, if it wasn't just the production of plastic that's giving people cancer and shortening lives and hurting poor kids at the point of use, at the point of disposal—once again—it's poor people who bear the burden. Often, we think we're doing a good thing. You're in your office and you're drinking your bottled water—or whatever it is—and

you think to yourself, "Hey, I'm going to throw this away. No, I'm going to be virtuous. I'm going to put it in the blue [recycling] bin." And then you look at your colleague and say, "Why, you cretin: You put yours in the white bin." And we use that as a moral tickle. We feel so good about ourselves.

Maybe I'll forgive myself. Not you, but I feel this way. And so we kind of have this kind of moral feel-good moment.

But if we were able to follow that little bottle on its journey, we would be shocked to discover that, all too often, that bottle is going to be put on a boat. It's going to go all the way across the ocean at some expense. And it's going to wind up in a developing country—often China. I think in our minds we imagine somebody's going to take the little bottle and say, "Oh, little bottle. We're so happy to see you, little bottle. You've served so well." He's given a little bottle massage, a little bottle medal. And we ask, "What would you like to do next?" The little bottle says, "I just don't know."

But that's not actually what happens. That bottle winds up getting burned. Recycling of plastic in many developing countries means the incineration of the plastic, the burning of the plastic, which releases incredibly toxic chemicals and, once again, kills people. And so poor people who are making these products in petrochemical centers like Cancer Alley; poor people who are consuming these products disproportionately; and then poor people, who even at the tail end of the recycling are having their lives shortened, are all being harmed greatly by this addiction we have to disposability.

> **Low-income people often are the ones who are buying the products that have those dangerous chemicals in them that their children are using.**

Now you think to yourself—because I know how you are—you say, "That sure is terrible for those poor people. It's just awful, those poor people. I hope someone does something to help them." But what we don't understand is: Here we are in Los Angeles. We worked very hard to get the smog reduction happening here in Los Angeles. But guess what? Because they're doing so much dirty production in Asia now, because the environmental laws don't protect the people in Asia now, almost all of the clean air gains and the toxic air gains that we've achieved here in California have been wiped out by dirty air coming over from Asia. So we all are being hit. We all are being impacted. It's just that the poor people get hit first and worst. But the dirty production, the burning of toxins, the lack of environmental standards in Asia is actually creating so much dirty air pollution it's coming across the ocean and has erased our gains here in California. We're back where we were in the 1970s. And so we're on one planet, and we have to be able to get to the root of these problems.

The root of this problem, in my view, is the idea of disposability itself. You see, if you understand the link between what we're doing to poison and pollute the planet and what we're doing to poor people, you arrive at a very troubling—but also very helpful—insight: *In order to trash the planet, you have to trash people. But if you create a world where you don't trash people, you can't trash the planet.* So now we are at a moment of the coming together of social justice and ecology as an idea. We finally can see that they are really, at the end of the day, one idea. And it's the idea that we don't have disposable anything. We don't have disposable resources. We don't have disposable species. And we don't have disposable people either. We don't have a throwaway planet, and we don't have throwaway children—it's all precious.

> One out of every four people locked up anywhere in the world is locked up right here in the United States. So that is consistent with this idea that disposability is something we believe in.

And as we all begin to come back to that basic understanding, new opportunities for action begin to emerge. Biomimicry, which is something that is an emerging science, winds up being a very important social justice idea. Biomimicry means respecting the wisdom of all species. Democracy, by the way, means respecting the wisdom of all people—and we'll get to that. But it turns out we're a pretty clever species. This big cortex: We're pretty proud of ourselves. But if we want to make something hard, we come up, "I know, I'm going to make a hard substance. I know, I'm going to get vacuums and furnaces and drag stuff out of the ground and get things

hot and poison and pollute, but I got this hard thing. I'm so clever." And you look behind you, and there's destruction all around you. But guess what? You're so clever, but you're not as clever as a clam.

A clamshell is hard. There's no vacuums, there's no big furnaces, there's no poison, there's no pollution. It turns out that our other species has figured out a long time ago how to create many of the things that we need using biological processes that nature knows how to use well. That insight of biomimicry, of our scientists finally realizing that we have as much to learn from other species—I don't mean taking a mouse and sticking it with stuff. I don't mean it from that way: abusing the little species—I mean actually respecting them, respecting what they've achieved. That's called biomimicry, and that opens the door to zero waste production, zero pollution production— that we could actually enjoy a high quality of life, a high standard of living without trashing the planet.

That idea of biomimicry, respecting the wisdom of all species, combined with the idea of democracy and social justice, respecting the wisdom and the worth of all people, would give us a different society. We would have a different economy. We would have a green society that Dr. King would be proud of. That should be the goal. And the way that we get there is to first of all recognize that the idea of disposability not only hurts the species we've talked about, but it even corrupts our own society as well.

In California, we lead the world in some of the green stuff; we also, unfortunately, lead the world in some of the gulag stuff. California has one of the highest incarceration rates of all the 50 states. We have a moral challenge in this moment. We are passionate about rescuing some dead materials from the landfill, but sometimes not as passionate about rescuing living beings, living people. And I would say that we live in a country—5 percent of the world's population, 25 percent of the greenhouse gases, but also 25 percent of the world's prisoners. One out of every four people locked up anywhere in the world is locked up right here in the United States. So that is consistent with this idea that disposability is something we believe in.

And yet, as a movement that has to broaden its constituency, that has to grow, that has to reach out beyond our natural comfort zone, one of the challenges to the success of this movement—of getting rid of things like plastic and helping the economy shift—is people look at our movement with some suspicion. And they ask a question, and the question is: How can these people be so passionate? A poor person, a low-income person, somebody in Cancer Alley, somebody in Watts, somebody in Harlem, somebody on an Indian reservation, might say to themselves, and rightfully so, "How can these people be so passionate about making sure that a plastic bottle has a second chance in life, or an aluminum can has a second chance, and yet, when my child gets in trouble and goes to prison, he doesn't get a second chance?" How can this movement be so passionate about saying we don't have throwaway stuff, no throwaway dead materials, and yet accept throwaway lives and throwaway communities like Cancer Alley? And so we now get a chance to be truly proud of this movement. When we take on topics like this, it gives us that extra call to reach out to other movements and to become more inclusive and to grow. And we can finally get out of this crazy dilemma that we've been in.

Most of you are good, softhearted people. When you were younger, you cared about the whole world, and at some point somebody said you had to pick an issue: You had to boil your love down to an issue. Can't love the whole world—you've got to work on trees, or you've got to work on immigration. You've got to shrink it down and be about one issue. And really, they fundamentally told you, "Are you going to hug a tree, or are you going to hug a child? Pick." Well, when you start working on issues like plastic, you realize that the whole thing is connected, and luckily most of us are blessed to have two arms. We can hug both. ⊕

Van Jones first presented "The Economic Injustice of Plastic" as a TED Talk in November 2010. Jones is founder of the Ella Baker Center for Human Rights, based in Oakland, California, and of Green for All, an NGO dedicated to "building an inclusive green economy strong enough to lift people out of poverty." He is the author of The Green Collar Economy: How One Solution Can Fix Our Two Biggest Problems. *Find this and more than 1,600 other TED talks at www.ted.com.*

See teaching ideas for this article, page 29.

Compare the six days of the Book of Genesis to the 4 billion years of geologic time. On that scale, one day equals about 666 million years. All day Monday until Tuesday noon Creation was busy getting the Earth going. Life began on Tuesday noon and the beautiful organic wholeness developed over the next four days. At 4 p.m. Sunday, the big reptiles; five hours later when the redwoods appeared there were no more big reptiles. At three minutes before midnight man appeared. One quarter of a second before midnight Christ arrived. At 1/40 of a second before midnight the Industrial Revolution began. We are surrounded by people who think that what we have been doing for 1/40 of a second can go on indefinitely. They are considered normal. But they are stark raving mad.

—David Brower
Founder, Sierra Club and
Friends of the Earth
From the film *Earth and
the American Dream*

ROGER PEET

Smarter Than Your Average Planet

Interconnections in the biosphere

BY DAVID SUZUKI

Interwoven, interlinked, joined, chained, bonded, coupled—no matter how you describe it, everything in nature seems somehow connected to everything else. From the symbiotic relationships between soil microorganisms and plants to marine nitrogen isotopes that end up in cedar trees via spawning salmon that are then eaten and distributed through the forest by bears, a precious few degrees of separation may be all that lie between any two parts of the planet, no matter how geographically isolated they may seem.

That puts humanity in a bit of a spot. We've relied on our pollution "just going away" for thousands of years. That doesn't really work anymore. It's finally caught up with us. Some of the modern pollutants we create stick around for a long time and can end up in our water, our food, and our blood. Greenhouse gases like carbon dioxide from our homes, factories, and industries linger in the atmosphere for hundreds of years, continuing to warm the planet long after we've stopped emitting them. Modern technologies that allow us to fish the deepest parts of the oceans, cut down the most isolated forests, and build cities that sprawl to the horizon are also gradually winking out the life-forms that have together made our planet habitable.

Although many of these actions may be economically efficient in the short term, they all have repercussions. And because our knowledge of how nature works is so incomplete, many of those repercussions are unforeseen and often unwelcome. We may think that eliminating one species is just a drop in nature's ample bucket, but we usually have no idea of how that creature fits into the web of life or what will happen when the web is broken.

In some ways, we are victims of the planet's own incredible generosity. Earth provided the ideal conditions, perhaps the only conditions, in which a species like ours might be able to evolve. We aren't exactly the most robust of creatures. We can't run terribly fast or jump very high. We lack fur to protect us from the cold. We drown if left in the water for very long. We even lack basic hunting and defensive attributes such as claws or sharp teeth.

But we do have big brains. And we used our big brains to take advantage of all that nature provided—the stable climate, the regularity of the seasons, the patterns of game migrations. We learned and took every advantage of every opportunity we could to get ahead.

So when it made sense to burn wood, then animal oils, then fossil fuels for energy, we did. And when we saw that building chimneys would send the smoke away from our homes, we built them. And when we realized that building them higher sent the pollution even farther away, we made them higher. We created a society that took full advantage of every single one of nature's services. Our success at doing this allowed our populations to soar and our economies to grow.

And when examined in isolation, each action seemed pretty small compared to the vastness of our planet. Our atmosphere seemed impossibly huge. Our oceans virtually limitless. Our soils endlessly fertile. What we didn't realize was that our air, our water, and even our soils and every living thing are all connected. These connections are what make our planet so very, very special—the only one like it in the known universe.

Living with the Legacy of "Away"

What do you do with waste? Why, you throw it away, of course. But think about the term "throw it away." Where exactly is this "away" place, and what happens to things when they get there?

For many North American cities, "away" could be a hundred miles or more by truck to a different country, state, or province. The city of Toronto actually ships its trash over the U.S. border to Michigan. And that's just household garbage being carted off to distant landfills. We also dump air pollutants up our smokestacks and out our tailpipes. We pour human and industrial waste into our rivers and oceans. We spray chemicals on our crops and hope that the residue goes away.

There's an old saying: "The solution to pollution is dilution." In other words, spread the pollution out and the problem goes away. Build your smokestack higher, flush the pollutants out with more clean water, or spread them out over more land.

Of course, we now realize that this solution wasn't much of a solution at all. It just pushed the problem over to our neighbors or onto the next generation. Not long ago the world seemed like a very large place. The atmosphere seemed vast, the oceans massive. There was nothing we couldn't dump out that wouldn't go away—eventually.

Today, we are just beginning to deal with a legacy of using nature as a dumping ground. Some countries are finally starting to reduce the heat-trapping emissions that are disrupting the climate, for example, but others are still ignoring the problem. Canada and the United States have continued to stall on making firm commitments to reducing their emissions, which continue to rise in both countries.

Meanwhile, research published in the journal *Environmental Science & Technology* in 2005 helped show how "away" may not exist at all. On the remote west coast of British Columbia, researchers found that grizzly bears eating a diet rich in salmon are accumulating toxins that build up in the fatty flesh of the fish. These toxins, a brew of persistent organic chemicals like PCBs and flame retardants, may come from as far away as Asia, but they don't just disappear. They can continue to exist for decades and be transported halfway around the world.

Researchers say that they don't know what effect these toxins are having on the bears, but the chemicals are known to mimic animal hormones and may pose a developmental risk to young cubs. The same chemicals are turning up in much higher concentrations in killer whales and polar bears.

> These connections are what make our planet so very, very special—the only one like it in the known universe.

Humans aren't immune to our legacy of hoping things go away either. Bulging landfills, smoggy skies, and a disrupted climate are some of the most obvious problems we face as a result. But this legacy often affects us in ways we can't even see. For example, in 2005, another paper published in *Environmental Science & Technology* reported that rice grown in the United States may contain up to five times more arsenic than rice grown in Europe or Asia.

In this case, the culprit was thought to be arsenic-based pesticides sprayed on cotton fields throughout the southern United States. Many of those fields have since been converted to rice crops, and arsenic left in the soil is finding its way into the grains. In 2008, researchers with Cornell University updated the study, confirming that the U.S. rice contained higher levels of arsenic. Fortunately for human health, it is in the form of methylated arsenic, which is considered less toxic than the inorganic form of the metal.

Still, such examples point to the ever-decreasing distance between "away" and human beings. We live in a disposable culture where it's easy to forget that things we throw away don't necessarily go away. And unless we start ensuring that the chemicals and other junk we release into the environment readily break down and don't build up over time, we will continue to build our legacy. "Away" may not be on the map, but it's now closer than ever.

Human Hormones Mess with Male Fish

Most people alive today were born after 1950. To these people, our modern world is just the way things have always been. Imagining life without TV, radio, telephones, and the internet is next to impossible. Teenagers probably have a hard time imagining life without text messaging.

> We live in a disposable culture where it's easy to forget that things we throw away don't necessarily go away.

And it's true: human reach is now profound. We are the most integrated, interconnected, and mobile species that has ever existed on this planet. Some of these interconnections produce marvelous results. We get to know other cultures. We can easily chat with friends and family on the other side of the world.

But we have to remember that, although we are connected with each other more than ever, we are also intimately connected to the rest of the natural world. These connections can manifest themselves physically, such as through global warming. But they can also manifest themselves biologically—and in some surprising ways.

In 2007, researchers writing in the journal *Proceedings of the National Academy of Sciences* reported that male fish became "feminized" when exposed to human hormones. Some of the fish, a type of fathead minnow, produced early-stage eggs in their testes, while others actually developed tissues for both reproductive organs.

How would fish be exposed to female human hormones? Through treated or untreated municipal wastewater, of course. It seems that widespread use of birth control pills has elevated the amount of estrogenic substances going into our waste stream. Remember, things that go down our toilets don't just disappear. They can actually survive simple sewage treatment processes and end up in our rivers, lakes, and oceans.

Reports of fish feminization as a result of human female hormones are fairly well documented—but no one had done long-term studies of the impact this can have on fish populations. For this study, researchers actually added the synthetic estrogen found in contraceptive pills to a remote lake in northern Ontario in amounts normally found in human wastewater. They did this for three years and monitored the results over a period of seven years.

The results were startling. As expected, the male fish developed some feminized characteristics, such as producing proteins normally synthesized in females. But what really disturbed the scientists: Populations of the fish crashed to near extinction levels by the end of the experiment. Feminization of the males combined with hormonal changes in the females apparently damaged their overall reproductive capacity to the point that the fish were unable to maintain their population.

The researchers conclude: "The results from this whole-lake experiment demonstrate that continued inputs of natural and synthetic estrogens and estrogen mimics to the aquatic environment in municipal wastewaters could decrease the reproductive success and sustainability of fish populations."

This spells trouble. Most of us have probably never heard of the fathead minnow, but these fish are a vital food source for well-known and popular sport fish that people have heard of—such as walleye, lake trout, and northern pike. They are also well studied and often used in toxicology testing because they have short life cycles, adapt well to lab conditions, and are representative of a large family of fish.

The authors of the report describe the fathead minnow as "a freshwater equivalent of the miner's

Interconnectedness—the Food Web

BY KATE LYMAN

The following activity was part of 3rd-grade teacher Kate Lyman's unit on food webs. See Lyman's "Lessons from a Garden Spider" on p. 48. —the editors

In studying food webs, I read aloud the book *Butternut Hollow Pond* by Brian Heinz. I asked for volunteers to act out the animals' actions as they captured and ate their prey. Next, I handed each student a card with a drawing of a living thing on it. (Another teacher and I drew some pictures or copied them from other materials.) My cards included a blue heron, a leopard frog, grass, fungi, a garden spider, and a grasshopper. I chose a student to hold a ball of yarn, and while holding on to the end of the yarn, threw the yarn to his or her "predator."

The game went on, as the yarn was tossed to frog, to heron, to grass (representing death), to fungi, to grasshopper, to spider, to frog, and so on, until all the students were holding pieces of yarn. The yarn crisscrossed the circle and formed a web of life.

When our web was completed, I asked students to predict what would happen if pesticides wiped out all the grasshoppers and other insects that spiders ate.

"The spiders would die!" several students responded.

"And then the frogs wouldn't have enough food," added another.

I asked all the students holding the grasshopper cards to drop their piece of yarn and they watched the web fall apart. 🌐

canary." In other words, what happens to the fish, as with the bird, could happen to humans in short order unless we are very careful. Cell phones and the internet aren't our only connections with each other and with the world. We are biological creatures too, and have to remember that our biological connections ultimately matter the most.

Ocean Life Makes Waves

Most people have heard of the "butterfly effect"—the idea that a small change, such as a butterfly flapping its wings in one part of the world, can set in motion a series of events that leads to a big event, such as a tornado, somewhere else. The term is largely used

as a metaphor, but science now shows that there's a literal aspect to the theory that has much broader implications.

To say that everything is connected to everything else has become a cliché, but it's true—especially in nature. Scientist and author James Lovelock uses the term "Gaia" to describe the Earth as a living, self-regulating system. According to this hypothesis, all of the planet's biological creatures are intimately connected with all of its physical systems, from the soils to the oceans and the atmosphere. Changes in any of these systems can affect everything else.

We can see how connected everything is when we release long-lasting substances into the atmosphere. Toxins, for example, can drift out of a smokestack in Hamilton, Ontario, or Mumbai, India, circle the Earth on the winds or ocean currents, and end up in seemingly pristine areas such as the far north. In fact, these toxins are now found concentrated in the fat of marine mammals and in human breast milk.

In an interconnected world, even a creature as small and seemingly inconsequential as the tiny, shrimp-like krill can have a big impact. These half- to three-quarter-inch-long creatures already play an important role in the ocean food chain and are a staple in the diet of some of the world's largest whales. But krill are so small that few people would have suspected they could play an important role in generating currents that help mix our ocean waters.

Yet that's exactly what a team of researchers from the University of Victoria found off the coast of British Columbia's Vancouver Island. The researchers looked at a deep ocean inlet where different layers of the water mix very little. They found that millions of krill, on their nightly upward migration from the deep water to the surface to feed, increased the mixing of water by three to four orders of magnitude. In other words, these tiny creatures actually cause quite a stir.

And it isn't just krill that cause this water stirring or "turbulence." All living organisms that exhibit similar behavior can cause turbulence, helping to bring cold, nutrient-rich water up to the surface. This exchange of cold and warm water is vital to the productivity of the oceans. It also helps break down human wastes, and it even plays an important role in the climate.

But turbulence has largely been thought to be driven almost exclusively by physical forces like the winds and the tides rather than by biological forces. The very idea that the behavior of individual organisms can affect entire systems seems fantastic. Yet the researchers in Victoria concluded that sea creatures themselves may be a critical, but overlooked, source of turbulence in the oceans.

Other researchers go even further. A 2006 article in the *Journal of Marine Research* calculated that, based on the math, swimming organisms may be one of the most important drivers of ocean turbulence. If this is the case, the authors concluded, then the overfishing that has caused fish stocks to plummet and the near extinction of many whale species as a result of hunting may have disrupted ocean turbulence enough to affect the planet's climate.

Seemingly small actions can have big consequences. More and more, we are finding that our world is not nearly as vast and limitless as we once supposed. Not only is it interconnected, but this very interconnectedness drives it. In this world, we are all butterflies, and we need to be mindful of what can happen when we flap our wings. ⊕

This article is excerpted from David Suzuki's book, The Big Picture: Reflections on Science, Humanity, and a Quickly Changing Planet *(Greystone Books). Suzuki is an award-winning scientist, environmentalist, and broadcaster. He is co-founder of the David Suzuki Foundation.*

 See teaching ideas for this article, page 29.

A Tunisian protester holding bread confronts riot police during a demonstration in Tunis on Jan. 18, 2011.

Reading the World in a Loaf of Bread

Soaring food prices, wild weather, upheaval, and a planetful of trouble

BY CHRISTIAN PARENTI

What can a humble loaf of bread tell us about the world?

Bread has classically been known as the staff of life. In much of the world, you can't get more basic, since that daily loaf often stands between the mass of humanity and starvation. Still, to read world politics from a loaf of bread you first have to ask: Of what exactly is that loaf made? Water, salt, and yeast, of course, but mainly wheat, which means when wheat prices increase globally, so does the price of that loaf—and so does trouble.

To imagine that there's nothing else in bread, however, is to misunderstand modern global agriculture. Another key ingredient in our loaf—call it a "factor of production"—is petroleum. Yes, crude oil, which appears in our bread as fertilizer and tractor fuel. Without it, wheat wouldn't be produced, processed, or moved across continents and oceans.

And don't forget labor. It's an ingredient in our loaf, too, but not perhaps in the way you might imagine. After all, mechanization has largely displaced workers from the field to the factory. Instead of untold thousands of peasants planting and harvesting wheat by hand, industrial workers now make tractors and threshers, produce fuel, chemical pesticides, and nitrogen fertilizer, all rendered from

Days of anger—the Egyptian people's revolution for bread, freedom, and human dignity.

change, now really kicking in, and increasingly the key destabilizing element in bringing that loaf of bread disastrously to market.

Marketing Disaster

When these ingredients mix in a way that sends the price of bread soaring, politics enters the picture. Consider this, for instance: the upheavals in Egypt during the Arab Spring, the democratic movements that spread through many countries in the Arab world in the spring of 2011. Egypt is also the world's single largest wheat importer, followed closely by Algeria and Morocco. Keep in mind that the Arab Spring started in Tunisia when rising food prices, high unemployment, and a widening gap between rich and poor triggered deadly riots and finally the flight of the country's dictator Zine El Abidine Ben Ali. His last act was a vow to reduce the price of sugar, milk, and bread—and it was too little too late.

With that, protests began in Egypt and the Algerian government ordered increased wheat imports to stave off growing unrest over food prices. As global wheat prices increased by 70 percent between June and December 2010, bread consumption in Egypt started to decline under what economists termed "price rationing." And that price kept rising all through the spring of 2011. By June, wheat cost 83 percent more than it had a year before. During the same time frame, corn prices surged by a staggering 91 percent. Egypt is the world's fourth largest corn importer. When not used to make bread, corn is often employed as a food additive and to feed poultry and livestock. Algeria, Syria, Morocco, and Saudi Arabia are among the top 15 corn importers. As those wheat and corn prices surged, it was not just the standard of living of the Egyptian poor that was threatened, but also their very lives, as climate change-driven food prices triggered political violence.

In Egypt, food is a volatile political issue. After all, one in five Egyptians live on less than $1 a day and the government provides subsidized bread to

petroleum and all crucial to modern wheat growing. If the labor power of those workers is transferred to the wheat field, it happens in the form of technology. Today, a single person driving a huge $400,000 combine, burning 200 gallons of fuel daily, guided by computers and GPS satellite navigation, can cover 20 acres an hour, and harvest 8,000 to 10,000 bushels of wheat in a single day.

Next, without financial capital—money—our loaf of bread wouldn't exist. It's necessary to purchase the oil, the fertilizer, that combine, and so on. But financial capital may indirectly affect the price of our loaf even more powerfully. When too much money moves through the global financial system, speculators start to bid up the price of various assets, including all the ingredients in bread. This sort of speculation contributes to rising fuel and grain prices.

The final ingredients come from nature: sunlight, oxygen, water, and nutritious soil—all in just the correct amounts and at just the right time. And there's one more input that can't be ignored, a different kind of contribution from nature: climate

> If you want to trace that near full-fledged crisis back to its environmental roots, the place to look is climate change.

A Deadly Drought

The world's first armed conflicts due to climate change may be just beginning. In a remote and arid part of northern Kenya, the Turkana nomadic community today brandish Kalashnikovs, ready to kill in defense of their depleting water sources. The district in which they live, herding goats, cattle, and camels, borders Uganda, Sudan, and Ethiopia. It is a landscape strewn with the carcasses of livestock following a prolonged and severe drought. With rains persistently failing to materialize, the traditional life of the Turkana stands threatened. Although droughts are not a recent phenomenon, those that used to happen—and were anticipated—every decade or so have now started ravaging the land every two or three years, throwing the community's migratory patterns into disarray.

Conflict is the most alarming symptom of climate change. Every drop of water is guarded fervently. Cattle rustling and tribal rivalry have a long history in East Africa and to some extent are intrinsic aspects of traditional pastoralist culture. A rivalry has existed historically between the Pokot, Samburu, and Turkana of Kenya, resulting in scuffles and encounters. But relatively small internal skirmishes over cattle have now become increasingly destructive and are spreading across country borders, with fatal armed warfare over diminishing water sources and grazing land.

Most weapons in the hands of the Turkana are thought to have permeated the porous borders of Somalia and Sudan. They are cheap, robust, need minimum maintenance, and can be used with little training. Akadaye, who lives in the village of Nasinuono, recently bought his 13-year-old son a gun for two bulls and a few goats.

"I had to invest in them," he explains, sliding his own gun down his shoulders. "My son has started herding cattle and these waterholes are so unsafe now." He pats his son's head as the boy proudly holds the gun, which looks bigger than him. The Toposa, feared rivals from Uganda and also desperate for water, are known to cross the border to attack the Turkana. The last skirmish left three dead and several wounded, with the Toposa making off with 200 cattle.

Eberhard Zeyhle of the African Medical and Research Foundation, who has worked tirelessly to reduce conflict and improve the general health of the Turkana, is clear as to where the fault lies. "The Turkana are a resilient lot. But now they need help. The world needs to wake up to the fact that actions taken in other parts of the globe directly affect the nomads' livelihood."

—Nash Colundalur

Reprinted from New Internationalist, *www.newint.org.*

14.2 million people in a population of 83 million. In 2010, overall food price inflation in Egypt was running at more than 20 percent. This had an instant and devastating impact on Egyptian families, who spend on average 40 percent of their meager monthly incomes simply feeding themselves.

Against this backdrop, World Bank President Robert Zoellick fretted that the global food system was "one shock away from a full-fledged crisis." And if you want to trace that near full-fledged crisis back to its environmental roots, the place to look is climate change, the increasingly extreme and devastating weather being experienced across this planet.

When it comes to bread, it went like this: In the summer of 2010, Russia, one of the world's leading wheat exporters, suffered its worst drought in 100 years. Known as the Black Sea drought, this extreme weather triggered fires that burned vast swaths of Russian forests, bleached farmlands, and damaged the country's breadbasket wheat crop so badly that its leaders (urged on by Western grain speculators) imposed a yearlong ban on wheat exports. As Russia is among the top four wheat exporters in any year, this caused prices to surge upward.

At the same time, massive flooding occurred in Australia, another significant wheat exporter, while excessive rains in the American Midwest and Canada damaged corn production. Freakishly massive flooding in Pakistan, which put some 20 percent of that country under water, also spooked markets and spurred on the speculators.

And that's when those climate-driven prices began to soar in Egypt. The ensuing crisis, triggered in part by that rise in the price of our loaf of bread, led to upheaval and finally the fall of the country's reigning autocrat Hosni Mubarak. Tunisia and Egypt helped trigger a crisis that led to an incipient civil war and then Western intervention in neighboring Libya, which meant most of that country's production of 1.4 million barrels of oil a day went off-line. That, in turn, caused the price of crude oil to surge, at its height hitting $125 a barrel, which set off yet more speculation in food markets, further driving up grain prices.

The world continues to experience climate-related environmental disasters that hurt grain crops. The global food system is visibly straining, if not snapping, under the intense pressure of rising demand, rising energy prices, growing water shortages, and most of all the onset of climate chaos.

And this, the experts tell us, is only the beginning. The price of our loaf of bread is forecast to increase by up to 90 percent over the next 20 years. That will mean yet more upheavals, more protest, greater desperation, heightened conflicts over water, increased migration, roiling ethnic and religious violence, banditry, civil war, and (if past history is any judge) possibly a raft of new interventions by imperial and possibly regional powers.

And how are we responding to this gathering crisis? Has there been a broad new international initiative focused on ensuring food security for the global poor—that is to say, a stable, affordable price for our loaf of bread? You already know the sad answer to that question.

> **Massive corporations are moving to further consolidate their control of world grain markets and vertically integrate their global supply chains in a new form of food imperialism.**

Instead, massive corporations like Glencore Xstrata, the world's largest commodity trading company, and the privately held and secretive Cargill, the world's biggest trader of agricultural commodities, are moving to further consolidate their control of world grain markets and vertically integrate their global supply chains in a new form of food imperialism designed to profit off global misery. While bread triggered war and revolution in the Middle East, Glencore made windfall profits on the surge in grain prices. And the more expensive our loaf of bread becomes, the more money firms like Glencore and Cargill stand to make. Consider that just about the worst possible form of "adaptation" to the climate crisis.

So what text should flash through our brains when reading our loaf of bread? A warning, obviously. But so far, it seems, a warning ignored. ⊕

Christian Parenti, author of Tropic of Chaos: Climate Change and the New Geography of Violence *(Nation Books), is a contributing editor at the* Nation *magazine, a Puffin Foundation Writing Fellow at the Nation Institute, and a visiting scholar at the City University of New York.*

See teaching ideas for this article, page 30.

The World Turned Upside Down

BY LEON ROSSELSON

In 1649
To St. George's Hill,
A ragged band they called the Diggers
Came to show the people's will
They defied the landlords
They defied the laws
They were the dispossessed reclaiming what was theirs

We come in peace they said
To dig and sow
We come to work the lands in common
And to make the waste ground grow
This Earth divided
We will make whole
So it will be
A common treasury for all

The sin of property
We do disdain
No man has any right to buy and sell
The Earth for private gain
By theft and murder
They took the land
Now everywhere the walls
Spring up at their command

They make the laws
To chain us well
The clergy dazzle us with heaven
Or they damn us into hell
We will not worship
The God they serve
The God of greed who feeds the rich
While poor folk starve

We work we eat together
We need no swords
We will not bow to the masters
Or pay rent to the lords
Still we are free
Though we are poor
You Diggers all stand up for glory
Stand up now

From the men of property
The orders came
They sent the hired men and troopers
To wipe out the Diggers' claim
Tear down their cottages
Destroy their corn
They were dispersed
But still the vision lingers on

You poor take courage
You rich take care
This Earth was made a common treasury
For everyone to share
All things in common
All people one
We come in peace
The orders came to cut them down

Reprinted with permission of Leon Rosselson.

Stealing and Selling Nature

Why we need to reclaim "the commons" in the curriculum

BY TIM SWINEHART

Most U.S. and world history textbooks teach students to ignore the role of nature in history. But as environmental crises—climate, water, soil, and biodiversity—threaten the viability of life as we know it for future generations, we can no longer afford a no-nature social studies curriculum. Our students need to know the environmental history of our current crises, including how nature was turned into a commodity to be bought and sold for private profit. They need to recognize that today's "enclosure of the commons" has a long history.

Early Resistance to Enclosure

In "The World Turned Upside Down," by Leon Rosselson, British singer-songwriter Billy Bragg sings about the 17th-century English Diggers:

> We come in peace they said, to dig and sow
> We come to work the lands in common

And to make the waste ground grow
This Earth divided, we will make whole
So it will be a common treasury for all.

In his excellent book, *The Value of Nothing*, Raj Patel defines "the commons":

> A commons is a resource, most often land, and refers both to the territory and to the ways people allocate the goods that come from that land. The commons has traditionally provided food, fuel, water, and medicinal plants for those who used it—it was the poorest people's life-support system.

If the commons is taught at all in history classes, it's likely as a passing reference to English enclosures—the process by which lands traditionally used in common by the poor for growing food, grazing animals, collecting firewood, and hunting game were fenced off and turned into private property. Some textbooks mention the peasant riots that arose in response to enclosures, or specific groups like the Diggers, which actively resisted enclosure by tearing down fences and reestablishing common areas for growing food. But to students reading their world history texts, this doomed fight by the rural poor must seem tragically misguided—especially since it is buried amid chapters that champion the innovation and progress brought on by the new economic order of industrial capitalism.

Some texts, like Holt McDougal's *Modern World History*, skip the peasants' resistance entirely, but sing the praises of innovative and enterprising wealthy landowners:

> In 1700, small farms covered England's landscape. Wealthy landowners, however, began buying up much of the land that village farmers had once worked. The large landowners dramatically improved farming methods. These innovations amounted to an agricultural revolution.

This disturbing historical summary leaves out as much as it gets wrong. Students could reasonably assume that English enclosures involved a fair exchange between "wealthy landowners" and "village farmers," not the forceful, sometimes violent evictions that removed peasants from land that their families had worked for generations. Take for example, the account of Betsy Mackay, 16, whose family was evicted by the Duke of Sutherland from the late-18th-century enclosures in Scotland referred to now as "the clearances":

> Our family was very reluctant to leave and stayed for some time, but the burning party came round and set fire to our house at both ends, reducing to ashes whatever remained within the walls. The people had to escape for their lives, some of them losing all their clothes except what they had on their back. The people were told they could go where they liked, provided they did not encumber the land that was by rights their own. The people were driven away like dogs.

The *Modern World History* narrative silences the voices of the poor, who struggled for centuries in England to maintain their traditional rights to common lands—rights enshrined in 1217 in the Charter of the Forest, the often overlooked sister document to Magna Carta. But don't bother looking for the Charter of the Forest in the textbook: The prologue, "The Rise of Democratic Ideas," praises Magna Carta for respecting "the individual rights and liberties" of nobles, but the idea of the commons didn't make the cut.

We need stories like those of Betsy Mackay, the Diggers, and the Charter of the Forest to get students to think critically about how the natural world—the source of subsistence for all people in all times, including today—has been appropriated historically to serve the interest of a few at the expense of the many.

This history is, of course, not limited to land enclosures during the British agricultural revolution. Around the world, European colonizers spent centuries violently "enclosing" the lands of indigenous peoples throughout the Americas, India, Asia, and

> 'The commons has traditionally provided food, fuel, water, and medicinal plants for those who used it—it was the poorest people's life-support system.'
>
> —Raj Patel,
> *The Value of Nothing*

Africa. And the process continues today, described by Indian scientist and activist Vandana Shiva in her essay "The Enclosure of the Commons":

The "enclosure" of biodiversity and knowledge is the final step in a series of enclosures that began with the rise of colonialism. Land and forests were the first resources to be "enclosed" and converted from commons to commodities. Later on, water resources were "enclosed" through dams, groundwater mining, and privatization schemes. Now it is the turn of biodiversity and knowledge to be "enclosed" through intellectual property rights.

The destruction of commons was essential for the industrial revolution, to provide a supply of natural resources for raw material to industry. A life-support system can be shared; it cannot be owned as private property or exploited for private profit. The commons, therefore, had to be privatized, and people's sustenance base in these commons had to be appropriated, to feed the engine of industrial progress and capital accumulation.

The enclosure of the commons has been called the revolution of the rich against the poor.

Missing Nature in U.S. History

Current U.S. history curriculum also contributes to an *ecological illiteracy* by glossing over the historical role of nature. And when we're not taught to understand the intimate and fundamental connections between people and the environment in our nation's history, it should come as no surprise that we struggle to make these same connections today.

> We need to tell stories that emphasize the vital relationships between people and the Earth.

One of the few places where nature shows up in U.S. history courses is an explanation of how Native American and European concepts of landownership differed. Here, for example, is Prentice Hall's *America:*

One item that Native Americans never traded was land. In their view, the land could not be owned. They believed that people had a right to use land and could grant others the right to use it, too. To buy and sell land, as other peoples have done throughout history, was unthinkable to them. Land, like all of nature, deserved respect.

By contrast, the Europeans who arrived on North American soil in the 1400s had quite a different idea about landownership. They frequently did not understand Indian attitudes and interpreted references to land use as meaning landownership. Such fundamental differences would prove to have lasting consequences for both Native Americans and European settlers.

No. Columbus, as just one example, did not misunderstand Taíno concepts of property—he came as a conqueror. He claimed Native American land before he knew anything about the people living there.

This is an opportunity to explore a different version of history with students—a history that begins with the naked truth that land inhabited by Native Americans had to be stolen before it could ever become private property. Instead, we have this later section of *America*, titled "Conflict with Native Americans":

Although the Native Americans did help the English through the difficult times, tensions persisted. Incidents of violence occurred side by side with regular trade. Exchanges begun on both sides with good intentions could become angry confrontations in a matter of minutes through simple misunderstandings. Indeed, the failure of each group to understand the culture of the other prevented any permanent cooperation between the English and Native Americans.

This is history of the worst kind, in which a seeming attempt at "balance" results in a morally ambiguous explanation for the dispossession and murder of millions of Native Americans. In *A People's History of the United States,* Howard Zinn offers a different version of the same story: "Behind the English invasion of North America, behind their massacre of Indians, their deception, their brutality, was that special powerful drive born in civilizations based on private property." For hundreds of years after the first contact between Europeans and Native Americans, the governments and laws of the colonizers worked, often violently, to imprint this notion of private property on the lands and peoples of the Americas.

Water and the Industrial Revolution

Although students often study the water cycle in biology, that very same water disappears from view when studying the industrial revolution in U.S. history. This historical omission creates at least two problems. First, it leaves us with the flawed impression that human ingenuity and technological advances built the industrial economy. It ignores the exploitation of rivers, land, fossil fuels, and the atmosphere that were the basis for industrial growth. The textile mills that heralded the birth of the industrial revolution in the United States would have produced few products and little wealth if they had not been powered by the river systems of the Northeast. Second and perhaps even more significant for teachers and students, the private appropriation—theft—of once public and common resources like rivers fundamentally changed the lives of people dependent on those resources for food and livelihood.

The story of early industrialization in the United States, as told in most textbooks, illustrates how the curriculum fosters ecological illiteracy. In Prentice Hall's *America*, the section titled "Inventions and Innovations" implies that industrialization was the result of witty and enterprising inventors using new technologies to revolutionize the way that goods, from cloth to guns, were produced in the nation's burgeoning capitalist economy. Nature is not entirely absent. Students read a description of New England's majestic rivers, which "gathered strength as they descended from the mountains, surged through valleys, and plunged over waterfalls." It's as if the rivers of the Northeast were *made* to provide power, and were simply waiting to be harnessed by the new industrialists. *America*'s discussion of nature and industrialization stops here, without a hint of how factory owners appropriated entire river systems, fundamentally transforming these ecosystems and decimating the fish populations that poor people depended on for food—a catastrophic story for those who weren't reaping the profits of the new industrial era.

America, like most texts, champions Samuel Slater, an immigrant who smuggled legally protected knowledge of textile machinery from Britain and built the first textile factory in the United States. Students read about how Slater "reproduced the complex British machinery" in a clothier's shop in Pawtucket, Rhode Island, and with the help of business partners, built "the nation's first successful water-powered textile mill in 1793":

> Textile producers soon began copying Slater's methods. By 1814, the United States boasted some 240 textile mills, most of them in Pennsylvania, New York, and New England. Slater and other mill owners grew wealthy by filling the needs of the growing American population for more and more cloth.

Thanks to the work of Zinn and other social historians, texts like *America* at least reference stories of class struggle between factory workers and owners. But students also need to learn the history of the struggle over how nature would be used and who had the right to use it.

In *Down to Earth: Nature's Role in American History,* Ted Steinberg uncovers the story of how nature—from rocks, trees, and rivers to climate and soil fertility—factored into U.S. history. He argues for a deeper understanding of how industrialization fundamentally changed the relationship between people and nature.

Steinberg explains that, before the rivers of the Northeast were transformed into the "surging, plunging" sources of power described in *America,* they served a different purpose in colonial American society:

> In the spring, when winter stores ran low, the colonists went fishing for shad, alewives, and salmon, species of fish that return from the ocean to freshwater streams to reproduce. Salmon were so plentiful during the colonial period that as late as 1700 they sold for only one cent a pound. The spring profusion of fish brought farmers descending on the region's rivers . . . securing an important supply of dietary protein at precisely the point in the seasonal cycle when they needed it most.

Learning about the colonists' dependence on these yearly fish runs is a valuable lesson in an age when most of our students are hard-pressed to say where any of their food comes from. But important lessons lie in the stories of what happened to these fish runs when factory owners began appropriating the rivers for industrial production.

The dams built by factory owners like Slater and corporations like the Boston Associates became a tangible focus of anger and protest. Farmers in 1792 petitioned the Rhode Island General Assembly to remove Slater's dam, but they were thwarted when one of Slater's partners preempted the petition through insider political connections. In 1859, people in Lake Village, New Hampshire, took a more hands-on approach, attacking and attempting to destroy a dam owned by the Boston Associates:

> The attackers included farmers angry over the flooding of their meadows to convenience out-of-state factories; upstream mill owners who resented being forced to follow the waterpower schedule of the lower Merrimack corporations; loggers who wanted the gates lowered to send timber downstream; and the poor and dispossessed, incensed that the economic transformation pulsing through the region had left them behind.

Water Becomes a Commodity

One of the most important innovations to create wealth for the Boston Associates was not a piece of mill technology but an idea—the "mill-power" concept. Steinberg explains the concept and its significance:

> A mill-power equaled the amount of water necessary to drive 3,584 spindles for spinning cotton yarn—the capacity of one of the Boston Associates' earliest factories—plus all the other machinery necessary for transforming

the yarn into cloth. The concept enabled the company to easily package water and put it up for sale. By the 1830s, companies at Lowell even purchased water without buying any land, breaking with past tradition. Water itself had become a commodity.

The story of how the Boston Associates turned water into a commodity represents one of the most important lessons environmental history can offer our students—the recognition that today's "private property" often hides a history of theft and dispossession. Steinberg sums it up well:

> The mills along the lower Merrimack incorporated the natural wealth available in the countryside into their designs for production and in so doing produced more than just cloth. They generated a chain of ecological and social consequences that spilled out beyond the factories, affecting places and people more than 100 miles away in a completely different state. Nothing better demonstrates the ways in which industrialization led to a major rationalization and reallocation of natural resources, enriching some at the expense of others.

It is important not to romanticize traditional natural commons as completely democratic or sustainable examples of how people can subsist from the environment—the dispossessed Native American inhabitants of the Northeast who had little or no access to traditional fishing grounds and the overfishing of some rivers by colonial farmers tell us otherwise. But we need to ask whether the new economy was always an improvement over the economy of the commons that preceded it.

Moving Nature to Center

When we look at the Diggers' resistance to enclosure, Native Americans resisting the land theft of European colonizers, or colonial-era farmers battling the damming of rivers in 18th-century New England, a different historical narrative emerges: The growth of industrial capitalism has been predicated on the private enclosure of the natural world. And these enclosures have met with resistance. This alterna-

> **The growth of industrial capitalism has been predicated on the private enclosure of the natural world. And these enclosures have met with resistance.**

tive narrative encourages critical conversation about the extent that "economic growth" has been used to justify the private seizure of the Earth's resources for the profits of a few while closing off those same resources—and decisions about how they should be used—to the rest of us. Even more important, this conversation can help us understand today's environmental crises—from the loss of global biodiversity to the melting of the Arctic—for what they really are: the culmination of 500 years of privatizing and commodifying the natural world.

The private enclosure of nature continues today; it's just hard to see. Like the proverbial fish surrounded by the water of the "free market," it's easy to assume that fossil fuel companies have some God-given right to profit from polluting our atmospheric commons. How will young people recognize this atmospheric grab when the school curriculum has erased all memory of our collective right to the natural commons?

Reclaiming the commons means fueling students' knowledge about a past that has conveniently disappeared. Educators did not create the environmental crisis, but we have a key role to play in alerting students to its causes—and potential solutions. ◉

> **Reclaiming these commons means fueling students' knowledge about a past that has conveniently disappeared.**

Tim Swinehart (timswinehart@gmail.com) teaches at Lincoln High School and Lewis & Clark College in Portland, Oregon.

 See teaching ideas for this article, page 30.

 Teaching Idea: **The Case for the Commons**

Have students compare this brief history of enclosure, written by Raj Patel, to the history in their textbook:

Sometimes piecemeal, sometimes sweeping, enclosure was the process by which land was once again taken out of public hands. Surveyors used chains to rope off areas of common land and formally assign title to a single individual. Not only fields but also forest and water were similarly enclosed—with lords preventing access to ponds and streams well stocked with fish, and to forests teeming with game that had provided the poor with meat. By 1500, 45 percent of cultivable land in England had been enclosed and took on a new logic—not only to provide private land for individual landlords, but also to drive up the price of rent for those landlords.

This theft was deeply unpopular and provided the backdrop for rebellions ranging from small-scale acts of insubordination to the 1381 Peasants' Revolt to the Diggers in the mid-1600s and beyond. The protests and resistance were always crushed, and because enclosure had seized the peasants' only means of survival, they had only two choices: work for their new landlords or try their luck in the cities. Adam Smith lamented the violence being done to the commons by the spread of private property, though the process was already over: "The wood of the forest, the grass of the field, and all the natural fruits of the Earth, which, when land was in common, cost the laborer only the trouble of gathering them. . . . [He must now] pay for the license to gather them; and must give up to the landlord a portion of what his labor either collects or produces." Within a generation, these displaced peasants were to become the proletarian backbone of the industrial revolution.

From "The Case for the Commons," by Raj Patel, published in Utne Reader.

The Commons

BY DAVID ROVICS

You build your fences and say there's nothing we can do
You say the world around us belongs fairly to the few
But about 6 billion people no doubt will agree
This world is our home, not your property

It's the commons, our right of birth
And to you who would enclose the land all around the Earth
Our future is your downfall, when we cut this ball and chain
You who'd sacrifice the public good for your private gain

With our sweat we built the railroads, built cities on these shores
But because you own the money you say that it's all yours
We laid the phone lines and the pipelines and then right before our eyes
You say these things our taxes paid for you now will privatize

Privatize the hospitals, privatize the schools
Privatize the prisons for all those who break your rules
And preparing for the day when all the wells run dry
You say you own the very rain that falls down from the sky

But it's the commons, our right of birth
And to you who'd own the water all around the Earth
Our future is your downfall, when we cut this ball and chain
You who'd sacrifice the public good for your private gain

You claim to own the harvest with your terminator seeds
You claim to own the genomes of every animal that breeds
You claim to own our culture and the music that we play
And with every song we download to your coffers we must pay

You would even own my name and you say it's for the best
Maybe you'll let us on your radio stations if our songs can pass your test
You own country, you own western, you say you've given us a choice
You may own the airwaves but you'll never own my voice

It's the commons, our right of birth
And to you who'd own the music all around the Earth
Our future is your downfall, when we cut this ball and chain
You who'd sacrifice the public good for your private gain

It's the commons, our right of birth
And to you who would own everything all around the Earth
Our future is your downfall, when we cut this ball and chain
You who'd sacrifice the public good for your private gain

Singer/songwriter David Rovics has toured four continents and has made more than 200 songs available for free. Find out more at www.davidrovics.com

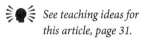 *See teaching ideas for this article, page 31.*

The war against the Earth began with the idea of separateness. Its contemporary seeds were sown when the living Earth was transformed into dead matter to facilitate the industrial revolution. Monocultures replaced diversity. "Raw materials" and "dead matter" replaced a vibrant Earth. Terra Nullius (the empty land, ready for occupation regardless of the presence of indigenous peoples) replaced Terra Madre (Mother Earth).

This philosophy goes back to Francis Bacon, called the father of modern science, who said that science and the inventions that result do not "merely exert a gentle guidance over nature's course; they have the power to conquer and subdue her, to shake her to her foundations."

Robert Boyle, the famous 17th-century chemist and a governor of the Corporation for the Propagation of the Gospel Among the New England Indians, was clear that he wanted to rid native people of their ideas about nature. He attacked their perception of nature "as a kind of goddess" and argued that "the veneration, wherewith men are imbued for what they call nature, has been a discouraging impediment to the empire of man over the inferior creatures of God."

The death-of-nature idea allows a war to be unleashed against the Earth. After all, if the Earth is merely dead matter, then nothing is being killed.

As philosopher and historian Carolyn Merchant points out, this shift of perspective—from nature as a living, nurturing mother to inert, dead, and manipulatable matter—was well suited to the activities that would lead to capitalism. The domination images created by Bacon and other leaders of the scientific revolution replaced those of the nurturing Earth, removing a cultural constraint on the exploitation of nature. "One does not readily slay a mother, dig into her entrails for gold, or mutilate her body," Merchant wrote.

Everything I Need to Know I Learned in the Forest

My ecological journey started in the forests of the Himalaya. My father was a forest conservator, and my mother became a farmer after fleeing the tragic partition of India and Pakistan. It is from the Himalayan forests and ecosystems that I learned most of

HILLARY SCHENKER

Two Views of Nature

BY VANDANA SHIVA

what I know about ecology. The songs and poems our mother composed for us were about trees, forests, and India's forest civilizations.

My involvement in the contemporary ecology movement began with Chipko, a nonviolent response to the large-scale deforestation that was taking place in the Himalayan region.

In the 1970s, peasant women from my region in the Garhwal Himalaya had come out in defense of the forests.

Logging had led to landslides and floods, and scarcity of water, fodder, and fuel. Since women provide these basic needs, the scarcity meant longer walks for collecting water and firewood, and a heavier burden.

> **The women declared that they would hug the trees, and the loggers would have to kill them before killing the trees.**

Women knew that the real value of forests was not the timber from a dead tree, but the springs and streams, food for their cattle, and fuel for their hearths. The women declared that they would hug the trees, and the loggers would have to kill them before killing the trees.

A folk song of that period said:

These beautiful oaks and rhododendrons,
They give us cool water
Don't cut these trees
We have to keep them alive.

In 1973, I had gone to visit my favorite forests and swim in my favorite stream before leaving for Canada to do my PhD. But the forests were gone, and the stream was reduced to a trickle.

I decided to become a volunteer for the Chipko movement, and I spent every vacation doing *pad yatras* (walking pilgrimages), documenting the deforestation and the work of the forest activists, and spreading the message of Chipko.

One of the dramatic Chipko actions took place in the Himalayan village of Adwani in 1977, when a village woman named Bachni Devi led resistance against her own husband, who had obtained a contract to cut trees. When officials arrived at the forest, the women held up lighted lanterns, although it was broad daylight. The forester asked them to explain. The women replied, "We have come to teach you forestry." He retorted, "You foolish women, how can you prevent tree felling by those who know the value of the forest? Do you know what forests bear? They produce profit and resin and timber."

The women sang back in chorus:

What do the forests bear?
Soil, water, and pure air.
Soil, water, and pure air
Sustain the Earth and all she bears. 🌐

Vandana Shiva is the director of the Research Foundation on Science, Technology, and Ecology. She is the author of numerous books, including Staying Alive: Women, Ecology, and Development, Soil Not Oil: Environmental Justice in an Age of Climate Crisis, Biopiracy: The Plunder of Nature and Knowledge, *and* The Violence of the Green Revolution. *This is excerpted from* Yes! Magazine.

See teaching ideas for this article, page 31.

RICARDO LEVINS MORALES

Principles of Environmental Justice

Delegates to the First National People of Color Environmental Leadership Summit in Washington, D.C., drafted and adopted 17 principles of Environmental Justice in 1991. —the editors

Preamble

We, the people of color gathered together at this multinational People of Color Environmental Leadership Summit, to begin to build a national and international movement of all peoples of color to fight the destruction and taking of our lands and communities, do hereby reestablish our spiritual interdependence to the sacredness of our Mother Earth; to respect and celebrate each of our cultures, languages, and beliefs about the natural world and our roles in healing ourselves; to ensure environmental justice; to promote economic alternatives that would contribute to the development of environmentally safe livelihoods; and, to secure our political, economic, and cultural liberation that has been denied for more than 500 years of colonization and oppression, resulting in the poisoning of our communities and land and the genocide of our peoples, do affirm and adopt these Principles of Environmental Justice:

1. **Environmental Justice** affirms the sacredness of Mother Earth, ecological unity and the interdependence of all species, and the right to be free from ecological destruction.

2. **Environmental Justice** demands that public policy be based on mutual respect and justice for all peoples, free from any form of discrimination or bias.

3. **Environmental Justice** mandates the right to ethical, balanced, and responsible uses of land and renewable resources in the interest of a sustainable planet for humans and other living things.

4. **Environmental Justice** calls for universal protection from nuclear testing; extraction, production, and disposal of toxic/hazardous wastes and poisons; and nuclear testing that threaten the fundamental right to clean air, land, water, and food.

5. **Environmental Justice** affirms the fundamental right to political, economic, cultural, and environmental self-determination of all peoples.

6. **Environmental Justice** demands the cessation of the production of all toxins, hazardous wastes, and radioactive materials, and that all past and current producers be held strictly accountable to the people for detoxification and the containment at the point of production.

7. **Environmental Justice** demands the right to participate as equal partners at every level of decision-making, including needs assessment, planning, implementation, enforcement, and evaluation.

8. **Environmental Justice** affirms the right of all workers to a safe and healthy work environment without being forced to choose between an unsafe livelihood and unemployment. It also affirms the right of those who work at home to be free from environmental hazards.

9. **Environmental Justice** protects the right of victims of environmental injustice to receive full compensation and reparations for damages as well as quality health care.

10. **Environmental Justice** considers governmental acts of environmental injustice a violation of international law, the Universal Declaration on Human Rights, and the United Nations Convention on Genocide.

11. **Environmental Justice** must recognize a special legal and natural relationship of Native Peoples to the U.S. government through treaties, agreements, compacts, and covenants affirming sovereignty and self-determination.

12. **Environmental Justice** affirms the need for urban and rural ecological policies to clean up and rebuild our cities and rural areas in balance with nature, honoring the cultural integrity of all our communities, and providing fair access for all to the full range of resources.

13. **Environmental Justice** calls for the strict enforcement of principles of informed consent, and a halt to the testing of experimental reproductive and medical procedures and vaccinations on people of color.

14. **Environmental Justice** opposes the destructive operations of multinational corporations.

15. **Environmental Justice** opposes military occupation, repression, and exploitation of lands, peoples and cultures, and other life forms.

16. **Environmental Justice** calls for the education of present and future generations, which emphasizes social and environmental issues, based on our experience and an appreciation of our diverse cultural perspectives.

17. **Environmental Justice** requires that we, as individuals, make personal and consumer choices to consume as little of Mother Earth's resources and to produce as little waste as possible; and make the conscious decision to challenge and reprioritize our lifestyles to ensure the health of the natural world for present and future generations.

The full document is available at www.ejnet.org/ej/ principles.html. More information on the environmental justice movement is available at the Environmental Justice Resource Center (EJRC).

See teaching ideas for this article, page 31.

Teaching Ideas
CHAPTER ONE: Everything Is Connected

Plastics and Poverty: Why the whole thing is connected—*page 4*
By Van Jones

Van Jones first presented "Plastics and Poverty" as a TED talk. You might want to watch the video of his talk with students instead of reading it. In discus-sion and writing, students can refer back to the text: www.ted.com/talks/van_jones_the_economic_injustice_of_plastic.html.

Prior to watching or reading Jones' talk, put a large plastic container in the front of the classroom and ask students to write as much as they know about plastic. Encourage them to "talk to" the plastic and ask it questions—what do they think is important to know about plastic?

Discuss with students:

• Why does Van Jones single out poor people as the ones hurt by disposable plastic products?
• Jones says that the poor are forced to buy the most dangerous products. What are some examples?
• Who does Jones blame for the poor being disproportionately harmed by plastic pollution? Do you agree?
• Jones says, "In order to trash the planet, you have to trash people." What's the connection that Jones sees? What examples can you think of?
• Why isn't recycling an adequate response to the problems that Jones describes?
• Jones uses the term "biomimicry" to describe how humans learn from nature. Can you think of examples of how humans could use biomimicry?
• Why does Jones believe the idea of disposability is harmful to people and the planet?
• Display the statistics that Jones offers:

The United States makes up about 5 percent

of the world's population, yet uses about 25 percent of the world's resources and has 25 percent of the world's incarcerated people.

Ask students to write or talk about whether these statistics are connected, and if so, how?

Jones uses the example of plastic to argue that the environmental movement and the social justice movement should not be separate. Ask students to think of other examples.

Smarter Than Your Average Planet—*page 8*
By David Suzuki

"Interwoven, interlinked, joined, chained, bonded, coupled—no matter how you describe it, everything in nature seems somehow connected to everything else." David Suzuki's metaphorical language suggests an assignment asking students to illustrate aspects of this interconnectedness—and how human practices can affect these webs of connection.

Ask students to notice as much natural life as they can in their community, and to list examples of it—for instance, oak trees, blackberry bushes, soil, spiders.

Suzuki writes: "Cell phones and the internet aren't our only connections with each other and with the world. We are biological creatures too, and we have to remember that our biological connections ultimately matter the most." Ask students: In what sense do biological connections "matter the most"? Why do you agree or disagree with Suzuki?

Suzuki writes that "all of the planet's biological creatures are intimately connected with all of its physical systems, from the soils to the oceans and the atmosphere. Changes in any of these systems can affect everything else." Assign each student a "biological creature" and to imagine that the creature has become extinct. Ask them to draw, chart, or write about how this loss would "affect everything else."

Suzuki closes by saying, "In this world, we are all butterflies, and we need to be mindful of what can happen when we flap our wings." Is Suzuki exaggerating? What evidence does he offer to support this conclusion? If he is right, what are the implications for each of us?

Reading the World in a Loaf of Bread
—*page 13*
By Christian Parenti

Before reading Parenti's article, brainstorm with students: How could climate changes in Australia, the

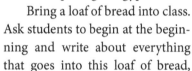

United States, and Russia contribute to an uprising in Egypt?

Bring a loaf of bread into class. Ask students to begin at the beginning and write about everything that goes into this loaf of bread, from start to finish.

Parenti's opening makes a good prompt for a student research project. "What can a humble _____ tell us about the world?" Give students other examples—chocolate bar, apple pie, pair of blue jeans, Barbie doll, etc.—and have them find other environmental and social connections.

Ask students to create a visual representation of the relationship between climate change, bread, and the uprising in Egypt.

Tell students that you would like them to solve a mystery: "What is the connection between climate change, bread, and the uprising in Egypt?" Cut up or rework parts of this article and distribute them to individuals in the class—for example, "One in five Egyptians live on less than $1 a day and the government provides subsidized bread to 14.2 million people in a population of 83 million." And, "In the summer of 2010, Russia, one of the world's leading wheat exporters, suffered its worst drought in 100 years." Have students circulate through the classroom hearing one another's clues in pairs as they try to "solve" the mystery.

Stealing and Selling Nature—*page 18*
By Tim Swinehart

Ask students to consider how private property first becomes "private." What social and legal customs do we use today to confer ownership of property? Has this always been the case? What other methods of "acquiring" property, including land or wealth, have people or nations used historically?

If you have access to a world history textbook, find the sections on English enclosure and/or the British agricultural revolution. Make copies that students can mark on, and ask them to read with a critical eye. Whose stories are told in this history? Who benefits from enclosure? Who loses?

Have students read the full text of Betsy Mackay's testimony about her family's eviction during the Highland Clearances in Scotland: www.chebucto. ns.ca/heritage/FSCNS/Scots_NS/Clans/MacKay/ History_MacKay/MacKay_Clr/Widow_Betsy_ MK.html

How would a textbook entry about enclosure be different if it told the story of Betsy Mackay's family and others like it? What are the implications of not including stories like Mackay's in most textbooks?

Introduce students to the "Discovery doctrine" as outlined by Chief Justice John Marshall in the 1823 U.S. Supreme Court case *Johnson v. M'Intosh*, which provides "legal" justification for the European theft of land from the indigenous peoples of the Americas.

Teachers seeking more background information on the commons and history of the commons can visit these websites:

http://onthecommons.org

www.utne.com/politics/commons-reclaiming-shared-resources-raj-patel.aspx

The Commons—*page 24*
By David Rovics

Ask students:

- Who is the "you" David Rovics addresses in this song?
- Can you think of some commons that you use and appreciate?
- Can you think of something today that is owned privately that was once part of our commons?

Two Views of Nature—*page 25*
By Vandana Shiva

Ask students:

- Is there a hierarchy in nature—an "empire," as Robert Boyle suggested—with "man over the inferior creatures of God"?
- What would it mean today if our society regarded Earth as our mother? What would we do differently? What would have to change in our world?

- Vandana Shiva says that it was the women in her region of India who came out in defense of the forests. Why was it especially the women?

Principles of Environmental Justice—*page 27*

Arrange students into pairs and ask them to come up with a definition of "environmental justice." If you think students need a prompt, give them the first principle as one example. You might tell them

that the authors of the declaration on environmental justice came up with 17 principles.

Divide students into small groups and give each group three or four of the principles. Ask them to think of as many examples of each environmental justice principle as they can. Then ask them to think of examples of when principles have been violated. If it's too challenging to do this without outside in-

formation, they might look online, or apply their principles to another article in the book—for example, Winona LaDuke's essay on uranium mining, or the Climate Change Mixer on page 92.

Find an example of something happening in the world that violates one or more of these principles. What would need to happen to bring this practice into line with the Principles of Environmental Justice?

Ask: What problems do you imagine gave rise to this gathering? Why would people come together to develop these principles?

Which of these principles is most important? Why?

"Fundamentally, the task is to articulate not just an alternative set of policy proposals but an alternative worldview to rival the one at the heart of the ecological crisis—embedded in interdependence rather than hyperindividualism, reciprocity rather than dominance, and cooperation rather than hierarchy."

—Naomi Klein, from *This Changes Everything: Capitalism vs. the Climate*

CHAPTER TWO: Grounding Our Teaching

DAVID McLIMANS

"Just imagine what our neighborhoods would be like if, instead of keeping our children isolated in classrooms for 12 years and more, we engaged them in community-building activities with the same audacity with which the Civil Rights Movement engaged them in desegregation activities 50 years ago! Just imagine how safe and lively our streets would be if, as a natural and normal part of the curriculum from K-12, schoolchildren were taking responsibility for maintaining neighborhood streets, planting community gardens, recycling waste, rehabbing houses, creating healthier school lunches, visiting and doing errands for the elderly, organizing neighborhood festivals, and painting public murals!"

—Grace Lee Boggs

BEC YOUNG

INTRODUCTION
Grounding Our Teaching

As Oberlin College professor David Orr famously wrote, "All education is environmental education. . . . By what is included or excluded we teach the young that they are part of, or apart from, the natural world."

Whether in biology, history, math, or PE, all curriculum teaches students about their relationship to the Earth—and specifically about their relationship to where they live. Unfortunately, too often students learn contempt for the Earth. Students learn that "real" education—the stuff that will make them "college and career ready"—takes place in a classroom, on a computer, or while listening to a teacher. Today, under the iron heel of standardized testing, many schools have even abolished recess. And budget cuts have erased field trips.

By contrast, the articles in this chapter demonstrate ways of fostering what early childhood educator Ann Pelo calls "an ecological identity." Part of "grounding our teaching"—literally—means getting students outside the classroom, to notice, to question, to listen, to smell, to feel, to "story" the world around us. This is the key insight of what has come to be called place-based education: We can center our teaching on the places we live—and not just in science class, but throughout the curriculum as well.

What happens when we focus our teaching about nature only on the far away—on the proverbial polar bear? In his article "Exploring Our Urban Wilderness," teacher Mark Hansen writes:

> Ask students what they know about the impact of deforestation on animal habitats and if they offer any answers at all, I'd wager they are more likely to bring up the Amazon than the impact of urbanization and industrial agriculture on the songbird populations in their own communities. This lack of grounding may help explain why critical ecological issues like global warming, groundwater pollution, and deforestation seem so remote to most children.

Too often, environmental education stops with knowledge of plants and animals. But the articles in this chapter on Turkey Creek in Mississippi and oil pollution in Milwaukee, Wisconsin, underscore that the environment also includes people. Grounding our students in their communities doesn't just connect them to nature; it also connects them to the ways their communities have been battered by powerful interests, and how those communities have been shaped by race and class.

The saying "know your place" is a command to not rock the boat, to not challenge one's position in a hierarchy. In this chapter, it means the opposite. We can teach students to know their place—to care enough about our communities that they are willing to defend them, and to make them better. ⊕

MIRIAM KLEIN STAHL

How My Schooling Taught Me Contempt for the Earth

BY BILL BIGELOW

I grew up hearing the phrase "no deposit, no return"—just throw stuff "away" and forget about it. The slogan was emblematic of a culture premised on unlimited extraction, production, and consumption—a culture living as if there were no tomorrow. Thanks to environmental activists, we now know that continuing on the present course means there won't be a tomorrow—or at least not one hospitable to life.

As a classroom teacher, I've thought and taught a lot about the hidden curriculum of race, class, and gender—and I continue to gain new insights as I probe the pervasiveness of white, middle-class, and male norms in schooling and the broader society. And that led me to consider the subterranean messages about the Earth that I learned in my own schooling. Although my classes included no *explicit* ecological curriculum, there was a profound *hidden* ecological curriculum—one that taught neglect and even contempt for the Earth. I don't mean to bash my elementary teachers for their ignorance of ecological issues. I never even heard the term "ecology" until my first year of college, 1969, when people were becoming more aware of environmental degradation. It would be disingenuous to shake a scolding finger at my late 1950s, early 1960s teachers. What I've discovered in this backward glance is how much my own teaching—as well as the school cultures where I've taught—resembles my early education, even though I talk more about ecological problems than my teachers did.

Bel Aire Elementary

When my family moved in 1957 from Los Angeles to Tiburon, California, a suburb of San Francisco, the area was still largely rural (I was 5). Cows grazed about a hundred yards from our house. Richardson Bay was just down the hill, across the railroad tracks; its rocky beaches stretched in both directions. Our neighborhood of 60 or so houses was nestled in rolling grass-covered hills dotted with eucalyptus trees. Blue belly lizards and horned toads inhabited huge rocks. Each spring the hills would bloom with brilliantly colored golden poppies and other wildflowers. Over the hill to the south, streams trickled into dense wetlands of cattails, frogs, and alligator lizards.

I spent every after-school moment—and every weekend or summer day—outside until it got dark. I knew where to dig the best underground forts, and how to avoid the toffee-like clay soil. I knew the places where, on rare occasions, I might find a salamander. From long observation at nearby ponds I knew the exact process of a pollywog's transition into a frog, and the relative speed of different kinds of snakes: garter vs. gopher vs. western racer. I knew the best climbing rocks. I was an expert on the properties of mud, and the precise kind of grass required for the fastest cardboard sledding. My playmates and I dug forts in the hills, built tree houses, hiked, explored, caught every reptile we could find, played

kick the can over great distances, and made rafts out of driftwood.

We had named key landmarks in the area: the Jungle, for a cavernous tangle of evergreen trees in a place that felt like a natural cathedral, but that we visited infrequently; the Trees, for a grove of huge eucalyptus where we often played; Naked Rock, (a name passed down to us by an older group of kids who claimed to have once danced naked around the rock); Eagle Rock; and Lizard Rock.

Nature surrounded us, but we were also surrounded by "development," by the continual construction of houses, the encroachment of new neighborhoods and roads crisscrossing the hills. My childhood was filled with the natural world, but also with the seeming inevitability of its commercial appropriation. We had a love/hate relationship with "development." We played hide-and-seek in the houses under construction, jumping off roofs, and rafting in basements when they flooded. But, inexorably, the builders seized and destroyed increasing amounts of our natural playground.

> **My schooling suppressed any notion that I would spend my life outdoors.**

How did our schooling extend or suppress our naive Earth knowledge and our love of place? Through silence about the Earth and the indigenous Miwok people of Tiburon, Bel Aire School, perched on the slopes of a steep golden-grassed hill, taught plenty. We actively learned to *not think* about the Earth, about the place where we were. We could have been anywhere—or nowhere. Teachers made no effort to incorporate our vast, if immature, knowledge of the land into the curriculum. Whether it was in the study of history, writing, science, arithmetic, reading, or art, school erected a Berlin Wall between academics and the rest of our lives. Although we spent our afternoons, weekends, and summers outdoors, aside from recess, school was an indoor affair—surrounded by metal, plastic, glass, brick, and linoleum. The hills above the school were a virtual wilderness of grasslands and trees, but in six years I can't recall a single "field trip" to those wide-open spaces. We became inured to spending days in manufactured space, accustomed to watching more Earth bulldozed and covered with yet more manufactured spaces.

My schooling suppressed any notion that I would spend my life outdoors. We were taught that the important work of society—which would be *our* work—occurs indoors, with books, and paper and pencils. The repetition of this indoor education taught us that the land beneath this structure was so much inert *stuff*—mere dirt on top of which happens real life. Outdoors was for play, for fun—but not for knowledge of self, culture, or the Earth. Real knowledge was "Egypt," arithmetic, report writing, the Civil War—even "Indians," but in a "let's-name-the-tribes-and-make-tepees" kind of way. School taught us our Earth knowledge was play/recess/other/trivia. Of course, there was a class component to this indoor education. By and large, we were the children of young professionals. We were being groomed for white-collar office work, not to be farmers or construction workers.

Maybe this heavy indoor bias is beginning to erode. For years, 6th graders from Portland Public Schools have spent a week at Outdoor School near Mt. Hood, learning about ecological issues—observing wildlife habitats, identifying plants, studying about how rivers and streams are formed. Kids explore the wilderness and learn rudimentary survival skills. I've never met a student who didn't cherish this one-week sojourn. But even this fine program has an unsettling subtext: In order to learn about the "outdoors," the Earth, one must travel away from the place where he or she lives. Nature is found in special places, well outside the city limits. The unintended message may be that urban areas are conquered territory, ecologically lost causes—and that the best we can hope for is an occa-

Bel Aire School, Tiburon, California, circa 1960.

sional escape to a pristine wilderness. Sadly, in the era of yearly budget cuts that have arrived with the regularity of Oregon rain, proponents of even this minimal environmental education have had to fight to keep it alive.

Who Was Here First?

We learned about "Indians" in elementary school, but not about the Miwoks who inhabited the land now parceled into neighborhoods with names like Little Reed Heights and Belveron Gardens. I had no way of knowing that First Nations peoples might have had different names and stories for the places where I played or how much I could learn from these stories about relationships between people and the land. And our teachers did not ask us to reflect on the place Bel Aire School occupied: Who owned this space where we were sitting? How did they come to control it? Who was here first? Why aren't those people here anymore? How did these other people teach their young? If our teachers had raised these questions we would have had to confront the contradic-

> **We may or may not have learned how to diagram a sentence, but we did learn to *not* question.**

tion that we were on the land we loved only because it had been twice stolen: First the Spaniards stole it from the native Miwoks, and then the United States took it from the Spaniards (by then, Mexicans).

Because we were not encouraged to reflect on the character of the land, we came to accept its transformation as "development" and "progress." Developers filled in the wetlands to build a new neighborhood and a junior high school. No one asked whether we agreed with this development; in fact, we weren't asked to consider it at all. I almost wrote that we watched helplessly as streams were buried, and the hills invaded by construction crews. But in truth, we didn't watch helplessly, we watched unconsciously. It never occurred to me to question the environmental justice of these actions. We may or may not have learned how to diagram a sentence, but we did learn to *not question*.

Nor was resistance in our conceptual vocabulary. When crews tore up the beach to build a four-lane highway just a couple of hundred yards from our house, no one protested. The kids in the neighborhood loved that beach, but the adults seemed to treat the land as empty space waiting to be done-to. I'm not saying that school created these notions of progress, but in numerous ways it legitimated them. (Ten years or so after demolishing about a mile of rugged beach, the powers that be changed their minds and decided not to continue the highway. With great fanfare, part of the bayfront land was turned into a soccer field.) One of my favorite pastimes when I was young was to imagine that the Russians had invaded and my friends and I were guerrilla soldiers defending our homeland. In real life, the Russians never arrived, but the bulldozers and dump trucks did. School had taught us to look for enemies in all the wrong places.

In school, we were never encouraged to think ecologically—to consider the interdependence of air, soil, water, plants, trees, animals, and humans. We lacked an ecological sensibility, so we regretted the loss of wetlands and forested areas to "development," but we couldn't critique this destruction in terms of the loss of the region's biodiversity. We were ecologically illiterate. Numerous species of plants and animals were wiped out on the Tiburon peninsula, but schooling offered us no conceptual framework to mourn the enormous loss.

This hidden ecological curriculum is politically useful for powerful interests in our society. Writer Wendell Berry notes that social elites "cannot take any place seriously because they must be ready at any moment, by the terms of power and wealth in the modern world, to destroy any place." Popular acceptance, if not support, for this destruction needs to be taught.

The hidden ecological curriculum at Bel Aire School encouraged students to *not think* about the Earth, to *not question* the system of commodification that turns the world, including the land, into things to be bought and sold. These are not merely curricular omissions, but active processes of moral anesthesia. The late poet Adrienne Rich wrote "lying is done with words, and also with silence." When the

> The late poet Adrienne Rich wrote "lying is done with words, and also with silence."

curriculum is silent about aspects of life—racism, sexism, global inequality, or the destruction of the Earth—that silence normalizes these patterns and implicitly tells kids, "Hey, nothing to worry about; that's just the way things are, the way they ought to be." And that's the lie.

Explorers, Discoverers, and the Earth

We learned contempt for the Earth not just in the *how* of schooling, but also in the *what* of schooling; harmful ecological messages were woven into the fabric of the curriculum. I recall social studies in 1st through 6th grades as one long celebration of the brave Europeans who carried civilization to the Americas. One year, my teacher assigned us each a different explorer. Mine was Coronado. We mapped their travels throughout the "New World"—new to whom?—and hung the maps around the room, commemorating the spread of European outposts in the supposed wilderness.

We studied "Indians" in 3rd and 4th grades, but in ways that reinforced a primitive-to-advanced continuum. We studied in depth the indigenous societies most like "us"—the Aztec and Inca, for example—with complex divisions of labor and networks of trade, powerful militaries, influential (i.e., imperialistic) states, and accumulations of great wealth. Other American Indian cultures dotted the primitive-to-advanced continuum at lesser points, and we discussed them more superficially. Highlighting "advanced" societies dismissed the wisdom of those cultures that lived in ecological balance for countless generations. If the latter were designated "primitive," we studied them as quaint artifacts, but not for what we could learn in order to reorient life today in our society. These curricular choices served to confirm that the society we lived in was the inevitable product of progress. Of course, as a 9- or 10-year-old, I wasn't conscious of any of this. That's what makes a hidden curriculum hidden.

Consumption as a Way of Life

The most effective aspects of any hidden curriculum are the ones that are hardest to see, the ones we simply take for granted. This includes the myth of the individual existing as independent agents in the world. In the book *Responsive Teaching,* authors C. A. Bowers and David Flinders ask teachers to consider "whether the culture is learned by students in a manner that leads them to view the 'self' as the basic unit of survival and progress or to recognize the interdependence of 'self,' culture, and the ecosystem." This is a vital concern. For me, Bel Aire School was both a symbol of and preparation for life in a society of essentially disconnected rational human beings seeking "success," which meant maximizing our material opportunities. Sure, we were taught to respect each other's property, not to hit one another, to cooperate on the playground and in sports. But the structure of being grouped by individual "ability," receiving individual grades, and the patterns of individual work taught us that our basic mission was to look out for number one. As early as elementary school we were conditioned to maneuver through the institution making rational choices that would enhance our ultimate salability as labor commodities—"If you want to get a good job. . ." The hidden ecological curriculum of the school structure highlighted "self," but failed to alert us to "the interdependence of 'self,' culture, and the ecosystem." The myth of the individual taught us to think about ourselves and our families but to *not think* about the Earth—or about cultural patterns that might be more ecologically responsible.

This curricular cult of the individual ensured that if and when students did become more aware of the ecological crisis, they would think about per-

sonal rather than systemic responses—for example, I should recycle more and buy less. But as John Bellamy Foster insists in *The Vulnerable Planet*: "The chief causes of the environmental destruction that faces us today are not biological, or the product of individual human choice. They are social and historical, rooted in the productive relations, technological imperatives, and historically conditioned demographic trends that characterize the dominant social system." A system premised on the commodification of nature and endless growth is inherently counter-ecological. But a curriculum that promotes an ideology of the autonomous individual fails to equip students to think systemically.

Toward an Ecologically Responsible Curriculum

I'd like to wax triumphant about how I've fundamentally "greened" my curriculum, but that process is ongoing. For now, I can offer broad principles I'm trying to effect as I construct an ecologically responsible curriculum:

- **As my critique of the hidden ecological curriculum at Bel Aire School suggests, place matters.** A concern for the Earth begins at home. Students ought to think about the history and character of the place they live: How has it changed and why? This means getting students outdoors, interrupting the traditional school-think that learning occurs primarily in classrooms.
- **An important component of this curriculum of place should be a focus on the ecological patterns of the original inhabitants of the land.** I'm not suggesting that we disable our critical filters when studying indigenous societies—some were sharply hierarchical, militaristic, and practiced slavery. But embedded in the traditions of many First Nations is a kind of ecological golden rule. Students should be exposed to cultures that honor the "voice" of the Earth.
- **Students need to develop an ecological literacy that alerts them to life's interconnectedness.** For example, in the Northwest, where I live, students should have an awareness of how deforestation pollutes

the rivers and affects the quality of drinking water and the viability of salmon spawning, etc. When they consider the possibility of the Northwest becoming a depot for coal to Asia, they should consider carbon dioxide, mercury, which social groups live closest to the railroad tracks, and the impact of diesel fumes on children—and not merely the potential for new jobs. Students should consider the Earth a living web of relationships that includes—and sustains—humanity.

- **An ecological curriculum doesn't merely entail studying nature.** It requires that we equip students to question the root concepts of Western civilization: "progress," "development," freedom for the autonomous individual, growth as goodness, private property as the basis of the good society. Throughout the curriculum, we need to ask how understandings of these ideas have helped or hindered ecological sustainability.

- **The power of a green curriculum lies in its "ecology"—the interdependence of social and environmental insights.** Just as there is no human epoch without ecological implications, no ecological issue exists without a social dimension. Earth-conscious teaching should prompt students to think about the intersection of race, class, gender, nationality, and the environment. This requires that we ask essential critical questions when studying the environment. To cite just one obvious example: Wealthy individuals with enormous carbon footprints will not be the ones turned into climate refugees as rising sea levels inundate the poor in places like Bangladesh. No "green curriculum" is worth the paper it is written on unless it addresses broader issues of social inequality.

In today's world, a deep ecological consciousness is a basic skill. The "buy-until-you-die" consumer orientation that bombards us from morning until night is not sustainable. The planet is in peril, and despite the conceit that suggests we humans are above it all, our fate is intimately coupled to that of the Earth, albeit unequally. It's about time the entire curriculum asks: What about the Earth? ⊕

Bill Bigelow (bbpdx@aol.com) is curriculum editor of Rethinking Schools *magazine.*

A s a teacher, I want to foster in children an ecological identity. I believe this identity, born in a particular place, opens children to a broader connection with the Earth; love for a specific place makes possible love for other places. An ecological identity allows us to experience the Earth as our home ground, and leaves us determined to live in honorable relationship with our planet.

We live in a culture that dismisses the significance of an ecological identity, a culture that encourages us to move around from place to place and that posits that we make home by the simple fact of habitation, rather than by intimate connection to the land, the sky, the air. Any place can become home, we're told. Which means, really, that no place is home.

This is a dangerous view. It leads to a way of living on the Earth that is exploitative and destructive. When no place is home, we don't mind so much when roads are bulldozed into wilderness forests to make logging easy. When no place is home, a dammed river is regrettable, but not a devastating blow to the heart. When no place is home, eating food grown thousands of miles away is normal, and it is easier to ignore the cost to the planet of processing and shipping it.

Finding a Place

Our work as teachers is to help children to braid their identities together with the place where they live by calling their attention to the air, the sky, the cracks in the sidewalk where the Earth bursts out of its cement cage. For me, teaching in a childcare program in Seattle located next to a canal that links Lake Union and Puget Sound, "place" means the smell of just-fallen cedar boughs and salty, piquant air, the sweet tartness of blackberries (and the scratch of blackberry thorns), the light gray of near-constant clouds, the rough-voiced calls of seagulls, and the rumble of boat engines. It is exhilarating to offer children this place as home ground.

Other places are less compelling as home ground. What does it mean to do this work of connecting children to place when the immediate environment numbs rather than delights the senses? What can we embrace in a school neighborhood dominated by concrete, cars, and convenience stores?

Children's worlds are small, detailed places—the crack in the sidewalk receives their full attention, as does the earthworm flipping over and over

A Pedagogy for Ecology

BY ANN PELO

on the pavement after rainfall. Children have access to elements of the natural world that many adults don't acknowledge. When we, like the children, tune ourselves more finely, we find the natural world waiting for us: cycles of light and dark, the feel and scent of the air, the particularities of the sky—these are elements of the natural world that can begin to anchor us in a place.

Rather than contribute to a sense of disconnection from place by writing off the environments around our most urban schools as unsalvageable or not worth knowing, teachers can instill in children an attitude of attention to the natural world in their neighborhoods. The sense of care for and connection to place becomes the foundation for a critical examination of how that place has been degraded. Rick Bass, in *The Book of Yaak,* describes his experience of the interplay between love of place and willingness to see the human damage done to that place: "As it became my home, the wounds that were being inflicted upon it—the insults—became my own."

Every child lives someplace. And that someplace begins to matter when we invite children to know where they are and to participate in the unfolding life of that place—they come to know the changes in the light and the feel of the air, and participate in a community of people who speak of such things.

Cultivating an Ecological Identity

Children know how to live intimately in place; they allow themselves to be imprinted by place. They give themselves over to the natural world, throwing endless rocks into a river, digging holes that go on forever, poking sticks into slivers of dirt in pavement, finding their way up the orneriest tree. They learn about place with their bodies and hearts. We can underscore that intuited, sensual, experiential knowledge by fostering a conscious knowledge of place.

How do we cultivate a love of place in young children's hearts and minds, moving beyond the tenets of recycling to intimate connection with their home ground? From my experiences as a childcare teacher, I've distilled a handful of principles.

- Walk the land.
- Learn the names.
- Embrace sensuality.
- Explore new perspectives.

- Learn the stories.
- Tell the stories.

My primary work is as a teacher in a full-day, year-round childcare program in an urban Seattle neighborhood that serves families privileged by race, class, and education. I've also worked closely with teachers and children in urban Head Start programs. The principles I suggest resonate in these widely varying contexts; all children deserve home ground.

Walk the Land

Contemporary U.S. culture is about novelty and fast-moving entertainment: a million television channels to surf, and news stories that flash bright and burn out fast. This disposition to move quickly and look superficially translates to a lack of authentic engagement with the Earth: Get to as many national parks as we can in a two-week vacation, drive to a scenic view, take some photos, and drive to the next place.

As teachers, we must be mindful of this cultural disposition to superficial knowledge. It's easy to fall into the habit of aiming for novelty, offering children many brief encounters with places, experiences that leave them familiar with the surface, but not the depths. Instead, we ought to invite children to look below the surface, to move slowly, to know a place deeply.

For many years, my emphasis in planning summer field trips was to get to as many city parks and beaches as I could. Each week, we'd head out to two or three different places, so that by the end of the summer we'd taken a grand tour of the city. I thought that by visiting a range of places in Seattle, the children would come to know their city. We had a hoot on those trips, but each place was a first encounter, and offered novelty rather than intimacy. The children came away from those summers not so much with a sense of place as with confusion about how these various places fit together to make up their home ground. We'd skimmed the surface of Seattle, but didn't know its depths.

> What can we embrace in a school neighborhood dominated by concrete, cars, and convenience stores?

Now, my emphasis has shifted. I plan regular visits to the same two or three places over the course of a year. Spending time at the same park and the same beach, we see it change throughout the year. I point out landmarks on the beach to help the children track the tide's movement up and down the beach. At the park, we choose a couple of trees that we visit regularly; we take photos and sketch them to help us notice the nuances of their seasonal cycles. From the top of a big rock at the park, the children play with their shadows on the ground below, noticing how shadow and light change over the year. The children greet the rhododendron bushes like dear old friends, and know the best places to find beetles and slugs.

My commitment to walking the land consciously with children has changed how I walk with them to the park in our neighborhood. I used to focus our walk on getting there efficiently and safely, and chose our route accordingly. Now, I've charted a longer route, one that takes us past a neighbor's yard full of rosemary and lavender and tall wild grasses. We take our time walking past this plot of earth, and I coach the children to point out what they notice about this familiar place. I worried that the children would become bored, walking the same path every day, or would stop seeing the land, so I developed several rituals for our walk. We pause at the rosemary to monitor changes in its fragrance, buds, and foliage, and to watch for the arrival of spit bugs, whose foamy nests delight the children. We pause at the wild grass to compare its growth to the children's growth, an inexact but joyfully chaotic measurement.

Learn the Names

When we talk about the natural world, we often speak in generalities, using categorical names to describe what we see: "a bird," "a butterfly," "a tree." We are unpracticed observers, clumsy in our seeing, quick to lump a wide range of individuals into broad, indistinct groups. These generalities are a barrier to intimacy: a bird is a bird is any bird, not this red-winged blackbird, here on the dogwood branch, singing its unique song.

Most of us don't have much of a repertoire of plant, insect, animal, tree, or bird names; I sure don't. For many years, I wasn't particularly interested in learning the names of the flora and fauna, and imagined that learning the names would be a chore,

a tedious exercise in memorization. When I turned 40 and visited Utah's red rock desert, it awakened me to a passionate love, born in my eastern Washington childhood, which I'd forgotten, or never consciously acknowledged: love for a spacious, uncluttered horizon, love for dirt, rock, and sage, for heat and dust and stars, for open sky. Being there taught me that learning the names is an exercise in love. I was in an entirely unfamiliar place, and had only the clumsiest of generic names for what I encountered: a bush, a rock, a lizard. As I began to fall in love with the red rock desert, I wanted to know everything about it, including the names it holds. I bought a field guide and began to learn the names—the identities—of the plants, the creatures, the types of rock. Each name was a step closer into relationship. The names helped me locate myself in the desert.

I carry a field guide to the Pacific Northwest with me now, when I'm out with the children in my group. We take it with us when we walk to the school playground around the corner, and when we go farther afield. We turn to it when we encounter a bug we don't recognize or find an unfamiliar creature revealed by a low tide. And I've created lotto and matching games from the field guide, photocopying images of familiar trees, birds, marine creatures. We use the images for matching games and bingo games: Together, we're learning the names of this place that is our shared home ground.

Embrace Sensuality

In a culture that values intellect more than intuition or emotion, typical environmental education too often emphasizes facts and information in lieu of experience. Plenty of plastic animals, nature games, videos, and books for children invite them to intellectualize—and commodify—the natural world. Teacher resource catalogues offer activity books and games that teach about endangered species, rain forest destruction, pollution, and recycling. These books and games keep the natural world at a distance.

To foster a love for place, we must engage our bodies and our hearts—as well as our minds—in a specific place. Intellectual and critical knowledge needs a foundation of sensual awareness, and, for very young children, sensual awareness is the starting place for other learning. How does the air feel on your skin? What birds do you hear on the playground?

A friend of mine taught in a Head Start program in a housing development that had been the scene of several shootings, and that had more graffiti than green. She wrestled with how to stir children's numbed senses awake in that harsh landscape where playing outdoors was dangerous. She decided to bring the sensual natural world into her classroom. She added cedar twigs to the sand table, and chestnuts, and stems of lavender. She included pinecones and seashells in the collection of play dough toys. She supplemented her drama area with baskets of rocks and shells, and included tree limbs, driftwood, stumps, and big rocks in her block area. She played CDs of birds native to the Northwest. And in early fall each year, she welcomed the children to her program with feasts of ripe blackberries, making jam and cobbler with the children, telling them about her adventures picking the blackberries in a wild bramble in the alley behind her apartment building.

Explore New Perspectives

Living in a place over time can breed a sense of familiarity, and familiarity can easily slip into a belief that we've got the land figured out. We stop expecting to be surprised, to be jolted into new ways of seeing; we become detached from the vitality of a place.

Our challenge is to see with new eyes, to look at the familiar as though we're seeing it for the first time. When we look closely and allow ourselves to be surprised by unexpected details and new insights, we develop an authenticity and humility in our experience of place, and wake up to its mysteries and delights.

Several years ago, one of the 4-year-old children in my group posed a simple question: Why do the leaves change color? Her question startled me awake: I saw the transformation of color through her eyes, a phenomenon consciously witnessed only once or twice in her young life, and one full of mystery and magic. Her question deserved my full attention, not a recital of the muddled information that I remembered from my science classes in school, and not a quick glance at an encyclopedia. Madeline's question launched our group on an in-depth study of the lives of leaves that carried us through the seasons.

My co-teacher, Sandra, and I took the children on a walk through the neighborhood to study the trees. Moving from one tree to the next, we began to see a pattern, and shared our observation with the

children: The leaves on the outermost branches began to change color before the leaves in the center of the tree. The children built on our observation, adding what they'd noticed: The leaves first changed color on their outermost edges, while the center of the leaves remained green. I suggested that we gather leaves to bring back to our room, where we could study them up close and record what we observed, sketching the details that we saw and adding nuances of color with watercolor paint. As we sketched the lines of the leaves, children pointed out the resemblance between the skeletal lines of leaves—the "bones" of a leaf, the children called them—and the tendons and lines on our hands: "The lines of the leaf feel like human bones." "The lines are like the lines on our hands." Excited by the children's observations, I suggested that we sketch our hands,

> I suggested that we sketch our hands, just as we'd sketched the leaves, knowing that our sketching would help us see ourselves in new ways, as cousins to leaves.

just as we'd sketched the leaves, knowing that our sketching would help us see ourselves in new ways, as cousins to leaves.

As we sketched, I asked the children to reflect on why the leaves change color in the autumn. "What is it about autumn that makes leaves change from green to red, orange, brown?" The children generated several theories: "In the fall, it's cold. Leaves huddle together on the ground to get warm. The trees are cold because they don't have any leaves to keep them warm." "The color is a coat to keep the leaves warm, because it's cold in the fall."

From this analysis, one child made a leap that deepened our conversation: "Leaves get sad when they start to die." From this decidedly unscientific conjecture, the children forged a potent connection to the leaves: "Like we give comfort to others when they're sad, the plant needs comfort." "I think a hug would help a leaf, and being with the leaf." "Maybe you could stay with it. You just give it comfort before it dies." "When it drops on the ground, that's when it needs you."

At Hilltop, we use an emergent pedagogy, developing curriculum from the children's questions and pursuits. In our study of the lives of leaves, I experienced the value of this pedagogy, as we lingered with questions, theories, and counter-theories, and with our not knowing. Our emergent curriculum framework allowed us to explore Madeline's question in the spirit in which it was posed: a question about the meaning of change and the identity of leaves. Through our exploration, we became intimates of leaves, anchored in our place.

Learn the Stories

To foster an intimate relationship with place, we need to know the stories and histories that are linked to that place, just as we do in our intimate relationships with people. In our work with young children, our focus in gathering these stories is as much about the children's imaginings as it is about scientific facts. We can invite their conjectures to complement the facts, opening the door to heartfelt connections.

Visiting a Head Start program one afternoon, I watched Natalie catch ants on the asphalt slab that served as the program's playground. She hovered over a crack in the pavement, carefully picking up each ant that crawled from the crack and dropping it into a bucket. Curious about her intention, I asked what she was planning for the ants: "They're bugs and we hafta kill them." I imagined contexts in her life in which this could be true: Had her family dealt with invasive insects at home? Had she experienced the pain of bee stings and itch of mosquito bites? I wanted respectfully to acknowledge these sorts of experiences, yet I didn't want them to become her only references for understanding and relating to the natural world. I said, "Sometimes, when bugs come into our houses, we have to kill them to keep ourselves healthy. And some bugs can bite us in painful ways. But sometimes we don't have to worry so much about the bugs we find. I'm curious about these ants. Where do you suppose they come from?"

Natalie was quick to imagine the ants' story: "The ants are in the hole talking. If they hear loud noises, they won't come out. We have to be very quiet! If they see us, they stay in because they're scared. When one ant wasn't looking, I got him! I'm faster than them—that's how I catch them."

"What's in the hole that the ants come from?" I asked.

"Maybe their family," Natalie mused. I offered her a clipboard and a pen, and invited her to draw what she imagined was in the hole. She began to sketch, talking aloud as she worked: "They're a family. They talk to each other and bring food to their baby. In the house, there's food and a table and a bed and a seat."

Natalie stopped drawing to look into her bucket: "There's 15 ants in the bucket! That's more than one family. That's a lot of families. They share one house in the hole. The ants come not fast because they're talking, saying their plan to come out to see what's outside. They want to find their family that's in the bucket. The ants in the bucket want to get out of the bucket and go to their family."

Natalie abruptly dumped the bucket upside down next to the crack in the pavement, and tapped it on its bottom. "Go home, ants! Go to your home. Go to your family."

When I invited Natalie to imagine the ants' story it helped her see her bucket from the inside as well as from above, and shifted her relationships with the ants. She moved from a defensive posture to that of being a protector. Particularly for children living in places where the natural world is degraded or dangerous, imagining the stories of a place can inspire new possibilities; it casts children into an active role as people who care about and take action on behalf of a place.

Tell the Stories

We're often encouraged to see the Earth as landscape, which is scenery—something to look at, but not to participate in. But when we collapse the distance between the land and ourselves and allow ourselves to become part of the story of a place, we give ourselves over to intimacy. This can be our work with young children—weaving them into the story of the place where they live.

One way I've begun trying to link the children to the land is by using observable markers anchored in place to measure our lives. "You'll start kindergarten in the fall, when the blackberries are ripe." "Christmas comes in the darkest part of winter, when the

> Natalie abruptly dumped the bucket upside down next to the crack in the pavement, and tapped it on its bottom. "Go home, ants! Go to your home. Go to your family."

sun sets while we're still at school, and the sun doesn't rise until we're back at school the next morning."

And I've been playing a game with the children that I learned from Richard Louv's book, *Last Child in the Woods*, "The Sound of a Creature Not Stirring." We listen for the sounds we don't hear (a leaf changing color, an earthworm moving through the soil, blackberries ripening)—a way to focus our attention on the Earth around us and to participate in what's happening in it.

A Foundation for Action

In *The Pine Island Paradox: Making Connections in a Disconnected World*, Kathleen Dean Moore writes, "Loving isn't just a state of being, it's a way of acting in the world. Love isn't a sort of bliss, it's a kind of work. . . . Obligation grows from love. It is the natural shape of caring." She writes: "To love a person or a place is to take responsibility for its well-being."

From love grows action. In my work with young children, I share stories of local environmental activists who have used their love of place to fuel their action. For example, I tell the story of a group of children and their families who launched a campaign to save the cedar tree at the school playground where we often play.

Children have loved the cedar tree at Coe School for a long time; children played at this tree even before you were born. One year, a mom was at a community meeting and learned that the city park department was planning to cut down the tree because it was damaging the asphalt on the playground with its big roots. She told the children in her daughter's kindergarten class, and those children and their families decided that they had to work to protect the cedar tree and to help the park department find another way to fix the problem of broken asphalt. The children and their families wrote letters to the city workers, telling them about how much they loved the cedar tree, and sharing their ideas for taking good care of the tree and the pavement on the playground. They had a meeting with the city workers, who hadn't known that the tree was important to the children. After the meeting, the city workers decided not to cut down the tree; they made

a plan with the children and their families and the other kids at Coe School about how they could work together to fix the asphalt and take care of the tree.

I watch for opportunities for the children to add their own chapters to the story of activism on behalf of beloved places. I want them to see themselves as part of a community of people anchored by fierce and determined love of place and who take responsibility for its well-being.

The poet Mary Oliver instructs us on how to open the natural world to children:

Teach the children. Show them daisies and the pale hepatica. Teach them the taste of sassafras and wintergreen. The lives of the blue sailors, mallow, sunbursts, the moccasin flowers. And the frisky ones—inkberry, lamb's quarters, blueberries. And the aromatic ones—rosemary, oregano. Give them peppermint to put in their pockets as they go to school. Give them the fields and the woods and the possibility of the world salvaged from the lords of profit. Stand them in the stream, head them upstream, rejoice as they learn to love this green space they live in, its sticks and leaves and then the silent, beautiful blossoms. Attention is the beginning of devotion.

And devotion is the beginning of action. ⊕

Ann Pelo focuses on Reggio-inspired education, social justice, and ecological teaching and learning. She worked as a teacher and teacher mentor for 16 years at Hilltop Children's Center, a full-day childcare program in Seattle, Washington. She is the author of five books, including The Goodness of Rain: Developing an Ecological Identity in Young Children.

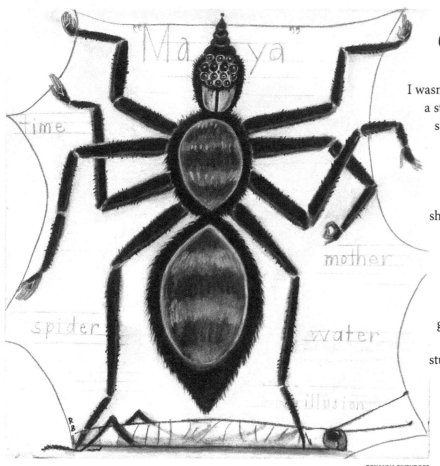

ROXANNA BIXENDORF

Lessons from a Garden Spider

How Charlotte transformed my classroom

BY KATE LYMAN

"**Y**ou're the best teacher in the whole world," said Maisee, as she hugged me. Actually, I wasn't "the best teacher"—it was our new guest, a strikingly beautiful black and yellow garden spider.

Juan had found the spider in the bushes outside our classroom door, and I had set her on a wooden frame on a bookcase in my 3rd-grade classroom. Overnight, she had made a dazzling symmetrical orb web. My students watched in amazement as the spider pounced on a grasshopper and quickly wrapped it in silk. "Awesome!" soon overtook "ewww!" as the students watched the spider feast on the grasshopper's blood.

In the few weeks that Charlotte (as my students named her) graced my classroom, the students learned many lessons from her. They learned about the web of life, the interdependence of predator and prey. Many changed their relationship to spiders from one of fear to one of respect.

My students learned about the sad, yet ever-renewing cycle of life. They also learned many facts about spiders, their body structures, and their different habitats and mechanisms of survival. And in a world that may lose 30 to 50 percent of all species by the middle of this century, I believe that observing and caring for one single spider helped us bridge the gap between the human and natural world.

Charlotte was an excellent team teacher. How could she have known that I had planned to do a lesson on food webs on the day that she first demonstrated her insect-trapping skills to us? As I watched my students hard at work creating food web drawings, I thanked Charlotte for providing a real-life demonstration of the concept. Teaching through words—even with the aid of illustrations in books—just didn't work with this class where 10 out of 14 of my students were learning English as a second language. (Six spoke Spanish as a first language, two spoke Hmong, one Khmer, and one Albanian.)

Reaching English Language Learners

My classroom reflects the increasing diversity and

pockets of poverty in areas of Madison, Wisconsin, a city commonly thought of as a white, well-off university town. About 65 percent of my school's families live on incomes below the poverty level, and 74 percent are from minority groups (mainly African American, Latino, and Hmong), with 33 percent designated as English language learners (ELLs). Some of the ELLs receive daily in-classroom support during the hour-long reading block and also for a few more hours a week during math and science instruction. However, the ESL teacher has to divide her time among three classrooms, all with a high percentage of ELLs, so the support cannot begin to meet every student's individual needs.

As the school year began, only five of my students were reading at grade level, and they all scored low on district tests. As I went over my class's test scores with my principal, I was reminded that more tests were looming: the 3rd-grade writing sample and the Wisconsin Third Grade Reading Test. I thought of all the standards that I was expected to meet. I felt overwhelmed by the challenge of meeting the standards and bringing my students' test scores up while at the same time making school meaningful and inspiring them to learn.

The challenge of teaching diverse learners is multiplied by the testing regimens and the standards that give teachers little freedom to adapt the content and methods of their teaching to meet their students' needs. For example, our district requires us to teach science through FOSS (Full Option Science System, developed by the University of California at Berkeley and published by Delta Education). It's a packaged program, complete with teacher training videos, and step-by-step lessons. We are supposed to teach science exclusively through FOSS and strictly adhere to the format of its lessons.

FOSS does have some advantages: Teachers find it helpful for teaching an area of science in which they may not have expertise. It also provides a structure and materials for beginning teachers. But many of us dislike the packaged, scripted manner in which FOSS is presented. It doesn't provide enough opportunities for the students to engage in genuine inquiry. I have found the FOSS units relating to natural science particularly objectionable.

For instance, the 3rd-grade FOSS "Structures of Life" kit teaches some of the same concepts that Charlotte and I were introducing, but it removes living creatures from their natural habitats and uses them as subjects of experimentation. One lesson I taught in this FOSS module explored the concept of habitat through crayfish that came shipped in boxes (many dead on arrival). We were supposed to store them in plastic dishpans, with plastic bowls for their "shelters," and feed them aquatic plants and cat food.

The plastic bowls proved to be a very unsuitable habitat; the crayfish proceeded to eat each other. This wasn't part of the planned FOSS curriculum, but the students and I discovered through further research a fact that was only briefly mentioned in my guide. When crayfish molt, they need to hide under rocks until their shells harden. Otherwise, they fall victim to predators, which, in our inadequate habitat, turned out to be other crayfish. This in itself would have been a great opportunity for further exploration. Looking back, I wish I had pursued the matter further. I could have encouraged the students to write to the creators of FOSS. At the time, I was happy only to separate the crayfish that had molted to spare my students from coming to school to discover more half-eaten bodies.

Further on in the module (which, after the crayfish disaster, I decided to skip), students are directed to "harness" snails with thread and duct tape and attach the thread to paper clips and metric weights. The expressed aim of the lesson—"to investigate the pulling strength of land snails"—seemed to conflict with the first goal presented in the introduction, to "develop an attitude of respect for life."

I read through this lesson several times with horror. I could not figure out its purpose. When students are finished with the snail experiments, teachers are advised (if they cannot keep the snails or find another teacher who needs to use them) to "euthanize" the snails by putting them in a freezer. ("Terminate" is the euphemism used in other FOSS kits containing living creatures.) The FOSS guide warns teachers that students might ask where the snails went. Its recommendation? "Tell them that you returned them to the place where they came from. Let them extract from that what they will."

A Respected Guest

Charlotte, on the other hand, was a respected guest in our classroom. My class knew that she was borrowed from the bushes outside our classroom and that we planned to return her. Never once did I see a student poke at her or disrupt her web creations. When students from other classrooms or my students' brothers and sisters came to visit Charlotte, my students taught them how to treat Charlotte with respect. My class became concerned about our spider's welfare at our school's Open House, when many families would be visiting with younger children.

Emily volunteered to write a warning note, which she taped to the bookcase:

> Charlotte here! This is a garden spider. The garden spider's name is Charlotte. Don't move the bookcase or the spider might fall. Charlotte is very kind. Do not touch her web. She made it to catch flies for our classroom. Don't take nothing from the bookcase. Don't touch the spider or she thinks you will hurt her. Please stay back. And her favorite food is grasshoppers.

In the next several weeks, as we continued to observe Charlotte, we learned about many different kinds of spiders through books and videos. Juan, a low reader and extremely reluctant writer, experienced a small transformation after he found Charlotte. Even the teacher across the hall remarked at how he came to school every day with a smile on his face. Every day during independent reading time, I'd find Juan reading spider books. And now he was writing four or five sentences at a time about spiders.

We learned about a garden spider's life cycle. Charlotte again came through by laying her eggs and weaving an egg sac. There was some controversy over what to do with the egg sac.

"Let's keep it in our room," suggested Chou. "Then in the spring, we'll have a whole bunch of baby spiders in our room!

"No!" protested several students. As much as they appreciated Charlotte's company, several objected to the idea of hundreds of spiders running around our classroom. Lilly suggested we place the egg sac back on the bushes where Juan found Charlotte so her babies would be born there. But other students worried that kids from the school might disturb the egg sac. Eventually, we agreed to bring the egg sac to the Aldo Leopold Nature Center, a preserve we had visited on a class field trip.

Charlotte's Exit

The students knew from our studies that after creating her egg sac, the garden spider dies. At the start of every day they ran over to the wooden frame to see how Charlotte was faring. She surprised us by weaving a few more webs and continuing to eat grasshoppers. But then she moved to the corner of the frame and stayed there for several days.

What Rosita wrote at that point proved that my students' connection with Charlotte transcended language barriers. Rosita is able to read and write in Spanish but knows only a few words in English. Because most of the instruction is in English, Rosita rarely appeared to be paying attention. But she was paying attention to Charlotte. This is what Rosita contributed to our class newsletter:

> Tenemos una araña en la clase que se llama Charlotte. Es una araña de jardín. Ella atrapó saltamontes y los envolvió con su telaraña. Ya tuvo sus huevos. Charlotte hizo una bolsita para que los metiera sus huevos. Charlotte ya no está en la tela que hizo. Charlotte está en la esquina. Charlotte ya no tiene el color amarillo. Ahora el color que tiene es gris. Charlotte no se mueve de su telaraña. Charlotte se va a morir cuando nacen los huevos.

> We have a spider in the class whose name is Charlotte. She is a garden spider. She caught grasshoppers and wrapped them with her web. Charlotte already laid her eggs. Charlotte made a little bag to put her eggs in. Charlotte is no longer in the web that she made. She is in the corner. Charlotte is no longer yellow. She is gray. She doesn't move from her web. Charlotte is going to die when her babies are born.

Rosita's observations were borne out the next morning. "Oh no, Charlotte is dying!" moaned Maisee.

"Look at her color. She is turning gray," noted Chou.

"She's getting smaller," said Juan. "Why is she getting smaller?" Several students gently touched Charlotte.

"She is officially dead," concluded Lilly. Several moans and many sighs followed.

"Our only spider!" said Maisee.

"She was a good trapper," added Emily. "We will need to bury her."

Emily put herself in charge of the funeral arrangements, labeling a small casket (an earring box) "Our best friend." Emily and Maisee drew pictures of Charlotte to place by her grave. Lilly wrote her gravestone inscription: "Here Lays Charlotte. Room 27's Spider!"

We discussed where to bury her. Juan said that she should be buried outside of our classroom, near the bushes where he found her. We planned the burial for the next day. At the end of the day, Chou ran to his locker and pulled a wilted bouquet of purple flowers from his backpack. I took the bouquet and thanked him. "Are they from your garden?" I asked. "Yes," he answered. "They're for Charlotte's grave."

The next day, we had Charlotte's funeral. Twice, by student request, we passed the box around the circle as students gravely inspected the dead spider. That afternoon, the class watched as I dug a hole and placed the box under the dirt. Lilly taped her grave marker to a tongue depressor and placed it in the dirt. Emily and Maisee added pictures they had drawn of Charlotte. "So we'll always remember her," said Maisee. Denitra thanked Charlotte for being part of our class. "Thank you for teaching us how you spin webs and how you catch grasshoppers. You were a nice spider. We will miss you." We went around the circle. One student at a time thanked Charlotte.

Denitra asked to speak again, "And I hope your babies will be born here and make their own webs."

I have no doubt that Charlotte helped teach my class our district's science standards relating to "characteristics of organisms" and "life cycles of organisms." She also engaged my students in learning and inspired them to expand their reading and writing abilities. Even after her burial, students continued to write about Charlotte.

Charlotte has even helped me teach multiplication; story problems about multiple legs have endless possibilities.

Beyond Charlotte

Our class moved on from studying spiders to insects, comparing the body structures, behaviors, and life cycles of a variety of insects. Although I used some lesson plans and student sheets from FOSS, we ob-served the insects in the schoolyard and only borrowed living things that we could return.

We did not limit our studies of insects to observations and worksheets. Instead, we learned and shared through literature and videos, visual arts, music, and drama.

As the weather got colder, I continued to have guest animals in my classroom. But they were homeless companion animals from the Humane Society instead of wild creatures. Most of my students lived in apartments and few could have pets, so they enjoyed the opportunities to meet and care for pets. (I was the only one with allergy problems.)

We further explored life cycles as we examined human beings from birth to death, and we shared and wrote about our own personal and cultural histories. We continued our examination of nature's cycles by studying the moon and its phases.

> I believe that observing and caring for one single spider helped us bridge the gap between the human and natural world.

Later in the year, we returned to topics relating to life and environmental science. Charlotte taught my class to care deeply about the fate of one spider. I saw how those feelings transferred to their research about and advocacy for endangered species.

Charlotte also taught me many lessons. She taught me not to underestimate my students. They did improve their reading and writing. They did make progress in learning multiplication and mastering science concepts.

With motivation, trust, and classroom experiences that touch them as deeply as Charlotte did, my students will learn. As our spider's namesake in E. B. White's book points out, spiders are "naturally patient." They know that if they construct a well-designed web and wait long enough, their efforts will pay off. Teachers have a lot to learn from spiders. ⊕

Kate Lyman has taught kindergarten through 3rd grade for more than 35 years in Madison, Wisconsin. Her professional interests are in the areas of multicultural and environmental education. Students' names have been changed.

SCOT BAKAL

"Before Today, I Was Afraid of Trees"

Rethinking nature deficit disorder

BY DOUG LARKIN

The last week of February turned out to be the peak of the maple-sugaring season that winter, and an inch of snow remained on the ground as the juniors in my chemistry class disembarked from the bus. Kevin Kopp, our guide, met us with empty buckets, and the students ran their hands over the bark of the oaks, beeches, and maples with an uncharacteristic quietness as Kevin talked about the different types of trees in downtown Trenton's Cadwalader Park.

For the moment, it was possible to imagine that we were somewhere other than urban central New Jersey. To many of my students it was a revelation that we were not actually standing in a natural forest, but in a place where someone had purposefully planted each of the towering trees decades ago. Their questions led to a discussion about how trees in urban environments not only look nice, but also help clean the air and lower energy costs by reducing the amount of sunlight absorbed by city surfaces on hot days.

When Kevin pointed out holes in one tree drilled by yellow-bellied sapsuckers, a group of students began racing from tree to tree, seeing who could be the first to find and run their fingers over undiscovered sapsucker holes. Students in another group raised their eyes to the treetops, looking for the birds. Kevin held up the metal spout and asked if anyone would like to hammer in the first tap.

"Ain't it gonna hurt the tree?" asked one of my more solemn students. Kevin assured him it wouldn't. We tapped two trees that day, and everyone who wanted a turn hammering got one. When the first bucket was finally hung from the tap, the students were clearly less than impressed with the leisurely *drip, drip, drip* of the watery tree sap. After explaining that the buckets would fill over the next few days, Kevin promised to bring them to us at school.

A Surprising Confession

We thanked Kevin, boarded our bus, and returned to school shortly before the end of the 80-minute block period. Back in the classroom, one of my African American students pulled me aside, saying, "I have to tell you something." At the beginning of the semester in January, she had loudly proclaimed to the whole class that she didn't like science. Though she dutifully completed her chemistry assignments, little in science class had seemed to hold meaning for her. So I was hoping for one of those minor teaching vindications, the kind of small victory that keeps weather-beaten teachers in urban schools coming back year after year. What she said, however, was more profound: "Before today, I was afraid of trees." This wasn't what I expected a popular high school junior—or anyone else, for that matter—to say, so I asked her to explain. She laughed and shrugged: "They're big and scary, and their bark is all rough-looking." She went on to describe how nature and the outdoors in general had always seemed somehow dangerous to her.

In thinking about what ideas my students might have about maple syrup making, a fear of trees had genuinely never occurred to me. I've looked over state and national high school science standards carefully since then, and none of them include anything about making students feel comfortable around trees. Yet there it was, as an unintended and valuable outcome of the trip.

Nature Deficit Disorder?

Science teachers in urban schools often serve students whose experiences with the natural environment are more obviously constrained by human factors than their suburban or rural peers. At the same time, parents and teachers are contending with an increase in sedentary indoor activities affecting young people of every demographic. Coupled with the shrinking opportunities many children have for experiencing nature in an unbounded form, these factors can lead to a lack of familiarity with the fundamental features of ecology and the natural world. Journalist Richard Louv has called the resulting situation "nature deficit disorder," a term that has resonated strongly with the environmental education community.

Yet those of us with an understanding of the history of multicultural education have good reason to be suspicious of any terminology that causes teachers to view specific students from a "deficit" perspective. Although Louv is careful to describe the deficit as applying to children's experiences with the natural world, like many diagnostic labels, it can become affixed to students themselves. Historically, some have used deficit language to marginalize students already struggling in schools. Rather than exploring options that seek to build upon students' strengths, experiences, and prior understandings of the world, people have used notions of cultural deprivation or deficiency to explain away the failures of teachers to be effective in teaching such students.

Environmental Education as Consciousness Raising

Administrators are fond of saying—particularly around testing season in the spring—that every teacher teaches reading and every teacher teaches math. I agree, but I will argue that we all must be environmental educators as well. This is especially important for science teachers because fundamental ideas in biology, chemistry, physics, and earth science are crucial to making sense of pressing environmental issues such as climate change, industrial pollution, radioactive waste, food safety, and the destruction and alteration of habitats.

> Those of us with an understanding of the history of multicultural education have good reason to be suspicious of any terminology that causes teachers to view specific students from a "deficit" perspective.

Three related goals frame most environmental education efforts. The first is fostering and sustaining a love of nature. The second is gaining a scientific understanding of the environment, with knowledge of the factors, processes, and interrelationships that describe and explain the living world. The last is helping students make intelligent and informed choices about where they live (defined as anywhere from the local neighborhood to the planet), which may or may not include a specific focus on social and environmental justice. These goals demand that we provide students with the attitudes, tools, and skills for action in their world. All are important, but schools and teachers prioritize them differently.

With these goals in mind, I often tried to push beyond the traditional sequence of topics in the discipline—which usually follows a predictable progression from properties of matter, through atomic theory and bonding, to the study of chemical reactions—to practical questions concerning the environment. Some chemistry curricula (e.g., Chem-Com) provide connections to environmental issues, and I often drew from these resources for teaching. There are countless possible links between chemistry and environmental justice, but like any curriculum, how far these go depends on the teacher.

For example, in my chemistry class that year, my students examined the suitability of Yucca Mountain as a repository for nuclear waste as part of our nuclear chemistry unit. Later, during a lesson on ions, we watched clips from the movie *Erin Brockovich* as a catalyst for discussion of both environmental justice and the electron configuration of hexavalent chromium (Cr+6).

In a high school class, being able to connect matters of environmental justice to chemistry is dependent upon students' ability to recognize their personal connections to these issues. That's why I'm glad I stopped ignoring Kopp's emails and got my students out tapping maple trees.

Not a single student passed on the opportunity to eat pancakes with syrup they had literally traced to its roots.

Tapping into Community Resources

Kevin Kopp is a well-known environmental educator in the central New Jersey area. I had dismissed his program announcements for many months because many of the activities seemed geared to younger students. One of his emails finally caught my attention:

> Bring your class to Cadwalader Park and learn about one of the true wonders of nature. This is a special program on maple sugaring and making maple syrup. Through stories and hands-on activities, students will learn about

the legends, history, and modern practices of making maple syrup. Each group will tap a tree, collect sap, and taste some Cadwalader Park maple syrup.

The idea that trees in our city could produce maple syrup was too good to pass up. Not only would this trip to Cadwalader Park—a place that had been publicized in recent years more for its violent crimes than for its environmental features—provide students with a solid connection between food, chemistry, and the environment, it would also be a way to develop their critical consciousness. Why was something like maple syrup, which could be gathered from trees in their neighborhoods, only a trace ingredient in the commercial packages of syrup that came with school breakfast? Pushing further, we could also ask who controls where our food comes from and why. If food exists naturally in places like our city park, where else does it grow? And what responsibility does that place on us to ensure a world that sustains these and other resources for current and future generations? *Not* thinking about these issues is an element of Louv's nature deficit disorder description, but we can just as well consider it a form of oppression. I thought that if students could see their own city differently, it would open up possibilities for further exploration.

Although I hadn't planned to teach solutions and molarity for another two months, it proved simple enough to develop an appropriate lesson on concentration that related to the task of boiling down syrup.

Kevin and I constructed a minimalist plan for the trip over email, agreeing to the approach of simply tapping the trees and telling stories along the way. He would focus on the history and process of maple sugaring, and I would inject science ideas into the narrative wherever appropriate. In preparation, I found myself reading widely about the science of maple sap, and reconsidering some of my own ideas about how it was produced. I would be remiss if I did not mention that I had to review most of the tree biology I had ever learned.

Kitchen Chemistry

The week after the trip, Kevin delivered two full five-gallon buckets to our class. According to our

plan, one was full of the maple sap and the other was tap water. We set students the challenge of identifying each without tasting. Within a short time, lab groups were scribbling possible procedures in their notebooks. One of the liquids was clear and the other was cloudy. I informed them that fact, in itself, wasn't enough to prove the liquids' identities. After all, these buckets of liquid had been sitting outside for a few days. That was one reason I ruled out tasting the unpasteurized liquids; also our chemistry lab equipment was not "food grade."

Once I had approved students' procedures for safety and feasibility, I let them proceed with their ideas for testing the two liquids. Though I had hoped that at least one group would examine the densities of the two liquids, as soon as the first lab group started a Bunsen burner with an evaporating dish, the rest of the class took the hint and quickly followed suit. Before the end of the period, many of the dishes contained a brown syrup in the bottom. There was still some skepticism on the part of the students that this was actually the same stuff they would put on waffles—especially because I wouldn't let them taste it.

The next week, I brought in some saucepans from home. Careful to keep the food preparation area in the back of the room clean, safe, and free of any lab equipment that could potentially contaminate the syrup, I tried to boil down the sap on an electric hot plate, but scorched my first test batch

during homeroom. I asked student volunteers from my morning physics classes to help me by stirring the sap and continually adding more whenever the mixture started to get too concentrated. By the time my chemistry class arrived in the afternoon, I had about 50 mL of Cadwalader Park syrup ready, and together we started another batch boiling down.

Although a formal study of chemical reactions was still weeks away, there would hardly be a more opportune time to discuss the chemistry of cooking. I did a brief demonstration of the differences between baking soda and baking powder, showing how each needed an acid in an aqueous solution in order to release the carbon dioxide. We examined the ingredients in a box of store-bought mix. Then, using a griddle from home, I whipped up a batch of pancakes, cutting them open to point out the role of the carbon dioxide in the rising of the batter as it cooked. In order not to break my own rules about food in the lab, we remained in the front half of the classroom as I passed out paper plates and plastic forks, and not a single student passed on the opportunity to eat pancakes with syrup they had literally traced to its roots. Later that week, a number of my students composed an email to Kevin. It read:

> Dear Mr. Kopp,
> This is Mr. Larkin's class.
> Bucket #1: Water
> Bucket #2: Sap
> We realized this after scientifically heating the evaporating dishes. After around five to 10 minutes the water began to evaporate from both dishes. Soon dish #2 began to turn a brownish color, while dish #1 just evaporated until nothing was left. When we first examined the contents of containers 1 and 2, container #2 had a cloudy color that differed from container #1. The sense of smell around the dishes was also different.
> Thank you for your time.
> Edwin, Kamal, Julio, Kenyatta,
> DeShawn, Yessenia, Sonia, and the rest of our period 6/7 chemistry class

Cultivating a Naturalist Intelligence

When I originally proposed the maple-sugaring trip to my students, they were eager to do something out of the ordinary, but they expressed doubts that we would really be doing chemistry. It was not unusual in their experience to go on a "fun" field trip with the loosest of ties to any learning goals.

Some teachers might see my student's admission about her fear of trees as deficient or even patholog-ical. Undoubtedly, her prior ideas about trees were influenced by the broad societal forces Louv described, which teach children to avoid having direct expe-riences with nature. Though Louv meant his critique to apply widely to modern childhood, in reality young people most likely to be considered deficient live in urban environments. Given the residential demographics of the United States, it is also true that these children are more likely to be children of color. Louv has giv-en voice to a vitally important idea, but his deficit language is ultimately harmful to students who could benefit from its implications. What's an urban envi-ronmental educator to do?

> **Developing naturalist intelligence entails more than just letting students loose in nature, though that isn't always a bad idea.**

Developmental psychologist Howard Gardner, who promoted the notion of "multiple intelligenc-es," revised his list to include the notion of a *natu-ralist intelligence*. He described it as the intelligence that "enables human beings to recognize, categorize, and draw upon certain features of the environment." Rather than viewing a student's understanding of the natural world as a deficit to be remediated, Gardner's work suggests that student ideas and ex-periences are a starting point.

Developing naturalist intelligence entails more than just letting students loose in nature, though that isn't always a bad idea. As teachers, we need to not only understand our students' prior knowledge about the natural world, but the ways they think about and experience it as well.

I recently observed a lesson in an urban middle school. The white teacher from a suburban back-ground had instructed her students (who were all African American) to go outside and look at the sky every night in order to collect data on the phases of the moon. Sitting in the back of the room, I heard one young girl say to her friend that she couldn't do the assignment because she was not allowed to go outside at night where she lived.

Although such a situation might be considered one more example of how nature deficit disorder has severed students' connections with the natural world, a different perspective might reveal this re-striction as her parents' rational response to living in a dangerous neighborhood. Had the teacher known more about her students' lives, she could have inten-tionally structured the activity differently, perhaps by anticipating times when students could observe the moon during daylight hours. Doing so would have built upon the opportunities students actually had for engaging with the natural world, rather than relying on the teacher's assumptions.

It's equally important, however, that urban ed-ucators don't assume students in their schools have minimal experience with the natural world. Some students and their families may come from or reg-ularly visit rural areas. Others, such as the Hmong students I taught in Wisconsin, may hunt in nearby forests, leveraging their rich cultural understanding of the natural world to a new ecosystem quite differ-ent from the hills of Southeast Asia.

Rather than viewing students as having nature deficit disorder, teachers can develop students' nat-uralist intelligence and critical consciousness by building on the ways they actually experience the world. Educators can extend Louv's ideas into for-mal science classrooms (without the deficit language baggage) by cultivating more opportunities for nat-ural experiences. The environmental knowledge a student uses to ride the subway system across town to school in New York City is quite different in char-acter from that of the suburban New Jersey youth who plays soccer every Saturday on a grassy field. It is also different from the knowledge a teenager in rural Pennsylvania uses to plant corn and milk cows. Yet all are forms of environmental knowledge that hold opportunities for teachers to meet students where they are in their thinking about the connec-tions between the human-influenced environment and the natural world.

They also give science teachers good reasons to make pancakes and syrup in class. ⊕

Doug Larkin teaches in the department of curriculum and teaching at Montclair State University in Montclair, New Jersey. Students' names have been changed.

My students' home terrain consists—at least on the surface—of houses, streets, schools, and stores. Like many urban kids, much of the unpaved, unfenced nature my 2nd and 3rd graders have access to is the domain of blown trash, sewage treatment smells, and "stranger danger."

But even the most beleaguered cities have hidden and not-so-hidden pockets of natural resilience. In Portland, Oregon, we have the Columbia Slough, a once-heavily polluted ancillary channel of the Columbia River that most of my students have been warned to stay clear of. It's not a pristine wilderness, but it's a wild space within our city that has provided many opportunities for my students to connect with their environment. Several years ago, I set out to explore both the concept of watersheds and the wild spaces in the slough with my students.

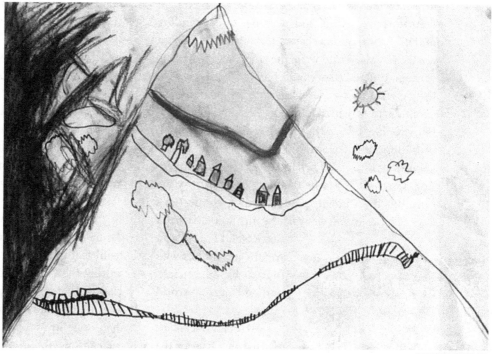

JULIAN ROHN-CAPELLARO

I hoped to bridge the gap between science curriculum and local communities. Ask students what they know about the impact of deforestation on animal habitats and if they offer any answers at all, I'd wager they are more likely to bring up the Amazon than the impact of urbanization and industrial agriculture on the songbird populations in their own communities. This lack of grounding may help explain why critical ecological issues like global warming, groundwater pollution, and deforestation seem so remote to most children.

I wanted my city kids to gain an organic connection to questions of environment and a concrete understanding of how ecosystems work. In previous years I felt I had squandered the learning potential of my school's patch of urban wilderness and reinforced my students' dislocated and disconnected sense of their natural surroundings. By studying the slough and the Columbia River watershed, my students have gained a sense of their impact on the world and how the choices they make—literally—trickle down.

Exploring Our Urban Wilderness

A 2nd- and 3rd-grade teacher helps his students discover the natural world in their urban environment

BY MARK HANSEN

Exploring "Place"

During my first years of teaching, I walked with students around the community, exploring the public housing project where many of them live. I wanted them to learn about their part of the city and how it fit into the larger urban landscape. I also wanted them to have a local and personal understanding of the forces that had shaped the community. I had them draw maps and write the stories of the busiest corners. I hoped they would find the links between their own stories—of moving into a new house, of climbing trees with their baby cousins, of getting into trouble, of arriving from a different country, of finding new friends—and the stories they read about the history of Portland. This kind of connection across time can be daunting for most 8-year-olds, but my students made startling connections between the recent arrival of some students from Mexico and the African American families who moved to our neighborhood during World War II.

We made 3-D plans of what our part of the city would look like if the students were in charge of the massive urban renewal project that was under way. They composed poems to their favorite plum trees and fire hydrants and wrote letters arguing for their preservation when the bulldozers came through. They paid homage to their favorite places and voiced what they thought was lacking (accessible shopping, a library, a clinic, rooftop gardens). Teams designed ideal neighborhoods, which we presented to city planners and families in the community. The activities asked students to think about the constituent elements of a healthy and happy community. What does a community need to function? Which features are absolutely necessary?

I evaluated the projects on the plausibility of the communities, whether or not people actually could and would want to live there. I invited architects and planners from the housing authority to explain the criteria they were using to redesign local public housing. My students debated whether a redesign was even necessary if it would mean that local residents would be displaced during the reconstruction. Fantasy plans for large houses with swimming pools began to give way to more community-centered features like parks, "old folks houses," and "a bigger school!"

Cecil even convinced his group to pare down the size of housing, because "you don't want to have to clean all that."

Mapping Flora and Fauna

As I biked around the neighborhood one day, I remembered LaTia arguing for "more nature, like wild animals," in our neighborhood. I'd agreed with her notion, but her planning team had ultimately dismissed nature as a low priority. Our focus was steadfastly "urban."

I decided students would map the flora and fauna of our neighborhood, to start identifying the wild features. We walked the neighborhood again, noting the kinds of trees that grew in parks and in people's yards, looking for wildflowers in the brambles by the railroad tracks. We revisited work they had done in 2nd grade on the animals that inhabited the neighborhood. I had them return to their maps and add places where plants grew untended and where animals might live. The crucial development in our ecological study of the area came when a blurb in the newspaper tipped me off that a new bike trail had opened a few blocks from our school, allowing access to the Columbia Slough.

I had taken for granted that my school lacked access to natural and wild settings. Early in my career, I had worked with suburban 5th graders monitoring a stream near their school, but it seemed impossible to expose my urban students to hands-on field science. I realized that there was no excuse for keeping my students cooped up in my class reading about the workings of some fictitious, abstracted watershed when they could walk five blocks to the wetland that drains our own watershed and get their hands dirty.

In their own backyard, students could learn about the ongoing negotiation between the world of trucks, houses, and pipes and the world of trails, nests, and sloughs. Students could see they lived in both of those worlds whether they stepped into the latter or not. I wanted them to apply their scientific

> **In their own backyard, students could learn about the ongoing negotiation between the world of trucks, houses, and pipes and the world of trails, nests, and sloughs.**

and civic understandings of their world in complex ways to address the notion of sustainability. I wondered if an understanding of the ecological processes taking place in the watershed would broaden and tie together their sense of the processes at work in our "built" neighborhood.

I wrote a letter/permission slip explaining to our families that we would use a paved and frequently used path, that we would not be in the water, and that children would be under direct supervision at all times. I also took the opportunity to invite them to join us in becoming slough scientists. Every single family responded positively, and a handful of brave parents said they would come along.

Before our first trip I explored what my students already knew about the slough. They thought it was stinky and polluted because it is so close to the sewer treatment plant in our neighborhood. They said the water was poisonous and that a lot of weeds grew there. Everything they said was negative. It was clear that part of our project needed to focus on rehabilitating the reputation of this little patch of wild space. I wondered how kids whose only immediate experience with nature was so grim could possibly maintain an optimistic outlook for the future of the environment. I hoped this curriculum could foster an ecological sensitivity and concern while staving off the pessimism that your average unit on the prospects of global rain forests can instill.

I asked what they wanted to know about the slough and they offered: "Is the slough safe?" "What will we see there?" "Do people live there?" "What animals live there?"

I added my own questions, letting them know I was learning, too: "How is our neighborhood connected to the slough?" "How does it feel to be there?"

The Urban Wilderness

Following the little map in the newspaper, we crossed the busy trucking lane that separates our school from the slough and entered a thriving urban wilderness. It's a complicated terrain, unlike any notion of wilderness I had ever entertained, completely consistent with the contradictory landscape I described above. Some kids were still reluctant to go, writing in their science notebooks that they anticipated it being "smelly and nasty."

The slough drains our watershed, and hosts a curious mix of occupants. It is home to both Portland International Airport and endangered western painted turtles. There are small mountains of stacked shipping containers and low, marshy islands that hide great blue heron rookeries. Coyotes and deer skulk along the railroad tracks, passing thick stands of cottonwood trees and acres of Subarus and Toyotas fresh from their berths in tanker ships.

The slough was also home to Vanport, Oregon's first large African American community, which was destroyed by a massive flood in 1948. I was aware of the many contradictions of the Columbia Slough's landscape, but I had yet to connect them to my school's neighborhood or to my own curriculum.

On our first few trips, fears about bad smells and danger abated as we spotted red-tailed hawks, woodpeckers, and raccoon tracks only a few yards from the railroad. We noticed tall Douglas firs that had a mysterious vine creeping up them and evidence of beavers chewing on some of the willows near the water. We saw a city landscaper working in a dirt field near the train tracks and asked him what he was doing. He explained that he was seeding the area with native grasses to keep the soil from washing into the slough when it rained. On the bridge over the slough we noticed an interesting bird in the water that "looked like a stork," according to Laney. Camille wasn't convinced so she consulted the bird guide her dad had stowed in her backpack. It turned out to be a great blue heron. Jabari and some others discovered coyote scat with evidence of rabbit fur. Mallard ducks swam away as we tromped down toward the shore one day, trailing some startled ducklings. Students began to fill their journals with lists of our sightings.

I was as astonished as the students were at the abundance of animal life. I started to ask them to reevaluate their connection to the watershed: What did it mean that an ecosystem in our area was healthy enough to support predators like coyotes? Was the area really even part of our "neighborhood" if it had

> I hoped this curriculum could foster an ecological sensitivity and concern while staving off the pessimism that your average unit on the prospects of global rain forests can instill.

so many animals in it? We were, after all, residents of a city. I was in no rush to answer these questions, because I believe healthy disequilibrium is at the heart of critical and scientific thinking. And frankly, the kids were not prepared to think that way without a lot of science instruction and time to mull over the issues.

I first asked kids to reflect in their journals about how they felt when they were in the slough. Some answers were negative:

"I was bored and I don't like walking so far," wrote Matthew.

"My mom said I can't get my clothes dirty, so I'm not going next time," Attiana wrote.

Some could not find enough exclamation points:

"I was really happy! I'm going to take my brothers there on our bikes!! And maybe I will show them the coyote scat!" wrote Angel. Some students began to connect their actions to the state of the watershed in the reflections I asked them to write. After one trip, Tiana wrote:

> Today it was raining and I am mad because Mr. Hansen said if it rained we can go back to school, but he lied. But I am having fun. Five minutes ago we saw some frogs and for now we saw one lizard. . . . Now Mr. Hansen is talking to those kids about our toilets, saying whatever you do in the toilet will go down to the slough sewer place. I didn't like it here [at] first but now it is like part of an interesting place. I think we should find a way to make sure the water stays nice, not toilet water. I think the slough shouldn't be so dirty from the city and the cars. These trucks are too loud. Where are they going? Where do all these animals come from? They better stay away from the trucks. We could clean this up. Bye.

Tiana's answer wandered, but I heard her trying to find herself in this new environment.

My students were beginning to understand the relationships that structure our environments, both built and unbuilt. They were beginning to think in terms of systems, a skill they can apply to other frameworks, as Ibrahim did in his science journal:

> The reason I like nature is I like to collect information and details when we go on trips. I think that the city is our habitat but instead

of rabbits there are cars. The cars need gas for their food and they need the streets and parking lots for their shelter. That's why there are gas stations in every part of the city.

I hope that fossil fuel-powered cars are only a temporary piece of our urban ecology. But Ibrahim was thinking of the city as a place with features that can be explained by the connections between them. I doubt he would have made those connections if he hadn't gotten so close to the rabbit warrens in the brambles of the slough and if I hadn't asked him to write about it regularly.

As the year drew to a close and we went on our last trip, I asked the kids to return to our original questions about the slough and reflect on how their thinking had changed. Some had little to say. A few said a little about seeing animals, which was fine, but didn't say much about themselves or the connections they made. Many kids echoed Marta, who wrote:

> I changed a lot. I like nature a lot. I learned about plants and birds and the city environment. I used to hate to get dirt on me, but now I don't mind. I will be a scientist.

I was most touched by Cortney, a girl who had been initially reluctant to explore our watershed and had shown little interest in finding her place within it. She wrote, "I learned that nature is not just a thing. It is something that you discover."

I wish Cortney had written more, because I was heartened to hear that she was more engaged than she had appeared. And she was right on target: Nature is not just a thing that can be learned in a year, and neither is a city. They are both about discovering relationships over time, many of them quite personal and local. This understanding of local, tangible relationships is the seed of critical insight into broader, more complicated environmental issues. As a teacher, I've learned the way to fuel a student's long-term passion for environmental justice is to deepen their relationship with their place. ⊕

Mark Hansen teaches at Lewis Elementary in Portland, Oregon. He is a co-director of the Oregon Writing Project and an editor of Rethinking Elementary Education. *The names of the children in this article have been changed.*

LOLITA PARKER JR.

As a freshwater paddler who had recently moved from Memphis to the Mississippi coast, I first encountered Turkey Creek and Mr. Flowers White from my canoe. White was clearly having more luck with his cane pole than my boys and I were with our spinning reels. He taught us some insider's tricks and then invited us back to his home for fried crappie. As he shared his fish and the recipe for a mean fry batter, he also shared stories about himself and his community. This was my introduction to the rich history that my students at Gulfport High School and I would later explore.

White traces his family history back to the founding of Turkey Creek in 1866. A group of freed slaves bought 320 acres of "swampland" north of what would later become Gulfport. They created a self-sufficient and socially isolated community of farms, residences, businesses, a church, and a school that flourished for more than a century. Then, beginning in the 1980s, Gulfport's growth nearly led to Turkey Creek's demise.

Gambling was legalized in our area in 1992. With the inrush of casino money, Gulfport became Mississippi's fastest growing city. City planners sought to annex only the affluent areas north of town, which would have created a dumbbell-shaped city. When a judge ruled against this plan, Turkey Creek was included in the annexation. Gulfport's

Looking for Justice at Turkey Creek

Students in Mississippi get out of the classroom and into the past

BY HARDY THAMES

planners then proceeded to make decisions with weighty implications for the fate of this low-income black community without including its members at the table. Acres of wetlands in the Turkey Creek watershed were filled in and new zoning laws passed to allow Walmart, Family Dollar, and other commercial buildings to go up along the section of Highway 49 (Gulfport's north-south corridor) that intersects Turkey Creek. As a result, the Turkey Creek community has shrunk precipitously and is now surrounded by an airport, concrete and electrical companies, and a strip mall business district.

According to Ella Holmes-Hines, Turkey Creek resident and longtime city councilwoman, the community has been "under attack" ever since its incorporation at the expense of Gulfport's growth. The environmental and political repercussions for Turkey Creek have been profound.

Hidden in Plain Sight

My students pass over Turkey Creek every time they go north of I-10 to get to the movie theater, the soccer fields, or any of the big-box stores and fast-food restaurants they frequent. In fact, the shortcut to Walmart takes them right through the heart of the Turkey Creek community. But I suspected that the community, creek, and watershed—their history and the controversy surrounding them—were as invisible to my students as they had been to me. I decided that Turkey Creek would be the next "text" for my elective contemporary issues course. In this history class, I attempt to empower students and foster empathy through case studies. The 21 students enrolled in the course were representative of our large public high school's demographics: 57 percent black and 40 percent white, with only a few Latina/o and Asian American students.

I introduced the unit to my students by reading an excerpt of the history compiled by Derrick Evans, founder of the Turkey Creek Community Initiative,

> **When my students read that Councilman Kim Savant had referred to Turkey Creek as a "drainage ditch" that would service a growing Gulfport, my students took offense.**

in response to the threat posed by the development of wetlands:

> The pioneers who settled the poorly drained "eight forties" [the eight original 40-acre parcels] were visionary, industrious, and innovative. With far less financial, political, or social capital than the celebrated founders of Gulfport, Turkey Creek's early settlers created arable land to practice sustainable agriculture, and developed a viable, self-sufficient African American community bound together by local customs and institutions.

My students were surprised to learn that such a community existed within our city. I challenged them to identify it: "Where is the African American community that was founded in 1866 by people who had been enslaved?" No response. "Where is the community, just a few miles from here, that predated the city of Gulfport by 30 years?" No response. "Which community's homes and businesses used to be where the airport now stands? Most of you cross the bridge running through it every day." Finally one student offered, "Oh, that bridge past the golf course on the way to Sonic?" I promised that, after our study, they would know the place's history and importance, and they would get to see the bridge from a canoe. With that, my students were on board.

Before we began our study of Turkey Creek, I searched the archives of the local *Sun Herald* newspaper, as well as the internet and academic journals, for relevant and accessible articles. Then I divided the published literature on Turkey Creek among small groups of students. I explained the overall driving question for the unit: "How is what happened to Turkey Creek a social justice issue?"

The students read the articles in their groups and presented the information to the class, which then generated lists of topics that interested them and questions they wished to pursue further. Their task was to formulate a driving question from an area of interest. For example, one group's question was "Is the water dirty?" When a student encountered the name of someone who might help them answer their questions, they added the name to the list of "follow-ups" on the back wall of our classroom. The students then called these community members, litigators, activists, and city officials—including the mayor—and invited them to come speak to the class. At times the students' attempts were more successful than mine. I had attempted without success to contact Turkey Creek activist Rose Johnson and Reilly Morse, president of the Mississippi Center for Justice. Then one day Armin announced that they had both accepted his invitation to come speak to us.

When my students read that Councilman Kim Savant had referred to Turkey Creek as a "drainage ditch" that would service a growing Gulfport, my students took offense. This "drainage ditch" flowed through the community that was home to Warren White (Flowers' brother), with whom Antonio would soon spend a day learning to make Warren's signature aluminum can airplane weather vanes. Antonio later reflected:

> It was like visiting my grandfather in the summer. Turkey Creek is a special place that should be appreciated by more people, not bulldozed or flooded. To say that it is a drainage ditch is just ignorant.

MISSISSIPPI HERITAGE TRUST / CHIP BOWMAN

As their knowledge grew, so did their appreciation for Turkey Creek and their ability to empathize with the residents.

A *Bill Moyers' Journal* segment on Leah Mahan's excellent documentary *Come Hell or High Water: The Battle for Turkey Creek* proved a rich source of material. We were moved by the accounts of residents whose relatives' graves had been "developed" into a parking lot for an apartment complex. We watched Eva Skinner clutching the 10-foot chain-link fence that prevents entry to the few marked graves left in the historic Turkey Creek cemetery and lamenting, "What's all that back there, they used to be graves. My son was buried on the corner. Oh, my God, it's sad, honey. Ain't no telling how many hundreds of people were buried here."

An article from our local newspaper archives stoked my students' outrage. When a reporter asked Savant about the legality of bulldozing and paving a graveyard, he replied, "We were going to rezone the property and I was told that there was a minority cemetery in that area." But when he looked at the written record, Savant told the reporter, he "couldn't find a cemetery . . . we couldn't find it." My students could not believe it. "It's down the street from here! Just take a walk and talk to people. Open your eyes." Later, in her culminating project on the Turkey Creek grave site, Raina would conclude: "This would not have happened to a 'white' graveyard."

Other students were moved by another moment in the video clip in which Derrick Evans floats along a bank that the city inexplicably clear-cut, saying, "It's like somebody died." These moments, juxtaposed against the disdain and dismissal expressed by Gulfport politicians, created a moral dissonance in my students, who were shocked that such things happened in their city, in their lifetime.

During Gulfport's frenzy of rezoning, clear-cutting, and indiscriminate building, Turkey Creek residents began to organize. When they successfully stopped a development that would fill in hundreds of acres of wetland, then-mayor Ken Combs called them "dumb bastards" in a meeting with the local newspaper's editorial board, noting that "those people" had not voted for him in the last election, in which his opponent was African American.

But no reading or video clip could compare to our field trips to Turkey Creek's homes and church, or measure up to class visits from community members whose passion and investment in the issue were infectious.

> To my students, Turkey Creek was a case of environmental injustice in which both environmental concerns and the residents of an historic African American community were being sacrificed in the name of development.

Final Projects: Research and Action

I asked students to choose a final project on the land, the culture, or the politics of Turkey Creek. My only requirement was that their projects should have both research and action components—a two- to three-page written research paper and a physical product or activity. So, for example, if students created photo essays, they would need to find a place to show them. When some students chose to focus on Turkey Creek's edible plants—the wild blueberries we picked from a bush inspired them—they documented their research as well as harvesting and preparing the plants. Some chose papers, and others shared presentations; still others, like Antonio, pursued the arts and crafts of Turkey Creek.

After students chose a topic or questions, we made a chart that helped keep their projects on track. Each group's project was different. I devoted one wall of my classroom to the groups' research, interviews, contacts, and daily activity.

Several groups invited guests into the school and recorded their meetings, creating an audio/video library future students can use. Julianne's group decided to study alternate routes for the city's proposed connector road—planned to link the port to the interstate at great cost to the low-income communities between the port and I-10. An environmental planner presented the city's plans. The proposed four-lane limited access highway (think 18-wheelers) would wind around low-income neighborhoods and an elementary school, and required the filling of scarce wetlands, which would increase flooding downstream at Turkey Creek. After tallying the costs of such a plan—blight, noise and air pollution, and downstream flooding—and comparing that to cheaper plans with far less environmental cost, Julianne was beside herself, demanding, "Where do these people live? It's just stupid. I bet they don't send their kids to the school that's going to be right next to an interstate. It's obviously a bad plan. I'm serious, I'm going to their houses."

Although the city council members and mayor were spared harassment, my students were beginning to take the perspectives of the Turkey Creek community as their own.

I asked my students to identify common themes in our research. One theme emerged clearly: To my students, Turkey Creek was a case of environmental injustice in which both environmental concerns and the residents of an historic African American community were being sacrificed in the name of development. Steven could not believe that "Beauvoir [Jefferson Davis' home overlooking the sound] has been rebuilt with a new library while this place is being wiped out."

We traced the logic of economic development: Poor people buy low-lying land because, being prone to flooding, it is cheapest. Gulfport, with the Gulf of Mexico to its south, could only grow northward. The Turkey Creek watershed encompasses North

Gulfport. For "development," the wetlands had to be filled in. This would degrade Turkey Creek and lead to flooding in communities downstream but would create taxable real estate for city coffers and—supposedly—jobs for all.

Just prior to our study of Turkey Creek, my students had spent two weeks making audio documentaries about how status and power operate in our high school by observing, interviewing, and surveying their peers in the cafeteria. Reading excerpts from Murray Milner Jr.'s *Freaks, Geeks, and Cool Kids* helped some of them see that power and status are still blue-chip commodities in high school. But most were uncritical, arguing that they cohabitated peacefully in a post–*Breakfast Club*, status-blind world. After seeing the unequal distribution of power in our city, my students became keen to see it in their school. Kendal's observations in particular factored into our discussions regarding power and status in Gulfport:

> People keep saying, "That's just how things are." I don't accept that. When you see the cafeteria you see mostly well-to-do white kids in the middle with the black kids kind of milling about on the walls. How can you ignore that we have a white-dominated past that is still visible here? Yeah, it's "how things are" but that doesn't make it right.

I played this section of her audio documentary back to my students and asked, "Does this apply to Gulfport?" The consensus was a resounding "yes." Beverly added, "Turkey Creek is definitely against the wall."

Such conclusions seemed warranted when my students asked incoming mayor Billy Hewes whether rich people deserve to be rich or whether the poor deserve their lot. He did "not like the tone of the question," and went on to admonish the offending student, saying that true wealth lies in having health and a family. Kendal later confirmed through her research that "many healthy families in Turkey Creek and the North Gulfport watershed area nevertheless know that they are not rich and suffer from poverty."

For students like Kendal, a young African American woman, this project was an opportunity to put research, observations, and feelings into one package. She would later say that these projects

helped her understand that inequalities exist, how they have come to be, and that a more equitable society is possible. She concluded:

> I saw that who had power and who doesn't depends on the history of the place. That doesn't mean blaming stuff on slavery or "The Man." It just shows our actions matter and make up what history is.

Connecting to the Community

To conclude our study, I asked my students to present their projects to the class. Although many projects focused on history, power, and status, others

Documentary: Come Hell or High Water

Directed by Leah Mahan / 56 min.
www.turkeycreekproject.org

Come Hell or High Water: The Battle for Turkey Creek is an intimate portrait of a community's struggle for survival. We follow Boston teacher Derrick Evans, who moves back to his Mississippi home after his community's ancestral graves are bulldozed to make way for Gulfport's "development." For years, Evans and longtime residents of this historic African American community resist powerful corporate interests and the politicians who do their bidding. And they survive the ravages of Hurricane Katrina and the BP oil catastrophe. This is the perfect classroom resource to introduce concepts of environmental racism and environmental justice. It's one of those small films that tells a gigantic story—about community, resilience, resistance, and hope.

—Bill Bigelow

celebrated the Turkey Creek community, its unique culture, and natural beauty.

For their comparison of two plans for the connector road, Julianne, Samantha, and Shelby researched eminent domain cases and spent hours with stakeholders, poring over maps and documents. They then made their own proposal. For future reference, they also created an issue profile guide that summarized the statements and viewpoints of everyone they researched or interviewed.

Steven and Beverly designed a community garden, wrote two grants for seeds and materials, and created an annual to-do calendar that covers gardening from planting to composting. They connected the school with the community by writing a proposal that was approved by the Mt. Pleasant Methodist Church, which owns the garden property.

Cecilia and Rileigh created photo essays. Some of the photos will hang in Mt. Pleasant Methodist Church's fellowship hall.

Philip and Avery created a historical map, using GPS as residents walked them through the area, pointing out where businesses used to flourish. This required spending time with residents during and after school. Pierce concluded: "These people are all pretty much a family, and it's pretty amazing that they are still having to fight to keep that place the way it is now. I truly feel that if we keep this project up we can help them out even more."

Emily made a meal using native plants, berries, and fish culled from the creek and its banks (baked mullet, dandelion salad, and blackberries) with an accompanying video documentary. I sat amazed as most of my students sampled it: "Not bad. Could use more salt," wrote one student.

"The Past Is Never Dead"

After jumping through many procedural hoops, we took our last field trip to Turkey Creek, canoes in tow (trailer and canoes donated for this project by a former outfitter). As my students floated from the cleared banks along the airport into the green canopy of what remains of the historic community, I reassured myself that at the least they had gained an appreciation of the creek's beauty. My hope is that they will be able to better perceive power relationships, see inequitable distributions of power and status, and find ways to counter racist historical legacies.

The world they inhabit is the result of human decisions—the landscape and people are the historical record. As my students peered six feet down into what is left of wetland in the Walmart parking lot, they got a 50-square-yard glimpse of what the land was 20 years ago, and a huge look at what it has become. The asphalt, concrete, and strip malls that rise above what used to be Turkey Creek's wetlands represent historical moments when leaders made decisions that had great social and environmental impacts.

We often talk of "bringing history to life," as though it ever died. My students realized, as William Faulkner famously put it, "The past is never dead. It's not even past." I am reassured that my students stand a better chance than previous leaders in countering a history that makes economic "progress" at the expense of the least powerful and most vulnerable. ⊕

Hardy Thames is a native Mississippian and teaches at Gulfport High School in Gulfport, Mississippi. Student names have been changed.

Resources

Brangham, William. "Turkey Creek: Watch and React," *Bill Moyers Journal.* This is an excerpt from Leah Mahan's documentary *Come Hell or High Water: The Battle for Turkey Creek.* Posted Nov. 16, 2007. Available at pbs.org/moyers/journal/blog/2007/11/watch_react_turkey_creek.html.

Miller Jr., Murray. *Freaks, Geeks, and Cool Kids: American Teenagers, Schools, and the Culture of Consumption.* Routledge, 2006.

"Turkey Creek," *Land Trust for the Mississippi Coastal Plain,* ltmcp.org/watershed-partnerships/turkey-creek-watershed-partnership.

"**M**r. Miller, Mr. Miller, you've got to see what we've found!" demanded Carl and Ken. The two students rushed into class, eager to share their findings from a civics class assignment to investigate areas named in a local environmental group's listing of toxic sites. They continued haranguing us until several teachers agreed to visit what Carl and Ken described as "a real mess."

When we got to the site, five blocks away from our school, we saw that Carl and Ken were not exaggerating. We stood on the edge of a smelly field, covered with pools of ooze. What was worse, the contamination had spread to an adjacent playground. We didn't have to look far to find the likely source of the ooze: the Moore Oil Company, whose building overshadowed the field.

That visit began a lesson in environmental activism that, before it was over, would have students videotaping toxic sites, collecting soil samples, preparing presentations for parents, engaging in discussions with the president of the company, speaking before the school board, and meeting with a

Students Blow the Whistle on Toxic Oil Contamination

BY LARRY MILLER AND DANAH OPLAND-DOBS

newspaper reporter who helped the students blow the whistle on a community health hazard.

Unlike some whistle-blowing incidents, this one had a happy ending. The students' activism prompted an agreement on a timetable for the cleanup, which the students monitored. Equally important, from our perspective as teachers, the students learned invaluable lessons in how civics involves far more than memorizing facts about the governmental separation of powers.

Leading Up to the Discovery

The civics class was part of the Career Certificate Program at Custer High School in Milwaukee, an interdisciplinary school-to-work program that also involved English, science, math, and technical classes. (Custer closed in 2011.)

Teachers in the program used the environment as a beginning theme, in part because 9th graders display an almost universal interest in the topic. In previous years, the environmental unit ended with field trips to a nature center, a sewage processing plant, and a landfill. That year, based on student input, we focused on issues closer to home. For example, students took part in science labs and field trips that included testing the water quality of a local creek in the Milwaukee River basin for turbidity, fungi growth, organism population, and oxygen content.

> **Unlike some whistle-blowing incidents, this one had a happy ending.**

In English classes, students corresponded with national and local environmental groups. They studied articles on citizen activism on environmental issues, and then gave group presentations. The main lesson they learned was that they could make a difference. In civics classes, students contrasted different perspectives on environmental issues. For example, students compared corporate statements on oil spills with the views of environmentalists.

The topic that led to discovering the pools of ooze involved an interdisciplinary study of the concept of density. Students traditionally study density in physical science, where students compare mass and volume to distinguish different materials. Our students did that in Jill Crowder's science class. But that was just the beginning. In David Caruso's algebra class, they mapped students' homes and figured out the density ratio of students per block within a mile of Custer. In Larry Miller's civics classes, they studied the density of toxic sites in Milwaukee County, using an environmental map of 400 sites developed by the local group Mobilization for Survival. The overall goal, for all the teachers, was to show students the complexity of a concept such as density, and show how it might be applied to different situations.

In Miller's class, students studied the issue of toxic waste for a month before he assigned them to investigate sites near their homes. (Every one of the 120 students in the program lived within eight blocks of an identified site.) Most of the sites didn't have much that students could observe: They consisted of underground storage tanks or hidden contamination. Carl and Ken made their find while scouting an area near a shortcut they took to school every day.

Looking for Answers

After the initial visit with Carl and Ken, we organized a formal field trip with eight students. We documented the site with video and photography, and took samples of the soil and liquid ooze. Back at school, students edited the video, reviewed the photographs, and tested the samples.

Next, we invited in a representative of the Wisconsin Department of Natural Resources. Scott Ferguson, a hydrogeologist, gave presentations to all 120 of our students on the history of Moore Oil and problems with getting companies to comply with governmental cleanup regulations. He also brought samples of the groundwater from the site, explaining that toxins like benzene, vinyl chloride, and cadmium were abundant in the groundwater. Always curious, our students asked how long it would take for the groundwater to clean itself naturally. The kids were shocked at Ferguson's reply: three generations.

It wasn't just the time frame that bothered the students. The park was in the neighborhood where many Custer students live, and their nephews, nieces, brothers, sisters, and cousins played at the park. The health hazard directly affected them and they wanted to take action.

Through our investigation we learned that Milwaukee Public Schools owned the playground. The district had been aware of the problem—to the extent that it had prohibited baseball games from being played in the park—but it had been slow to push for a cleanup. Following a brainstorming session, students decided to educate parents and students living in the area, approach the media to talk about their concerns, and ask the school board about the plan to get the park cleaned up. In hindsight, we realized that we made a mistake by not immediately contacting the management of Moore Oil, in the interest of accuracy, to get their version of events.

At the next parent-teacher organization meeting, Danah Opland-Dobs accompanied two students who gave a presentation on the oil spill. Two days later, at parent-teacher conferences, students wrote a leaflet asking parents to stop by a display where they showed a three-minute video of the toxic site. One couple that stopped by lived across the street from the spill. They could not thank the students enough for pursuing the issue.

Students were concerned that it was taking so long to start the cleanup at the site. To pressure all the parties involved, they decided to contact the media. The result was a front-page article in the metro news section of *The Milwaukee Journal* titled "Ooze and Ahhs: Students Find Oil at Playground." The students also told the reporter, Jack Norman, how the experience had been an eye-opener. "I didn't know anything about these types of experiences until I got involved in this program," said Samantha Piggee. Darius Bunch added: "Changes need to be made in our communities, and we are learning that we can do something about them."

Five days before the article was published, Norman had contacted the president of Moore Oil, Scott Haag, to ask for his view on the students' work. Haag, who was unaware of the students' investigation, was outraged. He immediately came to Custer and met with three teachers and the principal. Haag was concerned that teachers were not fulfilling our "educational duty" by talking with him before going to the press. We acknowledged that it would have been better to talk with him first, and arranged a more extensive meeting for the following week.

Teachers and seven students went to the meeting with Moore Oil. Haag made a presentation, as did the hydrogeologist representing the company Haag had hired to clean up the spill. The students

Ken, the Custer student who reported the oil spill.

took detailed notes on the hydrogeologist's explanation of cleanup plans and asked many questions. We were proud of the students and their ability to engage business-savvy adults. The students asked even tougher questions than we did. One student, Morgan Lampkin, demanded of Haag: "How could you know about the spill for so many years without making any real steps to clean it up?" Her questioning clarified that the company had known about the problem for years but had taken no action to clean up the mess due to legal wrangling over responsibility.

Going to the School Board

The next step in our campaign involved the school board and a request to meet with the committee that oversaw district facilities. As the meeting ap-

Classes observing the park affected by the oil spill.

proached, teachers and students became nervous. What exactly would we say to the committee? What would we demand?

At the meeting, one student said he was skeptical the bureaucracy would take action before winter, when freezing weather would make the cleanup impossible. School board members responded that the cleanup would start immediately. We later learned that Milwaukee city officials, Moore Oil, and the school board had come to an agreement that day on the timetable for a cleanup. The students were proud: Because of us, the cleanup began. We helped cut through the red tape.

> One student, Morgan Lampkin, demanded of Haag: "How could you know about the spill for so many years without making any real steps to clean it up?"

Moore Oil Company ended up paying the state of Wisconsin $133,000 in penalties for violating wastewater and hazardous waste laws at its two sites in the Milwaukee area. The company that did the cleanup, the U.S. Environmental Corporation, met with students to keep them informed of the progress of the cleanup. We took another field trip as the cleanup continued into the spring.

Some students involved in the project started an environmental club at school, worked on a science project doing cleanup of Milwaukee's streams, and helped rebuild a greenhouse at the school.

The most important lesson for the students, however, was that they saw they could make a change. They didn't have to just accept their community as a given—they could critique their natural surroundings. They saw the connection between school and their lives and became excited about what they were learning. ⊕

Larry Miller is a retired high school teacher and an elected member of the school board in Milwaukee. He is a Rethinking Schools *editor. Danah Opland-Dobs is a teacher mentor in Milwaukee.*

CHAPTER THREE: Facing Climate Chaos

ERIK RUIN

"To anyone outside who continues to deny and ignore the reality that is climate change, I dare them—I dare them to get off their ivory towers and away from the comfort of their armchairs. I dare them to go to the islands of the Pacific, the Caribbean, the Indian Ocean, and see the impacts of rising sea levels; to the mountainous regions of the Himalayas and the Andes, to see communities confronting glacial floods; to the Arctic, where communities grapple with the fast-dwindling sea ice sheets; the large deltas of the Mekong, the Ganges, the Amazon, the Nile, where lives and livelihoods are drowned; to the hills of Central America, that confront similar monstrous hurricanes; to the vast savannas of Africa, where climate change has likewise become a matter of life and death as food and water becomes scarce—not to forget the monstrous storms in the Gulf of Mexico and the Eastern Seaboard of North America, as well as the fires that have raged Down Under. And if that is not enough, they may want to see what has happened to the Philippines now."

—The Philippines' climate chief Naderev "Yeb" Saño, Nov. 11, 2013, at the U.N. Climate Change Conference, Warsaw, Poland

INTRODUCTION
Facing Climate Chaos

In January 2014, a poll conducted by the Pew Research Center found that people in the United States ranked global warming 19th on a list of the top 20 issues that they thought Congress and the president should address.

Schools share at least some of the responsibility for people's failure to recognize climate change as the life-altering issue that it is. For example, the adopted physical science textbook in our allegedly green city of Portland, Oregon—published by Pearson/Prentice Hall—buries its one-page coverage of climate change on p. 782. The few paragraphs seem designed to sow doubt. The section begins: "Human activities may also change climate over time." *May.* In textbooks, education journals, schools of education, and Common Core State Standards, one senses no curricular emergency when it comes to teaching about the climate.

We hope the articles in this chapter will help students recognize the significance of the climate crisis. And we also want them to see that the people being hit the hardest are the ones least responsible for it. Those people have names and stories. We focus especially on the impact of climate change on Indigenous peoples. As Matthew Gilbert of the Arctic's Gwich'in people writes in the chapter's first article, "Farewell, Sweet Ice": "Because nature is the fabric of our lives, we cannot really separate 'the climate' from our human selves. So when we talk about the environment and especially about the decline of caribou, we are talking about who we are and who we want to continue to be. It is a question of our very survival as a people."

Too often, explanations for the climate crisis are not really explanations at all; they are descriptions, as in "We are pouring too much carbon dioxide into the atmosphere." Yes. But why? And is it really accurate to say that "we" are doing this? Through activities like the Thingamabob simulation and the climate trial role play, students consider root causes of the crisis—an essential inquiry, if students are to think about the deep social changes we will need to respond fairly and decisively.

And that is a theme that runs throughout the chapter: We can take action to address this crisis, and people *are* taking action. Whether it is Rachel Cloues' 4th graders in "Polar Bears on Mission Street," the activists highlighted in the Climate Change Mixer, or the organizers of the Indigenous Peoples' Summit on Climate Change, people are acting for justice. And this activism is a much-needed antidote to despair. ◉

ERIK RUIN

Farewell, Sweet Ice

Melting ice is threatening the centuries-old society of the Gwich'in

BY MATTHEW GILBERT

Gwich'in elders long ago predicted that a day would come when the world would warm, and things would not be the same with the animals. That time is now.

My tribe, the Gwich'in of northeastern Alaska and northwestern Canada, are the northernmost Indian nation on the American continent. Our 8,000 tribal members live in 15 small villages dotted across a huge area of subarctic tundra and forest scattered with thousands of lakes and scores of rivers. Our home is also home to the Porcupine River Caribou Herd. For as long as anyone can remember, we have survived by hunting caribou. Despite the introduction of rifles, Christianity, a few snowmobiles and some electricity, we still make our living from subsistence hunting.

Now climate change has put our lives and livelihoods in immediate danger: The lakes, the rivers, the waterfowl and, most of all, the caribou that we depend on are under threat.

Because nature is the fabric of our lives, we cannot really separate "the climate" from our human selves. So when we talk about the environment and especially about the decline of caribou, we are talking about who we are and who we want to continue to be. It is a question of our very survival as a people.

Not far from Gwich'in land is the Arctic National Wildlife Refuge coastal plain. Potential oil and gas development there would further stress the caribou and fuel more greenhouse gas emissions, which would cause even more, very bad changes down the line. That is why we oppose plans to drill for oil there.

The biggest impact from climate change that we have seen recently has been two immense forest fires that laid waste millions of acres. When the giant fires of the Yukon Flats started, the marten were driven up to Arctic Village, north of their natural territories. Gwich'in, who trap the marten, are finding them scarce, and that means loss of fur for clothing and cash income from the sale of marten pelts.

We are even seeing a decline in snowfall to the point that it is making travel by sled and snowmobile difficult. Warmer weather means that creeks are taking longer to freeze, and when they do the ice is often too thin to hold heavy loads. This means that getting firewood in the early winter is becoming very difficult and dangerous.

Worst of all are the problems faced by the massive Porcupine River Caribou Herd. We rely on these animals for much of our food; their numbers used to be 178,000, but they have dropped to 129,000 in a mere decade. Calf mortality has risen because of unstable environmental conditions. When the Porcupine River thawed early, this was extremely serious, because it forced the pregnant cows to give birth south of the river rather than in their normal birthing grounds and summer pastures.

But the caribou's instinct to go north to the coastal plain of the Arctic National Wildlife Refuge is so strong that the cows then persisted in crossing the rushing water. Thousands of calves drowned as a result. A tragedy in itself, this also threatens our lives and culture.

The caribou's migration route has also been changing over the past 30 years. We have noticed how the herd is not as plentiful around Arctic Village as it was 30 years ago. My grandfather remembers vast numbers of caribou moving in waves near the village during the spring and summer. The herd was seeking shelter on glaciers and pockets of snow in the mountains. Those places are critical to the survival of newborn caribou: Mothers and calves rest on these cold spots to avoid voracious predatory insects that can kill the little calves. But now those spots have all disappeared. Since the ice and snow are gone in these pockets, the herd has had to move north of their usual area. Now they try to stay near the Brooks Range for its more plentiful snow, ice, and water.

With the herd shifting north, it's getting harder and harder for the hunters to locate them, and we find it much more difficult to provide for our families. Instead of being able to complete provisioning before the harshness of winter arrives, we have had to wait until winter so we can cross rivers and find hard snow surfaces. And not only is the herd smaller but the caribou are generally not as healthy.

The permafrost is melting everywhere and this is having a devastating effect on the Chandalar, Porcupine, and Yukon rivers: All three are getting very low.

> **Because nature is the fabric of our lives, we cannot really separate "the climate" from our human selves.**

It sounds contradictory to say that melting permafrost diminishes rivers, but the underground ice, the permafrost, held the water and regulated its flow. Only small amounts would melt gradually into the rivers. Now that the permafrost is melting, the water runs off quickly, washing away instead of trickling slowly through the filter of partially frozen ground. That leaves the rivers lower much sooner in the summer.

Water levels in the Chandalar River are so low that local moose hunters, who used to go upriver some 20 miles past Grandfather Mountain to hunt, now cannot. Gwich'in from the Yukon Flats travel to Old Crow, Canada, by river, but now they have to drag their boats in a lot of places because the water is so low.

Lakes are especially important to the Gwich'in diet because they have abundant wildlife. But these lakes are drying out. We have noticed that there is a definite decline in numbers of whitefish, for exam-

ple. Whitefish used to be the most common kind of fish, but they're getting scarce and we hardly make the winter harvest anymore. Because the fish feed on water plants all year long, they provide good nutrition from the waters to help us during a long winter without fresh produce.

> **My grandfather remembers vast numbers of caribou moving in waves near the village during the spring and summer.**

Global warming is affecting every facet of Gwich'in life. The Gwich'in elders saw it coming, a time when the weather would warm and change. The elders even foretold the problem of stratospheric ozone depletion. They had a name for what scientists call the ozone layer. In Gwich'in we call the shield against the sun's harsh rays "Zhee vee Luu." The elders knew something was happening to it. And just as the old stories were telling us: The climate is changing and the animals are changing their movements.

In 1988, at our first Gwich'in Gathering in 100 years, Gwich'in elders directed us to educate the world about the threats posed by oil and gas devel-

opment in the calving and nursery grounds of the Arctic National Wildlife Refuge. Now our whole way of life is threatened by global warming.

As Sarah James, Gwich'in Steering Committee spokesperson, has stated: "Protection of the refuge should not be treated as a separate issue from climate change. Let us connect Arctic Refuge with global warming. Drilling for more oil and burning it will contribute to climate change." We need solutions in our local communities, nationally and internationally to reduce greenhouse gas emissions that threaten not just us, here in the Far North, but all cultures everywhere. ⊕

Matthew Gilbert resides in Arctic Village, where he was raised by his grandparents, Reverend Trimble and Mary Gilbert. He earned a BA at the University of Alaska and recently completed a fellowship for the National Wildlife Federation documenting Gwich'in traditional knowledge of global warming.

 See teaching ideas for this article, page 174.

When the water receded after Hurricane Milo of 2030, there was a foot of sand covering the famous bow-tie floor in the lobby of the Fontainebleau Hotel in Miami Beach. A dead manatee floated in the pool where Elvis had once swum. Most of the damage occurred not from the hurricane's 175-mph winds, but from the 24-foot storm surge that overwhelmed the low-lying city. In South Beach, the old art deco buildings were swept off their foundations. Mansions on Star Island were flooded up to their cut glass doorknobs. A 17-mile stretch of Highway A1A that ran along the famous beaches up to Fort Lauderdale disappeared into the Atlantic. The storm knocked out the wastewater treatment plant on Virginia Key, forcing the city to dump hundreds of millions of gallons of raw sewage into Biscayne Bay. Tampons and condoms littered the beaches, and the stench of human excrement stoked fears of cholera. More than 800 people died, many of them swept away by the surging waters that submerged much of Miami

> **It was clear to those not fooling themselves that this storm was the beginning of the end.**

Goodbye, Miami

Rising sea levels will turn the nation's urban fantasyland into an American Atlantis

BY JEFF GOODELL

Beach and Fort Lauderdale; 13 people were killed in traffic accidents as they scrambled to escape the city after the news spread—falsely, it turned out—that one of the nuclear reactors at Turkey Point, an aging power plant 24 miles south of Miami, had been destroyed by the surge and sent a radioactive cloud over the city.

A City-by-City Forecast

The president, of course, said Miami would be back, that the hurricane did not kill the city, and that Americans did not give up. But it was clear to those not fooling themselves that this storm was the beginning of the end. With sea levels more than a foot higher than they'd been at the dawn of the century, South Florida was wet, vulnerable, and bankrupt. Attempts had been made to armor the coastline, to build sea walls and elevate buildings, but it was a futile undertaking. The coastline from Miami Beach up to Jupiter had been a little more than a series of rugged limestone crags since the mid-2020s, when the state, unable to lay out $100 million every few years to pump in fresh sand, had given up trying to save South Florida's world-famous beaches. In that past decade, tourist visits had plummeted by 40 percent, even after the Florida legislature agreed to allow casino gambling in a desperate attempt to raise

revenue for storm protection. The city of Homestead, in southern Miami-Dade County, which had been flattened by Hurricane Andrew in 1992, had to be completely abandoned. Thousands of tract homes were bulldozed because they were a public health hazard. In the parts of the county that were still inhabitable, only the wealthiest could afford to insure their homes. Mortgages were nearly impossible to get, mostly because banks didn't believe the homes would be there in 30 years. At high tide, many roads were impassable, even for the most modern semi-aquatic vehicles.

But Hurricane Milo was unexpectedly devastating. Because sea level rise had already pushed the water table so high, it took weeks for the storm waters to recede. Salt water corroded underground wiring, leaving parts of the city dark for months. Drinking water wells were ruined. Interstate 95 was clogged with cars and trucks stuffed with animals and personal belongings, as hundreds of thousands of people fled north to Orlando, the highest ground in central Florida. Developers drew up plans for new buildings on stilts, but few were built. A new, flexible carbon fiber bridge was proposed to link Miami Beach with the mainland, but the bankrupt city couldn't secure financing and the project fell apart. The skyscrapers that had gone up during the Obama years were gradually abandoned and used as staging grounds for drug runners and exotic-animal traffickers. A crocodile nested in the ruins of the Pérez Art Museum.

And still, the waters kept rising, nearly a foot each decade. By the latter end of the 21st century, Miami became something else entirely: a popular snorkeling spot where people could swim with sharks and sea turtles and explore the wreckage of a great American city. ◉

Jeff Goodell is a contributing editor to Rolling Stone *magazine and the author of* Big Coal: The Dirty Secret Behind America's Energy Future. *This article first appeared in* Rolling Stone.

 See teaching ideas for this article, page 174.

ERIK RUIN

I sat on a tall stool, facing the class of 9th graders, put a cigarette between my lips and flicked
on the lighter.

"Anyone mind if I smoke?"

Yes, they did mind: "That's disgusting." "It's
against the law to smoke here." "There's secondhand
smoke and it smells bad."

I hoped this opening to a unit on climate change
would underscore the idea that—even if students
don't have the vocabulary to express it—we are all
familiar with the concept of the "commons." In this

Teaching the Climate Crisis

BY BILL BIGELOW

classroom, we shared a breathing commons, and I didn't have to convince students that no one had an individual right to pollute it with cigarette smoke. I hoped the cigarette-in-the-classroom stunt would work as a metaphor: The Earth's atmosphere is just a bigger version of the classroom—a finite commons that none of us owns, but that each has a stake in.

I've become convinced that climate change—global warming, climate chaos; call it whatever you like—is the biggest issue facing humanity. As the renowned environmental activist Bill McKibben points out, this crisis

> represents the one overarching global civiliza-
> tional challenge that humans have ever faced.
> . . . The evidence gets worse by the day: Al-
> ready whole nations are evacuating, the Arc-
> tic is melting, and we have begun to release
> the massive storehouse of carbon trapped un-
> der the polar ice. Scientists figure the "safe"
> level of carbon dioxide in the atmosphere is
> about 350 parts per million. . . . Go beyond it
> for very long and we will trigger "feedbacks"
> that will result in runaway warming spiraling
> out of any human control and resulting in a
> largely inhospitable planet.

Tim Swinehart, co-editor of this volume and an active member of the Earth in Crisis curriculum workgroup of Portland, Oregon, Area Rethinking Schools, invited me to co-teach a unit on global warming to his 9th-grade global studies students at Lincoln High School. Tim and I teach social studies, not science. We knew that we were ill equipped to offer the kind of hard scientific instruction that would help students grasp exactly how and why the climate is changing. But just as all of us are responsible for the atmospheric commons, climate change falls into a curricular commons; Tim and I were committed to explore the social impact of global warming as well as some of its social roots. How the billions of metric tons of CO_2 we pump annually into the atmosphere affect the Earth's natural systems may be a scientific question. Why we do this, who it affects, and, at least in part, how we can stop it—these are social questions.

We especially wanted students to appreciate the inequality at the heart of climate change: Those who have the smallest carbon footprint are the ones most victimized by its consequences. We wanted students to probe beneath the glib "buy green" solutions to global warming. And Tim and I knew that in this unit we would toe a fine line between communicating the vast dangers of global warming and encouraging students to recognize their power to make a difference.

Global Warming Mixer

Where to start? Global warming feels so overwhelming and impersonal—something happening everywhere, yet nowhere in particular. It can have a kind of science fiction, someday-it's-gonna-get-really-bad feel.

I wrote a mixer activity (see the full activity on p. 92) to introduce students to the ways climate change affects individuals around the world—*today*, not in the distant future. In a mixer, students assume the roles of different individuals and, through meeting one another, learn about an issue that touches us all. For some of the people students portrayed, climate change is crashing through their lives right now. But for others, rising global temperatures have presented business opportunities—like the oil companies poised to exploit the Arctic, where an estimated 25 percent of the Earth's untapped fossil fuels beckons like buried treasure.

All of the 17 individuals in the mixer are based on real people. I wrote them in first person, and many of the roles incorporate the individuals' actual words. For example, Rinchen Wangchuk works with the Snow Leopard Conservancy, a grassroots habitat preservation organization:

> When I was a boy, after school ended for the
> summer, I remember slipping down the glacier
> that stretched far down the mountains near
> my village in the Nubra Valley—in Ladakh, the
> far northern part of India. Today, that glacier
> is almost gone. And I am watching the glaciers
> of the Karakoram Mountains disappear a little
> more every year. . . . Because it rains only two
> inches a year in Ladakh, we depend on the gla-
> ciers for 90 percent of our water . . . but what
> will happen if the glaciers disappear? How will
> we survive?

Students also meet Nobel Prize winner Wan-

Sea level rise threatens to drown the Pacific island nation of Kiribati,

CHRISTOPHER PALA/IPS

gari Maathai of Kenya, who describes unpredictable floods, drought, crop failures, and desertification afflicting huge swaths of Africa; and Enele Sopoaga of the South Pacific island of Tuvalu, who watches as rising sea levels threaten his land and people. Many of the individuals are activists, resisting mountaintop removal coal mining in West Virginia or blocking Bering Sea oil exploration.

Students also encounter climate change "winners," as Fox News calls them: Chris Loken, an apple grower in New York's Hudson Valley, where milder winters allow new crops of plums and peaches; and Russian oil man Roman Abramovich of Russia's Sibneft Oil Co.:

> It's simple: As temperatures rise every year, ice will melt and huge new areas will be open for oil and gas exploration in the Arctic. And as one of Russia's wealthiest men, and head of a large oil and gas company, this is the chance of a lifetime. Researchers tell us that one quarter of the Earth's untapped fossil fuels, including 375 billion barrels of oil, lie beneath the Arctic. . . . I'm a good businessman—a good *oil* businessman—so it's time to get to work.

Tim and I distributed roles to students and asked them to read these several times, to underline key information, and to list the three or four most important points about an individual's situation. The students circulated in the classroom and found a different individual to answer each of eight questions— for example, "Find someone who believes that he or she is hurt by climate change. How has this person been hurt?" "Find someone who believes that he or she might benefit from climate change. How might the person benefit?" A final question asked, "If possible, find someone here with whom you could take some joint action around global warming. What action might you take?"

Students seemed engaged in the activity, which took most of a 50-minute class. They collared one another to talk, grimaced at painful stories, delighted in shared experiences. Afterward, we asked students to write on three questions:

- Whom did you meet—or what situations did you hear about—that surprised you?
- Which themes came up in your conversations?
- Whom did you meet—or what situations did you hear about—that gave you hope?

Kaya wrote that she was surprised that "people from all over the world are being affected." Michael was surprised "that people were actually benefiting

from the global warming. Like Roman [Abramovich, the Russian oil man], who actually is getting richer off the melting glaciers and ice caps. So that was very surprising and a bit upsetting too since guys like mine [Anisur Rahman, from Bangladesh, whose lands are being washed away in horrible floods] are suffering so much." Abramovich may be a special case, but he exemplifies the basic inequality that I wanted students to grasp: We are all affected by global warming, but we're not all affected equally, and we're not equally responsible for its causes. As Carver wrote in his answer to this question, "It seems like the ones who couldn't afford to be affected were affected the most."

> We knew that, ultimately, students would act only from a place of hope and if it seemed possible to turn things around.

Alvaro, whose parents are immigrants from Mexico, in a class that was mostly white with a few African American students, was the only individual to note the link between climate and migration. "A lot of people were having problems and having to migrate," Alvaro told the class.

Students repeatedly identified one theme: water. "It was interesting how so many people all over the globe were affected by water that could either disappear or flood them because of global warming," Cole wrote. Aria noticed "how many of these involve water, both sea levels and glaciers," but added,

"I also think it's sad that everyone suffers but for some it's life and death and they see it more but don't have the influence or power to change it."

From the beginning, Tim and I wanted students to stay alert for signs that we might not be doomed—thus, the "hope" question we asked students to write on. The enormity of global warming has the potential to overwhelm and discourage. No doubt, Tim and I wanted to impress upon students the threat that climate change represents—a threat so huge that we literally don't have the words to express it. But we knew that, ultimately, students would act only from a place of hope and if it seemed possible to turn things around.

Ironically, some students found cause for hope in the problem's immensity. Selena wrote, "The more people [who are] affected badly, the more people want to help. This is good." And Adrian drew inspiration from the militancy of some activists: "I found out there were people who take the global warming fight to people's front doors. They break into oil company compounds to protest against drilling. I wish I could find these people so I could join them." (I had not included any activists breaking into oil company compounds, but one of the individuals in the role play—Stephanie Tunmore of Greenpeace—tried to physically block BP's oil drilling in the Beaufort Sea, north of Alaska.)

Obviously, I could have written hundreds of more roles, but the main aim of the activity was to introduce students to the breadth and inequality of global warming's impact. The media cast this as an "environmental issue," and it is. But global warming is also a racial and class issue. Those affected most profoundly are poor people and people of color. They are the most vulnerable to—and least responsible for—this catastrophe. Global warming also reflects the dynamics of empire, that so-called developed countries blast greenhouse gases into an atmosphere that the entire world depends on. As surely as Western colonial powers stripped the colonized world of resources, so too are the wealthy countries "mining" the atmosphere. Atiq Rahman, the Bangladeshi chairman of the Climate Action Network South Asia, captures the Third World's resentment when he warns, "If climatic change makes our country uninhabitable, we will march with our wet feet into your living rooms."

Textbook Disinformation

James Hansen, formerly of NASA, one of the world's leading climate scientists, said recently that the CEOs of large fossil fuel companies should be put on trial for high crimes against humanity and nature for spreading disinformation about global warming—disinformation that would end up in school textbooks. As Exhibit A, Hansen can use our textbook, *Modern World History* (Holt McDougal, 2012), adopted by Portland Public Schools for all required high school global studies classes. The book discusses global warming in three wretched paragraphs, buried on p. 679.

To provide students a bit more background before evaluating their textbook, we read and discussed an article from a special issue of *National Geographic* on the climate crisis. The article, "Proof Positive," describes the certainty of human-caused climate change and the relationship between carbon dioxide and global temperature changes. (See p. 106.) It lays out the science undergirding the Intergovernmental Panel on Climate Change's models that forecast dramatically rising temperatures this century, but also acknowledges the unpredictability of the many "amplifying feedbacks" and their consequences on Earth's natural systems.

Tim and I then asked students to look critically at their textbook. We had them read the three-paragraph passage and write about its adequacy and inadequacy, and to note which perspectives were missing.

Confirming Hansen's prediction that corporate-funded "scientific" research would find its way into the nation's textbooks, the second of *Modern World History's* three paragraphs on climate change begins "Not all scientists agree with the theory of the greenhouse effect." As science teachers have explained to me (and as "Proof Positive" confirms), this is inaccurate; the French physicist Joseph Fourier proposed the "greenhouse effect" in 1824. Today, no scientist disagrees with the "theory of the green-

house effect." Surely what the textbook writers meant to say is that the *human-caused* greenhouse effect is a theory, and this is what some students locked in on: Human-caused global warming is "just a theory." (After this article was first published in *Rethinking Schools,* Holt McDougal removed this one line from *Modern World History.*)

Maybe it shouldn't have surprised me that so many students doubted humans were changing the climate. It was a good lesson about how, even in an activity encouraging students to think critically about textbooks, the textbook's authority can undermine widespread scientific consensus. Tim and I explained that a scientific theory is more than a hunch. As Stephen Jay Gould wrote in his book *Hen's Teeth and Horse's Toes,* "facts and theories are different things, not rungs in a hierarchy of increasing certainty. Facts are the world's data. Theories are structures of ideas that explain and interpret facts." In other words, a scientific theory is a big deal; it must be consistent with all existing data. Tim and I had assumed we could proceed from the premise that human-caused global warming is real, not "just a theory." There was a lot of teacher-talk in this activity that we hadn't planned on.

> As surely as Western colonial powers stripped the colonized world of resources, so too are the wealthy countries "mining" the atmosphere.

Nonetheless, having gone through the role play, students recognized that the textbook does not mention a single individual, community, or culture affected by climate change. There's no story, no humanity, no appreciation of what's at risk. Nor does the text introduce students to activists like the people they met in the mixer.

Whether describing climate change, the Great Depression, or the Vietnam War, the biggest textbook bias is the failure to alert students to the power of organizing and collective action. When it comes to global warming, the textbook tells them to trust their leaders: "To combat this problem, the industrialized nations have called for limits on the release of greenhouse gases. In the past, developed nations were the worst polluters."

As one student pointed out, "This makes it sound like the rich countries are the good guys."

Exactly. And the bad guys? According to *Modern World History*, "So far, developing countries have resisted strict limits." Sadly, the textbook stands reality on its head.

Students also noted that the textbook communicates its disregard for climate change as a serious issue by burying its meager three paragraphs near the end of a 700-plus-page book. I suppose this is what we should expect when gigantic corporations produce our textbooks. These corporations are not neutral spectators to the social processes they describe; they benefit from today's global distribution of power and wealth, and whether it's climate change or free trade agreements, the textbook companies show no interest in helping students think critically about the world.

> As one student pointed out, "This makes it sound like the rich countries are the good guys."

Perhaps Tim and I should have introduced the textbook critique later in the unit when students had more background knowledge—and next time, we will. However, even this possibly misplaced activity alerted students to the fact that people make choices in how they describe the world, and students have a right to question those choices.

Thingamabobs and Climate Change

I've taught global studies on and off for more than 30 years. It's a class that exposes students to myriad forms of injustice: invasion and occupation, poverty and hunger, sweatshops and child labor—and now the causes and ravages of climate change. Lessons can turn into a litany of "people doing bad things to other people." Many students are not content to absorb injustice after injustice in the curriculum without wanting to make a positive difference. But how?

"Go shopping," George W. Bush told the country in the wake of 9/11. Even social justice organizations often encourage people to make things better through the marketplace: Buy fair trade chocolate, boycott Walmart. A consume- (or recycle-) your-way-to-justice orientation is especially prominent in today's responses to climate change. We're urged to buy compact fluorescent lightbulbs and hybrid cars; take a cloth bag to the grocery store. These actions can raise consciousness, reduce greenhouse gases, and help us identify with environmental justice on a daily basis. But if we respond to injustice only as consumers, we miss other potential responses.

We live with an economic system that distributes rewards on the basis of profit. Production decisions are largely privatized, even when we all bear the social and environmental costs. From the standpoint of greenhouse gas creation and the impact on the climate, this global capitalist system is, to put it mildly, problematic.

I decided to adapt a simulation I developed a number of years ago. "The Thingamabob Game" puts students in the position of executives in corporations that produce "thingamabobs"—which symbolize any manufactured good. I divide students into companies. Students name their companies and compete over five rounds to try to make the highest profit. The top four profit-makers get candy bars—the most for the top producers. The bottom three profit-makers get nothing. The catch is that the production of thingamabobs has environmental consequences; there is a trigger number of total thingamabob production that, if exceeded, leads to environmental catastrophe and everyone loses, no matter how much profit any particular company has amassed.

One trick for a successful game is having desirable goodies that students want to win. The best chocolate in Portland is Moonstruck, and so I bought lots of Moonstruck chocolate bars—organic dark chocolate, organic milk chocolate, ivory chocolate, and my personal favorite, dark chocolate espresso. Before I explained the rules, I told students, "I want to take a few minutes to tell you about the prizes we have for the winners." I have my teaching weaknesses, but I can describe the delights of chocolate with the best of them.

These chocolate bars were not BMWs, vacation homes in Maui, or home entertainment systems, but they might as well have been. After "meeting" the chocolate bars, students paid close attention as I went over the instructions:

> You are managers of a company that produces thingamabobs. You are in competition with other thingamabob companies. Even though you have highly paid managerial jobs, these are not necessarily secure. As with any capitalist company, you need to continually grow

and make a profit. Fail to return a sufficient profit and you'll lose your job.

Each company begins the game with $1,000 in capital. Each thingamabob costs $1 to produce. You will make $2 off of every thingamabob you produce and sell. So, for example, if you produce 100 thingamabobs in round one, you will spend $100, but you'll get $200 back, and end up with a total of $1,100.

To streamline the game, we assume that all the thingamabobs produced are also sold. Thus in the first round of the five rounds, a company that makes the maximum number of thingamabobs ends up with $2,000. With each successive round, companies have more capital and greater capacity to produce thingamabobs and profit. The catch is that the production of each combined 1,000 thingamabobs adds two parts per million of CO_2 to the atmosphere. The game begins roughly where the planet was in 2008, at 380 ppm of CO_2. The game's tension is that if the seven groups' total production exceeds a trigger number somewhere between 420 and 460 ppm

CO_2—between 20,000 and 40,000 thingamabobs— the Earth's environment is damaged beyond repair and everyone loses. In other words, no Moonstruck for anyone. On the wall I taped a folded piece of paper with the exact trigger number, 450 ppm—35,000 thingamabobs—but no one knew that number as they entered the game. (For more detailed instructions and teaching materials, see p. 147.)

It's possible that some, or even all, the groups can win this game. But it takes cooperation, lots of discussion, enforceable rules, alertness to the big picture, and collective restraint. Unfortunately, just like in the real world, the dynamics of a me-first, profit-driven economic system make this unlikely.

As we begin, I don't tell groups that they cannot cooperate, but I fan the flames of competition with disparaging comments about some groups' low production: "No way can you win this game with puny production like that." Around the second or third round, I pick up a Moonstruck bar and ask, "Did I tell you how great this chocolate is?"

I've played this game with high school freshmen and juniors, graduate students, teacher education cohorts, and groups of teachers. Alas, every group has destroyed the Earth. Tim's 9th graders went over the trigger figure by round four. The victorious company, Jellyfish, ended the game with a total of $12,000. But, as the saying goes, it was like winning at poker on the *Titanic*.

As the game concluded and students realized that I would keep my chocolate, Tim and I asked them to make sense of what we'd just gone through. We posed three questions:

1. Who or what was responsible for the Earth's destruction?
2. What are the real-world lessons of how you played the Thingamabob Game?
3. Suppose we were to play this game again. Give the class some advice on how you could or should approach it differently.

Most students tended to personalize blame for the Earth's destruction: "We were all reckless." "Greedy people." One student blamed me: "You said candy was the reward and it made us be competitive." Zak blamed "the companies," and added, "They would rather have money than help save the world."

Students recognized that this was not simply a game, that there was something happening in the classroom that had its counterpart in the world. As Selena wrote, "For me it made me think about all the greed in the world and how many think, 'We will be fine. I'm sure someone is doing something about it.'"

> "No one once said anything like 'remember we have to consider the environment.' It was still all about the prize."

Their writing was a fascinating stew of blame and insight. But no one identified the structure of the Thingamabob Game itself as the root of the problem. Ninth graders, along with all the rest of us, are not practiced in considering how the economic rules of the game can be identified and challenged. And another dynamic was also at play. No one stepped forward as a climate change organizer or activist, and said, "Look at what's happening. We're going to thingamabob ourselves to death."

Although Tim and I intended the game to raise questions about how our economic system is on a collision course with the Earth, we worried that the game might leave students with a sense of inevitable doom—that even as they saw what was happening, they still went ahead and destroyed the Earth. The next day, we gave students a second chance.

We began by reading aloud an article, "How to Be a Climate Hero," by Audrey Schulman, from *Orion* magazine. The article describes research on the "bystander effect"—that people tend to freeze when confronted by injustice or emergency if they're around other people who also don't respond. We asked students for examples from their own lives, and students shared moments such as walking by a possibly injured or ill person lying on the sidewalk, and not protesting when a classmate was treated unfairly because no one else spoke up.

The hopeful piece of this, according to Schulman, is that knowing about the danger of group inaction can prompt us to take action. Tim and I wanted students to see themselves as activists, to interrupt the insane spiral of greenhouse gases—in the real world as well as in Thingamabob's candy quest—and not allow others' passivity to lull them into a similarly deadly inactivity.

Before relaunching the game, we left them with some advice: "Remember, this is a game. The rules are made by humans and can be changed by humans. . . . Talk to each other. You control the game, the game doesn't control you."

Unfortunately, students returned to their "corporations" without having a conversation about how they wanted to replay the game. By the third round, they were only marginally kinder to the planet than the first time around. Various students called out to slow it down, to be "greener." And they were. However, ultimately, chocolate lust prevailed and production skyrocketed by 27,400 thingamabobs in the fifth and final round, leading to a game- (and Earth-) ending level of CO_2 production.

I was disappointed. I thought the group would find a way to resist the previous day's production war. Playing the game a second time with 11th graders across town at Franklin High School another year, one global studies class took to heart my "you control the game" message. They began the second game with a class meeting and essentially voted to abolish capitalism. They erased the corporate divisions, and the chaotic competition these created, and formed one big group. Relieved of competing against each other, they were able to collectively control thingamabob production at a sane and gradual pace.

Tim and I asked students to write about why, after they saw what happened the first time around, they destroyed the Earth a second time. They were pensive in their failure. Alvaro wrote that they lost "maybe because the competition still stayed in our minds. . . . It blinded us again." And Kaya pointed out that their class lacked climate hero/activists: "No one once said anything like 'remember we have to consider the environment.' It was still all about the prize."

During our discussion, many students seemed amazed that another class had created one big team, allowing everyone to "win." "You didn't tell us we could do that," one student complained. In fact, I did. Students simply couldn't hear it, because they are used to playing games by the established rules and not calling those rules into question. And, really, this was the game's punch line: We have to think systemically; we have to question the rules of the game; we have to work together to imagine new ways to "play" and "win."

The Thingamabob Game is still in process and it's not without its accompanying "fine print." All simulations are metaphors and highlight some aspects of social reality and distort other aspects. For one, these days we need to be going backward toward 350 ppm CO_2, not going up from the 2014 400 or so ppm. Thinking of the game's 450 ppm as the actual tipping point is wishful thinking. For another, not all thingamabobs are equally hostile to the atmosphere. Some, like solar panels or light rail cars or bicycles may be carbon-friendly. And, as the 2008 economic meltdown illustrates with devastating clarity, not all thingamabobs that are produced will be consumed. Finally, the simulation may communicate that all CO_2 pollution comes from the manufacturing of stuff, whereas much comes from transportation, agricultural practices, cutting down rainforests, and even how (and how much) we heat our homes.

Nonetheless, despite shortcomings, the Thingamabob simulation's essential insight remains: A fundamental incompatibility exists between an economy premised on an unquenchable drive for profit and the ecological imperative to reverse greenhouse gas pollution. Like the Franklin class that won the Thingamabob game, we have to recognize that our society's productive decisions are not private. What we produce, how we produce, and how much we produce all affect the atmospheric commons. These are decisions for us all.

Connection and Loss

One of the struggles in building a curriculum around climate change is searching for ways to help students grasp, *in a personal way,* what's at risk. Tim and I realized something was missing. We were hitting kids' heads but not their hearts.

Ultimately, climate change is about our connections to places and to people, and also about the potential loss of those connections. We decided to pause and ask students to explore these ideas. We asked students to write on one of two broad themes:

- Write a story about a special or "sacred" place. Describe an event or events that took place there that made it such an important place to you, that "rooted" you to the place.
- Write a story about a time in your life when something you cared deeply about was taken or stolen from you. This might be a precious possession, a place where you lived, a person you cared about, or something less physically tangible, like your innocence or your sense of hope.

Tim and I gave examples from our own lives and shared two stories students had written, one about a girl returning to a special place only to find that it had been "developed," and another about family turmoil and the loss of a parent. We wanted to offer powerful stories as models but also give students multiple points of entry, with different levels of risk in sharing. We told students we'd ask them to read their stories aloud in class but that they were free to pass or to have us read them anonymously.

After we brainstormed possible topics, I turned off the lights and asked students to close their eyes,

> One of the struggles in building a curriculum around climate change is searching for ways to help students grasp, *in a personal way,* what's at risk.

Connection and Loss

In a *Los Angeles Times* article, "Civilization's Last Chance," Bill McKibben quotes then-chief NASA climatologist James Hansen, who says that because of climate change, humanity is at risk of losing "a planet similar to that on which civilization developed and to which life on Earth is adapted. . . ." This should make us pause and consider what exactly this means to us personally—losing a planet that gave birth to civilization.

The only way we have to think about the future is in relation to things we've experienced in the past. Hansen's warning raises issues about our relationship to places that we are connected to, and that mean a great deal to us. It also suggests the threat of enormous loss.

That's what this writing assignment is about: connection and loss.

Assignment choices:

- Write a story about a special or "sacred" place. Describe an event or events that took place there that made it such an important place to you, that "rooted" you to the place.

- Write a story about a time in your life when something you cared deeply about was taken or stolen from you. This might be a precious possession, a place where you lived, a person you cared about, or something less physically tangible, like your innocence or your sense of hope.

put their heads down, and not talk. I led students through a guided visualization—a strategy I learned years ago from my wife and teaching partner, Linda Christensen. I asked students to visualize the place or person that they'd be writing about. I paused for 30 seconds to a minute after each instruction: What does the place look like? Let your mind's eye be a video camera. Try to capture as many details as you can. Which smells or sounds can you recall? . . . Now focus on the people involved. See their faces, hear their voices. Try to recall as many details about the people and place as you can. . . . Now try not to think in words. Let yourself be surrounded by the feelings you had about this place or incident. Just let the feelings wash over you. . . .

I told them that when I turned on the lights I didn't want to hear a single voice. They could keep their heads down and continue to recall details or they could begin writing. Tim and I would be available in the hall, if they had questions, but we didn't want any talking in the classroom as people wrote. The lights came on and they wrote in silence, with passion and determination.

The next day, we circled the desks and students read their writing aloud. Although the assignment gave students the option of writing about the loss of a prized possession, everyone wrote about the connection to and the loss of a beloved person or place. Several students wrote heartrending pieces about the death of parents and grandparents. After each student read, other class members offered positive comments about the writing, the content, or the reading. We also asked students to take notes on each other's pieces, and told them we'd be writing a "collective text" on two questions: (1) What common themes did you notice from our stories of connection and loss? (2) How do these themes connect to our study of climate change?

In the weeks I'd been in Tim's class, I'd never seen students reach so hard for ideas and discuss with such intensity. In her "collective text" answer to the first question, Kadee summed up what she'd heard: "Most stories started with a 'normal day,' and ended in 'I'll always remember.' There is kind of a theme of how the stories really impacted their life more than they felt at the time. No one wrote about objects or stuff. Shows how that 'new iPod' doesn't compare."

I thought we might be on thin ice asking students to relate these stories to climate change. But questions such as "What does this remind you of?" consistently generate valuable student insights, and this time was no different. Brandie wrote:

> Kadee's story was about losing her childhood. In her story, she had a line that went something like this, "In life, there are no rewinds or pauses, only play." In the real world and to global climate change, this line relates. There are no do-overs. We can't go back and try to change our mistakes, or pause so we can have time to fix things. . . . In my story, my grandma was like the world. She was like the world in the sense that she had a house (shelter) who everyone lived in. She provided food to them, shelter and happiness to people who needed it. Like the Earth. . . . People loved her, but took advantage of her, like the Earth. And one day, she wasn't there anymore, and left everyone in a mess. Like the Earth COULD.

This recognition of the power of our relationships to people and places, but also the fleeting character of these relationships—"we start out so naive and don't know what's going to happen next"—was a theme that ran through students' papers. Tim and I wanted to move back to the content of climate change, but fundamentally, this *was* the content: The "developed" world's patterns of production and consumption were putting at risk our relationships to everything and everyone that mattered.

We followed our read-around and discussion with the PBS NOW documentary *Paradise Lost,* about the South Pacific island nation of Kiribati, and asked students to think about the connection between the themes we identified in our own writing and what's happening to Kiribati. (See p. 118.)

Kiribati is the proverbial canary in the coal mine. It's spread over 33 tiny islands, with a population of 100,000. Its highest point is about 6.5 feet above sea level, if you don't count the coconut trees. Climate change—melting glaciers, warming seas— has doomed Kiribati. The ocean is rising and Kiribati's first climate refugees have already abandoned the islands for New Zealand.

Following our writing about connection and loss, the words in *Paradise Lost* of Ueantabo Mackenzie, who directs the local branch of the University of the South Pacific, resonated with students: "This country has been the basis of my being. And when it's no longer there, you know, it's unthinkable."

As with our opening mixer about stories of the global impact of climate change, we wanted to show students what these developments meant in people's lives. But it was the stories of their own lives that created the basis for a deeper bonding between "us" and "them." It was as if, after hearing each other's stories of connection and loss, students said to themselves, "This is what it must be like to lose your land and way of life."

> The lights came on and they wrote in silence, with passion and determination.

Students for Climate Action

We were running out of time. But we wanted to close this unit on a hopeful note and to prompt students to think of themselves as people who could take action. Young people from around the country were about to converge on Washington, D.C., for a week of lobbying, called Power Shift, and direct action: mass civil disobedience at the coal-fired power plant that serves the Capitol building. This seemed a good time to alert students to activism.

We created an activity that would allow them to take on the roles of student climate activists and to confront some strategic and tactical choices that actual campaigners encounter. We chose a role-play form that puts the entire class in the same position—in this case, members of a fictional Students for Climate Action—and offers six activities for students to debate and decide which to prioritize. Actions included helping climate refugees; lobbying Congress; engaging in civil disobedience; working on education in schools, churches, and community organizations; building climate coalitions with organizations in the Global South, like China and India; and getting people in the United States to reduce consumption. The actions certainly overlap, and we took some liberties mixing strategies and tactics, and intentionally left some open-ended. But the broad aim was to get students to imagine themselves as organizers. (A couple of weeks after this activity, several young climate activists who descended on Washington, D.C., were interviewed on Amy Goodman's radio-TV show *Democracy Now!* Tim showed a clip after our unit and said that students were rapt, listening to other young people tell why they traveled to D.C. for this activism.)

Students had to settle on three broad areas to tackle and to list them in order of preference. This meant deciding not to concentrate in three other areas. We seeded their discussions with pro and con arguments. For example, here is the civil disobedience option we presented:

Many Students for Climate Action members believe now is the time to increase pressure on the government to take dramatic action, and say that civil disobedience is the perfect tactic and coal the perfect target. They are concerned that the U.S. government isn't acting fast enough or boldly enough to reduce carbon emissions. They argue that big energy corporations, like ExxonMobil and Peabody Energy, still have too much power to influence the government, and that things won't change until the people act. One way that the people of the United States have exercised their power in the past is through civil disobedience—breaking the law in order to bring attention to a particular injustice. Many students at this conference are calling for widespread civil disobedience to bring attention to the injustice caused by the U.S. government's failure to act quickly to deal with climate change—but especially the injustice caused to your generation, which will bear the largest burden of slow action.

Other SCA members agree that coal is a huge problem and that these issues are urgent, but believe that civil disobedience is exactly the wrong thing to do. Why should we break laws and alienate people all over the country at the very time that more and more people are coming to agree that global warming is a serious problem? They argue that we'll look like a bunch of extremists or crazy people, if the police haul hundreds, or even thousands, of us off to jail. And at the very

moment that we finally have a president who agrees that climate change is real and needs to be dealt with, why would we choose a strategy of breaking the law? That would be nuts.

We explained to the class that we'd be available to answer questions of fact, "but this is your show. You need to figure out how to run your discussions and how to arrive at decisions." Although this group of 9th graders interrupted each other and fell to squabbling, many of their conversations were substantial and interesting. I took notes as students discussed. Here's a sliver of debate from the civil disobedience issue:

Cara: I vote yes. Last year we did the walkout against the war and many high school students did this. People did listen to us. We walked out of school and went downtown and talked to people. I think that civil disobedience works if a large group of people do it.

Daniel: I disagree. If we do this it will just get people angry and they will focus on us breaking the law.

Brandon: Obviously the whole idea of doing civil disobedience is that you're acting and you care. If you're going to act, why not break the law and help climate refugees?

Kadee: It is a good thing to bring attention to climate refugees. But this is such a big issue that breaking the law is not going to do anything. We have to act on it not just walk out.

Symara: I see it this way. Yeah, climate change is a huge issue, but segregation was an even bigger issue. It ended because people broke the law. We shouldn't worry if people get angry with us. We have to keep doing it.

Sam: If we do this we are going to come off like a bunch of radical crazies.

Brandie: No, we shouldn't do this. With Martin Luther King, they had rules that needed to be broke, and they broke those rules. But what rules do we break with climate change?

Yes, the discussion was pretty narrowly tactical and abstract, and perhaps we had not offered enough background about the coal plant issue. We'd also framed the question in a way that urged them to focus on civil disobedience rather than on the injustice the tactic might address. Still, Tim and I were encouraged that, on the whole, students took their activist roles seriously. (By the way, students voted 14 to 13 not to engage in civil disobedience.)

As Tim had predicted, students' top priority was to educate people on climate change issues. As Tim said, "No matter the role play, students always believe that education is the most important thing to do." Folded into that vote, however, was a decision to "educate" people to buy less stuff. "This is us educating us," Kadee pointed out. The SCA also voted overwhelmingly to reach out to grassroots environmental groups in the Global South with the aim of knitting together green alliances.

After about six weeks, my time at Lincoln had ended. Tim built on the activism we'd simulated in class, taking students to Salem, the state capital, where legislators were discussing global warming legislation. All along, we'd wondered whether our curriculum had found a balance between emphasizing the scope of the problem and highlighting how people can make a difference. In meeting with climate change activists from around the state, Tim wanted to connect students to individuals who grasp the civilization-threatening danger of global warming, but are not defeated by it. He also assigned students to interview and educate at least three people about climate change issues, helping students see themselves as teachers.

I'd done pieces of the activities in this unit over the years, but this was the first time I'd been part of creating a curriculum on global warming. I have a "this-should-have-been-so-much-better" feeling after almost any unit, but this sense was even more acute after this one. Ultimately, climate change is about every*thing* and every*one* on Earth. Deciding what to include and exclude is impossibly difficult—one more reason why all educators need to join the conversation about studying climate issues.

At times in this unit, I felt like a trespasser, a social studies teacher wandering through Scienceland. But if ever there were an issue that reveals how phony the divisions are between disciplines we call social studies and science, this is it. I ended my time at Lincoln feeling the urgency of teacher collaboration on curriculum that grounds students in a rich scientific understanding of Earth's natural systems as it exposes them to the social causes, consequences, and potential solutions to climate change.

This grassroots collaboration is all the more necessary, as textbook corporations become the de facto curriculum departments of U.S. schools. The gap couldn't be wider between our need for honest, critical curriculum on climate change and the pathetic materials available—witness the Holt McDougal/Houghton Mifflin global studies text described above. And, it's worth noting that Portland's physical science text adoption, *Physical Science: Concepts in Action* (Pearson/Prentice Hall, 2006), is as dismissive of human-caused climate change as the global studies adoption, hiding its misleading few paragraphs on p. 782.

That leaves curriculum development in our hands—teachers, environmental justice activists, scientists, communities affected by climate change. A billion people in Asia get their drinking water from glaciers that are disappearing, island nations like Kiribati and Tuvalu are drowning, huge swaths of Africa are becoming deserts, farmers in Australia are killing themselves in response to that country's worst drought on record, low-lying communities like New Orleans face the prospect of even more intense storms as oceans warm.

The threats are *so* dire that they have begun to prompt a profound social rethinking. Environmental justice movements are beginning to imagine a future that is greener, more cooperative, more democratic, and less oriented toward profit, consumption, and economic growth. This promises to be an era both terrifying and exhilarating. We have our work cut out for us. ⊕

Bill Bigelow (bbpdx@aol.com) is curriculum editor of Rethinking Schools *magazine.*

RINI TEMPLETON

Climate Change Mixer

BY BILL BIGELOW

No doubt, climate change affects everyone, everywhere. But not equally. Through role play, the Climate Change Mixer introduces students to 17 individuals around the world—each of whom is affected differently by climate change. For some, climate change threatens to force them to leave their land. For others, it is a business opportunity. In this activity, students meet one another in character and learn about the impact of climate change in their lives—and how each is responding.

Materials Needed:

- Mixer roles, cut up. One for every student in the class.
- Blank nametags. Enough for every student in the class. (Optional, but advised.)
- Copies of "Climate Change Mixer" questions for every student.

Time Required:

- One class period for the mixer. Time for follow-up discussion.

Suggested Procedure:

1. Explain to students that they are going to do an activity about the impact of climate change around the world. Distribute one role to each student in the class. There are only 17 roles, so in most classes, more than one student will be assigned the same individual. That's not a problem. You might point out to students that all of the roles describe actual people. In some cases, the roles incorporate these individuals' own words.

2. Distribute and have students fill out their nametags, using the name of the individual they are assigned. Tell students that in this activity you would like each of them to attempt to become these people from around the world. Ask students to read their roles several times and to memorize as much of the information as possible. Encourage them to underline key points. I ask students to list the three or four things they think are most important about their characters.

3. Distribute a copy of "Climate Change Mixer" to every student. Explain their assignment: Students will circulate through the classroom, meeting other individuals who also have some connection to global warming. They should use the questions on the sheet as a guide to talk with others about climate change and to complete the questions as fully as possible. Emphasize to students that they must use a different individual to answer each of the eight questions. (This is not *The Twilight Zone,* so students who have been assigned the same person may not meet themselves.) Tell them it's not a race; the aim is for students to spend time hearing each other's stories, not just hurriedly writing down answers to the different questions. Any role play risks stereotyping, so tell students not to adopt accents in an effort to represent an individual from another country. Encourage students to speak as if they are their assigned characters. Emphasize the use of the "I" voice, as sometimes students will begin by saying something like "My character lives in Bangladesh." It's important to the success of the activity that they attempt to *become* their characters—for example, to say, "I live in Bangladesh." Note that it's best to encourage students to meet one on one, as they circulate throughout the classroom. Some-

times students will cluster in groups, but this tends to allow some students to be passive and simply listen to others' conversations, rather than engaging in their own. Encouraging students to discuss the questions in pairs helps to address this potential problem. Finally, the last two questions ask students to begin to think about possible solutions. Tell students that answering both of these questions means they don't have to limit themselves to the information included in their role descriptions; they should try to propose ideas that are consistent with their characters' circumstances and concerns. For example, in one class, students playing two different individuals harmed by climate change decided they would make a film about the negative impact of rising temperatures throughout the world. This solution was not included in either role, but it was a creative response.

4. Ask students to stand up and begin to circulate throughout the class to meet one another and to fill out responses on the "Climate Change Mixer" questions student handout.

5. There is no set length of time for the mixer. I generally play a character myself so I can get a feel for how it's going and how much time students need. Allow at least a half hour for students to circulate.

6. After the students meet the other individuals, ask them to write briefly on some of what they learned from meeting people from around the world. Questions that I've used:

- Whom did you meet, or what situations did you hear about, that surprised you? Did you have any "aha's" while talking with people?
- Did anyone make you angry? Who?
- What themes seemed to come up in your conversations?
- Whom did you meet or which situations did you hear about that gave you hope?

7. Discuss these with students. See my article, "Teaching the Climate Crisis," on p. 79, for a description of how this played out with one group of students Tim Swinehart and I worked with.

Climate Change Mixer

1. Find someone who is hurt by climate change. Who is the person? How has this person been hurt? How might he or she be hurt in the future?

2. Find someone who might benefit from climate change. Who is the person? How might the person benefit?

3. Find someone who is affected by climate change in a way that is similar to how you're affected. Who is the person? How are your situations similar?

4. Find someone whose story involves a connection between water and climate change. Who is the person? What's the connection?

5. Find someone who will have to make life changes because of climate change. Who is the person? Why does this person have to make a life change? What might this individual do?

6. Find someone who lives on a different continent from you. How is this person affected by climate change? How is it different or similar to how you're affected?

7. Find someone who has an idea about what should be done to deal with global warming—or someone who is taking action in some way. Who is the person? What is the person's idea or action?

8. If possible, find someone here with whom you could take some joint action around global warming. Who is the person? What action might you take in common?

Larry Gibson

Mountaintop removal activist, Kayford Mountain, West Virginia

They say that to move away from oil we need to rely more on "clean coal," mined here in the USA.

Clean coal. That's a lie. That so-called clean coal comes from mountains in Appalachia that have been destroyed by coal companies, like Massey Energy. They blast mountains apart to get at the coal and dump everything they don't want in the valleys and streams, poisoning everything around.

When they talk about clean coal, they sure don't mean how they got it. They want you to focus on the fact that burning coal today produces less sulfur dioxide than it used to. That's the stuff that causes smog and acid rain. But burning coal still releases about twice as much carbon dioxide as natural gas, and a third more than oil—for the same amount of energy. And carbon dioxide is a greenhouse gas, the gases that cause global warming.

So mining coal is bad for the people of Kentucky and West Virginia, but it's also bad for the planet.

I've been fighting mountaintop removal of coal for more than 25 years. I'm not going to sit around and watch my home and the planet be destroyed. The coal companies care about the money. For me, it's not about the money. It's about the land. My mother gave me birth. The land gives me life. ⊕

Roman Abramovich

Sibneft Oil Co., Russia

Recently, I've seen a lot of articles asking whether global warming will be "good for Russia." This is a dumb question. Like anything, it will be good for some people and bad for some people. But I am doing everything I can to make sure that I am one of the people who benefits from global warming.

It's simple: As temperatures rise every year, ice will melt and huge new areas will be open for oil and gas exploration in the Arctic. And as one of Russia's wealthiest men, and head of a large oil and gas company, this is the chance of a lifetime. Researchers tell us that one quarter of the Earth's untapped fossil fuels, including 375 billion barrels of oil, lie beneath the Arctic. In the industry, we're talking about this opportunity as the new "black gold rush." Already our competitors in Norway—Statoil—are working on project Snow White, which will generate an estimated $70 billion in liquefied natural gas over the next 30 years. I'm not going to sit back and let the Norwegians or anyone else beat me out of this new business opportunity.

I'm sure that global warming is bad for a lot of people, but I'll leave that to the politicians and scientists. I'm a good businessman—a good *oil* businessman—so it's time to get to work. ⊕

Wangari Maathai

Green Belt Movement, Kenya

Africa is the continent that will be hit hardest by global warming. Unpredictable rains and floods, prolonged drought, crop failures, and fertile lands turned into deserts have already begun to change the face of Africa. The continent's poor and vulnerable will be hit the hardest. Already, some places in Africa are seeing temperatures rising twice as fast as world averages.

Wealthy countries will be affected, too. But for us, this is a matter of life and death. What makes this so outrageous is that our output of greenhouse gases is tiny when compared to the industrialized world's output. So the industrialized nations need to raise steady and reliable funds for the main victims of the climate crisis: the poor throughout the world.

For my part, I've been working in the Green Belt Movement for the last 30 years, since I was a young woman. We have mobilized millions of individual citizens in every country to plant trees, prevent soil loss, harvest rainwater, and practice less destructive forms of agriculture. We must protect the trees from the logging that is turning our continent into a desert. Our goal is to plant a billion trees. We will do our part to save the planet, but it is the rich countries that are most responsible. ⊕

Enele Sopoaga

Prime Minister, Tuvalu

Most people have never heard of my little island that is 400 miles from Fiji in the South Pacific. Tuvalu has 10,000 people in a place that averages just six feet above sea level. My people live on fish and fruit; everyone knows their neighbors and people don't even lock their doors.

Rising sea levels, caused by global warming, threaten the very existence of my land and people. Beginning in 2000, at high tide the water began covering places on the island that had never before been covered in the memory of even the oldest residents. In August 2002, the entire island flooded and the increased salinity [salt] has forced families to grow their root crops in metal buckets instead of in the ground. Many people believe that if current trends continue, there will be no more Tuvalu in less than 20 years.

The former prime minister of Australia said that if Tuvalu disappears, people should be relocated elsewhere. What incredible selfishness. How can anyone say that people in Tuvalu should suffer so that people in the so-called developed world can continue to fill our atmosphere with carbon dioxide by driving their big cars and buying stuff made halfway around the world? This is sick. That is why I have been speaking out. ⊕

Matthew Gilbert

Member of Gwich'in Tribe, Northern Alaska/Northwestern Canada

I am a member of the Gwich'in, the northernmost Indian nation on the American continent. There are about 8,000 Gwich'in. Because of global warming, we are threatened as a people.

We survive mostly from hunting caribou. Less snowfall is making sled and snowmobile transportation more difficult. Creeks are freezing later, and the ice is too thin to carry heavy loads. Lakes are drying up.

The worst threat is to the caribou. In 10 years, their number dropped from 178,000 to 129,000. Calves drown when they try to cross rivers that are usually frozen. My grandfather remembers vast numbers of caribou moving in waves near their village during spring and summer. No more. Our environment is in chaos. The hunters find it harder and harder to find the caribou that feed our people.

We must reduce greenhouse gases. My people are dying. ⊕

Chris Loken

Apple grower, Hudson Valley, New York

Everybody is saying awful things about global warming, and I know that it's bad for a lot of people. But recently Fox News did a report on climate change "winners" and they came to talk to me. As they said in their report, "There are some upsides to global warming."

Frankly, I saw this coming. I knew that things were going to get warmer and you know what they say about a crisis: It's also an opportunity.

I live in a beautiful place. Rolling hills. Good for apple trees. But I decided to diversify. Right next to the apples, I planted peach, apricot, and plum trees. Years ago. As I say, I saw this coming. These trees wouldn't have survived the winters of the old pre-global warming days. But our winters are getting milder, and I'm betting my trees will do just fine. As I told the Fox News people: "This farm here has been set up for the future." It's not easy running a farm these days, and if the weather decides to cooperate a little bit, who am I to argue? I'm sorry for those folks who are hurt by all this, but I've got to think of my family. ⊕

Stephanie Tunmore

Greenpeace climate campaigner

I joined the environmental organization Greenpeace because I felt like I had to do something to make the world a better place. To me, it seems that climate change is the most dangerous problem facing humanity and the environment. The consequences of global warming will be catastrophic, and we have to do something.

I've been working to save the Arctic. People think of the Arctic as just one big empty block of ice and snow. Either that, or where Santa Claus and the elves live. But it's an unbelievable place. There are species of birds and fish that are found only there and a few other places. Polar bears, musk oxen, and caribou reside there; and in the summer, snowy owls, ducks, and swans migrate there to nest. But already Alaska's North Slope has been taken over by 28 oil production plants, almost 5,000 wells, and 1,800 miles of pipes.

But the oil companies see global warming and the melting ice as an opportunity to drill for even more oil and gas. Haven't we learned anything? Why are we looking for more fossil fuels? The good thing is that more and more people are determined to stop oil development. We've taken direct action and have confronted the oil drillers in places like the Beaufort Sea, where we towed a fiberglass dome with two Greenpeace activists inside into a BP Northstar oil-drilling construction area. Two other activists unfurled a banner: **"Stop BP's Northstar, Save the Climate."** Direct action. That's what it will take to stop these oil-drilling criminals. ●

Rafael Hernandez

Immigrant rights activist, The Desert Angels, U.S.-Mexico border

In 1986, I crossed the border from Mexico to the United States, looking for a better life for my family. Now I am committed to helping migrants in need. My group, Los Angeles del Disierto—The Desert Angels—patrols both sides of the Mexican-California border near San Diego. We look for lost migrants and leave water, clothing, and food at key spots in desert locations to help people on their journey.

Recently, we rescued María Guadalupe Beltrán, a 29-year-old mother of four who had been burned severely in the huge Harris Fire on the border. Her father had died in Mexico and she had returned home to attend his funeral. She was caught in the fire coming back into the United States. But after suffering terribly, Beltrán died of her injuries. Afterward, I spoke to her husband, Rafael, who sat by her hospital bed for two weeks. He told me: "I asked the Virgin: 'Tell me whatever you want, please just don't take her.' But she did. At 11 in the morning my wife went away. She died at 11." Six migrants died in the fire and eight were injured.

The border patrol has pushed migrants to cross in unsafe desert areas. And global warming is making these areas even more unsafe, more deadly. Climate experts say that these wildfires, just like the awful ones in Greece, Australia, and Colorado, are going to happen more and more as the climate shifts. So María and other wildfire victims are also victims of global warming. ●

Rinchen Wangchuk

Snow Leopard Conservancy, Ladakh, India

When I was a boy, after school ended for the summer, I remember slipping down the glacier that stretched far down the mountains near my village in the Nubra Valley—in Ladakh, the far northern part of India. Today, that glacier is almost gone. And I am watching the glaciers of the Karakoram Mountains disappear a little more every year. One study found that each year, the glaciers lost between 49 and 66 feet, and another found that since the 1960s, more than 20 percent of the glaciers have disappeared. And as global warming increases, the glaciers will begin to melt faster and faster.

Glaciers are ice that has built up over thousands of years. Because it rains only two inches a year in Ladakh, we depend on the glaciers for 90 percent of our water. Farmers depend on this water to irrigate fields, and everyone depends on it for drinking. Ladakhis in the villages have worked out a cooperative system to share the water, but what will happen if the glaciers disappear? How will we survive?

In the rural areas of Ladakh, we have almost no cars. We pollute very little and release almost no greenhouse gases. It is unfair that the rich countries that produce so much carbon dioxide should be destroying the glaciers we depend on. ⊕

Moi Enomenga

Huaorani Indian, Eastern Ecuador

For years, the oil companies have invaded my people's lands and the lands of neighboring peoples—the Shuar, the Cofan, the Sequoya—in the rainforests of eastern Ecuador. First was Texaco. They left thousands of open pits that poisoned our rivers. Oil companies have spilled millions of gallons of crude oil and they continue to dump toxic chemicals into our rivers and streams. And oil development has also led to deforestation. When the oil companies build the roads, other "settlers" move in and chop down our forests and scare away our game.

With oil comes destruction. And now we learn that not only is oil development destroying our rainforest, it is destroying the world, through carbon dioxide pollution that leads to global warming. Oil kills the Huaorani through pollution and kills everyone through global warming. We say, "Leave the oil in the ground." Why do rich countries come here? People from the richest and most populated countries come to the poorest to take our resources, to live their life better, and leave us even poorer. But we are richer than they because we have the resources and the forest, and our calm life is better than their life in the city. We must all be concerned because this is the heart of the world and here we can breathe. So we, as Huaorani, ask those city people: Why do you want oil? We don't want oil. ⊕

Anisur Rahman

Mayor of Antarpara, Bangladesh

I am the mayor of Antarpara, a village in Bangladesh. Antarpara is on the Brahmaputra River that flows from the Himalaya Mountains in India. We are in the lowlands, and our village floods every year. We are used to it, and, in fact, the flooding is good because it leaves our land more fertile.

But now the floods are much worse. Now the floods are huge and each year they destroy our homes and carry off the land underneath them. My village used to have 239 families. Now we are 38 families. But where can we go when our homes are gone? Our country has 150 million people—the most densely populated in the world. I have an 18-month-old child. By the time she is grown, this village won't be here.

Where are we supposed to go? Do we all get tickets to America? ⊕

Steve Tritch

President and CEO, Westinghouse Electric

Before I became the head of Westinghouse I was senior vice president for Nuclear Fuel, providing nuclear fuel products and services to nuclear power plants throughout the world. Before that, I led the merging of the former ABB nuclear businesses into Westinghouse Electric, and was senior vice president of nuclear services. And before that, in 1991, I became manager of the Nuclear Safety Department, and later was appointed general manager of Westinghouse's Engineering Technology. Today, I belong to the American Nuclear Society and serve on the Nuclear Energy Institute's board of directors. I guess you could call me Mr. Nuke.

You might say that I'm a man on the hot seat these days. Not only are we running out of easy-to-find oil, but oil is also blamed for global warming. Coal is an abundant source of power, but it produces even larger amounts of greenhouse gases than oil—or natural gas. People are looking to my company, Westinghouse, for solutions. The solution is obvious: nuclear power. As I tell my employees, "What's good for the planet is good for Westinghouse."

Sure, the accident at the Fukushima nuclear plants in Japan was serious, and people were hurt. But the whole industry has learned from this accident, and even Japan still knows that nuclear power is the best way to go. The real threat is global warming. Global warming could destroy much of life on Earth. But nuclear power produces no greenhouse gases. They say nuclear power has dangers. Well, last year 5,200 Chinese coal miners died in accidents—and that's a lot more than have ever been hurt in a nuclear power accident. I see hope for the planet and Westinghouse is here to play our part. ⊕

Nancy Tanaka

Orchard Owner, Hood River Valley, Oregon

Our family has owned and operated fruit orchards in Oregon's Hood River Valley since my husband Ken's grandparents bought land here in 1917. Our family's only "time off" was when the U.S. government locked our family in internment camps during World War II. But that's another story.

Every generation of our family has farmed this land. And then we woke up to the front-page article in our local newspaper. It was a shocker. In fact, it scared us half to death. A study by Oregon State University found that 75 percent of the water during the summer months in the Upper Middle Fork of the Hood River comes from melting glaciers on Mt. Hood. And because of global warming, the glaciers are disappearing. That's our river. Well, we don't own it, but it's the river that irrigates our pears and cherries. Our family has grown fruit on this land since before we were born, and now they tell us that our irrigation water may be disappearing?

To tell you the truth, I never knew so much of the river's water in the summer came from glaciers. You see, glaciers on Mt. Hood are kind of small compared with glaciers on other mountains. Scientists say the problem is that glaciers have been shrinking because of global warming. I always thought global warming might affect the Arctic and the polar bears, but not the Upper Middle Fork of the Hood River. ⊕

Trisha Kehaulani Watson

Environmental lawyer, Hawaii

I was born and raised in the valley of Manoa, in the district of Kona (known today as Honolulu), on the island of Oahu. I am native Hawaiian. I am a lawyer specializing in environmental law—but much of my knowledge comes from talking with my family and *kupuna*, our elders.

Over the years, I have seen the beaches I played on my entire life steadily erode. In many places, the sand is disappearing.

My valley has always been very *waiwai* (wealthy, rainy, with much fresh running water), yet the waters have changed. We have far more unstable weather. When I was a little girl my grandfather used to take me down to the streams to watch the water rise when the heavy rains came. But things are much different today. The heavy rains are devastating. A few years ago we had a terrible flood wash through the valley. Since then, my street has been shut down numerous times due to dangerous flooding.

The seasons have also changed. It gets much colder than it used to, and also much hotter. The plants have changed because of it. Fruits come at unusual times of the year. Flowers bloom at different times of the year. Health problems also result from these weather changes.

The Earth is not well. ⊕

James Hansen

Former director, Goddard Institute for Space Studies, National Aeronautics and Space Administration (NASA), New York, City

I am a scientist, but I am also a grandfather. So that makes me especially interested in the future.

Recently, I was arrested at the White House in Washington, D.C., protesting the construction of the 1,700-mile Keystone XL Pipeline to send oil from the Tar Sands of Alberta, Canada, to Texas. Why would a scientist and a grandfather commit civil disobedience and get arrested? That's simple. If this pipeline is built and they continue to take this especially dirty and polluting oil from the Canadian Tar Sands, it makes it very unlikely that we will be able to stabilize the climate and avoid the disastrous effects that we are already beginning to see. As I've said, this pipeline is the fuse to the biggest carbon bomb on the planet.

Many years ago, I was one of the first scientists to warn that as we burn more fossil fuels—coal, oil, natural gas—the carbon dioxide created will heat the Earth to dangerous levels, with terrible, terrible consequences. I thought people would respond to scientists' rational arguments that we needed to end our addiction to fossil fuels. Now I know we need to take more drastic action.

So I volunteered to be arrested with 1,200 other people to draw attention to the importance of stopping this deadly pipeline from being built. I am more than 70 years old, but if need be, I will keep getting arrested. ⊕

Robert Lovelace

Ardoch Algonquin Indian leader, Ontario, Canada

In mid-February 2008, I was sentenced to six months in jail and ordered to pay a $15,000 fine. What was my "crime"? Trespassing on my own land—trying to block a uranium company, Frontenac Ventures, from prospecting on and polluting Algonquin Indian land. It began when we noticed people cutting down trees on land we had never ceded to the Canadian government. Someone had given Frontenac a prospecting license and they had gotten a court to issue an injunction against "trespassing." But this is our land, and Algonquin Indians and our non-Indian supporters organized a 101-day blockade to physically stop Frontenac from destroying the land. I was arrested and became a political prisoner.

Because of global warming, the nuclear power industry is claiming it is the "clean" alternative, because nuclear power does not generate greenhouse gases like coal or oil. The price of uranium shot from $43 a pound in 2006 to $75 a pound a couple of years after. It came down as a result of the 2011 nuclear disaster in Japan, but it will go back up. Canada is already the world's leading exporter of uranium, and our prime minister wants to increase exports and turn Canada into an "energy superpower."

There is nothing good about uranium mining. Uranium mining has no record other than environmental destruction and negative health issues. Mining companies clearcut the land and destroy the Earth to get at the uranium. Uranium can't be stored safely and other uranium mines around Canada have left land polluted with heavy metals like arsenic. And nuclear power itself is not clean. Nuclear waste stays radioactive for thousands of years and no one has found a safe way to store nuclear poisons that long. ⊕

Richard H. Anderson

CEO, Delta Airlines, Atlanta

I am CEO of Delta Airlines, and live in Atlanta. I'm a businessman and a lawyer, and have been in the airline business for more than 20 years. My job is to oversee Delta's long-term goals. Ultimately, I need to keep the company profitable for our investors and a secure and fulfilling place to work for our 80,000 employees.

I've been reading that air travel is bad for global warming. People say our jets produce a huge amount of carbon dioxide and other greenhouse gases that increase global warming. An article I read recently said, "Flying is one of the most destructive things we can do." This researcher concluded that "the only ethical option . . . is greatly to reduce the number of flights we take."

But ethics are complicated: Don't I have an ethical responsibility to my employees and stockholders—and to the 160 million customers who fly Delta every year, on more than 15,000 flights each day? And that means expanding air travel, advertising low fares, and trying to get people to take vacations to faraway places like Japan and China, to keep Delta profitable. Sure, we will try to pollute less, but we'll leave global warming to the politicians and scientists to figure out. I'm a businessman. ⊕

CLIMATE CHANGE TIMELINE

1890
Standard Oil Trust controls 88 percent of U.S. oil refining. By 1904, Standard Oil controls 91 percent of production and 85 percent of final sales.

1880s
Various European inventors create experimental gasoline-powered automobiles.

1859
Irish physicist John Tyndall confirms Fourier's theory that atmospheric gases like carbon dioxide trap heat. First oil well in United States is drilled in Pennsylvania.

1827
French mathematician Jean-Baptiste Fourier suggests that the Earth's atmosphere traps heat produced by the sun.

Circa 1750
Industrial Revolution begins in Manchester, England. Coal powers the mills.

1273
Wood- and coal-burning fires shroud English towns in smoke. Local regulations attempt to control the problem but fail. Atmospheric concentrations of CO_2 are about 280 parts per million (ppm).

1896
Swedish chemist Svante Arrhenius is the first to calculate that continued burning of coal and other fossil fuels will lead to a hotter Earth; later wins a Nobel Prize.

1899
The United States passes its first environmental law. The Rivers and Harbors Act makes it a misdemeanor to dump refuse into navigable waters without a permit.

1903
Henry Ford begins selling the gas-powered Model A automobile.

1904
The Willsie Sun Power Company builds the first U.S. solar-powered electrical plant in St. Louis. Soon afterward, the company builds another plant in the Mojave Desert at Needles, Calif. Within a few years, cheaper coal/gas facilities drive Willsie out of business.

1906
The city of Pittsburgh creates a smoke inspector's office and passes ordinances to regulate local air pollution. Air improves.

1920
There are 7.5 million gas-powered cars in the United States.

1948
Smog chokes the small industrial town of Donora, Pa. In five days, 20 people die, and 6,000 are sick or hospitalized. Air pollution becomes a national political issue.

1952
London smog, the product of a thermal inversion, kills 4,000 people in two weeks. Four years later, England's Clean Air Act becomes law.

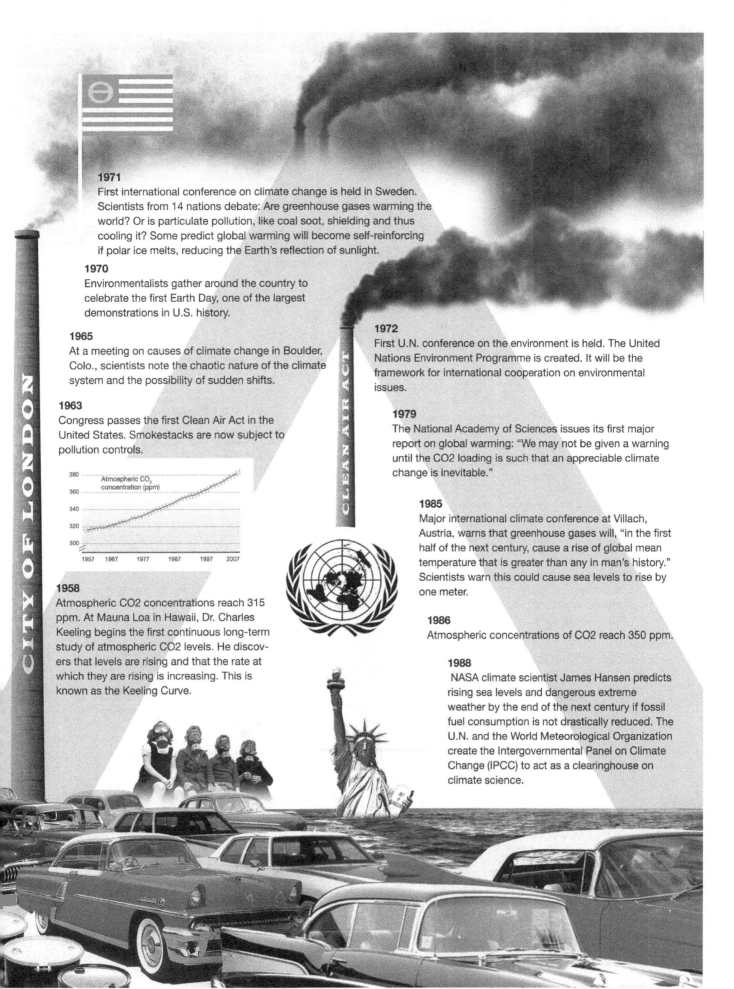

1971

First international conference on climate change is held in Sweden. Scientists from 14 nations debate: Are greenhouse gases warming the world? Or is particulate pollution, like coal soot, shielding and thus cooling it? Some predict global warming will become self-reinforcing if polar ice melts, reducing the Earth's reflection of sunlight.

1970

Environmentalists gather around the country to celebrate the first Earth Day, one of the largest demonstrations in U.S. history.

1965

At a meeting on causes of climate change in Boulder, Colo., scientists note the chaotic nature of the climate system and the possibility of sudden shifts.

1963

Congress passes the first Clean Air Act in the United States. Smokestacks are now subject to pollution controls.

Atmospheric CO$_2$ concentration (ppm)

1958

Atmospheric CO2 concentrations reach 315 ppm. At Mauna Loa in Hawaii, Dr. Charles Keeling begins the first continuous long-term study of atmospheric CO2 levels. He discovers that levels are rising and that the rate at which they are rising is increasing. This is known as the Keeling Curve.

CITY OF LONDON

CLEAN AIR ACT

1972

First U.N. conference on the environment is held. The United Nations Environment Programme is created. It will be the framework for international cooperation on environmental issues.

1979

The National Academy of Sciences issues its first major report on global warming: "We may not be given a warning until the CO2 loading is such that an appreciable climate change is inevitable."

1985

Major international climate conference at Villach, Austria, warns that greenhouse gases will, "in the first half of the next century, cause a rise of global mean temperature that is greater than any in man's history." Scientists warn this could cause sea levels to rise by one meter.

1986

Atmospheric concentrations of CO2 reach 350 ppm.

1988

NASA climate scientist James Hansen predicts rising sea levels and dangerous extreme weather by the end of the next century if fossil fuel consumption is not drastically reduced. The U.N. and the World Meteorological Organization create the Intergovernmental Panel on Climate Change (IPCC) to act as a clearinghouse on climate science.

MICHAEL DUFFY

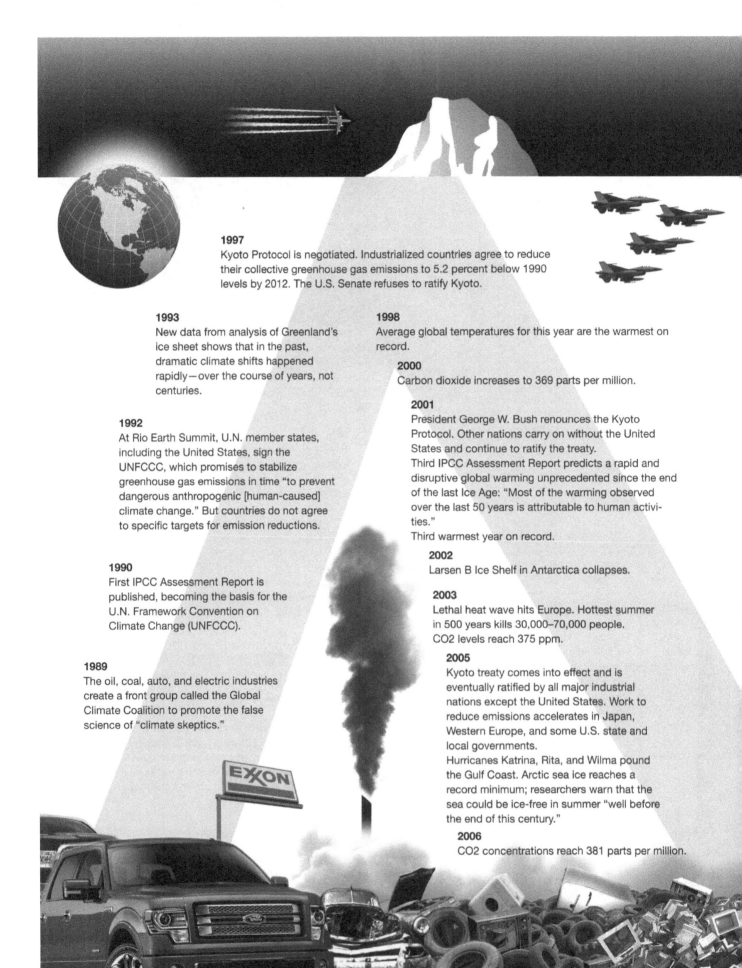

1997
Kyoto Protocol is negotiated. Industrialized countries agree to reduce their collective greenhouse gas emissions to 5.2 percent below 1990 levels by 2012. The U.S. Senate refuses to ratify Kyoto.

1993
New data from analysis of Greenland's ice sheet shows that in the past, dramatic climate shifts happened rapidly—over the course of years, not centuries.

1998
Average global temperatures for this year are the warmest on record.

2000
Carbon dioxide increases to 369 parts per million.

1992
At Rio Earth Summit, U.N. member states, including the United States, sign the UNFCCC, which promises to stabilize greenhouse gas emissions in time "to prevent dangerous anthropogenic [human-caused] climate change." But countries do not agree to specific targets for emission reductions.

2001
President George W. Bush renounces the Kyoto Protocol. Other nations carry on without the United States and continue to ratify the treaty.
Third IPCC Assessment Report predicts a rapid and disruptive global warming unprecedented since the end of the last Ice Age: "Most of the warming observed over the last 50 years is attributable to human activities."
Third warmest year on record.

2002
Larsen B Ice Shelf in Antarctica collapses.

1990
First IPCC Assessment Report is published, becoming the basis for the U.N. Framework Convention on Climate Change (UNFCCC).

2003
Lethal heat wave hits Europe. Hottest summer in 500 years kills 30,000–70,000 people.
CO2 levels reach 375 ppm.

2005
Kyoto treaty comes into effect and is eventually ratified by all major industrial nations except the United States. Work to reduce emissions accelerates in Japan, Western Europe, and some U.S. state and local governments.
Hurricanes Katrina, Rita, and Wilma pound the Gulf Coast. Arctic sea ice reaches a record minimum; researchers warn that the sea could be ice-free in summer "well before the end of this century."

1989
The oil, coal, auto, and electric industries create a front group called the Global Climate Coalition to promote the false science of "climate skeptics."

2006
CO2 concentrations reach 381 parts per million.

Sources: The Nation; Field Notes from a Catastrophe; 350.org; American Institute of Physics; PBS Frontline; Alternet.org; Ecowatch.com

2009

In October, the grassroots group 350.org coordinates a worldwide day of action on climate change, in which tens of thousands of people in 181 countries demand radical action to reduce emissions.

Largest civil disobedience action in U.S. history: 2,500 people blockade the gates of the Capitol Power Plant, which burns coal to provide heat to the U.S. Congress. Coal is the greatest contributor worldwide to carbon dioxide in the atmosphere.

Copenhagen climate conference fails to negotiate binding agreements. Atmospheric CO2 concentrations reach 390 ppm.

2008

The polar bear is put on the U.S. endangered species list because climate change is destroying its habitat.

Britain passes a climate change bill mandating 80 percent reductions in emissions by 2050.

Americans for Prosperity, funded by the Koch brothers, launch a "Hot Air" tour to oppose any regulation of carbon emissions. The tour's slogan is "Global Warming Alarmism: Lost Jobs, Higher Taxes, Less Freedom."

2007

IPCC Fourth Assessment Report sees possibility of abrupt and irreversible climate change.

At a U.N. conference in Bali, governments agree to a timetable for negotiating a successor agreement to the Kyoto Protocol. Negotiations will conclude in Copenhagen, December 2009.

The Supreme Court rules, in Massachusetts v. EPA, that the Environmental Protection Agency has the authority to regulate carbon dioxide emissions. Environmentalist Bill McKibben founds the climate action group, 350.org, recognizing that the atmosphere should not have concentrations of CO2 exceeding 350 parts per million.

2012

Level of CO2 in the atmosphere reaches 394 ppm.

Texas and Louisiana require teachers to teach "both sides" of the climate change issue as valid scientific positions—that humans may or may not be responsible for changing the climate.

NASA reports that nine of the 10 warmest years since 1880 have occurred since the year 2000.

350.org launches a Go Fossil Free: Divest from Fossil Fuels! campaign, urging colleges, universities, cities, religious institutions, and pension funds to withdraw investments from fossil fuel companies.

2013

May: Carbon dioxide concentrations in the atmosphere hit 400 ppm for the first time in human history.

More than a thousand protesters are arrested at the White House in civil disobedience actions to demand that the Keystone XL pipeline from Canada to the Gulf of Mexico not be approved.

2014

In a public letter, 93 faculty members at Harvard University demand that the university get rid of all its investments in the fossil fuel industry. Demonstrations against the fossil fuel industry spread to colleges and universities throughout the country. In May, students are arrested at Washington University in St. Louis protesting the university's ties with Peabody Energy, a giant coal company

September 21, 2014

An estimated 400,000 people marched through the streets of New York City for the People's Climate March. It was the largest environmental protest in history. Organizers declared: "We'll take to the streets to demand the world we know is within our reach: … a world safe from the ravages of climate change; a world with good jobs, clean air and water, and healthy communities."

MICHAEL DUFFY

Proof Positive

Charles David Keeling and the origin of climate science

BY ROBERT KUNZIG

May 19, 1955, 2:30 a.m. On a clear, nearly moonless night 27-year-old chemist Charles David Keeling stood alone on a footbridge in a forest near Big Sur, California. Starlight trickled through the redwoods; the Big Sur River rushed beneath his feet. Keeling held a spherical five-liter flask from which he had pumped all the air. He carefully opened the stopcock and let the cold, damp Big Sur air flood in.

If you had to pick a moment when the science of climate change was born, that would be as good a choice as any. Two weeks later Keeling was in Yosemite National Park, filling more flasks; in July he was standing in snow at nearly 13,000 feet in the Inyo Mountains; September found him on

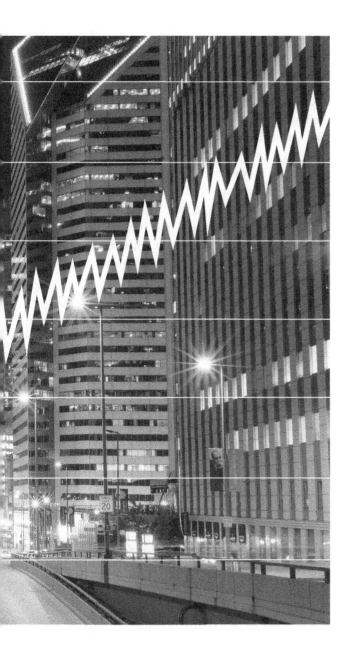

became a mission with global importance. Keeling died in 2005, but the measurements are still going on. In 2013, the atmospheric carbon dioxide level reached 394 parts per million. Judging from similar analysis of ancient air trapped in Antarctic ice cores, that is higher than it has ever been in the past 800,000 years.

The Keeling curve helps anchor the drifting debate about climate change in the hard rock of fact: thanks to Keeling, everyone agrees CO_2 is going up. Here's another indisputable fact: Earth's temperature is going up too. The global average surface temperature, compiled from thousands of locations on land and at sea, has risen by 0.74 degrees Celsius in the past century, a little over a degree Fahrenheit. Three-quarters of that increase happened in the past three decades; 11 of the 12 warmest years on record have occurred since 1995. "Warming of the climate system is unequivocal, concluded the Intergovernmental Panel on Climate Change (IPCC)—a panel of hundreds of the world's leading Earth scientists. In its Fourth Assessment Report, released in 2007, the group declared that it was more than 90 percent certain that the recent warming was caused by rising greenhouse gases, especially CO_2.

There is little question that human beings are responsible for the buildup. Humanity, largely by burning fossil fuels, pumps out more than 39 billion tons of CO_2 each year, and the amount continues to rise. A little over half of the human-generated emissions is enough to explain Keeling's curve of rising CO_2 perfectly. Most of the remaining CO_2 is dissolving into the ocean, making it measurably more acidic.

A few vocal skeptics continue to argue that climate change, if it's happening at all, has natural causes. There is little evidence to support such claims. Solar and volcanic activity, which can either warm or cool Earth's temperature, have been about the same over the past 50 years. Changes in the shape of Earth's orbit and the tilt of its axis, which are thought to trigger the

> There is little question that human beings are responsible for the buildup. Humanity, largely by burning fossil fuels, pumps out more than 39 billion tons of CO_2 each year, and the amount continues to rise.

a beach on the Olympic Peninsula. He was having fun, unburdened by any sense of mission. But back in his lab at the California Institute of Technology he would measure, more precisely and reliably than ever before, the tiny amount of carbon dioxide contained in his air samples. To his surprise, he found it was nearly the same everywhere, between 310 and 315 parts per million. That meant he could measure it in just one place, year after year, and see how it changed over the whole planet.

Which is what Keeling did, starting in 1958 on the treeless peak of Mauna Loa in Hawaii. Year after year he made them—even after he started boring people a bit, even after the funding agencies tried to cut him off. What had started as a fun project cycle of ice ages and warm interglacial periods, happen over millennia rather than decades. Moreover, when sophisticated computer models of climate include only natural causes, they don't reproduce the recent warming; when they include the increase in CO_2, they reproduce it quite accurately.

As long as the Keeling curve keeps rising, temperatures will rise too. We don't need 21st-century computer models to tell us that. Just old-fashioned undisputed 19th-century physics. It's been 187 years since the French mathematician Jean-Baptiste Joseph Fourier demonstrated that Earth's atmosphere traps heat. It's been 155 years since Irish physicist John Tyndall showed in laboratory experiments that carbon dioxide and water vapor do most of the trapping. And it's been 118 years since Swedish physicist Svante Arrhenius calculated that as CO_2 went up—which it was then just starting to do—the CO_2-warmed atmosphere would hold more water vapor too, amplifying the warming. In recent years that too has been confirmed: The amount of water vapor in the atmosphere really is increasing.

> In those four decades he had also been watching, from a hill near his home, as the city of San Diego sprawled east from the Pacific, covering farmland and wildland with houses, highways, and cars. "I have repeatedly asked myself how long these increases can go on."

When the same models that simulate the present climate so well are run into the future, they forecast a further warming by the end of this century of between 1.8 and 4 degrees Celsius, or 3.2 to 7.2 degrees Fahrenheit. The remaining scientific debate lies there—in just how hot it will get, and how fast, and what the impact will be. Part of the uncertainty is inherent to the chaotic climate itself; because it contains so many amplifying feedbacks, we will never be able to predict it with complete precision. And part of the uncertainty is due to us. Since 2000, emissions have been growing faster than the IPCC or anyone else thought possible—in part as a result of the explosive economic growth in China and India.

The IPCC reports, attacked as they sometimes are by climate skeptics, are actually deeply conservative and far from alarmist. Yet many climate scientists are alarmed these days. They see the surge in CO_2 emissions and in temperatures, they see ice melting rapidly in the Arctic and in Greenland, and they know that the uncertainty in the model forecasts cuts both ways. A spate of new research into the Earth's history fuels the fear that the models are more likely to be underestimating the severity of future climate change than overestimating it. In the past, it turns out climate has changed more dramatically and rapidly than is envisioned in the IPCC forecasts—even when we weren't there to give it a big kick.

There is no sign that CO_2 emissions are going to be cut soon. "The consumption of fossil fuel has increased globally nearly threefold since I began measuring CO_2," Keeling wrote in 1998, reminiscing on his career. In those four decades he had also been watching, from a hill near his home, as the city of San Diego sprawled east from the Pacific, covering farmland and wildland with houses, highways, and cars. "I have repeatedly asked myself how long these increases can go on."

According to the latest climate science, not much longer if we want to feel safe about the future. ⊕

Robert Kunzig is senior environmental editor for National Geographic *magazine and author of* Mapping the Deep *and co-author of* Fixing Climate: The Story of Climate Science—and How to Stop Global Warming. *Reprinted from* National Geographic.

Ask Yourself These Questions

BY GEORGE MONBIOT

THEA GAHR

To doubt, today, that man-made climate change is happening, you must abandon science and revert to some other means of understanding the world: alchemy perhaps, or magic.

Ice cores extracted from the Antarctic show that the levels of carbon dioxide and methane in the atmosphere (these are the two principal greenhouse gases) are now higher than they have been for 800,000 years. Throughout that period, the concentration of these gases has been closely tracked by global temperatures.

Carbon dioxide (CO_2) levels have been rising over the past century faster than at any time over the past 20,000 years. The only means by which greenhouse gases could have accumulated so swiftly is human action: Carbon dioxide is produced by burning oil, coal, and gas and by clearing forests, while methane is released from farms and coal mines and landfill sites.

Both gases let in heat from the sun more readily than they let it out. As their levels in the atmosphere increase, the temperature rises. The concentration of carbon dioxide, the more important of the two, has risen from 280 parts per million of air (ppm) in Marlowe's time to about 400 ppm today. Most of the growth has taken place in the past 50 years. The average global temperature over the past century has climbed, as a result, by 0.85° centigrade (1.5° Fahrenheit). According to the World Meteorological Organization, "the increase in temperature in the 20th century is likely to have been the largest in any century during the past 1,000 years."

If you reject this explanation for planetary warming, you should ask yourself the following questions:

1. Does the atmosphere contain carbon dioxide?
2. Does atmospheric carbon dioxide raise the average global temperature?
3. Will this influence be enhanced by the addition of more carbon dioxide?
4. Have human activities led to a net emission of carbon dioxide?

This is excerpted from Heat *by George Monbiot (Penguin Press).*

Carbon Matters

Middle school students get carbon cycle literate

BY JANA DEAN

It was the day after Hurricane Sandy and Eli and Krishna rushed into class with a question: "Ms. Dean, can it happen to us?" They'd seen images of buildings flooded to the second story and of cars that had floated to rest on top of each other. Our West Coast bay side town has early 20th-century buildings that look something like those flooded by Sandy in early 2013, and the students already knew that our streets flood when high tide, low air pressure, and heavy rainfall combine. The water backing up into storm drains in our streets has increased because of seas that have been rising due to heat trapped by a carbon-laden atmosphere. Eli and Krishna gave me an exciting opening to connect their up-to-the-moment concern to the biology, geology, and physics of climate change. I knew I could squeeze a lot of relevant science out of just one hurricane.

I was in for a surprise, however. The first time I mentioned climate change, Trevor's dad sent me an email insisting I "balance the science" and teach students that global warming may not be happening at all. Ultimately, the parent's persistent scrutiny, as uncomfortable as it made me, made my science even stronger and more specific. Instead of treating humanity's effect on the atmospheric system generally and broadly, I elected to isolate the carbon cycle from climate change and allow students to explore the difference between the short and long carbon cycles. As a result, they became more informed about the science behind climate change than most of the adults in their lives.

Carbon cycles, and as it does, it changes form by bonding with other elements. It hooks up with hydrogen, calcium, or oxygen as it finds its way through a planetary closed loop. It bonds with hydrogen to make the carbohydrates in plants and animals. It bonds with calcium to make the shells of sea creatures. It bonds with oxygen to make atmospheric carbon dioxide. In the short term, carbon shifts from CO_2 in the air through both photosynthesis and digestion to the starches and sugars that make up all living tissue. Then it shifts back to CO_2 and methane through respiration and decomposition. Over millennia, carbon-holding plants and animals are compressed by layers of sediment into coal and oil, becoming hydrocarbons. When burned as fuel, coal and oil break down into CO_2, which is suddenly released back into the atmosphere. What matters in climate science is not whether carbon enters the atmosphere but how fast the carbon cycles from the atmosphere to living organisms and to rocks and back again.

Both the short and long carbon cycles happen all the time. For example, eat a sandwich and you are participating in the short cycle: The plants and animals you are eating stored that energy in the last year at most. In contrast, drive your car to the sandwich shop and you are appropriating stored energy from deep time for a 10-minute excursion. Recognizing the difference between these two carbon cycles is a must for comprehending the link between burning fossil fuels and climate catastrophes like Hurricane Sandy.

The day after Krishna and Eli's questions about sea levels rising, I built interest for the rest of the class by showing students a cartoon from the "Week in Review" section of the *New York Times*. It depicts the Statue of Liberty with visitors arriving in dive suits. I asked students what they thought the cartoon was all about. Hands shot up. "Hurricane Sandy in New York City." "It's like New Orleans." "Hurricane Katrina." Although students had plenty of knowledge about big storms, no one mentioned the underlying phenomenon making such events more common and more catastrophic. They weren't yet connecting atmospheric carbon to the global rise in temperatures, storm frequency, and sea level. This seemed a perfect place to begin. If I proceeded step by step, I'd be able to build from what students knew to bring them to an understanding of one of the most important issues facing their generation. Usually I start a unit with a survey to find out what students know about a topic. This time, I relied only on that opening discussion and didn't anticipate the parent's coming challenge to the scientific basis of what I said next. I went ahead and told them that what had happened that week on the East Coast was due to climate change.

> **Drive your car to the sandwich shop and you are appropriating stored energy from deep time for a 10-minute excursion.**

I used a page on the *New York Times* website to connect rising sea levels to carbon emissions. The page uses models based on current carbon emissions as well as a reduction in carbon emissions to foretell future shorelines in U.S. cities. I asked my 6th graders to look for anything that surprised them. Miami turned out to be the star attraction: The entire city is expected to be underwater in a century or two. I wanted the class to move past the sensational, however, and think critically about what they were seeing. I asked them to begin to think about why some cities would be affected so much more than others. For example, why do the models predict that Seattle will lose only 13 percent of its land to the sea while places like Boston and Tampa Bay—37 and 50 percent, respectively—lose so much more? This put off discussion of carbon, but served to illustrate the potential effects of rising sea levels and broaden their perspectives beyond their own region.

Most students were able to infer that the differences are due to topography. If your city is flat, you flood. If you have hills, high ground will save you. This realization resulted in a heated discussion about the value of location. The neighborhood around our school sits on a ridge between two inlets. Sam said, "I'm going to be fine." He gestured to the northeast. "My house is right over there. I can walk to the water, but it's all downhill. I get tired coming home." Kathleen looked stricken. She and her mom live in a cottage by the bay. Inwardly, I cringed. My aim with such young students is to build urgency, not to scare them so much they quit being curious and clamor for a quick fix. I asked, "Do you have stairs to the beach?" She nodded gravely. I answered, "Then you probably are on high bank. Although we don't really know for sure, just like Seattle, that will likely give you more time." She relaxed, her immediate concern allayed. Then I drew cross sections showing the difference between high bank and low bank on the board. Students quickly saw that those desirable low bank beach properties came with a distinct future disadvantage. Isaiah didn't agree. "If you don't like where you're living, you can just move." I didn't want to leave the class there, thinking there wasn't anything to worry about because it's not happening tomorrow—or that the only consequence of climate change would be the arrival of the moving van.

When I teach about environmental issues I want students to realize that not everyone experiences the same impact. I asked students to talk to a partner about whether they thought sea level rise was fair. This launched a vigorous debate. Ava echoed Isaiah's earlier comment, saying people can move from endangered areas. But Lila stomped her foot and insisted, "There are places in New Orleans that are really beautiful. People live there and they love it. Their grandparents lived there and they do now and they don't want to move."

I temporarily cut off the debate to air a short video posted on YouTube by the United Nations:

> **Students quickly saw that those desirable low bank beach properties came with a distinct future disadvantage.**

Tuvalu: Sea Level Rise in the Pacific, Loss of Land and Culture (www.youtube.com/watch?v=L-gpH-gebunY). Students learned of the impending demise of the entire nations of Tuvalu and Kiribati, despite the fact that the remote islands have few carbon-emitting cars, factories, or power plants. These long-habitable atolls just happen to be low. I then directed students to take a stand on a continuum. If they thought sea level rise was fair, they would stand at one end. If they thought it wasn't fair, they'd stand at another. They could also stand anywhere else on the line if they saw that it was some balance of fair and unfair.

Students spread up and down the line and continued to voice their perspectives. Sam remained smug in his relative safety at the "fair" end. "I know the bay's close by but I live up here on this hill. I'm going to be fine." At the other end of the line, Keisha passionately spoke for the Tuvalans: "What are they going to do? Drive away? They live on an island!" Sam crossed his arms and exclaimed, "I don't know why we're even talking about this." Sam's resistance was nothing compared to what was quietly simmering in Trevor. Little did I know, I'd thrown Trevor into a moral crisis. He remembered that I had mentioned climate change as the cause of all these shifting shorelines. As class ended, he said, "But Ms. Dean, what if I don't believe in global warming?"

I asked, "Are you willing to learn more about it?"

He responded, "Yes?"

Even though Trevor's skepticism about climate change shouldn't have surprised me, it did. Our city council passed its first climate action plan way back in 1991 and in 2013 signed a resolution opposing the CO_2 disaster of new coal export facilities. The school district has a carbon emissions reduction plan and incorporates energy efficiency features in new construction. Transit runs on biodiesel and trumpets the reduction in carbon emissions on the back of every bus. People here have been publicly and actively tracking climate change for decades.

That afternoon I got the first email questioning my motives and asking for a balanced approach. Trevor's dad hoped his son was getting an understanding of both sides of the climate change issue. He went on to tell me that even though he didn't have a problem with the "theory" that climate change is a reality, he disagreed, and he wanted his side to be discussed. He asked that I present the argument that in fact it isn't happening after all. I took a deep breath and answered the parent, cc-ing my principal. I wrote about the sources I'd used thus far (the U.N. video and the *New York Times*) and that I'd next be using Climate Hot Map, a site curated by the Union of Concerned Scientists. I reassured Trevor's dad: "Everything on the site is backed by peer-reviewed research and presented in a form accessible to 6th graders."

I had chosen Climate Hot Map because a few students remained convinced that sea level rise was a problem

easily solved by human migration. If I were to get them engaged in the science of climate change, I'd have to persuade them that it mattered. Climate Hot Map provides links to global warming-related studies from around the world. It would allow them to see and understand crises like wildfires and water shortages. I provided a brief introduction for each of the organizational categories on the map: people, freshwater, oceans, ecosystems, and temperature. I told them to think about what interested them most and to activate those links and then to identify the three phenomena that most alarmed them. I added: "Everything on this map is a result of climate change. We'll learn more about what causes climate change later."

After using a class period to explore the site, students presented their greatest concerns. Forrest worried about wildfires in Australia due to increasingly low rainfall and high heat. Grant noted arctic amplification, which is an alarming feedback loop in which melting ice exposes less reflective ocean water that absorbs heat and in turn melts more ice. Sara passionately described the plight of Bordeaux vintners, although she would need to wait a few years to sample their wares. Afterward, I had them meet in geographically similar groups to discuss

their concerns in detail. After discussion, they wrote to the following prompt: "In a paragraph, describe what concerns you most about climate change. Explain what makes you care so much."

Students' passion surprised me. In my class, students draw an icon to represent the content on every page of notes or writing they create. For this assignment my 6th graders chose sad faces and stick figures with dark clouds over their heads. Isaiah ended his paragraph about aquacultural and agricultural issues in northern Europe with a wry "We need to take action on emissions (sorry SUV fans.)" Devi wrote about how rising water temperatures "contribute to all kinds of disasters like coral bleaching,

> First, I acknowledged the deep feelings in their sad-face and anger-cloud stick figures. I explained that passion makes you want to take action and science helps you know what action to take.

krill killing, and fish killing." He ended his paragraph by saying, "We can all stop global warming by using less emissions. Helping now is our only hope for survival." Then Meera offered her concern for forests, writing, "The mountain pine beetle eats trees and causes tree disease. Because of the increased temperatures from climate change, these tree-infecting beetles can survive through winter and multiply rapidly. This saddens me because when so many trees die (24.5 billion acres in British Columbia since 1994) the carbon levels in the atmosphere rise even more."

By then, Trevor's dad had begun transmitting an almost-daily influx of links to sources refuting any need for concern about climate change. I was a good sport. I spent entire evenings investigating the links he forwarded. As I tried to understand the science that called climate change into question, I didn't come up with anything I could share with my young scientists. Climate critics pointed to nonhuman sources of atmospheric CO_2, like volcanoes, or the vast stretch of prehuman geologic time during which temperatures fluctuated wildly. Some sources touted a slight dip in global average temperature in 2009. I did learn a lot from one site called "CO2 Science" about the power of photosynthesis, but its authors didn't consider the qualitative difference between the short carbon cycle (your sandwich) and the long one (the ancient hydrocarbons in your gas tank.) When I entertained the idea of bringing the "other side" to life so that students could decide for themselves whether climate change was happening, the string of lessons in graph reading and research analysis seemed endless, and I knew it would take me far away from my job of teaching science. Meanwhile, Trevor was increasingly silent and his dad's emails kept coming.

I went to my principal for help. His advice: "Drop the words climate change and global warming altogether and just teach the carbon cycle." When I answered that all the resources I'd found exist only because of the need to teach about climate change, he insisted, "Teach the kids the difference between politics and science. Politics is when we interpret the science and take action. We want people to make decisions based on science, right? So teach them the science and then connect the science to politics."

I wasn't sure I completely agreed with his take on the problem. I was concerned that teaching about the issue with a ban on terms so widely used in the media would make it far less relevant and meaningful to my students, not to mention contradict current scientific consensus. But I didn't know

what else to do, so I gave it a try. At the beginning of the next class I laid out the next few weeks.

First, I acknowledged the deep feelings in their sad-face and anger-cloud stick figures. I explained that passion makes you want to take action and science helps you know what action to take. Then I let them know that my job was to teach them science. I cited specific state standards having to do with systems and environmental issues and assured them that as a class we'd come to climate action based on the science. My doubter sat in the front seat. I've never had a 6th grader take such elaborate notes. He copied down every single word I spoke and every phrase I placed on the PowerPoint, and even asked me to repeat myself so he could get it all down.

That same day, students compared a windup watch (a closed system) to the digestive system of a dog (an open system) and identified open and closed systems they encountered in their lives and in the classroom. I ended the class by explaining that the Earth is a closed system. Except for inconsequential additions of matter in the form of errant asteroids, what's on Earth doesn't come or go, even though it continually shifts and changes form. "The movement of matter on the Earth, I told students, is called a cycle. This applies to carbon, which is what will matter once we have enough information to take action on climate change." It would be weeks before I uttered the phrase climate change again.

Next, students did a lab right out of our district-adopted FOSS Life Science materials. In it, students weigh water and celery stalks before and after leaving the stalks standing in water for 24 hours. The water seems to "disappear" from the vials. And it disappears the fastest from vials where the celery has the most leaves. Control vials eliminate evaporation as the cause. Given the relatively unchanged weight of the celery, the students determine that the plant must "use" the water for something. In fact, in photosynthesis, plants capture the C in CO_2 and the H in H_2O. In this way, they convert sunlight energy to chemical energy and store it as carbohydrates. As I explained the process to students, I was careful not to mention climate change. With the omission of the trigger phrase, Trevor relaxed in his note taking and the messages from Dad abated. It seemed Trevor could now happily learn some science along with his friends.

Then students investigated the way a finite

amount of carbon cycles through the Earth's closed system. Using lessons from the California Academy of Sciences (http://www.calacademy.org/teachers/resources/lessons), they identified objects containing carbon, including plant materials, plastic, soda pop, and seashells. They acted out the movement of carbon from the atmosphere (air), biosphere (living organisms), hydrosphere (water), and lithosphere (rocks). After a role play, in which dozens of donated tennis balls became flying carbon atoms, they created posters to illustrate the carbon cycle. On their posters they represented the processes by which carbon moves and changes form: photosynthesis, respiration, decomposition, calcification, and combustion. Even though the materials as written mention climate change at every turn, I didn't.

I finally had prepared students to think critically about the difference in carbon transfer between eating a sandwich and driving to the sandwich shop. I wondered how to broach the subject without inviting a fresh slew of emails. I knew the Keeling curve (which plots the increase in CO_2 in the atmosphere), might stir things up again. Instead, I chose to show students a YouTube video from a NASA scientist who describes the similarities and differences between coal and a banana. The friendly scientist, who introduces himself as Peter, explains that both coal and bananas capture atmospheric carbon via photosynthesis, but that the banana stores carbon in a short cycle involving digestion that lasts a few months at most, while coal stores carbon in a long cycle suddenly released when burned for energy. That afternoon, I got another email from Trevor's dad.

He suggested that I use a *Wall Street Journal* article about the growth of an isolated population of polar bears as proof that all was well in the Arctic. He thought I could use it to frame a class debate on whether or not climate change was happening at all. I responded: Students were learning about the carbon cycle in a way that would have them much more actively involved in learning science than they would be through a series of discussions about climate change. In my own mind I wasn't sure if banning discussion (and consequently the trigger phrase "climate change") was such a good idea, but since he didn't reply I figured no news was good news.

After comparing coal and bananas, our study of the carbon cycle culminated with a look at the human impact on the transfer of carbon. Students worked in groups to place the impacts on their posters. These included cement manufacturing, which moves carbon from the lithosphere to the atmosphere, and cattle farming, which shifts carbon from the biosphere to the atmosphere. They also included impacts of deforestation (biosphere to atmosphere), auto emissions, and burning coal for electricity (lithosphere to atmosphere.)

In order to know how well students were grasping the closed nature of the carbon cycle as well as human modifications to it, I asked them to respond in writing to the following questions:

- Are humans adding more carbon to the carbon cycle? Why or why not?
- What are human beings doing to change the carbon cycle? Give two examples.
- What is the difference between the short (fast) and long (slow) carbon cycle? Does the difference between bananas (fast carbon) and coal (slow carbon) matter? Why or why not?
- Do you think human alterations of the carbon cycle matter? Why or why not?

I was surprised by how difficult it was for my 6th graders to grasp that in spite of the increasing carbon in the atmosphere, the net planetary carbon remains stable. I suspect students struggle with the concept of a closed carbon loop because carbon alone is never visible. We see carbon only at rest in a seashell, the flesh on our bones, the leaf on a plant, or the dark rock of coal. The problematic carbon is in gas form, also invisible. My consistent message was "A finite amount of carbon cycles." I didn't say a word about the climate, and to my relief the doubting dad stayed quiet and Trevor remained relaxed.

After a role play, in which dozens of donated tennis balls became flying carbon atoms, they created posters to illustrate the carbon cycle.

Although a few students held onto the misconception that more carbon in the atmosphere means more carbon on Earth, I could see that they all understood that eating a banana and burning coal—while both transfer carbon to the atmosphere—affect the Earth in significantly different ways. Faye wrote, "We are releasing too much slow carbon from the lithosphere too fast for the biosphere to drink it up." Meera conveyed a nuanced understanding: "I'm not sure if humans by themselves are a problem. I don't think breathing is bad . . . but our factories and cars are really big problems since they pollute so much." And Trevor was learning too. He wrote, "I think that human alterations of the carbon cycle are a problem but not a problem. I believe that we shouldn't pollute and release bad gases but I think there is a reason we're able to breathe and have carbon go into the air and not 'destroy' the Earth. There are trees and an ocean and all the things that eliminate carbon from the air."

With students' emerging understanding of the short and long carbon cycles, I felt confident I could bring them back to climate change—the reason we were talking about carbon in the first place—and, given their strong feelings, what to do about it. We were only weeks from the end of the school year and I hadn't taught in depth about mining tar sands, coal mining, and the fracking process used to extract nat-

Helpful Links:

This series of maps on the *New York Times* website shows how rising sea levels may drastically affect areas in the United States.

www.nytimes.com/interactive/2012/11/24/opinion/sunday/what-could-disappear.html

YouTube video created by the United Nations: *Tuvalu: Sea Level Rise in the Pacific, Loss of Land and Culture* www.youtube.com/watch?v=L-gpHgebunY

The Union of Concerned Scientists' "Climate Hot Map" http://www.climatehotmap.org

YouTube video from a NASA scientist, describing the similarities and differences between coal and a banana: *Coal vs. banana: a two-minute explanation of the carbon cycle* http://www.youtube.com/watch?v=uStoBFtjy8U

ural gas. Sadly, I knew I wouldn't have time to do the topics justice and still plan an action. When I asked students what they'd like to do, the notion of fighting these sources of excess CO2 was conspicuously absent. We had spent so much time on the particulars of the carbon cycle, they didn't know nearly enough about what it had to do with 21st-century industrial living. I was satisfied, however, that everything on their list was connected to the carbon cycle. They included the impossible (hold your breath, stop eating, and the inevitable early adolescent "stop farting") as well as the possible (drive less, turn off our computers, and plant some trees.)

As the school year ended, the class participated in the joyful planting of a fast-growing carbon-absorbing maple tree, which they named Murphy Edgar Wood. On the last day of school, with lockers clean, books turned in, and desks pushed aside, my class held an appreciation raffle. Every student submitted an appreciation of someone in the class. With each draw, there were two winners: the one who wrote the appreciation and the one receiving it. Alec wrote his to the tree, "Dear Murphy Edgar Wood. Thank you for sequestering carbon on our campus. I am very appreciative of you reducing the greenhouse effect."

Although that little tree may not be much, I hope it symbolizes the fact that moving carbon from the long carbon cycle to the short one is not inevitable. I want my 6th graders to remember that we make choices as we interact with the Earth. I want them to remember that in planting Murphy, they took collective action and worked together to make a difference. They are young, and they will encounter climate change again and again in their lifetimes. Perhaps with their objective knowledge of the carbon cycle, they will engage the logic of those who say the wholesale transfer of carbon from Earth to sky is nothing to worry about. Moreover, when the coal trains threaten again to come through town—as they will—students will know why so many of our citizens picket and they'll have a scientific basis for choosing whether to join in on the action. ⊕

Jana Dean teaches 6th graders at Jefferson Middle School in Olympia, Washington. Students' names have been changed.

Atmospheric CO₂ at Mauna Loa Observatory

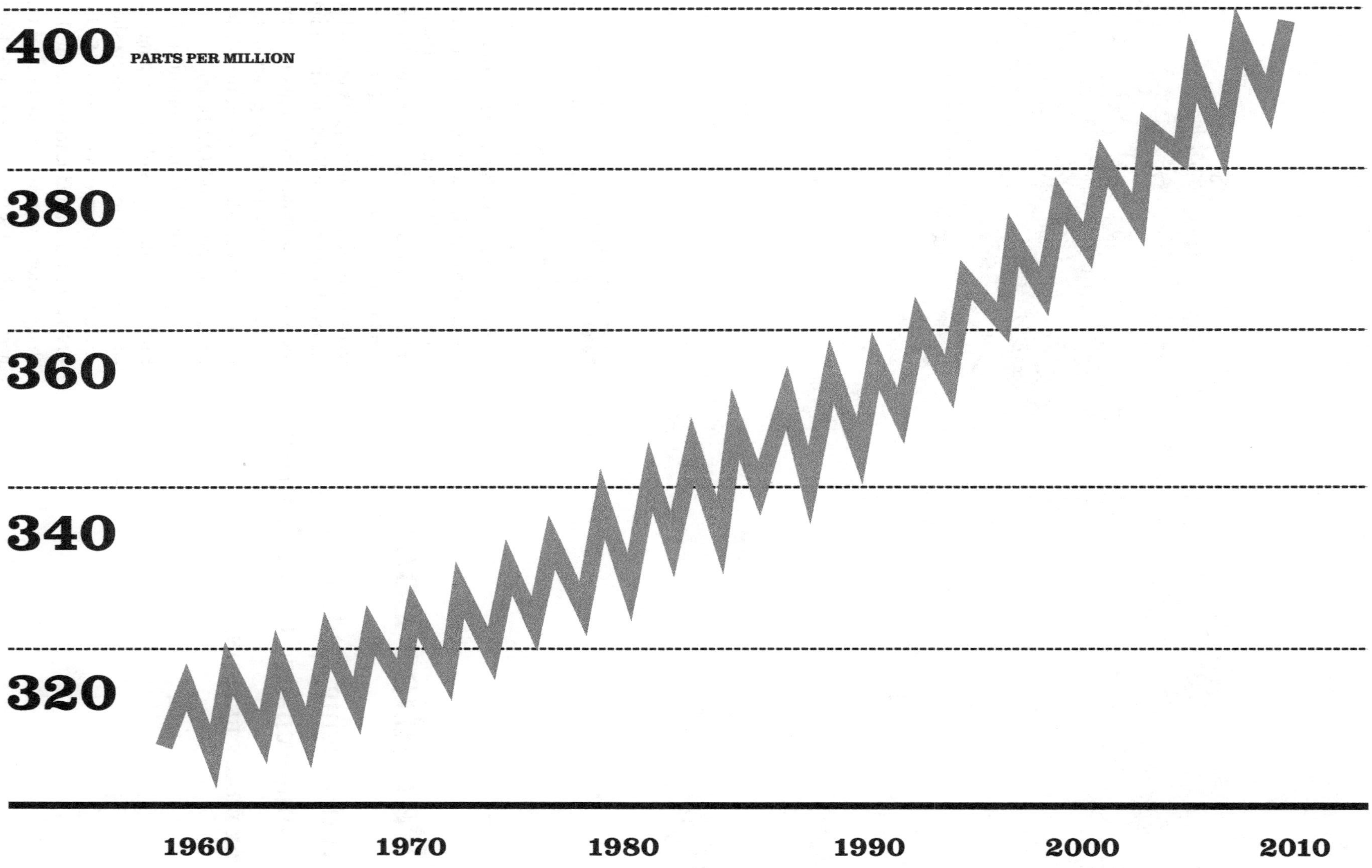

400 PARTS PER MILLION

380

360

340

320

1960　1970　1980　1990　2000　2010

Paradise Lost

Introducing students to climate change through story

BY BRADY BENNON

"This country has been the basis of my being. And when it's no longer there, you know, it's unthinkable." Ueantabo Mackenzie's haunting words in the PBS NOW documentary *Paradise Lost* shook me. I knew I wanted to teach a unit on global warming, especially after participating in the Portland-area Rethinking Schools curriculum group Earth in Crisis. I didn't have to be convinced that students need to learn about global warming; it's one of the defining issues of our time. But Mackenzie's message startled me: Global warming is here, right now, and it is uprooting people and destroying nations today, starting with Mackenzie's home on the island nation of Kiribati (pronounced KIRR-i-bas).

I grew up in Arizona, thousands of miles from Kiribati. My memories of the changing seasons usually boiled down to something like a shorts season and a jeans-with-sweatshirt season. So in the mid-1990s, when I started learning about the potential changes global warming might bring, I was vaguely nervous about the concept, but the danger of climate

change seemed remote and distant. What difference can a degree or two really make, I thought. I didn't grasp the rippling effects of global warming, including sea level rise, melting glaciers, and crop failures.

As I was planning my global warming unit—which I first taught to a freshman global studies class and later to a senior humanities class—it was important to ensure that my students didn't miss the point as I had. I didn't want them jumping straight into an investigation of the connections between carbon dioxide and rising temperatures, I didn't want them getting mired in the muck of political debates and international summits—without first hearing stories like Mackenzie's. I wanted them to see that, beyond the environmental damage, global warming is about people. Ultimately, I wanted them to care.

I decided that, before we watched *Paradise Lost,* I would help my students build empathy for climate change refugees and for people whose places are being altered by the changing climate. We started by reading the first chapter of Edward Abbey's *Desert*

Solitaire, "The First Morning." I asked students to highlight or underline words or phrases that were powerful, that spoke to them, or that were particularly descriptive. The piece begins:

> This is the most beautiful place on Earth. There are many such places. Every man, every woman, carries in heart and mind the image of the ideal place, the right place, the one true home, known or unknown, actual or visionary.

Abbey continues with a beautiful description of his first moments in his new home in a trailer near Moab, Utah, and the way in which the sights, sounds, and smells of the area fill him with a powerful sense of connection to that land as home.

After reading the chapter together, my students created a collective "found poem" by calling out the words or lines they highlighted when the moment felt right. The result was a poetic group effort that helped them explore Abbey's use of language to explain the power of place:

Landscape
the appeal of home
humps of pale rock
like petrified elephants
red dust and burnt cliffs
a cabin on the shore
known or unknown
such places
calling to me
apparent to me
beautiful

Words and phrases hung in the air and I brought it to a close. Then I explained that some of the places Abbey wrote about were under attack by developers and dam builders.

Afterward, I asked my students why they think places have such a profound effect on people.

Alex said: "Places are important because they can make us feel like we belong. We all need a home."

"I think a lot of my memories of people are mixed in with the places they took place. Like, when I think of my mom, I remember the house we lived in together and the food she made there," added Tori.

I loved that some of my students were beginning to personalize the idea of place. To help them go deeper, I borrowed an idea from our Earth in Crisis curriculum group and asked them to do some place writing of their own.

We started by brainstorming places they care deeply about—a special place in their home, at school, or even somewhere outdoors, where they feel they can be who they want to be and feel at peace. It could be someplace from their past or from their lives today. To prime the pump, I listed a few of my own examples: a trail through the forest, a particular soccer field where I enjoy playing pick-up soccer, the garden in my backyard. After picking one of the places on their list, students closed their eyes and imagined themselves in that place so they could absorb all the details like Edward Abbey did. I guided their visualization with prompts:

- Let your mind be a video camera. What do you see? What colors are there and how do those colors make you feel? Are there birds or animals? Maybe people you are close to?
- What do you smell as you look around you? Is it fragrant and floral like the flowers in the

background? Does it smell like fresh rain or fried chicken?
- What about the sounds that fill this place? A chirp? The wind blowing through the trees? The ocean lapping at the shore? Your friend or family member's soothing voice?
- Are there people who help make this place important? What do they look like or sound like? What feelings do they evoke in you?

Students opened their eyes and wrote for the next 15 minutes. Afterward, they shared their pieces with a partner. Then, I asked students to go back to their piece and write about what it would feel like to lose that place, to have that place taken or destroyed. From the sounds of their groans, I could tell this last task troubled many students, but they dove into their writing with intensity.

Then, we read our stories together as a read-around. Students' writing showed a deep connection to place, and they surfaced some powerful memories. Keesha, who was born in Liberia, wrote about a special tree from her village:

Every time I sit under this tree, it feels like nothing exists but nature, no cars, bikes, or smoke. This tree is not just the tree of peace, but the tree of home for many different animals, birds, and bugs. When it rains, I don't feel it much, so it's also the tree of protection. At age 9, I lost that tree. I came to the United States, hoping to find another tree like that one, but there's nothing like it.

Shana, who lives in foster care, wrote:

Every day I pass by it to and from school and I get a glimpse of my old place I used to call home. I see my mom and me laughing again. I can smell the incense. I can see us singing together as we cook dinner. At times, I just want to run and open the door to my home like I did as a kid. I am no longer that kid and that is no longer my home.

Erica's piece was eerily reminiscent of Mackenzie's words about Kiribati:

I look at that trailer and see my childhood,

my laughter, my fear, my balance, my thrills, my questions, my imagination, my safety, my saving grace, my youth. I see the place we ran to when the world wasn't what we wanted it to be, the place where everything was OK. Some of my heart sings to that trailer. To take it away would be to take part of me away from my soul.

When the places we love are destroyed or taken away, we lose more than land. We lose part of our identity.

Paradise Lost

Without delving into details about global warming, I had my students jump right into the 25-minute video *Paradise Lost*. I told them it was an introduction to global warming and would serve to help them understand why we're learning about the topic in social studies class in the first place. But it would not go deeply into the politics or science of the issue. I passed out a four-column chart and directed them to take notes about each of the four categories: people/things/objects, actions that happened, sensory details shown or described, and important words spoken. I mentioned that they would be using their notes to do some writing after the video. The video begins with a United Nations assessment that within the next 50 years, 6 million people a year will be displaced because of sea level rise and storms.

The film follows reporter Mona Iskander as she tours the island nation of Kiribati, which is less than 6.5 feet above sea level. Thirty-three separate islands make up this country east of Australia, which is disappearing because of rising sea levels. Iskander walks along sandy beaches surrounded by opalescent blue waters. Children play along the surf as fishermen throw their nets into the water nearby. The scene is idyllic and relaxed. Several students said that's a place they'd want to live. Noa, a recent immigrant from Tonga, said, "That looks like where I'm from."

The gentle voice of Anote Tong, president of Kiribati, breaks the spell of the enchanting island: "It's too late for countries like us. If we could achieve zero emission as a planet, still we would go down." The rising sea level from climate change is swallowing up the island right now.

In her first interview with Tong, Iskander exposes the injustice of the circumstances surrounding the island's demise: "What does it say to you that the poorest and the smallest countries, which are contributing the least to global warming, are the first ones to be affected by it?"

Tong's response is compelling: "Unless it hits you in the stomach, it means nothing to you." Later, he adds, "While the international community continues to point fingers at each other regarding the responsibility for and leadership on the issue, our people continue to experience the impact of climate change."

Tong's words cut to the essence of why I decided to lead off my unit with this film. I wanted to inspire my students to care about climate change and also to begin grappling with critical moral questions. For example, is it fair that per capita, the United States emits more than 17 tons of carbon dioxide annually, compared to 0.3 for the average Kiribati resident, but *their* land is among the first to disappear? Later in the unit, this foundation would enable us to investigate reasons why U.S. society is structured in ways that encourage and even require such fossil fuel use, to the benefit of multinational energy corporations.

> When the places we love are destroyed or taken away, we lose more than land. We lose part of our identity.

Throughout the film, Iskander documents the parts of the country already destroyed. By the time the film was made, Kiribati had to move "21 homes, a church, even their soccer field, or they would have been swallowed by the sea. The latest scientific reports say that within the century, the sea level will rise between 1 and 2 feet. That means much of this land will be gone."

The first-person accounts by Kiribati residents are disturbing. MacKenzie explains that the coconut trees, which are a crucial part of their diet and culture, are dying because of the saltwater intrusion. Linda Uan, a local tour guide, shows how the village water sources are now salty and unusable.

These interviews frequently lead to deeply moving statements about how profoundly sad it is to witness one's place get destroyed. My students were particularly moved by a brief interview with a Kiribati elder named Batee Baikitea, who says: "I love my land. If it is going to disappear, I will go with my land."

Uan adds: "It's our culture, our lands, it's everything. Everything's going to be lost. How would you take that? Losing one's land is emotional. There's no joking way about it; it is emotional."

The movie ends with Tong's decision to ask the wealthy nearby nations of New Zealand and Australia to take in thousands of Kiribati residents fleeing their island homes. The film raises questions about the obligations that wealthy countries emitting massive amounts of carbon dioxide have to climate refugees, as well as how climate refugees can prepare for the shock of such a move. It's a challenge the world will increasingly need to grapple with; 6 million refugees a year is no small matter. When I turned on the lights, I let the power of the film sink in for a few seconds as my students looked around at each other in astonished silence.

Kiribati Poems

While the impact of the film was still fresh, I wanted students to put words to Kiribati's situation, so I asked them to write poems. I discussed several types of poems they could choose to write. They could write a found poem, like the one we constructed as a class after reading Abbey's piece. Another option was a persona poem written from the perspective of someone or something in the film. "What are some examples of people or things whose perspectives you could write from?" I asked. Students listed options as I wrote them on the board. I was pleased when, after listing some of the Kiribati residents interviewed in the film, students made more creative suggestions: a palm tree, the ocean, the freshwater well, the *maneaba* (Kiribati village meeting center). As new ideas were added to the list, I noticed my reluctant writers getting more interested and less nervous about the assignment.

Wanting to help students think about potential structures for their poems, I provided two models: Linda Christensen's persona poem "Molly Craig" (from *Teaching for Joy and Justice*), and a student

> **The rising sea level from climate change is swallowing up the island right now.**

example of a found poem. I asked students to notice repeating lines, punctuation and line breaks, and sensory details. We talked about how each of those elements make the models interesting and add a sense of rhythm. I told students they didn't have to choose either of those poetry forms, as long as they conveyed the sense of place, the loss, and the injustice of climate change in Kiribati.

Finally, I provided students with a transcript of the film, so they could revisit parts of the video that spoke to them.

While students wrote, I projected my computer screen onto the board and wrote my own poem from the perspective of President Tong. Students who initially struggled seemed to settle into their own writing when they were able to watch me work on my own writing. I use this strategy for several reasons. My mid-thought pauses and occasional struggles demonstrate an element of solidarity with the students. Writing is not easy for anyone, me included, but it's worthwhile to work through the puzzling moments. It also sends a message to students that we are a community of writers, supportive and respectful of each other's words. Finally, it helps to establish a culture in which we share our work with each other so we can grow from each other's ideas and feedback.

When students finished writing their poems, we took turns reading them out loud. Veronica's poem, "Dancing as I Go," was written from the perspective of one of the traditional Kiribati dancers shown in the video:

Dancing as I go
With my feet buried in the sand,
I breathe in the rhythm of the wind and the
    waves
And wonder: how long?
As I perform the dance of my people to those
    who visit
I ask myself: will this be my last?
The last of my culture,
The last to pass on the traditions and ways of
    life,
The last to drink what is left of the remaining
    Kiribati water,
And the last to benefit from the bearings of the
    coconut tree.
Will those who I dance for return for me?

Save me?
Must I disappear with the land I love,
The land that is my home
Dancing as I go?

Lily's poem was both a celebration of Kiribati culture and a call to action:

Before we see the last of the tides come in,
Tell them we are here.

Before these waves beat down the last coconut
    tree,
Tell them we know. We knew all along.

Before our everything is flooding away,
Tell them about our home.

Tell them of its lush green palm leaves trailing
    across white beaches,
    Right up to the blinking blue sea.

Tell them our ways are simple.
We let the sun's rays walk down our backs,
And we are peaceful,
We are beautiful.

Before the read-around, I had asked the students to "write down lines that evoke emotions or cause you to want to take action, and write down themes that come up in the poems." After the readings, I asked the students: "So what? Your poems sound like you care about Kiribati. Why?"

Tyler said, "If my house was flooded because of something someone else did, I'd be mad."

Erica asked, "Is anyone trying to save Kiribati, or are they just going to drown?"

This was the place I wanted my students to get to before we dove deeper into global warming. During the rest of the unit, I hoped their concern for Kiribati would stay with them as they learned that pseudoscientists were paid to present false information about climate change. They would learn that corporations have tried to point the finger at individual consumers when, in fact, corporate practices are the lion's share of the problem. They would learn that oil companies spend far less than 1 percent of their budgets on renewable energy development while making billions in profit and painting them-

selves green. They would learn that a modern form of colonialism, climate colonialism, continues to exploit our atmosphere and impoverish and destroy countries like Kiribati. They would learn that, despite the warnings and urgent calls for help, nations have not passed binding treaties. In the end, through role plays, simulations, and community action with organizations such as Bill McKibben's 350.org, I hoped students would see themselves as truth-tellers and change-makers. But it all starts with caring.

"If my house was flooded because of something someone else did, I'd be mad."

The places where we live have a profound effect on our lives. They influence our ideas, beliefs, and how we see the world. Places give us meaning. Our memories make us who we are and are inseparable from the places where they are made. So what happens when our place gets destroyed? What happens to the people who are uprooted, ripped from their homes, torn from their place? We need to stop thinking of global warming as an abstraction. It is Kiribati. It is Katrina. It is Superstorm Sandy. Here in Oregon, it is a future of coastal towns inundated at high tide; increased wildfires, insect outbreaks, and tree diseases; and increased heat stress on crops. Global warming is you and me and all of us. Kiribati is just the beginning. ⊕

Brady Bennon teaches at Madison High School in Portland, Oregon.

Retreat of Andean Glaciers Foretells Global Water Woes

BY CAROLYN KORMANN

In 2009, the World Bank released yet another in a seemingly endless stream of reports by global institutions and universities chronicling the melting of the world's cryosphere, or ice zone. That report concerned the glaciers in the Andes and revealed the following: Bolivia's famed Chacaltaya glacier has lost 80 percent of its surface area since 1982, and Peruvian glaciers have lost more than one-fifth of their mass in the past 35 years, reducing by 12 percent the water flow to the country's coastal region, home to 60 percent of Peru's population.

And if warming trends continue, the study concluded, many of the Andes' tropical glaciers will disappear within 20 years, not only threatening the water supplies of 77 million people in the region, but also reducing hydropower production, which accounts for roughly half of the electricity generated in Bolivia, Peru, and Ecuador.

Chances are that many of Bolivia's Aymara Indians heard little or nothing about the report. But then the Aymara—who make up at least 25 percent of Bolivia's population—don't need the World Bank to tell them what they can see with their own eyes: that the great Andean ice caps are swiftly vanishing. Those who live near Bolivia's capital city of La Paz need only glance up at Illimani, the 21,135-foot mountain that looms over the city, and watch as its ice fields fade away. Their loss adds to a growing unease among the Aymara—and many Bolivians—who realize that the loss of the country's glaciers could have profound consequences.

The Aymara worship the ice-draped mountains as *Achachilas,* or life-giving deities, whose meltwater is vital to a region that suffers a five-month dry season and relies on agriculture to survive. Now, as greenhouse gas emissions heat the Earth, the Aymara are bracing for a future in which glaciers no longer can be counted on to supply life-sustaining water.

In recent decades, 20,000-year-old glaciers in Bolivia have been retreating so fast that 80 percent of the ice will be gone before a child born today reaches adulthood. So far this melting has brought temporary increases in stream flow and contributed to massive Amazonian floods that forced several hundred thousand people from their homes in 2008.

But within this decade, scientists predict that this torrent of meltwater will turn into a trickle as glaciers shrink, meaning that the age-old source of water during the dry season will steadily dwindle. Some highland farmers near La Paz already report decreased water supplies.

"Here you have precipitation only part of the year," said French glaciologist Patrick Ginot, standing at 16,500 feet next to Zongo glacier. "But it's stored on the glacier and then melting throughout the year, and so you have water throughout the year. If you lose the glacier, you have no more storage."

In effect, poor countries such as Bolivia are paying dearly for the massive energy consumption of the United States and the industrialized world. The so-called carbon footprint of the average Bolivian peasant is negligible, yet Bolivia's poor are not only among the first to feel the harsh effects of climate change, but also are sorely lacking the resources to adapt to it.

"The grand question here is, who compensates," says Oscar Paz, director of Bolivia's National Climate Change Program, "because we are not culpable for climate change. It's not fair that a country like Bolivia, which emits 0.02 percent of global greenhouse emissions, already has annual economic losses from the impacts of climate change equivalent to 4 percent of our GDP."

Bolivia is one of many countries, nearly all in the developing world, facing looming water shortages from melting glaciers. Up and down South America's western coast, Andean glaciers are the natural water towers to tens of millions of people, including those in the capital cities of Quito, Ecuador; Lima, Peru; Santiago, Chile; and La Paz.

Similarly, on the opposite side of the world, 2 billion people rely on meltwater from the Himalayas, which have lost 21 percent of their glacial mass since 1962. Himalayan glaciers are the main source of water for five major river systems whose flow irrigates much of China, India, and Pakistan's rice and wheat and that also supplies much of the region's drinking water. These river basins are the Ganges, with 407 million people; the Indus, with 178 million people; the Brahmaputra, with 118 million people; the Yangtze, with 368 million people; and the Yellow, with 147 million. Scientists predict that the Himalaya's smaller glaciers will be gone by 2035 and that many large ones will disappear by century's end, possibly leading to famine in a region whose population continues to soar.

Threats to Food Supply

"The world has never faced such a predictably massive threat to food production as that posed by the melting mountain glaciers of Asia," Lester Brown, president of the Earth Policy Institute, wrote.

Studies show glaciers melting at alarming rates throughout the world, yet unlike mountains in higher latitudes, ice melts year-round off tropical glaciers, which are found on peaks close to the equator and receive the sun's strongest rays.

"Glaciers, especially tropical glaciers, are the canaries in the coal mine for our global climate system," Lonnie Thompson, a preeminent glaciologist from Ohio State University, said during a climate change forum in Peru.

Bolivia's glaciated mountains are almost all in the Cordillera Real, or Royal Range, which soars from the northwest to southeast of La Paz and its adjacent slum city, El Alto, separating the arid, windswept expanse of the Altiplano (high plain) from the dripping verdure of the Amazon. Among these remote spires is a glacier that has become the most glaring symbol of Bolivia's rapidly transforming cryosphere.

Called Chacaltaya, which means "cold road" in Aymara, the glacier was once Bolivia's only ski resort and the world's highest. Now it is a barren, russet moraine studded with clues of its past: a lonely chunk of ice sticking out like an elongated diving board and a dirty white signpost with the fading graphic of a cartoonish condor on skis.

Looking down from Chacaltaya, the significance of its disappearance hits home. In the distance, the corrugated tin roofs of El Alto gleam across the endless Altiplano, which stretches like a placid brown ocean to the horizon. Water for the city's nearly 1 million residents comes mainly from the region's largest reservoir, situated at the base of a glaciated mountain cluster called Tuni Condoriri. Since 1983, the cluster has lost 35 percent of its ice mass. Glaciers Tuni and Condoriri, the two largest, are projected to disappear by 2025 and 2040, respectively, if not sooner.

> Edson Ramirez, Bolivia's leading glaciologist, published a study several years ago warning that water shortages would soon begin in El Alto and the outskirts of La Paz and worsen over the next decade.

Even closer is the glacier Zongo, the source for 10 cascading hydropower plants that provide a quarter of Bolivia's electricity. These days, Zongo is receding 33 feet a year. To the southwest stands Illimani, and though scientists have not monitored its glacial retreat, residents of nearby Palca say it is extreme.

The Andean Regional Project on Adaptation to Climate Change (PRAA) says that Palca and two other townships are most reliant on meltwater for survival and the most vulnerable rural districts to glacial loss. Pure geography, the areas' extreme poverty, and the lack of efficient irrigation methods are all factors.

Some residents already report decreases in flow, in part due to a drastic change in rainfall patterns. Worried about imminent water shortages, many Palca residents are migrating to the city or to other countries, such as Argentina. One irony of this migration is that many are moving from Palca to El Alto in hopes of a better life, yet water there also is running dry—the combined result of skyrocketing demand and diminishing natural reserves. A few decades ago, El Alto was just a small barrio next to the airport. In less than 20 years, the population has grown from 200,000 to 900,000, without any urban planning.

Edson Ramirez, Bolivia's leading glaciologist, published a study several years ago warning that water shortages would soon begin in El Alto and the outskirts of La Paz and worsen over the next decade. His team plotted a curve approximating when the water demand will surpass the amount that glaciers on Tuni and Condoriri will provide.

"Right now there is not a major problem in El Alto because the additional glacial melt has compensated for the demand, providing more water flow," says Germán Aramayo, the vice minister of water resources. "But we're going to begin to have problems."

In 1998, Ramirez and a team of French scientists presented Bolivian officials with the first results of their glacier-monitoring work, warning of the rapid retreat that was to come. No one believed them. Now there is little time to adapt before major water shortages begin. Nor does the government have the hundreds of millions of dollars needed to pay for these projects, which include building dams and reservoirs.

Many Bolivian officials believe that industrialized nations, the source of most greenhouse gases, have an obligation to help countries such as Bolivia mitigate the impact of climate change. Bolivia is planning to launch pilot projects in La Paz, El Alto, and four nearby communities that would, among other things, build more storage tanks to capture water in the rainy season; the World Bank will provide most of the funding. But far greater investment is needed to build larger reservoirs, help farmers acquire efficient drip irrigation technology, tap into underground aquifers, and rebuild municipal water systems, some of whose pipes leak half the water they carry.

Meanwhile, concern grows in places such as Palca.

"When I was a boy, the snows covered practically all these hills, but now year after year, they are melting away," says Roger Seja, leader of the Palca farmers' union. "It's very sad. How do we find a solution if nature herself, the universe itself, brings us this?" ⊕

Carolyn Kormann spent several months reporting in Ecuador, Peru, and Bolivia as a Middlebury Fellow in Environmental Journalism. This article first appeared in Yale e360.

 See teaching ideas for this article, page 175.

O n the Dec. 8, 2009, broadcast of *Democracy Now!*, Amy Goodman asked her guest, 15-year-old Mohamed Axam Maumoon, youth ambassador from the Maldives Islands to the U.N. climate talks in Copenhagen, for a message to young people everywhere about what climate change meant to him. Without hesitation, Axam turned to the camera and asked, "Would you commit murder . . . even while we are begging for mercy and begging for you to stop what you're doing, change your ways, and let our children see the future that we want to build for them?"

What does it mean to take Axam's question seriously? For many of us in the wealthy and so-called developed countries of the world, it means learning about the very real and life-threatening ways that climate change is affecting some of the world's poorest people. From the rapidly submerging islands of the Maldives, Kiribati, and Tuvalu, to the melting permafrost in native lands across the Arctic, Indigenous peoples around the world are confronting some of the worst effects of the climate crisis, despite having done so little to cause it. Axam's question prompts us to confront the injustice of a situation in which the wealthiest 20 percent of the world's popu-

"Don't Take Our Voices Away"

A role play on the Indigenous Peoples' Global Summit on Climate Change

BY JULIE TREICK O'NEILL AND TIM SWINEHART

lation has been responsible for more than 60 percent of global warming emissions.

The Indigenous Peoples' Climate Summit role play grew out of the Portland Area Rethinking Schools Earth in Crisis Curriculum Workgroup and the Oregon Writing Project. We designed it to introduce students to the broad injustice of the climate crisis and familiarize them with issues Indigenous groups around the world face as they confront climate change. The role play was inspired by the actual Indigenous Peoples' Global Summit on Climate Change, held in Anchorage, Alaska, in April 2009, when representatives from around the world exchanged experiences and observations from the front lines of climate change and agreed on a strategy for a worldwide campaign. The Anchorage summit highlighted how Indigenous peoples are combining traditional knowledge with new practices to adapt to climatic changes and the important role that Indigenous perspectives can play as the rest of the world attempts to respond and adapt to the realities of a quickly changing climate.

> **Following the example of the Anchorage summit, we wanted to model a collaborative decision-making process and give voice to groups that are often left out of international climate talks.**

We wanted to give our global studies students—9th graders at Lincoln High School, a large public school serving Portland's predominantly white, relatively prosperous west side—the opportunity to educate one another about how Indigenous peoples are confronting the effects of climate change. (Although we were not team teaching, we met daily to plan the unit and to share student responses.) Following the example of the Anchorage summit, we wanted to model a collaborative decision-making process and give voice to groups that are often left out of international climate talks. (Even the name of the most prominent climate monitoring organization, the *Intergovernmental* Panel on Climate Change, marginalizes Indigenous peoples, who rarely have their own national governments.) So we developed a role play in which students are divided into small groups, each of which represents an Indigenous group that attended the Anchorage summit. We wrote a profile sheet (see p. 134) for each group that details how they are being affected by climate change. The groups have an opportunity to discuss their own situation, teach and learn from the other groups, and, finally, agree on a common list of demands.

The role play includes six groups of Indigenous peoples: the Kiribati People of the Pacific Islands, Yup'ik People of Alaska, Bambara People of Sub-Saharan Africa, Aymara People of Bolivia, Indigenous Peoples of the Amazon, and Diné (Navajo) People of the American Southwest—people most of our students knew nothing about. In each of the roles, Indigenous people, as farmers or hunter-gatherers with intimate ties to the land, are validated and honored as legitimate observers of climate change.

Many nature shows, environmental groups, and even our own Oregon Zoo highlight the plight of the polar bear. And, like a word-association test, when we first brought up the issue of climate change in class, it was inevitably followed by "those-poor-bears" comments from our students. We built on that association by putting it in a larger perspective, with an exploration of the overall environmental consequences of climate change and the impact of these changes on the survival of peoples and cultures. For example, the Yup'ik role includes the following passage:

> The permafrost is dying and that means your way of life is threatened. You depend on hunting and trapping on the tundra for polar bears, walrus, seals, caribou, and reindeer, and harvesting fish from the sea. Less snowfall is making sled and snowmobile transportation more difficult. Creeks are freezing later, and the ice is too thin to carry heavy loads. Lakes are drying up. In 10 years, the number of caribou in one part of Alaska has dropped from 178,000 to 129,000. Calves drown when they try to cross rivers that are usually frozen. Your elders remember vast numbers of caribou moving in waves near their villages during spring and summer. No more. The environment is in chaos. The hunters find it harder and harder to find the caribou that feed your people.

In Portland we, too, deal with climate change issues. The glaciers on Mt. Hood are receding, threatening water supplies in the Hood River Valley and the apples, pears, and cherries the region provides. Like climate change itself, this feels abstract to many of our students, for whom food comes from the store and climate change happens somewhere else. The situations presented in the roles introduce students to people who face climate change directly; we want students to recognize the urgency and intimacy of climate change for many of the world's people. As the Bambara role, from sub-Saharan Africa, illustrates:

> The Sahara Desert is growing—you know, because you've seen it with your own eyes. Some measurements show the desert growing by up to 30 miles per year, taking over grasslands and trees in its path. It's starting to feel like you might be next. Your ancestors have lived near the desert for hundreds of years, farming special varieties of maize, millet, and sorghum adapted to the warm temperatures and dry climate of your homeland. But as temperatures all over sub-Saharan Africa get warmer, farming that was already difficult to begin with has gotten much worse. . . .

Learning to Empathize

We divided the class into six groups and asked them to form small circles around the classroom. As we distributed the roles to our students (students within each group received the same role), we asked them to read carefully, highlighting information they felt was vital to understanding the particular climate challenges their characters faced. The roles are packed with information, so we asked students to read them aloud in small groups and discuss the situations confronting their groups.

"This sucks—we are *dying!*" Jeyonna announced to the rest of the Bambara convened around her. The handouts we distributed to the six groups asked two questions: First, what did the group need the rest of the world to know about how climate change is affecting their region? Second, what actions—in order of priority—would they like to see the world take to address the problems facing Indigenous peoples as a result of climate change? We urged them to be imaginative. We reminded them that in some instances, their roles described actions their group was already taking, but they should feel free to propose other

JENNA POPE

actions that could help Indigenous people in their group and in other groups.

"What we need is nuclear power!" Quinn announced confidently to no one in particular, as the class settled into their task.

"If you want it so bad, you go work in the mines," Amanda called out, in character as a Diné. "If there are nuclear reactors," she explained, "you have to live with the waste."

The roles deal not only with how climate change affects Indigenous peoples but also with the impact of supposed alternatives—what's called "mitigation" in climate change jargon. For example, as the Diné role explains to students:

> **It was important to us that our students encounter activists who were not defeated by the scale of these problems, but were knitting together alliances to address them.**

> Farming is not the only way that the Diné are connected to climate change. As energy companies look for ways to make electricity that releases less greenhouse gas, some people are talking about nuclear power as a perfect solution to our climate problems. They say that nuclear power can produce all the electricity we need, and not release greenhouse gases into the atmosphere.

But where is the uranium that fuels the nuclear power plants mined? From your land. Diné people are some of the poorest in the United States, but Diné land is rich with uranium resources. You grew up hearing stories of the Diné men who worked in the yellowcake uranium mines, from the 1940s to the 1980s. You want others to hear these stories about family and friends who came home each day with clothes covered in yellow uranium dust. The companies that ran the mines told workers not to worry about the dust—that it was safe—but people now know that the mines exposed workers and their families to high levels of radiation. . . . So Indigenous peoples need to speak with one voice and say that not only do we have to support *real* solutions to climate change, we have to oppose *false* solutions—like nuclear power—that just lead to more poisoning of Indigenous people.

Finding "One Voice"

Finding this "one voice" was the impetus for the actual Anchorage summit. We wanted students to simulate the knowledge exchange and solidarity-building that took place at that summit. It was important to us that our students encounter activists who were not defeated by the scale of these problems, but were knitting together alliances to address them.

So for our next step, we explained to the students that they needed to find out about the situations of other groups, learn from their expertise, and look for commonalities, allies, and possible shared strategies for action. We asked students to choose half their group (generally three students) to move around the room as rovers/ information sharers. The other half would serve as stationary representatives, receiving the groups moving around the room. In this way, in a class of 30 to 36 students, each group had the opportunity to interact with the other five groups in three rotations, of about 10 to 15 minutes each, moving clockwise around the classroom.

In their meetings with other groups, we asked students to focus first on the specific ways they were experiencing climate change in their homelands and how these changes made life more difficult. (We provided graphic organizers so they could keep track of what they learned.) Many discovered common themes. "You have too much water and we don't have any," Jeyonna observed, as a member of the Bambara. "Too bad we can't just take your excess."

"We can't hunt; the ice is receding," Jake added, speaking as a Yup'ik. "People are going hungry."

"Us too! It's food. We can't grow it in the desert."

We told them that once they understood the problems faced by the other group they were speaking with, they should take notes on the actions the other group proposed to deal with those issues: "We need to stop burning fossil fuels—50 percent within five years, 100 percent within 10." "We need to mandate zero-emission power plants in all major cities." "We need to stop cutting trees in the Amazon." "We need to create more local economies."

Representing Indigenous cultures freed students from their own perspectives and their own limited experiences with climate change. It allowed them to suggest radical solutions, to envision a much different world from one ruled by the mighty dollar—because the true bottom line, for many of the roles, is that people and cultures are dying. As Usaia, reflecting on his role as a resident of Kiribati, wrote, "It affects a part of your body to see that the place you were born is going under water."

Once they returned to their original groups, the students compared notes with other group members to prepare for the full summit by deciding on their top two priorities for action.

Building Consensus

To get the larger summit conversation started, we asked each group to propose one of its top two priorities. We wrote the priorities on the board and the groups began negotiations. Students in each group had an opportunity to make the case for their respective priority and then the class discussed its merit. We explained and instituted a consensus model for decision-making; this encouraged the kids to really evaluate the issues, looking for com-

> "We keep fighting with big mining companies who don't seem to care about us at all. All they want is our uranium, but my whole family is sick from mining."

monalities and opportunities for compromise. Instead of the basic majority vote that we use to make decisions in other role plays, all of the groups had to express support for a proposal before adding it to the final list of action items—or at the very least, any hesitant group had to agree to "live with" a proposal by not blocking its addition to the list. To infuse some tension in the deliberations and to make sure that students didn't simply produce a laundry list of possible strategies, we asked them to agree on the three most important action points. Students struggled a little with the consensus approach, but it encouraged them to think beyond their individual groups.

Most discussion centered on whether to pursue long-term or immediate actions. "People are starving right now!" Kristi reminded the summit, speaking as a Bambara. "People need to know that—it has to be a priority!"

"That's true," Helene agreed, representing the Indigenous peoples of the Amazon, "but we also want a voice in the future. We need to be included and valued." Back and forth the discussion went as students suggested combining certain points, clarifying others, and eliminating a few.

Mary and Keegan, speaking as Diné representatives, made a convincing plea for moving toward more local economies. "We keep fighting with big mining companies who don't seem to care about us at all. All they want is our uranium, but my whole family is sick from mining," Mary argued.

"Yeah, that's why we need to let the rest of the world know that nuclear power is not a good solution to fossil fuels," Keegan added. "*We* should control what happens on Diné land, not some big company. We want to keep building solar panels on our land, not make more uranium mines."

Even if their arguments were simplistic at times, we were impressed that, within the first weeks of school, our 9th graders seemed fully invested in a multiday discussion that ranged from plans to cut emissions to limiting deforestation to funding programs to meet the needs of climate refugees.

In the end, most classes decided on a top action priority that combined the demands of some groups for wealthy countries to begin immediate and drastic reductions of greenhouse gas emissions with other groups' focus on proactive measures to help with a global transition to renewable sources of en-

ergy. You could almost see the lightbulb go off above Sami's head when she raised her hand to suggest, "Why don't we combine the first three suggestions: Stop using fossil fuels, cut emissions, and create more alternative energy? They all work toward the same goal, so we can see them as one action item. We can all support that, right?"

The Anchorage Declaration

After two days of intense conversation, we drew the summit to a close. Then we read the action items listed in the Anchorage Declaration from the actual Indigenous Peoples' Global Summit on Climate Change. Because students had invested so much energy in our own classroom discussions, they seemed eager to read the declaration from the real summit. For homework, we asked them to compare their list and the actual list, and to reflect on the similarities and differences.

In our discussion the next day, Keegan offered, "I think we did pretty well. I mean, our first action item is almost exactly the same as what they came up with at the real conference." When we asked about anything that we might have missed in our discussion, Sonya pointed out that "number three of the Anchorage Declaration talks about the historical debt that wealthy nations owe because they have burned fossil fuels for the last 100 years. Even though this means we wouldn't be as wealthy, it seems fair since fossil fuels have made our lives so much better."

Another item from the Anchorage Declaration that didn't make it into our action items was a plan to send representatives from the Indigenous Summit to the U.N. Climate Conference in Copenhagen. The stated goals in the Anchorage Declaration are to recognize the importance of traditional knowledge, to include Indigenous peoples' observations of climate change alongside those of scientists, and to fully include Indigenous voices in international negotiations to create global climate change policy.

On the day President Obama was to receive his Nobel Peace Prize, a coalition of North American Indigenous groups marched on the U.S. embassy in Copenhagen. Clayton Thomas-Muller from the Canadian-based Indigenous Tar Sands Campaign had this to say:

So we're here in Copenhagen at the United Nations international climate negotiations . . . to say we want a just and clean future, a new economic paradigm that doesn't sacrifice our communities at the altar of irresponsible policies for the economic benefit of the select few who pull the economic strings.

As it turned out, the Copenhagen climate talks relied on a particularly exclusive, undemocratic process to produce the weak and unenforceable "accord" heralded as progress by Obama and the leaders of a handful of nations. Subsequent climate conferences have yielded no enforceable agreements to cut greenhouse gas emissions.

But even if Indigenous voices were not given a prominent official role in the U.N. climate talks, Indigenous people did play a crucial role in the protests in the streets of Copenhagen and the incredible gathering of thousands of civil society groups that came to Denmark to advocate for a fair and just climate treaty. By December of that year, our classes had moved on to new topics, but as we took time out to watch some of the coverage of the Copenhagen talks on *Democracy Now!,* students recalled their connections to the Indigenous leaders they saw on the screen—the real people on whom their roles had been based.

The quietest voices
Have the loudest meaning
Every word said is like
An earthquake.

"My name is Johnson Cerda. I am a Quechua Indian from the Ecuadorian Amazon, and I grew up in the rainforest."

"That's me!" Helene shouted, as Cerda continued:

We are here because we understand that . . . in the climate negotiation, we need at least to put our voice first. Second, we want to insert some safeguards for Indigenous peoples. And the third thing is that we need also to say here that we have knowledge, and we can share our knowledge.

As the students were watching the Indigenous activists, Julie leaned over and whispered to Amanda Henderson, who is a member of the Warm Springs tribe of Central Oregon, "That might be you up there one day." Julie was unprepared when Amanda came to class the next day and handed her a poem:

The quietest voices
Have the loudest meaning
Every word said is like
An earthquake.
It sends a big movement
It moves the biggest barriers down
It can open a new state of mind.
The quietest voices
Can join and become
A million voices.
For what we say can
Be pushed aside
Forgot about.
But when we come together,
We are heard
We do count
We are ready to stand up
We won't take no for an answer
We will speak until
Everyone hears us
We will not be quiet anymore
We are important
We do count.
Don't take our voices away. ⊕

Julie Treick O'Neill and Tim Swinehart teach social studies at Lincoln High School in Portland, Oregon. Dianne Leahy and Bill Bigelow contributed to writing the Indigenous Peoples' Climate Summit role play.

Indigenous Peoples' Global Summit on Climate Change* Anchorage, Alaska

Welcome people of the Arctic, of the Americas, of the Pacific, of Africa, of the Caribbean, to the Indigenous Peoples' Global Summit on Climate Change. We thank the Ahtna and the Dena'ina Athabascan peoples in whose lands we are gathered.

Mother Earth is no longer in a period of climate change, but in climate crisis. Rising oceans, thawing permafrost, larger storms, expanding deserts, dying coral reefs, and declining forests threaten the stability of Indigenous peoples around the world.

While the wealthier nations of the world debate the best way to limit the effects of climate change in the future, we face the immediate destruction of our homes, our lands, our traditions, and our cultures. As Patricia Cochran, chair of the Inuit Circumpolar Council, said, "Indigenous peoples have contributed the least to the global problem of climate change, but will almost certainly bear the greatest brunt of its impact."

We meet here today, because we will not stand by and witness the destruction of our way of life. We represent only a few of the 5,000 groups of Indigenous peoples in more than 70 countries. Together we have a global population of 350 million, representing about 6 percent of humanity. We have persisted through 500 years of invasion, genocide, land theft, and now environmental destruction—we will outlast this threat as well.

We honor our solidarity as Indigenous peoples living in areas that are the most vulnerable to the impacts and root causes of climate change. We reaffirm the unbreakable and sacred connection between land, air, water, oceans, forests, sea ice, plants, animals, and our human communities as the material and spiritual basis for our existence.

In fact, the world desperately needs the wisdom of Indigenous perspectives, perhaps now more than ever. Because of our long cultural and spiritual connection to the land, oceans, and wildlife, Indigenous peoples have a lot to offer the rest of the world as it considers how best to deal with the climate crisis.

The goal of this summit is to develop a list of action items to present to the rest of the world at the next United Nations climate change conference. Because of the stakes posed by climate change, the conference could be the site of decisions that affect all of humanity, so it is essential that Indigenous peoples be given full participation. Today we start that process of participation. ●

*Adapted from the original, www.indigenoussummit.com

Role Play

Role-Indigenous Peoples' Global Summit
Kiribati People of the Pacific Islands

You represent the people of Kiribati (pronounced KEER-ih-bahs), a group of islands located in the tropical Pacific Ocean, to the north and east of Australia. The chain of 33 islands is home to more than 100,000 people, all of whom live only a few feet above sea level.

Kiribati is expected to be one the world's first countries to lose its territory as a result of sea level rise from global climate change. Within the last 10 years, two of the smallest islands have already disappeared underwater. As people in the United States debate whether rising sea levels will someday affect cities like New York and San Francisco, you're watching the effects right outside your front door. This is not some future scenario for the people of Kiribati; your neighbor must decide soon whether or not to move her grandmother's grave when the rising seas force her to move her home.

Your ancestors have lived here for thousands of years, but now you might lose your home because of a problem you didn't cause. But you know who's to blame: People in the wealthy countries of Europe and the United States have caused greenhouse gases to increase in the atmosphere, by burning so many fossil fuels—like coal, oil, and natural gas—for the last 200 years. These people claim the right to build their big factories, but how can this compare to your right to a secure home? These wealthy countries must immediately begin to dramatically reduce their use of fossil fuels. They need to agree to binding agreements that require them to reduce their carbon emissions.

As sea levels rise, salt water from the ocean has polluted the freshwater sources that the people and plants of Kiribati depend on for survival. As a symbol of things to come, one Kiribati island, Tepuka Savilivili, no longer has any coconut trees—killed off by increasing levels of salt water in the ground. Coconut trees are the backbone of traditional Kiribati culture, ranging in use from food to building materials. Because of its importance, your people refer to it as "the tree of life."

Kiribati's land and coconut trees are not the only things threatened by climate change—the traditional island culture is also at risk of disappearing with the rising ocean. For thousands of years, your ancestors have lived here, building their culture in relationship with the natural environment. If you are forced to leave Kiribati, forced to migrate to Australia or New Zealand, certain parts of Kiribati culture will be lost forever. This is why many Kiribati will choose not to leave their homes.

However, some Kiribati residents believe that they will have to leave, and have sought refugee status in other countries. They consider themselves refugees because through no fault of their own, their land—your land—is becoming uninhabitable. Your people don't want to leave Kiribati, but you may not have a choice. But as of now, there is no official status for climate refugees.

Kiribati president Anote Tong is already planning for the migration of your people. Here is part of a speech he gave to the United Nations last year:

> The relocation of the 100,000 people of Kiribati cannot be done overnight. It requires long-term forward planning and the sooner we act, the less stressful and the less painful it would be for all concerned. We must provide them with the education and training to make them competitive and marketable in international labor markets. This strategy provides our people with an option so that when they choose to migrate, they will migrate on merit and with dignity. They will be received by their adopted countries not as burdens, but as worthwhile members of the community.

But so far, the so-called developed nations have agreed to no plan for climate refugee migration. Do they expect you to drown in the ocean as they continue to pollute the atmosphere? ⊕

Role Play

Role-Indigenous Peoples' Global Summit

Yup'ik People of Alaska

You are a Yup'ik Eskimo and live in the Arctic. Your people have always had a close relationship with nature and notice even small changes in the environment. Recently, you have worried because the climate has become unpredictable and the landscape is changing in dramatic ways.

The permafrost is melting—and that means your way of life is threatened. You depend on hunting and fishing walrus, seals, and salmon. Autumn freeze-up occurs up to a month later than usual and the spring thaw seems earlier every year. The multiyear sea ice is smaller and now drifts far from your community in the spring, taking with it the seals your community relies on for food. In the winter the sea ice is thin and broken, making travel dangerous for even the most experienced hunters. In the fall, storms have become more frequent and severe, making boating difficult. You have seen thunder and lightning for the first time.

Hotter weather in the summer is melting the permafrost and causing large-scale slumping on the coastline and along the shores of inland lakes. The melting has already caused one inland lake near you to drain into the ocean, killing the freshwater fish. Even the foundations under buildings are shifting. And with the melting permafrost, buried toxic waste left by the U.S. military is increasingly polluting water and poisoning fish—and people.

The village of Newtok, about 800 kilometers west of Anchorage, is one of several Yup'ik villages in need of relocation due to climate change. Because of higher average temperatures, intensifying river flow and melting permafrost are destroying homes and infrastructure. More than 300 residents have been forced to relocate to a higher site 15 kilometers west on Nelson Island. This will cost tens of millions of dollars. What will become of your people? And why should your people suffer so that other nations can continue burning more and more coal and oil and living comfortably?

The Arctic is warming at twice the rate of the rest of the world. Scientists think that summer ice could vanish in the next 10 to 20 years. Oil companies say that one quarter of the Earth's untapped fossil fuels, including 375 billion barrels of oil, lie beneath the Arctic. They can't wait for the ice to melt. They call this the new "black gold rush." So as your people try to hang on to your land and culture, oil companies profit from your suffering and become richer and richer.

It's urgent that you join with other Indigenous peoples to stop climate change and to demand justice for your people. ⊕

Role Play

Role-Indigenous Peoples' Global Summit

Bambara People of Sub-Saharan Africa

The Sahara Desert is growing—you know, because you've seen it with your own eyes. Some measurements show the desert growing by up to 30 miles per year, taking over grasslands and trees in its path. It's starting to feel like you might be next. Your ancestors have lived near the desert for hundreds of years, farming special varieties of maize, millet, and sorghum adapted to the warm temperatures and dry climate of your homeland. But as temperatures all over sub-Saharan Africa get warmer, farming that was already difficult has gotten much worse.

The Bambara live in countries like Mali, Senegal, Burkina Faso, and Niger—all countries with farmers who face greater challenges as temperatures increase from global warming.

The Bambara have been good at adapting to the changing climate of the sub-Saharan region, but these changes may be too much. Even the hardiest varieties of the region's three main crops—maize, millet, and sorghum—would probably not tolerate the temperature increases forecast for the coming decades. Droughts in the 1970s and 1980s killed 100,000 people in the region, but this would be much worse. Without immediate solutions, food security across all of Africa could be threatened.

Malnutrition has become worse among your people. It is common to see children with a reddish fuzz on top of their heads—a sign of malnutrition. Half the Bambara children die before their first birthday. For some, the climate crisis is in the future, but for your people it is happening *now*. Poverty and malnutrition have many causes, but the changing climate is a big one.

Urgent measures must be taken to stock seed banks and help farmers to stay a step ahead of Africa's shifting agricultural map. More than 40 percent of Africa's population lives on less than a dollar a day, and 70 percent of these poor are located in rural areas and largely dependent on agriculture for survival.

One of the reasons you made the trip to this summit is to make sure that the rest of the world is aware of the food security issues your people face as a result of climate change. The rich nations of the world are still debating the terms of an international climate change agreement, but they seem to focus mostly on how to regulate global carbon emissions. That's important. It is long overdue especially for Americans and Europeans to start limiting their carbon emissions. But you want to be sure that upcoming climate talks also consider how to help people like the Bambara continue to feed themselves. Cutting emissions is important for the long run, but your people need assistance right now. ⬤

Role-Indigenous Peoples' Global Summit

Aymara People of Bolivia

From the capital city of La Paz, you need only to look toward the mountains for a reminder that climate change isn't just an issue that will face future generations. The mountains surrounding La Paz, like the towering 21,135-foot Illimani, are home to glaciers that have provided life-sustaining water to the Aymara people for more than 2,000 years. In fact, glacial meltwater is the main reason that farming was possible in this high plains region of Bolivia, which experiences a five-month dry season each year. The Aymara have historically worshiped the mountains as *Achachilas*, or life-giving deities, but one glance up at the quickly vanishing ice caps foretells a future with very little water.

Here are just a few of the startling facts: A recent World Bank study of Andean glaciers concluded that if current warming trends continue, many of the glaciers of Bolivia and Peru will disappear within 20 years—threatening the water supplies for 77 million people and greatly reducing hydropower production, which provides about half of the region's electricity. Glaciers in Bolivia that are 20,000 years old are retreating so fast that 80 percent of the ice will be gone before a child born today reaches adulthood. The glacier Zongo, near La Paz and the source for 10 hydropower plants that provide 25 percent of Bolivia's electricity, is receding at 33 feet a year.

Meanwhile, halfway around the world, *2 billion people* face similar shortages of glacial meltwater from the Himalayas, where glaciers have lost 21 percent of their mass since 1962. Himalayan glaciers provide water for river systems that reach *almost a third* of all humans on the planet: the Ganges, with 407 million people; the Indus, with 178 million people; the Brahmaputra, with 118 million people; the Yangtze, with 368 million people; and the Yellow, with 147 million.

Ten years ago, the Bolivian city of Cochabamba was ground zero in the global fight to keep water—a free, life-sustaining resource provided by Mother Earth—in the hands of the people and out of the hands of multinational companies (like Nestlé and Bechtel) that are in the business of selling water for profit. People from around Bolivia descended on Cochabamba to fight the privatization of the city's water supply, and won: Bechtel left Bolivia and hasn't been back since. Bolivia's president, Evo Morales—the country's first Indigenous president—nationalized Bolivia's water supply in 2007, but as Bolivia's water supplies diminish, Bechtel and Nestlé will likely be back, offering to provide badly needed water at a cost.

As glaciers continue to melt, the people of Bolivia will call on their government to build reservoirs and to help farmers develop more efficient irrigation techniques—in fact, some of these projects are under way. But Bolivia is one of the poorest nations in the Western Hemisphere. Should your people be forced to pay for large-scale climate adaptation projects when the greenhouse gas emissions of industrialized Northern countries are causing the glaciers to melt? The population of the industrialized North makes up only 20 percent of the global population, but it is responsible for 60 percent of historic carbon emissions. The wealth of the North was built on a "climate debt"—an unequal use of the atmosphere as a dumping ground for carbon—so it only seems fair that some of this debt be repaid now. ⊕

Role-Indigenous Peoples' Global Summit

Indigenous Peoples of the Amazon

Your people live in rainforests in the Amazon basin—the Yanomami, Ashéninka, Huaorani, Secoya, Achuar, Yawalapiti, and many, many others. The Amazon region is the largest rainforest in the world and consumes about one-fifth of the carbon dioxide produced from burning fossil fuels. The Amazon's role as a "carbon sink" that sucks global-warming carbon dioxide out of the atmosphere may be hurt. During a major drought in 2005, the fires that broke out in the western Amazon region put more carbon into the atmosphere than they soaked up. This is likely to occur again as savannas (grasslands) replace rainforest, which will have a huge effect on your livelihood and the other 60 million Indigenous people in the region.

Climate scientists predict that higher worldwide temperatures caused by increased greenhouse gases in the atmosphere will reduce rainfall in the Amazon region, which will cause widespread local drought. With less water and tree growth, "homegrown" rainfall produced by the forest itself will decrease as well, as this depends on water passed into the atmosphere above the forests by the trees. The cycle continues, with even less rain causing more drought, and so on. This cycle puts the whole Amazon region at risk, which puts the whole world at risk.

Tropical rainforests have long been home to Indigenous peoples who have shaped civilizations and cultures based on the environment in which they live. Living from nature, Native peoples have learned to watch their surroundings and understand the intricacies of the rainforest. Over generations, your people have learned the importance of living within your environment and have come to rely on the countless renewable benefits that forests provide. Like other Indigenous peoples, you believe that the Earth is our historian, our educator, the provider of food, medicine, clothing, and protection. This is something Indigenous peoples have to teach the entire world.

When most people think about greenhouse gas pollution and climate change, they think about burning fossil fuels like coal, oil, and natural gas. But every year, cutting down forests is responsible for as much as 15 percent of worldwide greenhouse gas emissions. You know the unique role the Amazon basin plays in regulating global climate. You also know how loggers, miners, oil companies, pipeline builders, and ranchers have continued to steal your land and kill your people. If the so-called developed world really wants to stop climate change, it has to take serious action to protect the rainforests where you live. The survival of Indigenous peoples in the Amazon will contribute to the survival of everyone on Earth.

Moving out of your forest homes is not an option. Your people have lived in and with the forests for millennia. You're not going anywhere. ⊕

Role Play

Role-Indigenous Peoples' Global Summit

Diné (Navajo) People of the American Southwest

You come from the "painted" red-rock deserts of the U.S. Southwest, homeland of the Diné People (in English, known as Navajo). The Diné/Navajo Nation is home to the largest Native American population in the United States, more than 180,000 people. In some ways, your hometown of Shiprock looks like any other small town in America—gas stations, motels, a few restaurants and grocery stores—but it's quite different in other ways. More than half the Navajo population lives below the U.S. poverty line.

Many people on the reservation still speak the traditional Diné language (in addition to English), and across the desert, farmers still practice traditional dryland methods of agriculture. But the entire Southwest is 10 years into a deep drought that shows no sign of letting up, and even the traditional methods that Diné have used to grow food in the desert for the last thousand years are starting to fail. As the rate of global climate change increases, farming will likely get even harder for your people. One Diné farmer, Charles Chi, describes it this way: "People are messing with Mother Nature. And just like my forefathers—my grandpa used to say one of these days, there's only going to be two weathers—fall or summer. Today, I think that he was telling the truth."

Farming is not the only way that the Diné people are connected to climate change. As energy companies look for ways to make electricity that releases lower levels of greenhouse gases, some people are talking about nuclear power as a solution to climate problems. They say that nuclear power can produce all the electricity we need without releasing greenhouse gases into the atmosphere.

But where is the uranium mined that fuels the nuclear plants? From your land. The Diné people are some of the poorest in the United States, but Diné land is rich with uranium resources. You grew up hearing stories of the Diné men who worked in the yellowcake uranium mines, from the 1940s to the

1980s. You want others to hear the stories about family and friends who came home each day with clothes covered in yellow uranium dust. The companies that ran the mines told workers not to worry about the dust—that it was safe—but people now know that the mines exposed workers and their families to high levels of radiation.

The contaminated dust made its way into people's lives in other ways too—hundreds of abandoned uranium pit mines filled with water and then became watering holes for Diné sheep herds. When families like yours butchered their sheep, they ate the uranium-infected meat. Today, the Diné suffer from high cancer rates and respiratory problems. One study found that cancer rates among Diné teenagers living near mine tailings are *17 times* the national average.

In 2005, the Navajo Nation banned uranium mining on its territory. But Hydro Resources Inc. has been working with the U.S. Nuclear Regulatory Commission to try to get approval for mining near Navajo communities in New Mexico. The company estimates that nearly 100 million pounds of uranium exists on those sites, worth billions of dollars. The group Eastern Navajo Diné Against Uranium Mining has fought Hydro for more than a decade.

So Indigenous peoples need to speak with one voice and say that not only do we have to support *real* solutions to climate change, we have to oppose *false* solutions—like nuclear power—that just lead to more poisoning of Indigenous people. ⊕

Discussion Questions—Indigenous Peoples' Global Summit on Climate Change

From the perspective of your role, answer the two questions below. As you meet with other groups and learn about their situations, you will likely want to add information to your answers.

1. What do you want the rest of the world to know about how climate change is affecting you and other Indigenous people?

2. What actions would you like to see the rest of the world take to address the problems facing Indigenous peoples as a result of climate change? **Put these in order of priority.**

Pre-Conference Meetings

Group Names	What the world needs to know.	What the world needs to do —Action Items

The Anchorage Declaration

24 April 2009

1. We call for mandatory greenhouse gas emissions limits for industrialized nations of the world. In recognizing the root causes of climate change, we call for these nations to also decrease dependency on fossil fuels (coal, oil, natural gas). We further call for a just and fair transition to local, renewable energy systems—owned and controlled by our local communities.

The conference could not reach consensus on the development of fossil fuels on Indigenous lands. Two possible action items are:

- An eventual phaseout of fossil fuel development, recognizing and respecting the right of Indigenous nations to use their fossil fuel resources as they see fit.
- A halt to new fossil fuel development on or near Indigenous lands.

2. We call on the U.N. conference participants in Copenhagen to recognize the importance of Traditional Knowledge and practices shared by Indigenous Peoples to address the problems of climate change.

3. We call on the wealthy, industrialized nations to recognize their contributions to greenhouse gas emissions over the last 100 years as the primary cause of climate change. We call on these nations to pay this historical and ecological debt.

4. We call on the United Nations' scientists to recognize and include Indigenous Peoples' climate change assessments (such as Inuit knowledge of changing ice conditions, Taíno knowledge of changing oceans, etc.)

5. We call upon the U.N. to fully involve Indigenous Peoples as active and equal participants in all the decision-making processes to address global climate change.

6. We challenge nations to abandon false solutions to climate change that negatively impact Indigenous Peoples' rights, lands, air, oceans, forests, territories, and waters. These include nuclear energy, large-scale dams, geo-engineering techniques, "clean coal," agrofuels, plantations, and market-based mechanisms such as carbon trading.

7. We call for a fund to be created to enable Indigenous Peoples to deal with the challenges created by climate change and to foster our empowerment, capacity-building, and education—with a specific focus on Indigenous youth and women.

8. We call on financial institutions to provide risk insurance for Indigenous Peoples to allow them to recover from extreme weather events.

9. We call upon the United Nations Food and Agriculture Organization (FAO) and other relevant U.N. bodies to establish an Indigenous Peoples' working group to address the impacts of climate change on food security and food independence for Indigenous Peoples.

10. We call on nations to recognize and respect traditional ownership of traditional lands, air, forests, waters, oceans, sea ice, and sacred sites, including those guaranteed by treaties. In particular, nations must ensure that Indigenous Peoples have the right to mobility and are not forcibly removed or settled away from their traditional lands and territories. In the case of climate change migrants, appropriate programs and measures must address their rights, status, conditions, and vulnerabilities.

11. We call upon nations to return and restore lands, territories, waters, forests, oceans, sea ice, and sacred sites that have been taken from Indigenous Peoples, limiting our access to our traditional ways of living, thereby causing us to misuse and expose our lands to activities and conditions that contribute to climate change.

We offer to share with humanity our Traditional Knowledge, innovations, and practices relevant to climate change, provided our fundamental rights as intergenerational guardians of this knowledge are fully recognized and respected. We reiterate the urgent need for collective action. ⊕

Agreed by consensus of the participants in the Indigenous Peoples' Global Summit on Climate Change, Anchorage Alaska, April 24, 2009. (Adapted by Julie Treick O'Neill and Tim Swinehart.)

"**S**o, is her house actually sinking?"

"Yes, Heather, it is."

"But, that's so sad! I want to do something about that!"

No doubt my preservice secondary education student, Heather, is familiar with the topic of climate change. Everywhere we look, we see media coverage. But there still seems to be something missing. There still appears to be a disconnect, for my preservice teachers, anyway, between what they read about online and what they see in their day-to-day lives. And this has huge implications for their futures as public school teachers. One way to address this disconnect has been to put a face to the topic of climate change. By connecting all of my "Heathers" to students who live in places where climate change is having actual, observable effects, a topic that was once theoretical becomes real.

Kwigillingok, Alaska, vs. Bellingham, Washington

My teacher ed students at Western Washington University in Bellingham, come from multiple walks of life, are at different points in their educational and

Climate Change in Kwigillingok

First-person narratives bring climate change closer to home

BY LAUREN G. McCLANAHAN

working careers, and have different goals for their futures as middle and high school teachers. However, one commonality my students tend to share is geography. Most hail from western Washington state—up and down the "I-5" corridor. Take the freeway north, and in 15 minutes, you're in Canada. A few hours south, and you've crossed into Oregon. On a daily basis, my students don't give much thought to climate change. No doubt, many claim to be "green" through and through. They recycle, use compact fluorescent bulbs, and buy local whenever possible. And these efforts are important; but as for the big changes—the catastrophic ones happening in our circumpolar regions—my students just don't see it. In contrast, the students of Kwigillingok, Alaska, see these changes every day and can document firsthand how their village is changing because of them.

Kwigillingok is a small Yup'ik fishing village in western Alaska that sits along the Bering Sea. With a population of about 400, the residents depend on a subsistence culture to survive, much as they have done for thousands of years. Fishing, hunting, and creating and selling crafts are as integral today as they have been for centuries. However, our warming Earth is now threatening that culture.

I began working with the students of "Kwig" several years ago, when one of my former students was hired to teach in the Lower Kuskokwim School District. What started as a simple pen pal relationship between her high school students and my college students slowly transformed into the project described here. And although the students have changed over the years, the questions that they were asking of one another became more focused, until we decided that the topic of climate change was the main issue that everyone wanted to discuss.

The biggest challenge faced by the residents of Kwig is the melting of the permafrost, that layer of frozen ground that lies just below the Earth's surface and that is supposed to stay frozen year-round. Recently, that permafrost has begun to melt, and as a result, major changes are taking place. Many homes and other structures in the village are beginning to sink, leaning to one side as the permafrost they were built upon begins to shift. In addition to sinking homes, new, invasive species of plants are beginning to take root and grow, which in turn is slowly changing the migratory patterns of big game such as the local musk ox populations. Fishing, too, has been affected by the warming trend, and fish camps have had to relocate depending on the changing location of the fish. These are big changes that Kwig's high school students can see and feel. They wanted to tell us about these changes, to tell the future teachers "down there" to share with their future students. They wanted to let everyone know that climate change is real and has a face and a name—hence the "First Person Singular" project. This was a project to create a warning for the rest of us, those of us who do not have to prop our houses up with sandbags or who do not have to go hungry due to a lack of fish in our rivers. Or at least not yet.

The "First Person Singular" Project

As mentioned, the relationship between Kwig and WWU began when a former student of mine was hired into the district. The more the students shared with us about their culture and their harsh, yet beautiful landscape, the more I felt as if I had to visit. In an initial visit, I met with the teachers and students at several village schools. I saw firsthand what the students of this region had to share with my students, not only from a cultural perspective, but also a scientific one as we delved deeper into the issue of climate change and its effect on their culture.

Recently, one of my current students approached me about completing his student teaching internship at the Kwig school. "I just want a very rural, very challenging school setting," he told me. Well, did I have the place for him. Luckily, my intrepid student would not mind hauling his own water, which is what he would have to do, since many of the buildings in Kwig have no running water. He did, however, have the luxury of a newly installed incinerator toilet in his cabin. Before he left, my student (a future English teacher) and I had talked about doing a project with his students that would combine disciplines and allow the students' own voices to be heard. The concept of place-based ed-

> Many homes and other structures in the village are beginning to sink, leaning to one side as the permafrost they were built upon begins to shift.

ucation, of focusing curriculum on local issues, had been an important part of our university classes, and my student wanted to try it out. He liked the combination of using the local setting as the classroom, and letting his students "direct" their learning—two of the main components of place-based education. So, with his students' input, we decided upon a project, and I made plans to help facilitate the project after he eased into his new role as student teacher. I figured that another visit would give me an opportunity not only to formally observe my student teacher, but also to work in person with the project and students I had been thinking about for some time.

Before I traveled to Kwig for the second time, I asked the high school students (with whom I communicated by email) to photograph any evidence of climate change they noticed in their village. Then, when I arrived, my student teacher and I sat down with the students to talk about the photos they had taken. This technique of using "auto-driven photo elicitation" (as it is called in the field of visual studies) proved to be beneficial. Auto-driven photo elicitation is simply when people involved in a research study take their own photographs, and use those photos as the basis for later interviews. The photos gave us a starting point—something on which to focus our conversation. Otherwise, I was afraid that the conversation might become too abstract, or even too uncomfortable (seeing as how the students had never met me face-to-face, but only through email). However, by focusing on the photos, we were able to get to the heart of what was important to the students. After all, we were talking about their photos, of evidence of climate change in their village.

After we spent time talking about the photos (individually and as a group), I asked students to pick a favorite photo and write about why it was the best choice to illustrate the effects of climate change. Because we had talked about the photos first, the writing part was easy. They could describe, in detail, why their photos mattered, and why their audience, my preservice teachers, needed to know about them. Then, after they had written their paragraphs, I asked students to read their paragraphs (or parts of their paragraphs) into a digital voice recorder so that we could incorporate their own voices (literally) into our final product. One of the students even volunteered to play the piano so our project would have a soundtrack.

One of the students photographed a leaning building. He described it this way:

> The world is changing. It's getting warmer and warmer. Ice is melting everywhere, even underground. The melting of the permafrost causes hills, houses, and other buildings to sink. Permafrost is a section in the ground where everything is frozen. It melts and refreezes around the year, but lately, there has been more melt than freeze. If we don't do something, we could lose this beautiful land that we lived in for thousands of years, forever.

He then wrote the same paragraph in his native Yup'ik language, and read them both aloud. This was powerful. Another student photographed seagulls that were hanging around later in the season than usual. "It's unusual for them to still be here [in October]," she explained, which suggests that [the ground] is not as cold as it looks.

Once they wrote the paragraphs and made the recordings, students responded to several prompts that they had generated, such as "What is worth preserving in Kwig?" One student responded, "We don't have a lot of money. We need to stay near the ocean so we can fish. We don't want to have to move farther and farther back every few years. We can't leave, but we can't stay, either." When asked what message they wanted to send to the preservice teachers in Washington, one student said, "Please understand that what you do down there has a great impact on us up here. Understand that we're all in this together. Climate change doesn't just affect polar bears—it affects people, too."

The project's final phase was to put our photos, words, and voices into a very short iMovie. The students helped plan the sequencing, and then we put it together. And although the movie was only four and a half minutes long, it sent a strong message to the preservice teachers it was meant to educate. After viewing the movie, one of my preservice teachers wrote, "Now that I know this—now that I have seen these kids' faces and heard their stories—I can't un-

> "If we don't do something, we could lose this beautiful land that we lived in for thousands of years, forever."

know it. Now I have to decide what I can do about it, both in my classroom, and in my everyday life." This short film is online at www.rethinkingschools.org/earth.

Larger Implications

Place-based education, while not a new concept, is particularly well suited for the inclusion of student voices. By grounding learning in local phenomena and students' lived experience, it can be easily adapted to fit any number of school curricula. For example, nearly every city or town has local issues that can be studied in greater depth, be they environmental issues (toxic wastewater), social justice issues (migrant workers' access to health care), or issues dealing with the economy (how city taxes are used to fund local schools). In the case of our project, climate change was an obvious topic for exploration, given our fortunate connection with students in the Far North. Plus, the topic fits nicely into the definition of "sustainability education." Within WWU's Woodring College of Education, our underlying assumption is that education for sustainability (as opposed to education about sustainability) will result in citizens who are more likely to engage in personal behavior or contribute to public policy decisions in the best interest of the environmental commons and future generations.

> "Now that I know this—about the challenges facing Kwig due to climate change—I feel obligated to do something about it."

Personal Implications

When students take control of their learning, and take control of how they demonstrate their learning, amazing things can happen. The Kwig high school students learned that they not only had some important things to say, but they also discovered an audience that was receptive to and respectful of their words and ideas. My preservice teachers learned that they are not the experts on everything and sometimes they have to step aside to let the experts step forward (in this case, the students themselves). The idea of relinquishing power in the classroom can intimidate a new teacher, but it is an important lesson, especially regarding student engagement.

After viewing the high school students' movie, my preservice teachers had a lot to say about place-based education, and how this project connects students to their local communities, and society as a whole. One student commented, "Obviously, the kids in the movie care about what is happening to their homes and land. We need that heart in schools, or what they are learning means nothing."

Many students also commented on the topic of climate change. Said one: "Now that I know this—about the challenges facing Kwig due to climate change—I feel obligated to do something about it." Climate change now has a name and a face. It's personal.

The more we know about others through their stories, in their own voices, the more inspired we might be to recognize those voices in our own. ⊕

Lauren G. McClanahan is a professor at Western Washington University in Bellingham, Washington.

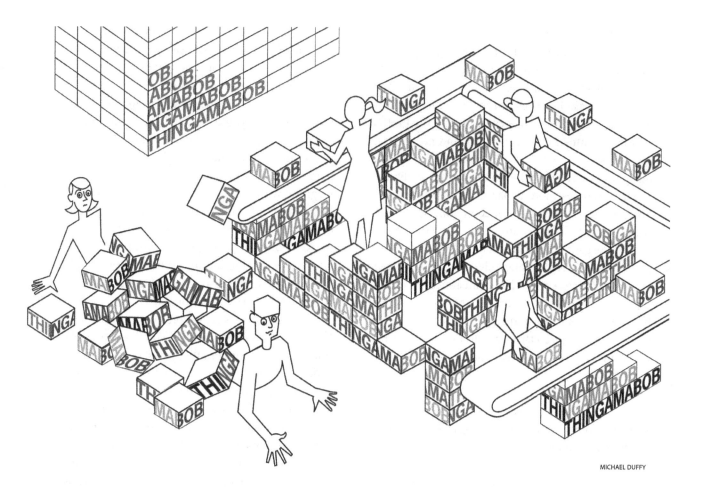

MICHAEL DUFFY

The Thingamabob Game

A simulation on capitalism vs. the climate

BY BILL BIGELOW

The premise of this activity—and a premise of this book—is that an economic system driven by the profit motive inevitably collides with the health of the planet in general, and with climate stability in particular. A challenge for educators is finding ways to help students *experience* this fact—and wrestle with its implications. The Thingamabob Game helps students grasp the essential relationship between climate and capitalism better than anything else in my curriculum. And coming to this realization is not merely academic. How we think about solving the climate crisis depends, in large part, on what we think is causing it. [See this game explained in context in "Teaching the Climate Crisis," p. 79.]

In the Thingamabob Game, small groups of students represent competing manufacturers of "thingamabobs"—goods that, as in the real world, require natural resources to produce and whose production creates greenhouse gases, especially carbon dioxide. In the game, as in the real world, the more we consume and produce, the more carbon dioxide is released into the atmosphere, and the more we put at risk life on Earth. As of this book's 2014 publication, CO_2 in the atmosphere is increasing at the rate of more than two parts per million per year. (The *rate* of increase has risen throughout the years, in addition to overall CO_2 concentrations.) The Earth's ecology is immensely complicated, and no one can say the precise effect of, say, 450 ppm CO_2 in the atmosphere, or 550 ppm. But as the impact of climate change becomes more apparent, scientists paint a grim future. In this activity, competing groups of students/manufacturers are equipped with knowledge of the gathering calamity. They then determine if they can be environmentally responsible, given the rewards offered by—and punishments exacted from—a profit-based economy. The game can be fun, frenetic, and frustrating. But the activity helps students gain more clarity about strategies for responding to the climate crisis and saving the planet.

Materials Needed:

1. Copies of the Thingamabob Game Role Sheet; one for each student.

2. Five Thingamabob Production Round slips per group. I generally have seven groups, depending on class size. (Cut these up beforehand.) Even groups as small as two students work fine.

3. Several candy bars or other desirable food products. (See the Thingamabob Game Role Sheet for the exact number you could end up needing.)

Suggested Procedure:

1. Before beginning the game, put the chart on page 152 on the board or overhead.

2. Distribute copies of the Thingamabob Game Role Sheet to students. Read it aloud. You'll notice that for simplicity's sake, each "company" doesn't have to worry about developing markets for its goods. In this game, whatever the companies produce will be sold. Thus with each round, each company increases its capital and can produce even more thingamabobs, if the student "managers" of that company so choose. (See #10, below for caveats on the game's limitations.)

The math is simple: Each company starts out with $1,000. Thingamabobs sell for $2 a piece and cost $1 to produce. So a company makes a dollar for every thingamabob produced. (Companies may not borrow money in the game—although that would be a twist you could introduce—and cannot spend any more money than they have.) So for the first round, an individual group's maximum production is 1,000 thingamabobs, which would leave the group with a total of $2,000 after the round, if the group decided to produce as much as possible.

The carbon math is equally simple. For every thousand thingamabobs produced by *all the groups put together*, CO_2 in the atmosphere increases by two parts per million. So with each round, as in the real world, CO_2 rise is inexorable.

Note that the role sheet promises candy for all the winners. It's important that you have desirable candy awards ready, and that you show these to students to motivate them to try to win. However, every class I have ever done this with—maybe 30 or 40—has produced so many thingamabobs that it triggered climate disaster, so students aren't likely to earn the reward you select.

The concluding paragraph of the instructions warns students that at the end of the five rounds if the *total* number of thingamabobs produced (i.e., by all seven groups in all five rounds) produces carbon dioxide concentrations over the trigger figure, every company loses the game. Students don't know the precise trigger point of environmental destruction. *I set the figure at **450 parts per million—production of 35,000 thingamabobs**—and I write that on the board and cover it up. Students know only that the figure is somewhere between 420 and 460 ppm—between 20,000 and 40,000 thingamabobs. Emphasize the tension in the game—as in real life: *They will be rewarded based on how much profit they produce for their company; but the more thingamabobs they produce, the closer they bring the planet to climate catastrophe and environmental devastation.*

3. Divide the class into seven groups. It's fine if groups have only a couple of students in a group, or as many as five or six (if you're in one of the many school districts where class sizes are ballooning due to budget cuts.) Tell each group to come up with a thingamabob company name.

4. Distribute five Thingamabob Production Round slips to each group. Ask them to make their first production decisions. They should discuss these within each small group, complete the information on the slip, and hand them to you without revealing their numbers to their competitors.

5. Begin by writing all the company names on the board or overhead. Then post the first round production figures. Be sure to add up the number of thingamabobs produced in each round and to keep a running total of all the thingamabobs produced in the game, and for each round, the carbon dioxide ppm, beginning with the game's starting point of 380 ppm. Point out the "loser" companies whose profits don't match those of their competitors. Tell these companies that their stockholders are getting restless because their competitors are so much more successful, even though they began with the same amount of capital. If one company decides that it wants to carry the banner for the Earth and produce no or few thingamabobs, I may declare that company bankrupt and distribute those students to other groups so that they get the message that failing to compete has consequences. I always remind students how good that chocolate will taste for the students in the winning companies.

6. Continue round by round, indicating the most and least profitable companies. Also emphasize how the total carbon dioxide count is getting dangerously high. (Remember, *you* know that the trigger figure of climate no-return is 450 ppm; *they* know only that it is between 420 and 460.) Finish all five rounds, even if they exceed the 450 figure. One year my students went over the 460 number in the fourth round, and I told them that new "scientific evidence" found that the trigger point is higher, between 460 and 480 ppm. My students in that class still went over, topping out at somewhere around 500 ppm.

7. If by some miracle, at the end of round five, thingamabob production of all companies has not exceeded the 450 ppm trigger number, award the prizes to the groups as indicated on their role sheet.

8. Afterward, before discussing, ask students to write about the activity. Keep it simple, something like "Why did you destroy the Earth? You saw it coming, but you did it anyway." Perhaps ask students to comment on the observation by ex-financier James Goldsmith that "winning" in a game like this is like "winning at poker on the Titanic." You might also ask them to reflect on what, if anything, they could have done that would have led to a more positive outcome.

Or you might have students write about the game's main lessons. Propose some that they might react to:

- Human beings are inherently greedy and competitive. People will never be able to co-operate to solve the climate crisis. It's human nature.
- What's missing from this game is a government. To keep corporations' greed in check, we need a strong government that enacts and enforces regulations on carbon dioxide emissions.
- The cause of the climate crisis is the capitalist system itself. So long as we have an economic system driven by profit, corporations will be rewarded for endangering the planet. If we want to deal with the roots of the climate crisis we need to create an economic system that does not reward greed.

9. Begin the discussion by getting students to talk about what they wrote. Some further discussion questions include:

- Describe what went on in your group. What pressures did you feel?
- What prevented you from being more ecologically oriented?
- How does the game resemble real life? What was unrealistic about the game?
- Is the game "rigged"? Could the rules be changed in ways that would not lead to climate ruin?
- Invent a new set of rules for the "game" that would not lead to environmental destruction. What different behaviors could be rewarded?

10. Now for the fine print. Simulations are metaphors. They're useful in illuminating aspects of reality, but they can obscure or miss other important aspects. The Thingamabob Game effectively highlights how the capitalist market has no built-in alarm system to protect the Earth. As social critic David Korten writes, "There are no price signals indicating that the poor are going hungry because they have been forced off their lands; nor is there any price signal to tell polluters that too much CO_2 is being released into the air, or that toxins should not be dumped into soils or waters." But the simulation may imply that we all suffer equally as the Earth deteriorates. Nothing in the activity suggests that the consequences of climate change reverberate unevenly through the global landscape, affecting the Third World, the poor, and people of color more harshly than those with more privilege. And the game's cataclysmic end may distort the way things are likely to play out in real life, as pockets of the world become unlivable but global elites exist relatively comfortably in gated communities, shielded from the worst aspects of the climate crisis.

As mentioned earlier, to underscore the pressure on corporations to maintain high production, even at great ecological peril, vagaries of the market do not factor into the Thingamabob Game. Obviously, in real life everything that is produced is not necessarily purchased. This is one of the central contradictions of market economies: They tend toward overproduction and boom-and-bust cycles. And the production of every "thingamabob" will not be equally unfriendly to the climate. Some production—for example, solar panels—may soften humans' effect. You can discuss these and other limitations with students. Still, the game's essential caution—that a profit-oriented economic system is incompatible with climate stability and environmental responsibility, more broadly—is one that is hard to deny. ⊕

Bill Bigelow (bbpdx@aol.com) is curriculum editor of Rethinking Schools *magazine.*

The Thingamabob Game Role Sheet

Y ou are managers of a company that produces thingamabobs. You are in competition with other thingamabob companies. Even though you have important and highly paid managerial jobs, these jobs are not necessarily secure. As with any capitalist company, you need to continually grow and make a profit. Fail to return a sufficient profit and you'll lose your job.

But the threat of global warming raises some questions about your thingamabob business. Here is what the best science tells us: Over the past 350 million years or so, the sun's energy has been stored on Earth in the form of carbon—especially in oil, coal, and natural gas. Burning carbon-based fuels provides energy that runs our cars and heats our homes. This carbon-based energy also produces and transports your thingamabobs. But here's the problem: Burning carbon-based fuel releases carbon dioxide gas. Carbon dioxide (CO_2) traps the sun's warmth within the atmosphere—which is why it's called a "greenhouse gas." The main threat is that as CO_2 and other greenhouse gases build up in the atmosphere, Earth's temperatures rise. And as temperatures have begun to warm, the Earth faces dire situations: Glaciers are disappearing; permafrost in Alaska and Siberia has started to melt; corral reefs in the Indian Ocean and South Pacific are dying; species are going extinct at an increasingly rapid rate; weather patterns are changing, leading to more intense storm activity. And the seas are rising, which jeopardizes every coastal area in the world.

Since the 1700s, CO_2 alone has increased in the atmosphere from about 275 parts per million (ppm) to around 400 ppm. Most of this increase has taken place since 1950. And today, it's rising by 2.1 ppm every year. In this period, according to NASA, global average temperatures have risen 0.8° Celsius (1.4° Fahrenheit). No one can predict for certain the impact of, say, 450 ppm CO_2 or 550 ppm CO_2. Continuing on this course could have catastrophic consequences.

Naturally, the production of thingamabobs is not the only cause of rising greenhouse gases. How we heat our homes, how we get to work, even how our food is raised plays a role. But production of thingamabobs definitely increases the concentration of CO_2 in the atmosphere. Some of this is from the mining and shipping of raw materials to make the thingamabobs; some is from thingamabob production itself, which requires a great deal of energy; some is from the shipping of thingamabobs from China, where most of your factories are located.

Rules of the Game

Each company begins the game with $1,000 in capital. Each thingamabob costs $1 to produce. You will make $2 off of every thingamabob you produce and sell. (So, for example, if you produce 100 thingamabobs in round one, you will spend $100, but you'll get $200 back, and end up with a total of $1,100.) Of course, with every thingamabob produced, the Earth comes one step closer to ecological disaster. **In the game, production of each 1,000 thingamabobs adds an estimated 2 ppm carbon dioxide to the atmosphere.** The world in the Thingamabob Game begins at 380 CO_2 ppm.

To simulate the real-life consequences, here's how scoring will work. There will be five "production" rounds. At the end of the fifth round, you will be rewarded not on how nice you are to each other, or to the Earth, but on how much profit you've made for the company:

Rewards:

Top two groups:	Candy for every group member
Group 3	Two candy bars to split between group members
Group 4	One candy bar to split between group members
Group 5	Nothing
Group 6	Nothing
Group 7	Nothing

If all groups tie, each group will receive one candy bar to share.

Here's the catch: If the total production of thingamabobs for all groups produces CO_2 concentrations over the trigger number—somewhere between 420 and 460 ppm (that is, between 20,000 and 40,000 thingamabobs)—the Earth's environment will be damaged beyond repair, and no one will receive any candy. ⊕

Company Name	Round 1 Avail. Capital	Round 1 Things Produced	Round 2 Avail. Capital	Round 2 Things Produced	Round 3 Avail. Capital	Round 3 Things Produced	Round 4 Avail. Capital	Round 4 Things Produced	Round 5 Avail. Capital	Round 5 Things Produced
1.	$1,000									
2.	$1,000									
3.	$1,000									
4.	$1,000									
5.	$1,000									
6.	$1,000									
7.	$1,000									
Running total "things"										
CO₂ ppm										
Game begins 380 ppm–2 ppm per 1,000 things										

Thingamabob Production

Round # _____

Company name: _____Available capital ($): _____

Number of thingamabobs produced this round: _____

Total available capital after production: _____

Thingamabob Production

Round # _____

Company name: _____Available capital ($): _____

Number of thingamabobs produced this round: _____

Total available capital after production: _____

Thingamabob Production

Round # _____

Company name: _____Available capital ($): _____

Number of thingamabobs produced this round: _____

Total available capital after production: _____

Thingamabob Production

Round # _____

Company name: _____Available capital ($): _____

Number of thingamabobs produced this round: _____

Total available capital after production: _____

Remember the Carbon Footprint of War

BY BRUCE E. JOHANSEN

Compared to the all-too-obvious death and environmental mayhem caused by warfare, the long-range toll of war's carbon footprint is less visible but hardly harmless. Modern war waged at long distances is hugely carbon dioxide intensive. The U.S. armed forces, which maintain as many as 1,000 bases in other countries, consume about 2 million gallons of oil per day, half of it in jet fuel. Fuel economy has not been a priority in modern fossil-fueled warfare. Humvees average 4 miles per gallon, while an Apache helicopter gets half a mile per gallon.

Consumption of fossil fuels has increased over time, with great waste. The Air Force alone uses half the oil consumed by the Department of Defense. For example, it burned through 2.6 billion gallons of fuel during one six-month period in Iraq and Afghanistan in 2006. At that time the armed forces consumed as much fuel per month in limited wars as they did during World War II between 1941 and 1945.

Growing Carbon Production

At the beginning of World War I in Europe, just over a hundred years ago, the main motive force in battle was the horse and shoe leather, as troops in Europe marched off to battle on foot or horseback. The advent of aerial bombardment and increasing use of tanks caused a dramatic escalation in war's carbon dioxide production. War is often a powerful technological motor, and carbon-consumption innovator. World War II began with quarter-century-old biplanes and ended with jet-propelled fighters, resulting in a massive increase in fuel consumption.

The mechanization of the military provided many more opportunities to ramp up carbon dioxide production during the world wars of the early 20th century. World War II's Sherman tank, for example, got 0.8 miles per gallon. Seventy-five years later, tank mileage had not improved: the 68-ton Abrams tank got 0.5 miles per gallon. Typical fuel consumption of a fighter jet was 300 to 400 gallons per hour at full thrust, or 100 gallons per hour at cruising speed during hundreds of hours of training and combat missions. Blasting to supersonic speed on its afterburners, an F-15 fighter can burn as much as four gallons of fuel per second. According to Gar Smith in *Earth Island Journal,* the B-52 Stratocruiser, with eight jet engines, consumes 86 barrels of fuel per hour.

Individuals are told to reduce our "carbon footprint," and we should. But how many years of riding a bike to work would it take for me to offset one F-15 flying for an hour? Assuming that my bike replaces a car that gets 25 miles per gallon, my daily commute of five miles would use a gallon a week. That's 350 weeks, roughly seven years, to fuel a fighter jet at full thrust for one hour.

During the 1950s and 1960s, the U.S. military flew B-52s at all times, on the theory that an airborne fleet would prevent the Soviet Union from obliterating the entire U.S. nuclear-armed armada on the ground. Each of these B-52s burned thousands of gallons of fossil fuel per hour while aloft. That's 73 bike commuters' annual fuel savings for every hour a B-52 is in the air.

Size, Scope, and Complexity

The carbon footprint of war is important and intriguing, but impossible to calculate precisely because of its size, scope, and complexity. The carbon footprint for a bag of potato chips, a carton of milk, or a pair of athletic shoes may be calculated by adding up each unit's proportion of manufacturing and transportation energy inputs along the entire life cycle of a product. Calculating the carbon footprint of a single consumer product is complex, but we can do it. When we become really serious about carbon footprints, we will know the amount of greenhouse gases generated by each platoon sent to war, each bomb dropped, each tank deployed. However, today we know the carbon footprint of a bag of potato chips from a Safeway grocery store in California, but war—that elephant in the greenhouse—remains unmeasured. If the Pentagon has ever done such a thing, no one seems to be bragging about it to civilians.

> Individuals are told to reduce our "carbon footprint," and we should. But how many years of riding a bike to work would it take for me to offset one F-15 flying for an hour?

The United States launched the Iraq war on the pretext of protecting vital oil supplies, even as it consumed oil at a phenomenal rate. At the start of the Iraq war, in 2003, the United Kingdom Green Party estimated that the United States, Britain, and the minor parties of the "coalition of the willing" were burning the same amount of fuel in the Iraq war (40,000 barrels a day) as all 1.1 billion people living in India. The U.S. Air Force uses 2.6 billion gallons of jet fuel a year, 10 percent of the U.S. domestic market. By the end of 2007, according to a report from Oil Change International, the Iraq war had put at least 141 million metric tons of carbon dioxide equivalent into the air—as much as adding 25 million cars to the roads. The Iraq war added more greenhouse gases to the atmosphere than 60 percent of the world's nations.

So how would one begin to sketch the carbon footprint of a war? Here is a preliminary sketch:

- First, add all the energy used to produce the weapons, transport, and other provisions consumed in the war.

- Add the emissions produced getting soldiers, supplies, and civilian contractors to the theater of war, and home again—in the case of a war pursued thousands of miles from home, often by air transport. Add the cost of running armed personnel carriers, heating and cooling soldiers' lodgings, and so forth, as well as the greenhouse gases caused by the conduct of combat itself.

- Add the carbon and other greenhouse gases added to the atmosphere by fires initiated by bombings and other explosions. In Iraq, pay special attention to intentional sabotage of oil pipelines and suicide bombings, as well as improvised explosive devices.

- Add the carbon cost of tending the wounded. Iraq's emergency room spanned nearly half the world, from airborne surgery to the Landstuhl Regional Medical Center in Germany and hospitals in the United States.

The Urgency of Solutions

Peacemakers in our time are often assumed to be naive dreamers. Given the environmental crisis, however, a timely end to war is not naive, but necessary. Armies of the future must study the best ways to solve international conflicts without armed conflict and the monumental pollution that accompanies their death and destruction.

The greenhouse gas emissions of war should be regulated on a worldwide basis, and the United States, the world's premier military power, should take the lead in de-carbonizing international relations. With the carbon footprint of war adding to its cost in blood and treasure, this tally of greenhouse gas emissions should convince us that the Earth can no longer afford fossil-fueled war.

Bruce E. Johansen is the Jacob J. Isaacson University Research Professor in the School of Communication and the Native American Studies Program at the University of Nebraska at Omaha.

See teaching ideas for this article, page 175.

DEFENSE.GOV NEWS PHOTOS

In June, when my 4th-grade students and I reflected on the highlights of their year, I was moved to learn how deeply the class had been affected by our Polar Bear/Global Warming Project. They felt a strong sense of accomplishment about this community service project, and for me, the project provided a big lesson on activism and inspiring hope.

"The project was a lot of work, but it was fun too," one student told me. Several asked if they could keep working through the summer. One of the students, Cecilia, explained that at first her dad did not believe in global warming, but by the end of our project she felt she had convinced him it's an important—and real—problem.

It all began with "Feeling the Heat," an article in *Time for Kids* magazine about the sad plight of the polar bears as they face melting summer ice. I read the article out loud to my class, hoping to familiarize them with the issue of global warming. Not surprisingly, the text brought up many complex concepts for my 9- and 10-year-old students, all English language learners, some of them newcomers. (My school, Sanchez Elementary, is located near San Francisco's Mission District and is comprised of 63 percent English language learners, predominantly Spanish-speaking.) I struggled to find clear ways to answer their questions and respond to their comments, realizing that although I understand the basics of climate change, I wasn't prepared—emotionally—to talk about it with children. I felt completely hopeless about the possibilities of slowing global warming or doing anything that could help the polar bears.

"I know!" one student, Maria Luisa, offered enthusiastically. "Let's collect money to save the polar bears!"

"This isn't really the kind of problem money can just fix," I responded curtly. I'll admit I was over-

DAVID MCLIMANS

Polar Bears on Mission Street
4th graders take on climate change

BY RACHEL CLOUES

whelmed by the immensity of the problem and my students' earnest, yet understandably naive, desire to find a quick solution. How could I explain to 4th graders that the world, especially this country, has a life-threatening addiction to fossil fuels? Or that the oil industry dominates government decisions? Or that policy change moves slowly and often backward? I couldn't figure out how to convey hope to my students, or in any way be positive about the looming crisis of global warming.

> I couldn't figure out how to convey hope to my students, or in any way be positive about the looming crisis of global warming.

Of course, we had a brief discussion of simple, proactive steps offered in the article (turn off lights, drive less) but after that, I pretty much dropped the topic of global warming. I didn't feel like I had the tools or the inspiration to tackle the issue. But it kept coming up in my classroom, over and over, during the next several weeks; my students were fascinated by the polar bears.

"Polar Bear SOS"

In the springtime, my principal asked all the teachers at our school to carry out a community service project with our classes to honor César Chávez, whose birthday is a state holiday in California. He suggested simple possibilities such as cleaning the schoolyard, working with younger students, or collecting garbage at a local park. Our timeline: one week to carry out the project and present it to the school for César Chávez Day.

Around that same time, a friend forwarded me an email from the nonprofit Natural Resources Defense Council (NRDC) asking for contributions to fund an educational advertisement for national television. NRDC 's "Polar Bear SOS" campaign supported a proposal by the U.S. Fish and Wildlife Service to list the bears as "threatened" under the Endangered Species Act, and it was also raising awareness about global warming. The poignant ad, narrated by children, showed photos of polar bears struggling to survive in their melting habitat. Viewers would be encouraged to submit comments online at NRDC.org in support of the Endangered

Species Act proposal. If the bears became officially categorized as threatened, then the government would be required to take steps to protect them. And since melting arctic ice is now indisputably linked to global warming, this opens up possibilities for legal support requiring car manufacturers and industries to reduce carbon emissions. I decided to tell my students about the campaign, retract my statement about money not being helpful in solving the problem of global warming, and see if they wanted to raise funds to send to NRDC.

Students had a few other ideas for community service projects, but the 4th graders voted unanimously to take on a fundraiser for the newly named Room 18 Polar Bear Project. They really liked the idea of doing something in response to the article we had read, which seemed to have created significant empathy for the polar bears. Taking into account that my school's families are generally poor (71 percent of students receive free or reduced lunch), I adopted a very modest attitude about the project. "If you want to ask your parents for a couple of dollars to donate, that would be great," I told them. "Bring in what you can." I also told my students that we would spend the next few days learning as much as we could about global warming.

That evening, I spent some time thinking about and planning a few lessons to get us started. I was aware of the short timeline, but I felt like breaking away from our nominally scripted curriculum and desk work and focusing on an issue relevant to our school community and to the global community as well. In the spirit of César Chávez, who confronted injustice directly and thought out of the box, it seemed worth tackling. A key point I wanted to help students comprehend is that the polar bears are in danger of losing their unique habitat due, in large part, to the actions of humans. By understanding and teaching others about the causes and effects of global warming, we could begin to help solve the problem. A different part of this project, I realized, was teaching students that another way to create change is to support organizations that work toward the same goal on a larger scale.

I found helpful resources on San Francisco's excellent Department of the Environment website (sfenvironment.org). I downloaded and copied short, kid-friendly articles with graphics in both English and Spanish. I asked students to work in small groups and use highlighters to pick out and discuss important facts and concepts. Then, as a large group, we talked about the questions that came up. I tried to clarify confusing sections, for example those involving the atmosphere, by drawing pictures and making diagrams on the whiteboard. The next day I took my class to the computer lab where we explored NRDC's website, including the polar bear ad. We read about how cars and industries overload the atmosphere with carbon dioxide, exacerbating the greenhouse effect. Because we live in an urban community, it was easy for my students to see how too many cars on the road create air pollution, and understand the benefits of public transportation. My students underlined these important facts and other information in their copies of the articles, and each of them went home that day with the information in Spanish. I asked them to share what they were learning with their families.

In an attempt to better convey the concept of the greenhouse effect, which was difficult for my students to grasp, I designed a simple, age-appropriate model. I used a small, stuffed cloth Earth ball and put it inside a plastic ziplock bag. In front of the students, I blew air into the bag and sealed it, to show them how the atmosphere is a thin layer of air around our planet. I took the Earth ball out, dampened it with water, and put it back inside the bag. We put the model in the sun for an hour or so, and while the water heated up, evaporated and condensed on the inside of the bag, the children began to see how the Earth stays warm enough to sustain life.

"It's just like the windows on our car," one student said. "When it's sunny outside, it gets really hot inside!"

"Exactly," I replied. "Some heat feels good and is important so animals and plants can live on Earth. But pollution in the atmosphere, like carbon dioxide, traps heat and makes the planet too hot."

"That's why glaciers and ice caps are melting!" My students seemed to make the connection back to the polar bear article. The plastic bag did, indeed, function as a real greenhouse. It was a rough model, but it was better than nothing.

I suggested that the class figure out a way to share their learning with other students in the school, so our project would have a bigger impact. An artistic group, the students quickly decided to make bilingual posters to put up around the school,

in the shape of polar bears, with tips to reduce global warming. All over the school we posted these messages: "Take the bus instead of driving," "Ride your bike and walk more," "Turn off your lights when you're not using them," and "Donate money to NRDC.org in Room 18 to help the polar bears!" With more time, it would have been great to send pairs of students from our classroom to visit each classroom in the school. As it was, a couple of students wrote letters to all the teachers and asked them to read the letters out loud in their classes. The letter explained our project and asked for contributions to our campaign for the polar bears. "Dear Teachers and Students," it began. "We are learning about Global Warming in Room 18. You can help by not using your cars because the pollution is going into the atmosphere. . . ."

Esteban's Success

On day three of our project, an amazing thing happened. One of my students, Esteban, arrived with his parents at our classroom door, holding fistfuls of money, which he eagerly thrust toward me— crumpled dollar bills and coins, all spilling onto the floor. He explained that the previous evening he had walked down Mission Street—the bustling, lively center of the predominantly Latino Mission District where he lives—and talked to passers-by about our Polar Bear Project. Esteban had carried around his copies of the bilingual articles, which explained the basics of global warming, and he shared them with the strangers who stopped to talk with him on the street. "If they spoke English, I showed them the article in English, and if they spoke Spanish, I used the one in Spanish," Esteban explained to me when I asked him how he had approached people. "I just told people about why it's important to help the polar bears, and how we can stop global warming. I told them our class is raising money to send to NRDC."

We counted out Esteban's money: It added up to more than $100, and nobody could have been prouder than Esteban (though his parents were very proud, too). His classmates and I were impressed, and everybody was energized to try to raise more money. "I'm going to ask everybody in my apartment building to donate money!" shouted several students.

By the end of the week, Room 18 had raised almost $300, and it seemed everybody in our school was talking about the Polar Bear Project. Children from other classrooms came to our room to drop their quarters, dimes, and nickels into our can. Teachers, volunteers, and our school principal all contributed money.

What thrilled me the most was the way the term "global warming" rolled off the tongues of my 4th graders. They were able to articulate concepts like what produces greenhouse gases; how the atmosphere traps heat; why polar caps and glaciers are melting; and, most importantly, many of the simple ways we can all help curb the problem. I was relieved that the project had moved beyond a "Save the Polar Bears" campaign and toward genuine understanding of a complex global problem. When I looked around at my class—100 percent engaged in making posters, writing letters, reading more information, and counting money—I also realized they were getting an authentic taste of activism. Esteban, who is a smart but sometimes socially awkward kid, was the obvious leader and hero of our campaign.

César Chávez Day

The Polar Bear Project culminated on the day before César Chávez Day, when our school welcomed Rita C. Medina and Lydia Medina, the sister and niece of César Chávez, who spoke to students and staff. These women, along with other family members, have established nonprofit organizations to promote the legacy of César Chávez. Rita Medina, well into her 80s, continues to march each year in support of farmworkers. After some inspiring words from these great activists and leaders in the farmworkers' movement, the classes presented their community service projects.

My students stood together, holding their posters and talking about the issue they had taken on and the money they had raised to help the polar bears, who, they explained, are part of our global community. "The ice where polar bears live is melting, and we want to help save them. We need to stop global warming," they

> I was relieved that the project had moved beyond a "Save the Polar Bears" campaign and toward genuine understanding of a complex global problem.

told our school community. Rita and Lydia Medina congratulated me and my students and told us how impressed they were with our example of activism. They said their brother and uncle, César Chávez, would be proud. They pulled out their pocketbooks in front of everybody and contributed their own donations to our can of funds. It was an incredible honor to be recognized publicly by two such inspiring social justice leaders.

That evening, I sent our donation to NRDC. My attitude toward teaching about global warming had completely changed. I still feel overwhelmed by the problem, of course, but the feeling of hopelessness has been replaced with a real sense of hope. I am inspired that a classroom of kids cared enough to go out and ask for money to help save the polar bears. By deciding to walk down Mission Street and talk about global warming, Esteban inspired many of his classmates to go out and talk about what they were learning. Their ability to speak two languages was clearly an important tool.

The students felt empowered by the information they learned from reading and discussing, as well as the sizable fund they collected. We all felt energized after stepping away from textbooks, workbooks, and standardized tests and focusing on a hands-on project. My students acted, spoke, and felt like activists. Then, fortunately, their hard work was acknowledged by adult activists who look like them and share their linguistic and cultural background.

Embarking on the project also helped me realize that I can—and should—teach students about global warming, even if it overwhelms me. As a thank-you gift, the NRDC sent polar bear bookmarks and a handwritten card to my class. They ran their ad on television and thousands of people responded in support of placing polar bears on the Threatened and Endangered Species list (the bears were listed as threatened in 2008).

I sense a growing awareness and, perhaps, a shift in consciousness in our communities regarding global warming. The example set by César Chávez and his family reminds me that turning knowledge into action is powerful—and it creates the potential to overcome what at first seems impossible. ⊕

Rachel Cloues is a teacher-librarian at a K–8 school in San Francisco.

I was googling myself recently (in an attempt, if you must know, to locate an essay that I had published somewhere), and I managed to misspell my own name. So I was directed to the one source that had mangled my name in the same way. And that is how I was confronted, in an obscure blog, with the question "Why isn't Sandra Steingraber [with dyslexic spelling] talking about climate change?"

It was unsettling. As the days went by, I began an imaginary argument.

Look, I first wrote about receding glaciers in 1988. I was assigning Al Gore to college students in 1992. Not long ago, I made climate instability the centerpiece of a commencement address I gave at a rural college in coal-is-king Pennsylvania. And if you think all the trustees were pleased with that theme, I invite you to give it a try. So the question is not "Why is S. S. not talking about climate change?" The question is "Why is S. S. not talking about it AT HOME?"

OK. Why don't you talk about it at home?

Because I have young children and because I believe that frightening problems need to be solved by adults who should just shut up and get to work.

So, how long are you going to keep hiding the truth from your kids?

That's as far as I got before three other notable things happened. First, Elijah asked to be a polar bear for Halloween. As I pinned the chenille fabric, it occurred to me that his costume might well outlast the species. I decided not to tell him that.

A month later, Elijah asked his sister for a weather report. Faith walked out onto the porch, spread out her arms in the manner of Saint Francis, and came back in. "It's global warmingish," she said and went back to her cereal. No comment from me.

And then I overheard a conversation on the playground. One child said, "I know why it's hot. Do you?"

Another said, "It's because the Earth is sick." They all nodded. I said nothing.

It's time to sit down with my kids and have the Global Warming Talk. I carried off the Sex Talk—and its many sequels—with grace and good biology. Surely, I can rise to this new occasion.

On the surface, procreation and climate change seem opposite narratives. Sex knits molecules of

BEC YOUNG

The Big Talk

How to tell a 6-year-old where all the birds and bees have gone

BY SANDRA STEINGRABER

air, food, and water into living organisms. Climate change unravels all that. The ending of the sex story is the birth of a family. The climate change story ends with what biologist E. O. Wilson calls the Eremozoic Era—the Era of Loneliness.

But then I realized that the two stories share a common epistemological challenge. Both are counterintuitive. In the former case, you have to accept that your ordinary existence began with an extraordinary, unthinkable act (namely, your parents having intercourse). In the latter case, you have to accept that the collective acts of ordinary objects—cars, planes, dishwashers, iPods—are ushering in things extraordinary and unthinkable (dissolving coral reefs, daffodils in January). So, I reasoned, perhaps the same pedagogical lessons apply: During the Big Talk, keep it simple, leave the door open for further conversation, offer reading material as follow-up.

Of which there is no shortage. In fact, a veritable cottage industry of children's books on climate change has sprung up almost overnight. These range from the primer, *Why Are the Ice Caps Melting?*, in which lessons on the ravaging of ecosystems also offer plenty of opportunities to practice silent *e*, to the ultra-sophisticated *How We Know What We Know About Our Changing Climate: Scientists and Kids Explore Global Warming*, by foremost environmental author Lynne Cherry, in which middle school readers are cast as co-principal investigators. This new literary subgenre is impressive. Reading its various offerings, I found myself admiring the respectful tones and clear explanations. These books describe global warming as a reality that no longer lingers in the realm of debate. And yet, they are not, for the most part, scary. Indeed, the first sentence in the inside flap of *How We Know What We Know* is "This is not a scary book."

And here is where the pediatric versions of the climate change story depart from their adult counterparts. The recent crop of books on global warming intended for grown-ups focuses on the surreal disconnect between the evidence for rapidly approaching, irreversible planetary tipping points (overwhelming) and the political response to that evidence (mostly zilch). The children's books profile heroic individuals fighting to save the planet—in ways that kids can get involved. To read the children's literature is to see the world's people working ardently and in concert with each other to solve a big problem—and enjoying a grand adventure while they're at it.

Is this the fiction we all should be laboring under? I don't know. I do know that a fatalistic mindset, which afflicts many adults but almost no children, is a big part of what's preventing us from derailing the global warming train that has now left the station. On this, I wholly agree with sociologist Eileen Crist, who argues that fatalism, masquerading as realism, is a form of capitulation that strengthens the very trends that generate it. I do know that we grown-ups need visions of effective challenges and radical actions that can turn into self-fulfilling prophecies.

I also know that I needed something to say to my 6-year-old when we walked home from the library in April—no leaves to offer shade, the bank's LED sign reading 84 degrees—and he turned his ingenuous face to mine to ask, "Mama, is it supposed to be so hot?"

So I am working on my talk. For inspiration, I have arranged on my desk three documents. One is an essay that Rachel Carson published in *Popular Science* in 1951—eight years before my birth. It's titled "Why Our Winters Are Getting Warmer," and it includes a drawing of Manhattan deluged by seawater. Another is Carson's essay "Help Your Child to Wonder," published five years later. The third is a book by poet Audre Lorde that includes the sentence "Your silence will not protect you."

My talk features a story about a boat in which we all live—people, butterflies, polar bears. A storm starts to rock the boat. The waves are chemical pollution, habitat destruction, industrial fishing, and warfare. Now along comes a really big wave. Global warming. The already-rocking boat is in danger of flipping over.

Then what happens? I don't know. For the first time in my life, I have writer's block. Somebody help me out here. ⊕

Sandra Steingraber is an ecologist and author who explores the links between human rights and the environment, with a focus on chemical contamination. She is author of Living Downstream *and* Raising Elijah. *This article originally appeared in* Orion *magazine.*

 See teaching ideas for this article, page 176.

O ur response to the climate crisis depends on who or what we think is causing it. This trial role play helps students understand how complicated it is to assess "guilt." It's an engaging activity in which students defend entities charged with causing the climate crisis, as they point the finger at others. Ultimately, it asks them to consider how different factors are linked together, and, depending on their assessment of blame, to propose what justice looks like. The climate crisis is no laughing matter, but this activity provides a lively and playful way to explore the issues.

Materials needed:

1. Construction paper for making name placards, one for each of the six groups.

2. Colored markers.

3. Copies of the indictments. Members of each group get copies of their own indictment—for example, all those in the U.S. Government group receive the U.S. Government indictment. Sometimes students like to see the indictments of other groups, and this can lead to a more fully argued trial, so you may want extra copies of indictments for all the groups.

Who's to Blame for the Climate Crisis?

A trial role play on the roots of global warming

BY BILL BIGELOW

Suggested procedure:

1. In preparation, list the names of all the "defendants" on the board: The "Market"—the System of Global Capitalism; the U.S. Government; U.S. Consumers; the Governments of China, India, and Other "Developing" Countries; and Oil and Coal Companies.

2. Explain that in this trial role play, each group is charged with causing the climate crisis and all its ravages: the destruction of cultures, extinction of species, putting at risk the lives of people all over the world, and even threatening the lives of people in future generations. Groups of students will prepare a defense for their role and you as teacher will play the prosecutor. (When I taught this lesson at Madison High School in Portland, I was joined by two colleagues, Adam Sanchez and Brady Bennon. Adam shared prosecutor duties with me, and Brady played the judge. Having three teachers in the classroom at once is a luxury, and is not essential to the activity, but it's an excellent opportunity for this kind of collaboration.) Explain that students' responsibility will be both to defend their own group and to explain which group or groups they think are guilty, and why.

My rule in trial role plays is that students in a group can choose to plead guilty, but they cannot claim sole "guilt"; they must accuse at least one other defendant. If students are puzzled by any of the defendants, tell them not to worry: The charges are explained in each indictment and, in any event, it's your job as the prosecutor to argue the charges against them. Tell them each group will receive a written copy of their indictment.

3. Describe the order of the activity:

    a. In their groups, students will read the indictments and will prepare a defense against the charges. I ask students to write up their defenses, as they'll be presenting them aloud and may want to refer back to them. A written defense helps focus the group.

    b. Before the trial begins, you will select several students to be jurors, and will swear them to neutrality.

    c. As prosecutor, you will begin by arguing the guilt of one group.

    d. The students in the accused group will defend themselves and will offer their thoughts on who is the guilty party or parties. (Another option here is to allow or encourage students to call witnesses from other groups to be questioned. For example, U.S. Consumers call Oil and Coal Company executives to the stand: "Are you aware that burning coal is the single greatest source of carbon dioxide in the atmosphere?")

    e. The jury then questions the accused group, and others may question the group and offer direct rebuttals.

    f. This process is repeated until all groups have been accused and have defended themselves. The jury then deliberates and makes its decision.

4. Ask students to count off one through six and to form groups throughout the class. (My "trick" to get students into their groups as quickly as possible is to give first choice to the first group that assembles its members into a tight circle and is ready to work.) Distribute the indictment sheets to each group. Remember, each group should read these indictments carefully and begin to prepare its defense. I've included possible lines of defense on each indictment, just to get them started; tell students that they should use these only if they find them helpful.

As students read and discuss, I circulate from group to group, raising questions, helping them think through possible defenses, but also playing devil's advocate to help them sharpen arguments. At this point, I distribute a placard and marker to each group and ask them to write their group name and display it for others to see.

Also, as mentioned, I tell students that I'm happy to share the indictments for the other groups if they think that would be helpful. It will also be useful to have copies of materials from the climate change unit available, as students may want to incorporate information from them into their defense.

5. When each group appears ready—and this will vary from class to class, and will also depend on the time you are able to commit to the activity—choose a jury; in a large class, one member from each group, or a total of three students in a smaller class. Publicly swear them to neutrality, as they will no longer represent the groups to which they previously belonged. And, as I mentioned, if you can enlist other adults to participate, that's ideal.

6. The order of prosecution is up to you. Most recently, I prosecuted, in order: the Oil and Coal Companies; the U.S. Government; the Governments of China, India, and Other "Developing" Countries; U.S. Consumers; and the "Market"—the System of Global Capitalism. I save the "Market" for last. Because it is a nonhuman entity, it is the most challenging to prosecute, and it helps if students have heard all the previous prosecutions. As prosecutor, I don't read the indictments aloud, because this can become tedious. I use each indictment as an outline to improvise, embellish, and basically, come at each group as hard as I can.

7. After each group has been prosecuted and has had ample opportunity to defend itself, I ask the jury to step outside the classroom to deliberate. Jurors can find one group guilty and acquit all the others, but I encourage them to assign percentage blame—for example, one group may have 25 percent guilt, another 20 percent, and so on. I ask them to be prepared to explain their verdict. As the jury deliberates, I ask everyone else to step out of their roles and to write out what I've asked the jury to do. Years ago, in a different trial role play, a student of mine rebelled against the idea of assigning percentage blame. She argued that what was important was not how each group was separately responsible for a given social/environmental crime, but how each group's blame connected to every other group's blame. Of course, she was right, and so now I encourage people to take a more holistic approach to assigning guilt, if students find this helpful.

8. The jury returns to deliver and explain its verdict and then we discuss. Here are some possible discussion questions:

- What were some of the most effective arguments you heard in the trial?
- Were there any "innocent" groups? Why didn't you assign them any blame?

- Were there any groups whose guilt was directly connected to the guilt of other groups? In what way?
- Are there other groups that were not indicted that should have been included?
- How much responsibility for the climate crisis did you assign to the System of Global Capitalism?
- What would it mean to "convict" any of these groups? How would you sentence them?
- When Ken Gadbow, a teacher in Portland, uses trial role plays, he introduces students to the concept of restorative justice—the idea that what we should aim for in a justice system is not punishment, but the repairing of harm that has been done and ensuring that the injustice stops. Another piece of restorative justice is that the victim must be actively involved in effecting justice. Introduce this concept to students and ask them to apply restorative justice to the global warming trial. What changes need to be made to deal with the crimes highlighted in the trial?

> **Who or what is responsible for global warming and all the misery it brings—and will bring?**

9. Especially if this trial role play is taught at the end of a unit, it lends itself to a writing assignment. The trial deals with the question: Who or what is responsible for global warming and all the misery it brings—and will bring? Ask students to write an essay that "answers" this question and that also proposes what should be done about it.

Bill Bigelow (bbpdx@aol.com) is curriculum editor of Rethinking Schools *magazine.*

U.S. Government

The Indictment: You are charged with the destruction of cultures, species, and putting at risk the lives of countless millions of people around the world. But your crime is also about the future. You are destroying the lives of people throughout the world who are alive today. *And* you are destroying the lives of people throughout the world who are yet to be born.

- Until 2007, the United States was the leading emitter of greenhouse gases, and it is still by far the largest emitter per capita of any major nation. It emits twice as much carbon dioxide as Great Britain, Germany, or Italy; and three times as much as Mexico, France, Sweden, or Switzerland. And the U.S. government has done almost nothing to promote alternative energy. Your pushing of petroleum-based products—which cause a huge amount of carbon dioxide emissions—is a case in point. You could try to discourage gas use by taxing it, but instead you keep taxes low. Here are some recent examples of taxes on a gallon of gas in Europe: $3.92 in France; $4.10 in Germany; and $4.40 in Great Britain. And the United States? The average combined federal, state, and local taxes are 38 cents a gallon.

- The U.S. government continues to give huge subsidies to fossil fuel industries. Instead of giving gifts to the oil and coal companies, why not tax their windfall profits and use the money to fight climate change. In just 10 years—from 2001 to 2011, the five largest oil companies made a staggering $1 trillion in profit, that's $1,000,000,000,000. Imagine if you—the U.S. government—taxed these profits at a higher rate? Think of all the money that could be used for good.

- It's not only the oil industry that is the problem. It's also coal—which emits even more carbon dioxide than oil. The United States sells coal that we the people own at just over $1 a ton. Companies can sell this at $100 or more per ton. And the coal companies now want to export millions of tons of this U.S.-owned coal to Asia, which will produce even more carbon dioxide when it's burned.

- You could implement a mandatory carbon cap, and begin reducing emissions dramatically. Or you could heavily tax production of oil, coal, and natural gas. You could lead the world by example. Instead, you do the opposite. The U.S. government refused to sign the Kyoto Protocol—the most significant effort taken by governments around the world to begin to reduce greenhouse gases—and you have refused to support binding international agreements to reduce greenhouse gas emissions.

- Instead of real measures, you push ethanol, which is a gift to the agribusiness companies. Creating ethanol from corn is immoral. First, it raises the cost of food around the world. And, when everything is taken into account, ethanol creates as much greenhouse gas pollution as petroleum.

- Many groups may be responsible for climate change and the disasters it is unleashing and will unleash in the future. But those who are most responsible are those who: (1) knew about the crime; (2) had the capacity to stop the crime; and (3) failed to act to stop the crime.

Possible defenses:
- You are elected by the people. If any "crimes" have been committed, blame the people; they put you in power.
- The government itself is not a huge producer of greenhouse gases. You don't drill for oil, you don't dig coal. Those cars on the road aren't yours.
- You would support an international agreement on climate change if the developing countries would agree that they're part of the problem. It's unfair for the biggest polluters—like China—to have no restriction on their greenhouse gas emissions.
- You have to move slowly and carefully or you could create economic chaos. That could lead to violence on a global scale.

U.S. Consumers

The Indictment: You are charged with the destruction of cultures, species, and putting at risk the lives of countless millions of people around the world. But your crime is also about the future. You are destroying the lives of people throughout the world who are alive today. *And* you are destroying the lives of people throughout the world who are yet to be born.

- At the core of the global warming crisis lies U.S. consumer culture—which you enjoy and perpetuate. Buy, buy, buy. That's you. The United States has about 5 percent of the world's population, but consumes more than 25 percent of the world's resources and creates more than 25 percent of the world's waste. If everyone in the world consumed like you—a typical U.S. consumer—we'd need three to five more planets. The average U.S. person consumes twice as much as they did 50 years ago. Your houses are twice as large as they were in the 1970s. You even spend between three and four times as much time shopping as people in other so-called developed countries in Europe. Underlying all this consumption is the burning of fossil fuels that creates carbon dioxide that heats the planet and is leading to environmental catastrophe.

- Your addiction to the automobile and to oil creates a huge amount of greenhouse gas pollution that causes global warming. You drive everywhere, forgetting that when you get into the car, you hurt people all over the world as well as people not yet born. And you're greedy. The average person in the U.S. creates *four times* the carbon dioxide emissions as the average Chinese person. China as a whole may release more greenhouse gases than the United States, but if every Chinese person lived like an American, we'd be in even worse trouble than we are now.

- Which leads us to the next point in the indictment. People in China—and India, Indonesia, Nigeria, and Mexico—look to the United States as the model for what life ought to be like. Big houses, big televisions, fast cars, freeways, lots of stuff—that's the image that gets beamed around the world. Your lifestyle has become the standard that everyone in the world strives for. And because of this, the planet is dying. In fact, more people in the world are living like Americans. (More are starving, too; growing inequality is a byproduct of more people getting rich.) People around the world look at you and say, "I want that, too." If not for you, people around the world would not be acquiring habits that emit increasing amounts of greenhouse gases.

- You can try to blame the coal companies, the oil companies, the government, other countries. But in all the world, you're the biggest contributor to global warming. Without consumers, no one can sell anything. It all starts with your idea that consumption equals happiness.

Possible defenses:
- You have broken no laws.
- It's the government and energy companies that have failed to invest in more renewable energy. *You* didn't refuse to sign an international treaty to limit greenhouse gases; your government refused.
- No one is forcing people around the world to strive for your lifestyle. People have to take responsibility for their own actions.
- It's not the individual's responsibility to think about big global problems like climate change. It's the government's responsibility; that's what governments are for.
- You live in a capitalist system. And the whole point of capitalism is profit. As consumers, you are just playing your part. If consumption is the problem, blame capitalism, not the consumer; you have no choice but to buy from capitalist companies.

The Governments of China, India, and Other "Developing" Countries

The Indictment: You are charged with the destruction of cultures, species, and putting at risk the lives of countless millions of people around the world. But your crime is also about the future. You are destroying the lives of people throughout the world who are alive today. *And* you are destroying the lives of people throughout the world who are yet to be born.

- The fuel that releases the most carbon dioxide into the atmosphere is not oil—it's coal. And no one in the world is producing as much coal as China. China burns more coal every year than the United States, Japan, and the European Union *combined*. In the next eight years, China plans to add 562 coal-fired plants. And these aren't small plants; each could power a city the size of Dallas. India plans to build 213 plants during this period. We know that coal is the worst greenhouse gas polluter and yet governments of so-called developing countries continue down this path.

- Everyone wants to blame the United States and rich countries for greenhouse gases, but the leading emitter of carbon dioxide is China, which produces more than 9 billion tons of carbon dioxide, compared to about 5 billion tons by the United States. China and other developing countries are doing the greatest damage to the environment.

- What is so sad is that countries like China and India could lead the world in developing alternatives. China is the world's largest country. By 2030, India will have more people than China. Think of the difference it would make if each of these countries pursued "green" technologies instead of coal. Instead of choosing the best of the world's technology—like solar and wind power—China and India choose coal, the most polluting source of energy. Instead of relying on conservation, sharing, and renewable energy, developing countries imitate the worst aspects of the so-called developed countries like the United States.

- It's like your governments are trying to murder your own people. Because of global warming, the glaciers are melting at a frightening pace. During the dry season, glacial melt provides 70 percent of the water for the Ganges River in India. The Ganges is the leading source of water for more than 400 million people.

- Your countries are not just burning coal and oil; you're burning your forests, too. Cutting down trees and burning rainforest land is responsible for up to 15 percent of greenhouse gases every year; transport and industry account for only 14 percent each. People think that you poor countries aren't to blame for climate change, but you are. Which countries were the largest emitters of greenhouse gases in 2013? China, the United States, and India.

Possible defenses:
- The per-capita carbon dioxide emissions are still less in every developing country than in the United States and most other developed countries.
- How hypocritical of the developed countries. They already built their cities and destroyed their forests. But they want to deny countries like China and India the same right to pull themselves out of poverty. The atmosphere is filled with carbon dioxide over long periods of time. What matters is not *today's* carbon pollution; it's carbon pollution throughout modern history.
- Rich countries could do the world a big favor by transferring technology to the developing world without worrying about patent protection.
- If the rich countries don't want poor countries to destroy their forests or drill for oil, they should pay fees to poor countries to make up for lost revenue.
- One of the reasons that China uses so much energy is because it makes just about everything that's used in the West. Stop using Chinese-made goods if you're going to complain about our carbon dioxide emissions.

Role Play

Oil and Coal Companies

The Indictment: You are charged with the destruction of cultures, species, and putting at risk the lives of countless millions of people around the world. But your crime is also about the future. You are destroying the lives of people throughout the world who are alive today. *And* you are destroying the lives of people throughout the world who are yet to be born.

- The oil companies are well known for poisoning the planet. But one of your biggest crimes has been poisoning people's minds. Precious time has been lost addressing global warming because you have spent millions and millions lying to and confusing people around the world. Between 1998 and 2005, ExxonMobil gave $16 million to groups that denied global warming was a problem. As the Union of Concerned Scientists reported, you spent money trying to make it appear that there was scientific debate and doubt about human causes for global warming when there was no scientific doubt. To spend millions lying to the world is a crime against humanity and nature. As Dr. James McCarthy, Harvard professor of biological oceanography and former chair of the Intergovernmental Panel on Climate Change, said: "It's shameful that ExxonMobil has sought to obscure the facts for so long when the future of our planet depends on the steps we take now and in the coming years."

- Everyone knows that burning coal and oil creates carbon dioxide, a greenhouse gas, and contributes to heating up the planet. But still you act like we "need" more coal and oil. You continue to destroy mountains in Appalachia to get more coal, and you want to mine coal on public land in Montana and Wyoming to sell in Asia. And you continue to push for opening up the Arctic National Wildlife Refuge to oil drilling.

- Despite more indications of a global climate catastrophe, you're laughing all the way to the bank. In a recent five-year period, the largest five coal companies made almost $8.5 billion. But that's nothing compared to the oil companies. In one year alone, 2013, the top five oil companies made $93 billion. To put that in perspective, that is $254 million *every single day*. You coal and oil companies don't produce energy, you produce profits. You could put money into solar and wind power—or better yet, into encouraging people to use less energy. But, hey, you wouldn't make as much money, so you'll never do this.

- And you continue to fight consumer attempts to make automobiles more fuel efficient. Why? Because the more oil consumed in gas-guzzling cars, the higher your profits. It's as if you care nothing for the future. In your ideal world, you just drill and drill and drill without thinking about anything besides where we can drill for oil next.

Possible defenses:
- A corporation—whether it produces oil or furniture—is legally obligated to look out for its stockholders, not the public. It's the government's job to look out for the welfare of society, not private, for-profit companies.
- Coal produces almost half the electricity in the United States. Don't like coal? Turn off your lights. Turn off your TV.
- If consumers didn't buy oil, we would have to find something else to sell. Don't blame the suppliers, blame the customers. We are simply playing by the rules of capitalism. We try hard to make our expenses less than our revenue. Those are the rules of the game. Don't like it? Blame capitalism, not coal and oil companies.
- Everything we've done has been perfectly legal. If it's so bad, pass a law and we'll stop doing it.

"The Market"—the System of Global Capitalism

The Indictment: You are charged with the destruction of cultures, species, and putting at risk the lives of countless millions of people around the world. But your crime is also about the future. You are destroying the lives of people throughout the world who are alive today. *And* you are destroying the lives of people throughout the world who are yet to be born.

- It's tough to accuse a system of being responsible for a crime. We tend to think of individuals or maybe groups of people committing crimes, but a system? In capitalism, the rules are simple: Make profits and you survive and prosper. Fail to make a profit and you don't survive. What doesn't show up as profits is invisible to you. Here's an example. If it is most profitable for landowners in Indonesia to cut down the forests and burn them to make way for palm oil plantations, then that's what will happen. However, burning the forests releases huge amounts of greenhouse gases, which has a devastating impact on climate change. But the system of capitalism is blind to anything except profit.

- If there is a profit in stopping global warming, fine; capitalism will work to stop it. But the climate crisis and all the disasters that it is unleashing can be very profitable. Security companies like Blackwater promise to protect the rich people from the poor people—as they did in New Orleans following Katrina. Or as they will do as the glaciers melt and push farmers off the land and into cities; or as food prices skyrocket as landowners find it more profitable to grow crops for ethanol than for food, and the poor protest and threaten the rich. Another disaster? Hmmm. Sounds like a source of profit.

- Sure, capitalism will produce technologies to limit greenhouse gases. It will produce solar panels, windmills, bicycles, light rail cars—whatever. But only so long as there is a profit. At the same moment capitalism is producing solar panels, its oil companies are off in the Arctic Ocean, taking advantage of climate change to explore for more oil—which will be burned to produce still more greenhouse gases.

- The market, capitalism, the profit system, neoliberalism. These days you go by lots of names. But whatever we call you, the facts remain: You have no heart, no empathy, no conscience. Profit is the only objective. And the problem is that all the benefits from polluting the atmosphere with greenhouse gases are private, and all the costs will be borne by people—mostly poor people—all around the world.

- Blame the transnational oil companies or the coal companies if you like. They are the ones that extract fossil fuels, which get burned for energy and result in carbon dioxide emissions. Or blame the auto companies or the logging companies. But they all behave the way they do because of the rules of the capitalist system: Do whatever yields the greatest profit. At the root of this climate crisis is a system that demands grow, grow, grow; buy, buy, buy; sell, sell, sell; profit, profit, profit. As a result, the atmosphere fills with more heat-trapping gases.

Possible defenses:
- Only humans can be charged with crimes.
- It will be the profit system—global capitalism—that will save humanity. People will develop green technology not because they are different kinds of people, but because they can make a profit from saving the planet.
- No one forces anyone to buy a polluting car or to use polluting energy. Human choice is at the root of what happens in the world, not the profit system.
- Your job is simply to get things produced. The government is responsible for protecting the environment and preventing climate change, not you.

Bali Principles of Climate Justice

The International Climate Justice Network, an international coalition of groups, gathered in Johannesburg, South Africa, in 2008 for the Earth Summit. The group released a set of principles aimed at "putting a human face" on climate change. The Bali Principles of Climate Justice redefine climate change from a human rights and environmental justice perspective.

Preamble

Whereas climate change is a scientific reality whose effects are already being felt around the world;

Whereas if consumption of fossil fuels, deforestation, and other ecological devastation continues at current rates, it is certain that climate change will result in increased temperatures, sea level rise, changes in agricultural patterns, and increased frequency and magnitude of "natural" disasters such as floods, droughts, loss of biodiversity, intense storms, and epidemics;

Whereas deforestation contributes to climate change, while having a negative impact on a broad array of local communities;

Whereas communities and the environment feel the impacts of the fossil fuel economy at every stage of its life cycle, from exploration to production to refining to distribution to consumption to disposal of waste;

Whereas climate change and its associated impacts are a global manifestation of this local chain of impacts;

Whereas fossil fuel production and consumption helps drive corporate-led globalization;

Whereas climate change is being caused primarily by industrialized nations and transnational corporations;

INDYMEDIA.ORG.UK

Whereas the multilateral development banks, transnational corporations and Northern governments, particularly the United States, have compromised the democratic nature of the United Nations as it attempts to address the problem;

Whereas the perpetration of climate change violates the Universal Declaration of Human Rights, and the United Nations Convention on Genocide;

Whereas the impacts of climate change are disproportionately felt by small island states, women, youth, coastal peoples, local communities, Indigenous peoples, fisherfolk, poor people, and the elderly;

Whereas local communities, affected people, and Indigenous peoples have been kept out of the global processes to address climate change;

Whereas market-based mechanisms and technological "fixes" currently being promoted by transnational corporations are false solutions and are exacerbating the problem;

Whereas unsustainable production and consumption practices are at the root of this and other global environmental problems;

Whereas this unsustainable consumption exists primarily in the North, but also among elites within the South;

Whereas the impacts will be most devastating to the vast majority of the people in the South, as well as the "South" within the North;

Whereas the impacts of climate change threaten food sovereignty and the security of livelihoods of natural resource-based local economies;

Whereas the impacts of climate change threaten the health of communities around the world—especially those who are vulnerable and marginalized, in particular children and elderly people;

Whereas combating climate change must entail profound shifts from unsustainable production, consumption, and lifestyles, with industrialized countries taking the lead;

We, representatives of people's movements together with activist organizations working for social and environmental justice resolve to begin to build an international movement of all peoples for Climate Justice based on the following core principles:

1. Affirming the sacredness of Mother Earth, ecological unity, and the interdependence of all species, Climate Justice insists that communities have the right to be free from climate change, its related impacts, and other forms of ecological destruction.
2. Climate Justice affirms the need to reduce with an aim to eliminate the production of greenhouse gases and associated local pollutants.
3. Climate Justice affirms the rights of Indigenous peoples and affected communities to represent and speak for themselves.
4. Climate Justice affirms that governments are responsible for addressing climate change in a manner that is both democratically accountable to their people and in accordance with the princi-

ple of common but differentiated responsibilities.
5. Climate Justice demands that communities, particularly affected communities, play a leading role in national and international processes to address climate change.
6. Climate Justice opposes the role of transnational corporations in shaping unsustainable production and consumption patterns and lifestyles, as well as their role in unduly influencing national and international decision-making.
7. Climate Justice calls for the recognition of a principle of ecological debt that industrialized governments and transnational corporations owe the rest of the world as a result of their appropriation of the planet's capacity to absorb greenhouse gases.
8. Affirming the principle of ecological debt, Climate Justice demands that fossil fuel and extractive industries be held strictly liable for all past and current life cycle impacts relating to the production of greenhouse gases and associated local pollutants.
9. Affirming the principle of ecological debt, Climate Justice protects the rights of victims of climate change and associated injustices to receive full compensation, restoration, and reparation for loss of land, livelihood, and other damages.
10. Climate Justice calls for a moratorium on all new fossil fuel exploration and exploitation; a moratorium on the construction of new nuclear power plants; the phaseout of the use of nuclear power worldwide; and a moratorium on the construction of large hydro schemes.
11. Climate Justice calls for clean, renewable, locally controlled, and low-impact energy resources in the interest of a sustainable planet for all living things.
12. Climate Justice affirms the right of all people, including the poor, women, rural, and indigenous peoples, to have access to affordable and sustainable energy.
13. Climate Justice affirms that any market-based or technological solution to climate change, such as carbon trading and carbon sequestration, should be subject to principles of democratic accountability, ecological sustainability, and social justice.
14. Climate Justice affirms the right of all workers employed in extractive, fossil fuel, and other

greenhouse gas-producing industries to a safe and healthy work environment without being forced to choose between an unsafe livelihood based on unsustainable production and unemployment.

15. Climate Justice affirms the need for solutions to climate change that do not externalize costs to the environment and communities, and are in line with the principles of a just transition.

16. Climate Justice is committed to preventing the extinction of cultures and biodiversity due to climate change and its associated impacts.

17. Climate Justice affirms the need for socioeconomic models that safeguard the fundamental rights to clean air, land, water, food, and healthy ecosystems.

18. Climate Justice affirms the rights of communities dependent on natural resources for their livelihood and cultures to own and manage the same in a sustainable manner, and is opposed to the commodification of nature and its resources.

19. Climate Justice demands that public policy be based on mutual respect and justice for all peoples, free from any form of discrimination or bias.

20. Climate Justice recognizes the right to self-determination of Indigenous peoples, and their right to control their lands, including subsurface land, territories, and resources, and the right to the protection against any action or conduct that may result in the destruction or degradation of their territories and cultural way of life.

21. Climate Justice affirms the right of Indigenous peoples and local communities to participate effectively at every level of decision-making, including needs assessment, planning, implementation, enforcement and evaluation, the strict enforcement of principles of prior informed consent, and the right to say "No."

22. Climate Justice affirms the need for solutions that address women's rights.

23. Climate Justice affirms the right of youth as equal partners in the movement to address climate change and its associated impacts.

24. Climate Justice opposes military action, occupation, repression, and exploitation of lands, water, oceans, peoples and cultures, and other life forms, especially as it relates to the fossil fuel industry's role in this respect.

25. Climate Justice calls for the education of present and future generations, emphasizes climate, energy, social, and environmental issues, while basing itself on real-life experiences and an appreciation of diverse cultural perspectives.

26. Climate Justice requires that we, as individuals and communities, make personal and consumer choices to consume as little of Mother Earth's resources, conserve our need for energy; and make the conscious decision to challenge and reprioritize our lifestyles, rethinking our ethics with relation to the environment and the Mother Earth; while utilizing clean, renewable, low-impact energy; and ensuring the health of the natural world for present and future generations.

27. Climate Justice affirms the rights of unborn generations to natural resources, a stable climate, and a healthy planet.

Adopted using the "Environmental Justice Principles" developed at the 1991 People of Color Environmental Justice Leadership Summit, Washington, D.C., as a blueprint.

Endorsed by:
CorpWatch, U.S.; Friends of the Earth International; Global Resistance; Greenpeace International; GroundWork, South Africa; Indigenous Environmental Network, North America; Indigenous Information Network, Kenya; National Alliance of People's Movements, India; National Fishworkers Forum, India; OilWatch Africa; OilWatch International; Southwest Network for Environmental and Economic Justice, U.S.; Third World Network, Malaysia; and World Rainforest Movement, Uruguay. 🌐

 See teaching ideas for this article, page 176.

Farewell, Sweet Ice—*page 74*
By Matthew Gilbert

Ask students to consider the following questions:

- What's your gut reaction about what's happening to the Gwich'in?
- Gilbert writes: "Because nature is the fabric of our lives, we cannot really separate 'the climate' from our human selves." What does he mean by that?
- List the ways the Gwich'in have been affected by climate change.
- Some of the changes affecting the Gwich'in will be permanent. Who or what is responsible for what is happening to the Gwich'in? Who should pay for the negative climate consequences affecting the Gwich'in?

Ask students to create a drawing (or another visual representation) of an aspect or a number of aspects of the impact of global warming on the Gwich'in.

Consider pairing "Farewell, Sweet Ice" with the film *Chasing Ice*. The film follows photographer James Balog on a quest to document the world's  vanishing glaciers. The story of Balog and his team setting up photography equipment in these extreme weather locations is engaging for students, and the beautiful and horrific images of disappearing ice make the film worth showing.

Before watching the film, ask students to write about or discuss the emotions they associate with climate change. During the film, have students take notes on the scenes or photographic images they feel most strongly about.

After the film, ask students:

- Which scenes or images stood out to you?
- What about the film do you find most hopeful?
- At the end of the film Balog says, "When my daughters look at me 25 years from now and say, 'What were you doing when you knew that global warming was happening and you guys knew what was coming down the road?' I want to be able to say to them, 'Guys, I was doing everything I knew how to do.'" What would you like to be able to say in response to the same question from your children?

Goodbye, Miami—*page 77*
By Jeff Goodell

Ask students:

- What are some similarities and what are some differences between what is happening to the Gwich'in in the Arctic and what Jeff Goodell imagines will happen to Miami?
- Does Goodell's portrayal of Miami in 2030 seem believable? Does anything seem exaggerated?
- If Goodell's portrayal were to come true, which social groups would be hurt the most? Are there any groups that might benefit, at least in the short term?
- Gilbert and Goodell offer two portraits of the consequences of climate change—one that is happening and one that *might* happen. Choose a community you are familiar with and imagine the possible effects of severe climate change at some point in the future. You might use Goodell's piece as a model, or you might use more of a story format.

Climate Timeline —*page 102*

Ask students to study the timeline, and consider the following questions/activities:

- Choose five events from the Climate Timeline that you believe are significant. For each of these, write about why you think they are important.
- Through research, try to add at least five more important dates to the timeline. These might

include events after the final 2014 entry.

- Choose at least three points on the timeline that seem to you to be "choice points"—times when things might have gone in a different—or better—direction if people had acted differently. What are these, and how might things have gone differently?
- What themes or trends can you identify on the timeline? Identify several themes and use colored markers to highlight them.
- Choose three events on the timeline and research them. What important or interesting information are you able to add to the items on the timeline?
- What "big ideas" do you take away from looking at this timeline? What did you find surprising? What upset or troubled you? What gave you hope?

Retreat of Andean Glaciers Foretells Global Water Woes—*page 124*
By Carolyn Kormann

Pick one glacier somewhere in the world that is melting as a result of climate change. Write the glacier's name on the board and circle it. With students, construct a "spillover chart" of as many plausible consequences as the class can imagine—for example, each consequence gives rise to other consequences, which give rise to still more consequences, etc.

Ask students to read "Retreat of Andean Glaciers Foretells Global Water Woes" and to highlight some of the consequences they predicted in the spillover chart. Are there potential consequences students identified that are not included in the article? Are consequences mentioned in the article that students did not predict in their spillover chart?

"The grand question here is, who compensates," says Oscar Paz, director of Bolivia's National Climate Change Program, "because we are not culpable for climate change. It's not fair that a country like Bolivia, which emits 0.02 percent of global greenhouse emissions, already has annual economic losses from the impacts of climate change equivalent to 4 percent of our GDP." Paz believes Bolivia should not have to pay for the economic losses created by climate change. Why do you agree or disagree? List with students the specific costs that people in Bo-

livia will incur as the Bolivian glaciers melt. Who should pay these costs? What recourse does Bolivia have? If students believe the countries most responsible for greenhouse gas emissions over time should pay, what if these countries refuse?

Remember the Carbon Footprint of War—*page 154*
By Bruce E. Johansen

The term "carbon footprint" has worked its way into the popular lexicon, and numerous websites help  students calculate their personal carbon footprints. One book on global warming targeted at young people, *The Down-to-Earth Guide to Global Warming*, tells students: "Your carbon footprint comes from normal, everyday activities like using your computer, turning on the light in your bedroom, taking a bath (heating water uses energy!), and riding in a bus or car to school." The concept of a carbon footprint *individualizes* carbon pollution, and draws students' attention to their personal patterns of consumption. There is nothing wrong with that. But if that's where it ends, students may miss other important sources of carbon pollution in the atmosphere, like war and preparation for war. The idea of a personal carbon footprint may narrow our thinking about climate activism to individual consumer choices.

Before reading "Remember the Carbon Footprint of War," ask students to list some of the things they do in life that contribute to their personal carbon footprint. Or better yet, have them use one of the online carbon footprint calculators. Afterward, ask students to list other human-caused ways that carbon enters the atmosphere—ways for which they are not personally responsible. How many students listed the military or warfare? Ask students to list the ways that war contributes to greenhouse gases. Read "Remember the Carbon Footprint of War," and have students compare Johansen's findings with their thoughts prior to reading the article. Ask whether or not the fossil fuel use of the U.S. military is part of their *personal* carbon footprint. Why or why not?

Ask students to use examples from this article to construct mathematics word problems.

The Big Talk—*page 161*

By Sandra Steingraber

In "The Big Talk," Sandra Steingraber wrestles with why she has not talked with her children about global warming. By the end of the article, she commits herself to have the "Global Warming Talk" with her kids. In the article, she begins the talk, but then trails off. "For the first time in my life, I have writer's block. Somebody help me out here."

Ask students:

- When did you first hear of the dangers of climate change? How old were you? Was that too young to hear this story? Should you have been told earlier?

- How did learning about climate change affect you? At what age should we talk to young children about the significance of global climate change?

Assign to students:

- Imagine that you are describing climate change to a 6-year-old. What will you focus on? Climate change is such a huge, complicated problem that you'll need to simplify the issues in a way that a child can understand.

- Will it be helpful to use a metaphor, like the boat in the story that Sandra Steingraber begins? Feel free to start a different story with your own metaphor.

- Will your story be hopeful? Many children's stories have a larger lesson or moral—will your story convey a larger message?

Students might write this assignment as a children's story, or as a conversation that includes the imagined comments and questions of young children. One option would be for students to illustrate their stories and approach teachers of young children to see whether it might be appropriate to read their stories aloud in class.

Bali Principles of Climate Justice

—*page 171*

Working with this document might feel redundant if your students have done the Indigenous Peoples' Summit on Climate Change role play (see p. 127), but if they have not, this is a valuable document to

teach. Ask students in pairs to define the term "justice." See if they can come up with a succinct one- or two-line definition. Ask them to share some of the definitions aloud and to identify commonalities. The specific language is not important, but see if students can agree on a list of concepts that are central to the idea of justice.

In small groups, ask students to apply this definition to the climate: What is *climate* justice? Encourage students to think of all the ways that people throughout the world are affected by climate change. What would it mean for these people to have justice?

Does justice also apply to other species? After students have answered these questions, distribute the Bali Principles of Climate Justice. Draw their attention to the second part of the document, which begins: "We, representatives of people's movements together with activist organizations working for social and environmental justice resolve to begin to build an international movement of all peoples for Climate Justice . . ." Ask students to locate "climate justice" items on this list that they included on their lists. Ask them to identify at least five items on the list they did not identify but that they feel are important.

As an alternative, you might distribute just the "Whereas" statements from the Bali Principles. Explain to students that a resolution like this is a kind of inverted essay: The opening "Whereas" statements are the evidence, and the demands/resolutions that follow are like thesis statements—conclusions that flow from the "Whereas" statements. Ask students to choose a specific number of "Whereas" statements and to pair these with statements about climate justice. You might give an example or two from the document.

Ask students to choose a certain number of statements about climate justice and to "translate" these into language that can be more easily understood.

CHAPTER FOUR: Burning the Future

GLOBAL WARMING/GREEN PATRIOT POSTERS

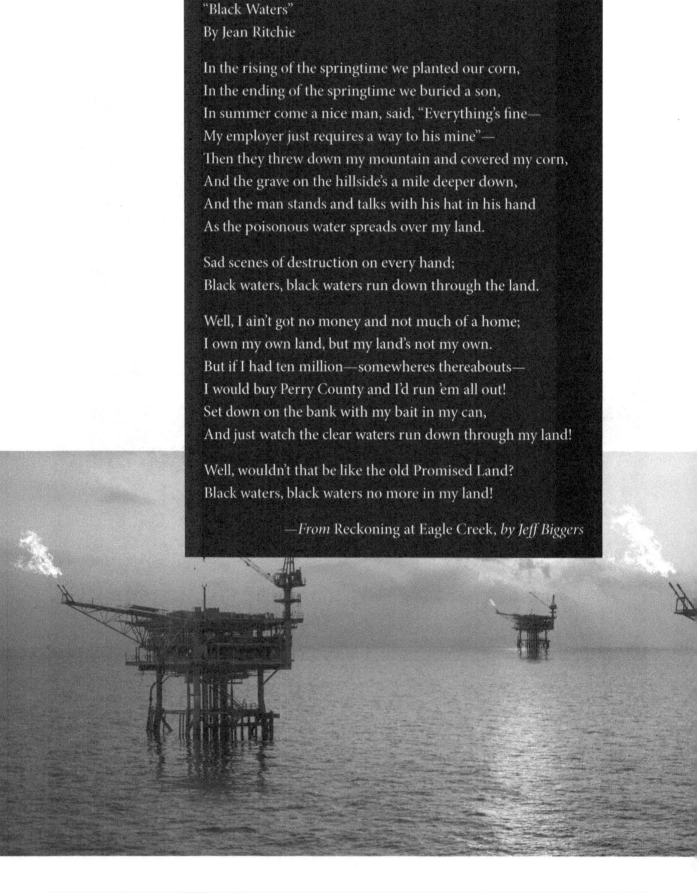

"Black Waters"
By Jean Ritchie

In the rising of the springtime we planted our corn,
In the ending of the springtime we buried a son,
In summer come a nice man, said, "Everything's fine—
My employer just requires a way to his mine"—
Then they threw down my mountain and covered my corn,
And the grave on the hillside's a mile deeper down,
And the man stands and talks with his hat in his hand
As the poisonous water spreads over my land.

Sad scenes of destruction on every hand;
Black waters, black waters run down through the land.

Well, I ain't got no money and not much of a home;
I own my own land, but my land's not my own.
But if I had ten million—somewheres thereabouts—
I would buy Perry County and I'd run 'em all out!
Set down on the bank with my bait in my can,
And just watch the clear waters run down through my land!

Well, wouldn't that be like the old Promised Land?
Black waters, black waters no more in my land!

—*From* Reckoning at Eagle Creek, *by Jeff Biggers*

INTRODUCTION
Burning the Future

The title of this chapter, "Burning the Future," is borrowed from a film of the same name, about coal and mountaintop removal mining in Appalachia. It has a metaphorical ring, but it's no metaphor—it's literally true: We are burning the future.

This chapter deals with the overall issue of fossil fuels and then is divided into three sections: coal (p. 198), oil (p. 230), and natural gas and fracking (p. 247).

The chapter is an invitation to teachers not only to teach about fossil fuels, but also to teach against fossil fuels. No doubt, this sounds partisan, and it is. But an anti-fossil fuels curriculum is based on science: 80 percent of known reserves of coal, oil, and natural gas need to stay in the ground if we are to avoid a climate catastrophe.

According to the latest report of the Intergovernmental Panel on Climate Change, the atmosphere has a ceiling of about 1 trillion tons of carbon, to preserve a reasonable hope of keeping climate change to a 2 degrees Celsius (3.6° Fahrenheit) increase over preindustrial times. The problem is we've already deposited well over 500 billion tons of carbon in the atmosphere, and at the rate humanity is burning fossil fuels, we'll reach the trillion-ton limit around 2040. In 2013 alone, we released 36 billion tons of carbon into the atmosphere.

We use the "we" voice here, and the collective term "humanity," but, as we emphasize in this chapter and throughout the book, a vast inequality exists between the carbon dumpers and those who suffer from the dumping. The same holds true for fossil fuels extraction. The coal companies clearcut forests, shove mountaintops into streams, and then call it an "act of God" when floods devastate communities. But activists like Maria Gunnoe know better: "I didn't see God up there in those haul trucks, filling in the creeks." (See "They Can Bury Me in These Hills, but I Ain't Leavin'" p. 210.) Those who order the land's blasting and scraping live at a comfortable distance from the victims of their decisions.

Fossil fuels do not just burn the future. They wreak havoc on the air we breathe, the water we drink, the oceans we depend on, our bodies' chemistry, and the democracy we need. This may be a grim future we confront, but the way students learn about these dynamics needn't be deadening. This chapter features mysteries, role plays, simulations, games, mixers, storytelling, poetry, and imaginative writing. Implicit in all these activities is the promise that people can make sense of what is happening around us, and we can take action on behalf of a different future. And, without minimizing the dangers we face, we can have fun doing it.

Yes, we are burning the future. But nothing is inevitable. This is a key lesson from history, and—we hope—from the activities included in this chapter. ⊕

The Mystery of the Three Scary Numbers

BY BILL BIGELOW

Every now and then an article comes along that takes such a novel approach to an issue, I feel like I'm seeing something with new eyes. Such was the case when I read "Global Warming's Terrifying New Math," Bill McKibben's 2012 *Rolling Stone* article. It made me see our climate predicament with such clarity that I knew immediately I had to figure out how to turn this article into curriculum.

The "terrifying new math" is pretty simple. McKibben, founder of 350.org and the world's most prominent climate campaigner, proposes that there are just three numbers that we need to pay attention

ADAM MILLS ELLIOTT

bottom lines," he writes. The first scary number.

The second scary number is **565 gigatons**—or 565 thousand million tons. That's humanity's carbon "budget"—how much carbon dioxide we can pour into the atmosphere with a reasonable chance of keeping global temperatures to a two degrees Celsius increase. That 565 gigatons sounds like a lot until we hear that global carbon dioxide emissions rose by 31.6 gigatons in 2011, and that projections call for humanity to blast through our 565-gigaton quota in less than 16 years.

Which brings us to the final number that makes the other two numbers so frightening: **2,795 gigatons.** This number represents the stored carbon in reserves held by coal, oil, and gas companies, and the countries—Saudi Arabia and Kuwait, for example—that act like fossil fuel companies. McKibben notes that this number was first highlighted by the Carbon Tracker Initiative, a group of London financial analysts and environmentalists. In other words, the fossil fuel industry already has plans to exploit five times as much carbon as can be burned without exceeding the two degrees ceiling. Burning these fossil fuels would enter the world into a dystopia of climate science fiction—a rise in sea levels not seen in human history, species extinction, droughts, superstorms, heat waves from hell, coral kill-offs, and consequences we cannot yet imagine.

"Here's another way of saying it: We need to leave at least 80 percent of that coal and gas and oil underground," McKibben writes. "The problem is, extracting and burning that coal and oil and gas is already factored into the share prices of the companies involved—the value of that carbon is already counted as part of the economy." This would be the equivalent of these companies writing off $20 trillion.

Not only is the fossil fuel industry not planning to write off any of this $20 trillion, the industry is using its immense wealth to add new reserves. Just as an example, in 2012 Exxon's CEO Rex Tillerson bragged that his company planned to spend $37 billion through 2016 on increasing oil production.

The simplicity of McKibben's "three scary numbers" helped me put into perspective some of the "softer" responses to global warming. So many environmentalists, teachers, and students want to concentrate on alternatives: everything from buying locally to stepping up recycling, planting more trees, and developing greener sources of energy. No doubt,

to in order to reach some radical conclusions about the future of fossil fuels.

The first number is **2 degrees Celsius**, or about 3.6 degrees Fahrenheit. In the 2009 Copenhagen Climate Accord, 167 countries, including the United States, pledged that "deep cuts in global [greenhouse gas] emissions are required . . . so as to hold the increase in global temperature below 2 degrees Celsius." The Copenhagen Accord was a timid, inadequate document. According to McKibben, even a two-degree rise in global temperatures is fraught with danger, but it's the only international consensus on a climate target—"the bottomest of it's crucial to imagine and work for alternatives. But for any of this to make a difference, we need to recognize fossil fuels—and those who exploit them—as immediate and staggering threats to life on Earth. One clear implication is that we cannot nice our way out of this. We have to educate and enlist our students in imagining a very different future in terms of energy use and fighting to make that happen.

Yes, a full curricular treatment of climate chaos needs to do more than merely frighten students with scary numbers. But these numbers of McKibben's invest our thinking about the climate with a two-plus-two-equals-four certainty. It's not probable that the route we're on leads to catastrophe—it's for sure.

The Mystery Activity

I love the structure of mixer activities that get students up out of their seats and talking with one another to figure out a bigger picture. Rather than asking students to assume the roles of individuals in history or around the world, I decided to write clues drawn largely from McKibben's *Rolling Stone* piece. I wanted students to solve the "mystery of the three scary numbers" by talking with one another. Well, maybe not solve, but at least come to recognize why these numbers are, in fact, so scary and begin to reflect on their implications. Further activities or discussion about the climate crisis would build from a common recognition of the mathematical fact that we are on an unsustainable trajectory.

My colleague and co-editor, Tim Swinehart, who teaches at Lincoln High School in Portland, invited me into his economics class to teach a couple of "Three Scary Numbers" sessions with him.

I held up a copy of *Rolling Stone*. "Anyone familiar with this magazine?" Maybe a third of the students raised a hand. "I began reading *Rolling Stone* in 1968, when I was about your age. Last year, the magazine published an article that generated more interest, more likes, more shares, more Tweets, than any article *Rolling Stone* had ever published. And here's the thing: The article is about just three numbers, three very scary numbers.

"So we're going to do an activity that we call the 'Mystery of the Three Scary Numbers.' Basically, you have two tasks: figure out what the three numbers are and why they're scary. Afterward, we'll talk about the meaning of these scary numbers and what

we can do about them."

We distributed a question sheet to everyone (see p. 190), and each student also received a clue. There were 29 students in the class and I'd written exactly 29 clues. In the clues, each of the three scary numbers was in 16-point bold type so students were sure to spot these "this-is-a big-deal" numerals.

The handout asked questions like:

- Find someone who has one of the three "scary" numbers (in large, bold type). What is the number?
- List as many details as you can find out about this number (at least three).
- Find three other numbers about climate change. What is the number and why is it important?

Some of the clues stuck faithfully to describing something about one of the three scary numbers, for example:

Two degrees Celsius is about 3.6 degrees Fahrenheit. In 2009, 167 countries signed on to the Copenhagen Accord. These 167 countries are the biggest polluters in the world, responsible for 87 percent of all greenhouse gas emissions. The accord states that we cannot raise the Earth's temperature more than two degrees Celsius without risking planetary disaster. All 167 countries, including the United States, pledged: "We agree that deep cuts in global [greenhouse gas] emissions are required . . . so as to hold the increase in global temperature below two degrees Celsius."

Other clues focused on different numbers:

Over the past 30 years, permanent Arctic sea ice has shrunk to half its previous area and thickness. As it diminishes, global warming increases. This is due to several things, including release of the potent greenhouse gas methane trapped under nearby permafrost, and because ice reflects the sun's energy whereas oceans absorb it. Oil companies see the disappearance of Arctic ice as an opportunity to make more profit by drilling for more oil—which will create even more glob-

Demonstration against fossil fuel exports from the Northwest. Columbia River between Oregon and Washington, July 27, 2013.

ADAM MILLS ELLIOTT

al warming. For example, Royal Dutch Shell has spent $4.5 billion preparing to drill in the Arctic. One of the world's leading environmentalists, David Suzuki, calls this "insane."

One clue featured the "Keeling curve"—the graph that depicts the inexorable rise in atmospheric carbon dioxide that Charles David Keeling began measuring in 1958 at Mauna Loa in Hawaii. When Keeling first began his measurements there, he recorded 313 parts per million; in 2013, it passed 400 parts per million.

Obviously, the more one knows about basic climate science, the easier time one will have with this activity. When teacher (and *Rethinking Schools* editorial associate) Adam Sanchez did this activity with 9th graders across town at Madison High School, he realized that his students needed a bit more initial familiarity with the concept of greenhouse gases and the relationship between burning coal, oil, and gas and releasing carbon into the atmosphere. (Of course, all these numbers are moving targets, as carbon dioxide in the atmosphere increases yearly, carbon reserves are on the rise, and the two degrees limit seems far too generous with every new climate study. The numbers included in the activity described here will change over time, but the essential problem captured by the mystery activity will remain.)

I introduced the mixer by reminding students that each of them had a different clue and that each clue offered important information that would help them figure out the mystery of what makes these three numbers so scary. Tim and I encouraged the students not to wait until the very end to begin making sense of this, but to talk with one another about the big picture as they circulated throughout the class. The rules of the game were simple: They had to share clues verbally—no handing over clues to anyone—and conversations had to be one-on-one (to encourage maximum participation). Finally, this was a get-up-and-mingle activity, so no just hanging out at one's desk waiting for "callers" to arrive.

Students wasted no time: "I need a bold number! Who has a scary number?"

One student encountered 2,795 gigatons. "That's a lot," she said in a quiet voice.

"So what are these numbers saying?" another asked.

"The numbers are important because we only have a couple of years."

"Well, we're already in trouble."

Toward the end of the activity, I watched a student cock her head and ask no one in particular: "So is this saying that we're going to die?"

Students were fully engaged throughout the half hour or so of the clue hunt. When conversations began winding down, Tim and I asked everyone to form a large circle and continue to discuss the final assignment sheet questions, which asked: Why are these numbers important? What actions should be taken?

We wanted students to feel free to share whatever occurred to them, so we did not overdirect the conversation. After this activity, students would continue to study the climate crisis with Tim and connect real human beings with these numbers. For now, we simply wanted to hear how they made sense of this new information.

"These are insane numbers," Matt said. He mentioned the potential species extinction and the rising seas. Cory pointed out that, at current rates, "We're on track to hit two degrees quickly, it's not some far-off endpoint"—which was exactly the sensibility we hoped students would draw from the activity: Climate change is not about the future, it's about now.

Michele was struck by the possibility of widespread desertification. Even James, a confirmed libertarian, argued that there was no reason to think that the market would somehow on its own be moved by these numbers: "I've never had it quantified like this, or had such a grim picture painted. . . . This has to be a shift that *we* make."

Not surprisingly, when it came to what "we" should do, students were all over the map. One student offered a techno-fix: "NASA is thinking about Mars." Sonia and many other students thought as responsible consumers: We should "use more local products and make permanent changes, not just 'I rode the bus one day'"; we should recycle and compost more; we should cut down on meat and travel; we should walk more; we should stop wasting water. And for some students, "we" extended to what the government should do: Start taxing coal; find alternative sources of energy; "the government should lead a 'war on climate change.'"

Interestingly, the more students talked, the more distant their solutions became. When a couple of students began criticizing the Chinese government for its alleged climate crimes, I pointed out how the conversation had drifted from changes that were more in our power to influence to those that weren't.

For homework, we gave the class an abbreviated version of McKibben's "Global Warming's Terrifying New Math" article to reinforce the information they encountered in the mystery activity. McKibben does not write with high school readers in mind, but having encountered much of his argument in the mystery, we knew students would find it more accessible. McKibben's strategic punch line is the need to launch a campaign to demand that colleges, retirement systems, and cities divest from holdings in fossil fuel companies—borrowing from the important divestment activism of the anti-apartheid movement during the 1980s. In May 2014, for example, after pressure from Fossil Free, Stanford University announced it would no longer invest in coal mining companies.

Given the terrifying math McKibben presents, Tim and I did not seek to suggest that there was a single "do-this" answer. We wanted to raise the question of what we should do—not answer it. So, in addition to McKibben's divestment proposal, we introduced students to a *Huffington Post* critique by Christian Parenti, author of *Tropic of Chaos,* who argues that attacking fossil fuels through the stock market is misguided for a host of reasons and that we need to focus our energies on "the important things government can do, right now, if pressured by grassroots action."

We weren't looking for students to take sides. But we did want them to recognize the urgency of activism. Maria wrote: "The three scary numbers are very scary. What scares me the most is how well this information is known without any action."

Of course, people are acting, and more study would introduce students to a range of strategies and actions. For now, we were content simply to have students "do the math," in the words of the 350.org campaign that built from McKibben's *Rolling Stone* article. Do the math, and recognize the profound immorality of leaving the future of life on Earth to the profit-driven choices of the fossil fuel industry.

As Matt wrote, "This made me want to change how this country functions. We are past the time of oil and coal." ⊕

Bill Bigelow (bbpdx@aol.com) is curriculum editor of Rethinking Schools *magazine.*

Clues to "The Mystery of the Three Scary Numbers"

Two degrees Celsius is about 3.6 degrees Fahrenheit. In 2009, 167 countries signed on to the Copenhagen Accord. These 167 countries are the biggest polluters in the world, responsible for 87 percent of all greenhouse gas emissions. The accord states that we cannot raise the Earth's temperature more than **two degrees Celsius**, without risking planetary disaster. All 167 countries, including the United States, pledged that: "We agree that deep cuts in global [greenhouse gas] emissions are required . . . so as to hold the increase in global temperature below **two degrees Celsius**."

Former NASA scientist James Hansen, the world's most prominent climatologist, believes that the Copenhagen target of keeping global warming under **two degrees Celsius** is not good enough. He says: "The target that has been talked about in international negotiations for two degrees of warming is actually a prescription for long-term disaster."

At the 2009 Copenhagen climate summit, a spokesman for small island nations warned that many island nations would not survive if the planet warmed by **two degrees Celsius**: "Some countries will flat-out disappear."

Many scientists believe that allowing the Earth to warm by **two degrees Celsius** could be a disaster. "Any number much above one degree involves a gamble," writes MIT's Kerry Emanuel, a leading authority on hurricanes, "and the odds become less and less favorable as the temperature goes up." Thomas Lovejoy, once the World Bank's chief biodiversity advisor, says this: "If we're seeing what we're seeing today at 0.8 degrees Celsius [for example, Superstorm Sandy], two degrees is simply too much."

To prevent a planetary catastrophe—rising sea levels, melting glaciers, disrupted food production, a scarcity of freshwater, more violent and deadly storms, more frequent droughts, increased warfare over scarce resources, etc.—the climate may not be allowed to rise more than **two degrees Celsius**. This is the *only* number that the vast majority of the world's nations have agreed to about the climate.

Scientists estimate that humans can pour about **565** more **gigatons** of carbon dioxide into the atmosphere by 2050 and still have some hope of staying below two degrees Celsius. [A gigaton is 1 billion tons—that is, a thousand million tons.] The **565-gigaton** figure was derived from one of the most sophisticated computer simulation models that have been built by climate scientists around the world over the past few decades.

Computer models calculate that even if we stopped all CO_2 (carbon dioxide) releases now, the temperature would likely still rise another 0.8 degrees Celsius, as previously released carbon continues to overheat the atmosphere. That means we're already three-quarters of the way to the two degrees Celsius limit—because we've already heated the planet 0.8 degrees Celsius.

In late May 2012, the International Energy Agency published its latest figures of how much carbon dioxide is being released into the atmosphere: CO_2 emissions in 2011 were 31.6 gigatons, up 3.2 percent from emissions the year before. [A gigaton is 1 billion tons—that is, one thousand million tons.] Study after study predicts that carbon emissions will keep growing by roughly 3 percent a year—and at that rate, we'll blow through our **565-gigaton** allowance in 16 years, around the time today's preschoolers will be graduating from high school.

Fossil fuel companies—and countries like Venezuela or Kuwait that act like fossil fuel companies—already have a huge amount of coal, oil, and natural gas in the ground that they own or have access to. The amount of these "reserves"—when burned for energy—would release an estimated **2,795 gigatons** of carbon dioxide into the atmosphere. [A gigaton is 1 billion tons—that is, one thousand million tons.] That is the number calculated by the Carbon Tracker Initiative, a team of London financial analysts and environmentalists.

2,795 gigatons is higher than **565 gigatons.** *Five* times higher.

If all the fossil fuels owned by just two giant oil companies, Russia's Lukoil and ExxonMobil, were sold and burned, it would release more than 80 gigatons of carbon dioxide into the atmosphere.

Energy corporations and big energy producing countries like Saudi Arabia and Kuwait, have estimated reserves of coal, oil and gas that—if burned for energy—would release **2,795 gigatons** of carbon emissions. John Fullerton, a former managing director at JP Morgan, who now runs the Capital Institute, calculates that at today's market value, the **2,795 gigatons** of carbon emissions are worth about $27 trillion—that's 27 thousand billion dollars: $27,000,000,000,000.

Ken Salazar, President Obama's former secretary of the interior, opened up a huge area of Wyoming for coal extraction. The total basin contains 67.5 gigatons worth of carbon, if all that coal is burned for energy.

According to NOAA, the National Oceanic and Atmospheric Administration, the average temperature in the lower 48 United States in 2012 was the hottest ever recorded. It was 55.3 degrees, one degree above the previous record and 3.2 degrees higher than the 20th-century average, NOAA scientists said.

According to James Hansen, former climatologist with NASA, the U.S. National Aeronautics and Space Administration, the "tar sands" of Alberta, Canada, contain as much as 240 gigatons of carbon—which, if burned, would take up almost half of the available atmospheric space if we take the **565 gigatons of carbon** limit seriously. The company, TransCanada, has proposed that it build a pipeline from Canada through North Dakota, South Dakota, Nebraska, Kansas, Oklahoma, and Texas to export oil around the world.

"Lots of companies do rotten things in the course of their business—pay terrible wages, make people work in sweatshops—and we pressure them to change those practices," according to writer and journalist Naomi Klein. "But these [three] numbers make clear that with the fossil fuel industry [coal, oil, and natural gas], wrecking the planet is their business model. It's what they do."

If the oil reserves of just six companies—Exxon, BP, Chevron, ConocoPhillips, Shell, and the Russian firm Gazprom—were burned for energy, this would use up more than a quarter of the **565 gigatons of carbon** limit that is needed to keep the planet from warming more than two degrees Celsius. (Each of these companies continues to search for more oil.)

In early March 2012, Exxon CEO Rex Tillerson told Wall Street analysts that the company plans to spend $37 billion a year through 2016 (about $100 million *a day*) searching for *more* oil and gas.

Two-thirds of wheat grown in poor countries, and almost a quarter of the wheat grown in rich countries—nearly half the world's total crop—is at risk from global warming. In order to keep up with the world's growing population, global wheat production needs to rise 50 percent.

Around the world, the Earth's average temperature has risen more than one degree Fahrenheit (0.8 degrees Celsius) since 1880, and about twice that in parts of the Arctic. That may not sound like much, but we're already starting to see more intense rainstorms; severe droughts and heat waves are becoming more frequent. Rising seas are damaging homes near the water. Some populations of animals are starting to die out.

This graph shows the increase over time of the carbon dioxide concentration in the atmosphere. Every major scientific organization in the world, and 97 percent of climate scientists, attribute this increase to human causes—mostly burning fossil fuels like coal, oil, and natural gas. The higher the concentration of carbon dioxide in the atmosphere, the warmer the planet becomes.

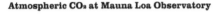

Atmospheric CO₂ at Mauna Loa Observatory

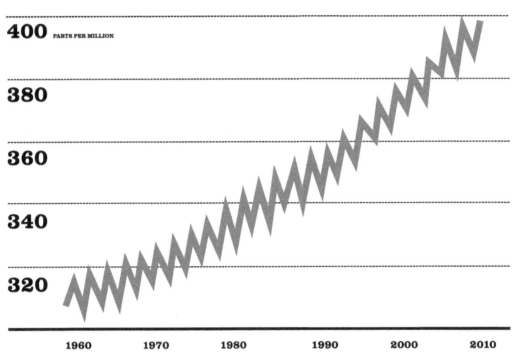

Over the past 30 years, permanent Arctic sea ice has shrunk to half its previous area and thickness. As it diminishes, global warming increases. This is due to several things, including release of the potent greenhouse gas methane trapped under nearby permafrost, and because ice reflects the sun's energy whereas oceans absorb it. Oil companies see the disappearance of Arctic ice as an opportunity to make more profit by drilling for more oil—which will create even more global warming. For example, Royal Dutch Shell has spent $4.5 billion preparing to drill in the Arctic. David Suzuki, one of the world's leading environmentalists, calls this "insane."

There is overwhelming evidence that our climate is warming due to pollution from human activities. That's the conclusion reached by 97 percent of climate scientists and every major national academy of science in the world. When we burn dirty fossil fuels like oil and coal, and when we cut down forests that store carbon, we pollute our atmosphere with greenhouse gases and warm our planet. This is not controversial among scientists.

Because of global warming, the world's glaciers are melting. All scientific organizations and the vast majority of climate scientists (97 percent) believe that global warming is caused by human activity. Here is how *National Geographic* magazine describes it: "Everywhere on Earth ice is changing. The famed snows of Kilimanjaro have melted more than 80 percent since 1912. Glaciers in the Garhwal Himalaya in India are retreating so fast that researchers believe that most central and eastern Himalayan glaciers could virtually disappear by 2035. Arctic sea ice has thinned significantly over the past half century, and its extent has declined by about 10 percent in the past 30 years. NASA's repeated laser altimeter readings show the edges of Greenland's ice sheet shrinking. Spring freshwater ice breakup in the Northern Hemisphere now occurs nine days earlier than it did 150 years ago, and autumn freeze-up 10 days later. Thawing permafrost has caused the ground to subside more than 15 feet (4.6 meters) in parts of Alaska. From the Arctic to Peru, from Switzerland to the equatorial glaciers of Man Jaya in Indonesia, massive ice fields, monstrous glaciers, and sea ice are disappearing, fast." The results include rising sea levels and putting at risk the freshwater supply of billions of people.

The U.S. Department of Defense has said that global warming will create more instability and warfare around the world. Global warming is already creating more violent storms, drought, lack of food and water, mass migration, and the spread of disease. All these will create tension between people around the world and lead to increased military conflict. According to the *New York Times,* Secretary of State John Kerry (then a U.S. senator) has argued that the continuing conflict in southern Sudan, which has killed and displaced *tens of thousands* of people, is a result of drought and expansion of deserts in the north. "That is going to be repeated many times over and on a much larger scale," he said. Global warming is killing people in many different ways.

According to climate scientists at Oxford University in Great Britain, humanity could probably keep the Earth's average temperature rise below **two degrees Celsius** in the future if we cut carbon emissions every year by 2.4 percent. For true safety, scientists estimate that humanity would need to cut carbon emissions by twice that rate.

There are currently several proposals to export coal through the Columbia River Gorge to Asia. The Sightline Institute, an environmental think tank in Seattle, estimated that if just *two* of these coal export terminal proposals were approved—in Longview and Bellingham, Washington—it would add 199 million tons of carbon dioxide to the atmosphere, every single year. And this includes just the actual burning of the coal: not the "mining, processing, rail shipping, storing, maritime shipping, constructing new port or rail facilities, or any other related activities." Over 10 years, the coal burned in Asia from the coal exports would be equal to two gigatons of carbon dioxide. (A gigaton is a billion tons.)

According to an estimate of the Congressional Budget Office, the top 20 percent of the wealthiest people in the United States are responsible for consumption that releases three times the carbon dioxide—the main greenhouse gas—as the bottom 20 percent of the population.

The Mystery of the Three Scary Numbers

1. Find someone who has one of the three numbers (in boldface). What is the number?

2. What are as many details as you can find out about this number? Try to find at least three.

3. Find someone who has a different one of the three numbers (in boldface). What is the number?

4. What are as many details as you can find out about this second number? Try to find at least three.

5. Find someone who has a different one of the three numbers (in boldface). What is the number?

6. What are as many details as you can find out about this final number? Try to find at least three.

7. Find three other numbers that are not directly connected to one of the Three Scary Numbers. What is each number and why is it important?

8. Once you have finished questions 1 through 7, find someone who has also finished and discuss why these numbers are important, and what actions should be taken. Write your thoughts here:

MICHAEL DUFFY

A Matter of Degrees

The arithmetic of a warming climate

BY BILL MCKIBBEN

Three hundred years ago, when we started burning coal, there was no reason not to. We had no idea yet that it could change the climate—and, more importantly, we were burning such tiny quantities that it didn't really matter. The ease and mobility it provided for human beings, especially as smoky coal turned into much cleaner oil, came with few discernible costs; it seemed almost like magic.

But over the course of the last generation, we've all learned that this potion carried the most powerful hangover possible. We've already burned enough coal and gas and oil to push us out of the Holocene; we've raised the planet's temperature about one degree, and that's been enough to cause massive outbreaks of drought and flooding, enough to make seawater 30 percent more acidic. The (literally) burning question for the Earth is, how fast can we get off this stuff?

Which is where the arithmetic comes in. Even though our heating of the climate has already caused some truly ugly results, scientists calculate that if we pass two degrees Celsius, we're entering the guaranteed catastrophe zone. That's why the world's nations—even the recalcitrant ones, like the United States and Saudi Arabia—are theoretically committed to making sure we stop short of two degrees. "I cannot negotiate on the two degrees," said German Chancellor Angela Merkel, in 2007. In July 2009, the Major Economies Forum, which includes all the countries that burn vast quantities of carbon, agreed to limit temperature increases to below two degrees. In December 2009, as the Copenhagen conference was fizzling, President Obama attempted to cover its failure with a nonbinding agreement he and others drafted on the fly. It opened with this:

> To achieve the ultimate objective of the convention to stabilize greenhouse gas concentration in the atmosphere at a level that would prevent dangerous anthropogenic interference with the climate system, we shall, recognizing the scientific view that the increase in global temperature should be below two degrees Celsius, on the basis of equity and in the context of sustainable development, enhance our long-term cooperative action to combat climate change.

The poorest nations at Copenhagen were demanding 1.5 degrees as a target, and even that is probably too high given recent data—but for now let's just say this: The one thing about global warming that the world agrees on is that two degrees is too much.

> **The (literally) burning question for the Earth is, how fast can we get off this stuff?**

Scientists also agree that to stand a reasonable chance of avoiding a two-degree rise, we can't emit more than 565 gigatons of CO_2 over the next 40 years. It's like saying, if you want to keep your blood alcohol level legal for driving, you can't drink more than eight beers in the next six hours. It's a limit: 565 gigatons. On the other side is a world where much of what we call civilization gets undermined, beginning in the poorest places that have done the least to cause the trouble.

Last year some analysts in the U.K. decided to add up how much carbon all the world's fossil fuel companies (and the countries like Venezuela that are essentially fossil fuel companies) have listed as reserves. That is, how much they have already found and are planning to dig up and burn. The total was enough to generate 2,795 gigatons of CO_2—five times as much as scientists say it's thinkable to emit.

Here's another way of saying it: We need to leave at least 80 percent of that coal and oil and gas underground. The problem is, extracting and burning that coal and oil and gas is already factored into the share prices of the companies involved—the value of that carbon is already counted as part of the economy. John Fullerton of the Capital Institute calculated last year that choosing not to burn that carbon would require a $20 trillion write-off—a figure equivalent to about 40 percent of the planet's GDP. If we took the threat to our climate seriously, Fullerton wrote, "fossil fuel-intensive economies and investors would be severely damaged."

Now you know, in hard numbers, why the fossil fuel industry has fought so hard to prevent meaningful regulation of its emissions; it is planning on burning far more carbon than the atmosphere can safely absorb. There are two choices here: a healthy balance sheet for the richest industry on Earth, or a healthy Earth. Or consider this: *If you've got oil, coal, and gas stocks in your portfolio, your investment can pay off only if the planet tanks.* And that's what markets are betting will happen. The price of these stocks stays high precisely because investors assume the company's lobbyists will continue to succeed in preventing real change.

You can do the math right down to the level of individual companies. ExxonMobil, according to the U.K. analysts, has about 40 gigatons of CO_2 emissions in its reserves. That means that a single

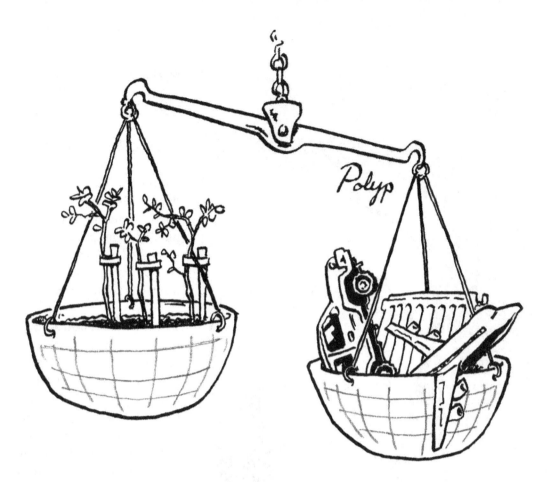

company could use up more than a 15th of the planet's safety margin. BP has about 35 gigatons in its reserves. Chevron and ConocoPhillips about 20 gigatons each.

Coal giant Peabody? Ten gigatons. And it's not as if they're content with what they've got. Exxon-Mobil, for instance, boasts that it spends $100 million every day looking for more gas and oil—that is, it spends $100 million a day looking for carbon that scientists say simply can't be burned. In 2008, it spent just $4 million on renewables research—*for the entire year.*

Given that ExxonMobil and its brethren have purchased large fleets of congressmen, this political battle is going to be an uphill one at best. We'll wage it, and sometimes we'll win small, temporary victories—the fight against the Keystone pipeline, for instance. But as the carbon math makes clear, we've got to address the heart of the matter, questioning the legitimacy of business models wherein financial success equals global catastrophe. For instance, given the numbers, how can any college or university justify having fossil fuel stocks in its portfolio? They are educating young people to be productive members of our society for the next 60 or 70 years, but they're paying for it in a way that guarantees

those kids have no future. The same goes for the parents saving up to put those kids through school. It's ironic that many "ethical investors" avoid gambling stocks but load up on oil; these guys aren't just gambling our future, they're hell-bent on guaranteeing it doesn't exist.

Figuring out how to write off those carbon reserves without tanking the economy is going to be every bit as hard as figuring out how to retool our energy grid for renewables. If you're a college trustee seeking to sell all those fossil fuel shares, your advisor will insist you're increasing risk and reducing returns by not being properly "diversified." But the math tells us what we have to do. In their black-and-white way, the numbers make it clear that today's fossil fuel enterprise is profoundly immoral. ◉

> **Given the numbers, how can any college or university justify having fossil fuel stocks in its portfolio?**

Bill McKibben is Schumann Distinguished Scholar at Middlebury College and founder of 350.org.

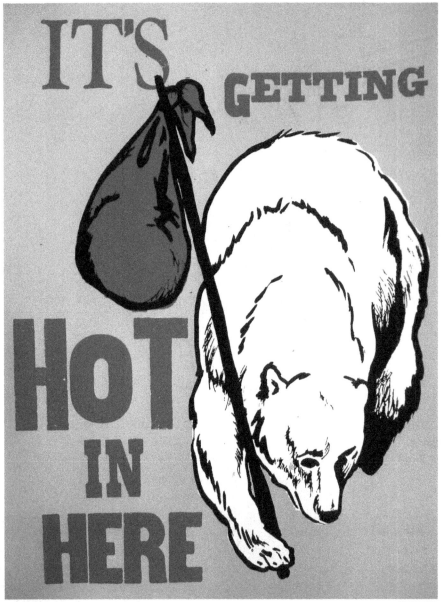

IT'S GETTING HOT IN HERE

MARY TREMONTE

A Short History of the Three Ages of Carbon—and the Dangers Ahead

BY MICHAEL T. KLARE

In today's world, one thing is guaranteed: Global carbon emissions will soar far beyond our current worst-case assumptions, meaning intense heat waves will become commonplace and our few remaining wilderness areas will be destroyed. Planet Earth will be a far—possibly unimaginably—harsher and more blistering place. In that light, it's worth exploring in greater depth just how we ended up in such a predicament, one carbon age at a time.

The First Carbon Era

The first carbon era began in the late 1800s, with the introduction of coal-powered steam engines and their widespread application to all manner of industrial enterprises. Initially used to power textile mills and industrial plants, coal was also used in transportation (steam-powered ships and railroads), mining, and the large-scale production of iron. What we now call the Industrial Revolution was largely comprised of the increased use of coal and steam power for productive activities. Eventually, coal would also be used to generate electricity, a field in which it remains dominant today.

This was the era in which vast armies of hard-pressed workers built continent-spanning railroads and mammoth textile mills as factory towns proliferated and cities grew. It was the era, above all, of the expansion of the British Empire. For a time, Great Britain was the biggest producer and consumer of coal, the world's leading manufacturer, its top industrial innovator, and its dominant power—and all of these were connected. By mastering the technology of coal, a small island off the coast of Europe was able to accumulate vast wealth, develop the world's most advanced weaponry, and control the global sea-lanes.

The same coal technology that gave Britain such global advantages also brought great misery. As noted by energy analyst Paul Roberts in *The*

End of Oil, the coal then being consumed in England was of the brown lignite variety, "chock full of sulfur and other impurities." When burned, "it produced an acrid, choking smoke that stung the eyes and lungs and blackened walls and clothes." By the end of the 19th century, the air in London and other coal-powered cities was so polluted that "trees died, marble facades dissolved, and respiratory ailments became epidemic."

For Great Britain and other early industrial powers, the substitution of oil and gas for coal was a godsend, allowing improved air quality, the restoration of cities, and a reduction in respiratory ailments. In many parts of the world, of course, the Age of Coal is not over. In China and India, among other places, coal remains the principal source of energy, condemning their cities and populations to a 21st-century version of 19th-century London and Manchester.

The Second Carbon Era

The Age of Oil began in 1859 when commercial production began in western Pennsylvania, but only truly took off after World War II, with the explosive growth of automobile ownership. Before 1940, oil played an important role in illumination and lubrication, among other applications, but it was not as important as coal; after the war, oil became the world's principal source of energy. From 10 million barrels per day in 1950, global consumption soared to 77 million in 2000.

Driving the global rise of petroleum was its close association with the internal combustion engine (ICE). Due to oil's superior portability and energy intensity (that is, the amount of energy it releases per unit of volume), it makes the ideal fuel for mobile, versatile ICEs. Just as coal rose to prominence by fueling steam engines, oil came to prominence by fueling the world's growing fleets of cars, trucks, planes, trains, and ships. Today, petroleum supplies about 97 percent of all energy used in transportation worldwide.

The rise of oil was also because of its use in agriculture and warfare. In a relatively short period of time, oil-powered tractors and other agricultural machines replaced animals as the primary source of power on farms around the world. A similar transition occurred on the modern battlefield, with oil-powered tanks and planes replacing the cavalry as the main source of offensive power.

These were the years of mass automobile ownership, continent-spanning highways, endless suburbs, giant malls, cheap flights, mechanized agriculture, artificial fibers, and—above all else—the global expansion of U.S. power. Because the United States possessed mammoth reserves of oil, was the first to master the technology of oil extraction and refining, and was the most successful at using petroleum in transportation, manufacturing, agriculture, and war, it emerged as the richest and most powerful country of the 21st century. Thanks to the technology of oil, the United States was able to accumulate staggering levels of wealth, deploy armies and military bases to every continent, and control the global air and sea-lanes—extending its power to every corner of the planet.

> Thanks to the technology of oil, the United States was able to accumulate staggering levels of wealth, deploy armies and military bases to every continent, and control the global air and sea-lanes—extending its power to every corner of the planet.

However, just as Britain experienced negative consequences from its excessive reliance on coal, the United States—and the rest of the world—too has suffered in various ways from its reliance on oil. To ensure the safety of its overseas sources of supply, the U.S. government has established tortuous relationships with foreign oil suppliers and has fought several costly, debilitating wars in the Persian Gulf region. Overreliance on motor vehicles for personal and commercial transportation has left the country ill equipped to deal with supply disruptions and price spikes. Most of all, the vast increase in oil consumption—here and elsewhere—has increased carbon dioxide emissions, accelerating planetary warming (a process begun during the first carbon era) and exposing the country to the ever more devastating effects of climate change.

The Age of Unconventional Oil and Gas

These were the hallmarks of the exploitation of conventional petroleum: the explosive growth of automotive and aviation travel, the rise of suburbs around the world, the mechanization of agriculture and warfare, the global supremacy of the United States, and the onset of climate change. Today, most of the world's oil still comes from a few hundred giant onshore fields in Iran, Iraq, Kuwait, Russia, Saudi Arabia, the United Arab Emirates, the United States, and Venezuela, among other countries; some additional oil is acquired from offshore fields in the North Sea, the Gulf of Guinea, and the Gulf of Mexico. This oil comes out of the ground in liquid form and requires relatively little processing before being refined into commercial fuels.

But such conventional oil is disappearing. According to the International Energy Agency, the major fields that currently provide most of the world's oil will lose two-thirds of their production over the next 25 years, with their net output plunging from 68 million barrels per day in 2009 to a mere 26 million barrels in 2035. The IEA assures us that new oil will be found to replace those lost supplies, but most of this will be of an "unconventional" nature. In the coming decades, unconventional oils will account for a growing share of the global petroleum inventory, eventually becoming our main source of supply.

> **The production of unconventional oil and gas turns out to require vast amounts of water.**

The same is true for natural gas, the second most important source of world energy. The global supply of conventional gas, like conventional oil, is shrinking, and we are becoming more dependent on unconventional sources of supply—especially from the Arctic, the deep oceans, and shale rock via hydraulic fracturing—or "fracking."

In certain ways, unconventional hydrocarbons are like conventional fuels. Both are largely composed of hydrogen and carbon, and can be burned to produce heat and energy. But in time the differences between them will become more important. Unconventional fuels—especially heavy oils and tar sands—tend to have a higher proportion of carbon to hydrogen than conventional oil, and so release more carbon dioxide when burned. Arctic and deep offshore oil require more energy to extract, and so produce higher carbon emissions.

"Many new breeds of petroleum fuels are nothing like conventional oil," Deborah Gordon, a specialist on the topic at the Carnegie Endowment for International Peace, wrote in 2012. "Unconventional oils tend to be heavy, complex, carbon laden, and locked up deep in the Earth, tightly trapped between or bound to sand, tar, and rock."

By far the most worrisome consequence of the distinctive nature of unconventional fuels is their extreme impact on the environment. Because they are often characterized by higher ratios of carbon to hydrogen, and generally require more energy to extract and be converted into usable materials, they produce more carbon dioxide emissions per unit of energy released. In addition, the process that produces shale gas, hailed as a "clean" fossil fuel, is believed by many scientists to cause widespread releases of methane, a particularly strong greenhouse gas.

All of this means that, as the consumption of fossil fuels grows, increasing, not decreasing, amounts of CO_2 and methane will be released into the atmosphere and, instead of slowing, global warming will speed up.

And here's another problem with the third carbon age: The production of unconventional oil and gas turns out to require vast amounts of water—for fracking operations, to extract tar sands and extra heavy oil, and to facilitate the transport and refining of such fuels. This is producing a growing threat of water contamination, especially in areas of intense fracking and tar sands production—and a decrease of available water supplies for other purposes, like irrigation and household use. As climate change intensifies, drought will become the norm in many areas and so this competition will only grow fiercer.

Along with these and other environmental impacts, the transition from conventional to unconventional fuels will have unknown economic and geopolitical consequences. For example, the exploitation of unconventional oil and gas reserves from previously inaccessible regions involves the introduction of untested production technologies, including deep sea and Arctic drilling, hydrofracking, and tar sands upgrading.

It's clear that, for the giant energy firms, unconventional energy is the next big thing and, as among the most profitable companies in history, they plan to spend huge amounts to ensure that they continue to be so. If this means that they fail to invest much in renewable energy like wind or solar, so be it.

In other words, energy firms, banks, lending

agencies, and governments will be biased in favor of next-generation fossil fuel production, making it more difficult to establish national and international curbs on carbon emissions. This is evident, for example, in the U.S. government's continued support for deep offshore drilling and shale gas development. And, more and more, we see growing international interest in the development of shale and heavy oil reserves, at the same time that new investment in green energy is being cut back.

The transition from conventional to unconventional oil and gas will also have a large, if still unknown, impact on political and military affairs.

U.S. and Canadian companies are playing a huge role in the development of many of the new unconventional fossil fuel technologies; and some of the world's largest unconventional oil and gas reserves are located in North America. The effect of this is to strengthen U.S. global power at the expense of rival energy producers like Russia and Venezuela, and energy-importing states like China and India, which lack the resources and technology to produce unconventional fuels.

For the time being, no other country is capable of exploiting unconventional resources on such a large scale as the United States.

At the same time, other countries will seek to develop their own capacity to exploit unconventional resources in what might be considered a fossil fuels version of an arms race. This will require a great deal of effort, but such resources are widely distributed across the planet and in time other major producers of unconventional fuels will emerge, challenging the United States' advantage in this realm. Of course, this will increase the global destructiveness of the third age of carbon. Sooner or later, much of international relations will revolve around these issues.

Surviving the Third Carbon Era

Unless something dramatic happens, the world will depend more and more on the exploitation of unconventional energy. This, in turn, means an increase in the buildup of greenhouse gases with little possibility of preventing catastrophic climate effects. Yes, we will also witness progress in the development and installation of renewable forms of energy, but these will not be as significant as the development of unconventional oil and gas.

Life in the third carbon era will benefit some. Those who rely on fossil fuels for transportation, heating, and the like will still have plenty of oil and natural gas. Banks, the energy corporations, and other economic interests will make enormous profits from the explosive expansion of the unconventional oil business and global increases in the consumption of these fuels. But most of us won't benefit. Quite the opposite. Instead, we'll experience the discomfort and suffering accompanying the heating of the planet, the scarcity of contested water supplies in many regions, and the destruction of the natural landscape.

What can be done to cut short the third carbon era and avert the worst of these outcomes? Calling for greater investment in green energy is essential but not enough—given the power of those pushing the development of unconventional fuels. Campaigning for curbs on carbon emissions is necessary, but will no doubt be difficult, as so much is invested in unconventional energy.

In addition to these efforts, we need to expose the unique and terrible dangers of unconventional energy and to demonize those who choose to invest in these fuels rather than in green alternatives. Some efforts are already under way, including student-initiated campaigns to persuade or compel college and university trustees to divest from any investments in fossil fuel companies. We need more. We need a systemic drive to identify and resist those responsible for our growing reliance on unconventional fuels.

To survive this era, humanity must become much smarter about this new kind of energy and then take the steps necessary to compress the third carbon era and hasten in the Age of Renewables before we burn ourselves off this planet. ⊕

Michael T. Klare is the Five College Professor of Peace and World Security Studies at Hampshire College in Amherst, Massachusetts. His newest book, The Race for What's Left: The Global Scramble for the World's Last Resources, *has just recently been published. His other books include* Rising Powers, Shrinking Planet: The New Geopolitics of Energy *and* Blood and Oil: The Dangers and Consequences of America's Growing Dependency on Imported Petroleum. *A documentary version of that book is available from the Media Education Foundation.*

 See teaching ideas for this article, page 261.

BURNING THE FUTURE: Coal

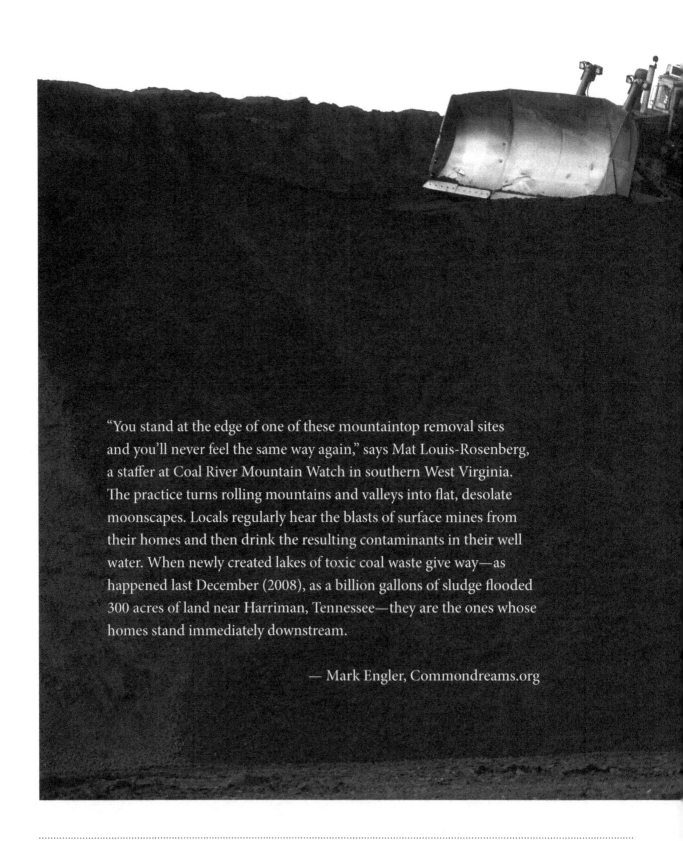

"You stand at the edge of one of these mountaintop removal sites and you'll never feel the same way again," says Mat Louis-Rosenberg, a staffer at Coal River Mountain Watch in southern West Virginia. The practice turns rolling mountains and valleys into flat, desolate moonscapes. Locals regularly hear the blasts of surface mines from their homes and then drink the resulting contaminants in their well water. When newly created lakes of toxic coal waste give way—as happened last December (2008), as a billion gallons of sludge flooded 300 acres of land near Harriman, Tennessee—they are the ones whose homes stand immediately downstream.

— Mark Engler, Commondreams.org

PAUL K. ANDERSON

Coal, Chocolate Chip Cookies, and Mountaintop Removal

BY BILL BIGELOW

In 30 years of teaching, I'd never taught explicitly about coal. Coal appeared in my social studies curriculum solely as a labor issue. We read passages about the 1914 Ludlow Massacre of striking coal miners and their families in Colorado, and watched John Sayles' excellent film *Matewan* when we studied early 20th-century labor struggles. But coal was mostly invisible in my history classes.

At the risk of sounding melodramatic, the world cannot afford this kind of curricular invisibility today. Forty-five percent of the main greenhouse gas produced in the United States, carbon dioxide, comes from burning coal for electricity; so does two-thirds of all the sulfur dioxide pollution. According to the American Lung Association, coal is responsible for thousands of premature deaths every year—"more hazardous air emissions than from any other industrial pollution sources," according to *Toxic Air*, an American Lung Association report. Forty percent of this country's electricity comes from burning coal: more than a *billion tons* of coal every year—almost 15 pounds of coal burned *each day* for every person in the United States. And most coal mining in the

BEEHIVE COLLECTIVE

United States these days is strip mining—the Earth is essentially skinned alive to get at the coal seams within. Coal companies have sliced the tops off 500 mountains in Appalachia and dumped the waste in the valleys, burying 1,200 miles of streams and poisoning residents' water. The term for this is mountaintop removal, and it's not a metaphor.

So I decided that it was time to break my curricular silence on coal. Now that I no longer have my own high school classes, my friends and colleagues Tim Swinehart and Julie Treick O'Neill invited me to help them co-teach a piece of a unit on climate change and energy to their 9th-grade global studies classes at Portland's Lincoln High School. (Tim and Julie teach separate classes but plan together.)

Coal is hidden—in the curriculum, but also literally: We encounter coal every time we turn on a light, but—at least in urban areas—our students almost never see it. I wanted to begin our brief introduction to coal by showing some of the stuff to students, but it was almost impossible to find. I sent an email to the 60 teachers on our Earth in Crisis curriculum list asking if anyone had any coal. No

luck. So I called Portland General Electric, which runs Oregon's only coal-fired power plant, in eastern Oregon. They promised to send me some coal but never did. Finally, I called several coal companies around the country and left messages. A few days later a package postmarked St. Louis arrived from Peabody Energy, the world's largest coal company. Eureka. It contained a little baggie of chunks of coal.

In class, I began by telling students that we were going to play a game, but first I wanted to show them something. I brought out the coal on a piece of paper on a stool and placed it in front of the room: "I'd like you to describe this. Feel free to come up and take a closer look, but don't touch it and please don't talk with each other about it. Just write what you see. There's no right or wrong description." I didn't say what it was.

Most students stayed seated and wrote, but four or five came up and carefully studied the little hunks of coal. After a few minutes, I asked students to turn to the person next to them and read their description aloud, and then asked for a few volunteers to read to the whole class. Students' accounts were observant and straightforward. Emma described it as "pitch black . . . bits crumbling off, sharp edges, no one looks like the other. Some look shiny from different angles, others look dusty and charcoal gray. . . . All the pieces are huddled together, not a single one left alone." Students described "burnt chunks," "kinda crumbly," "chalky rocks," a "weak rock that you could probably crush into powder with your foot." A number of students mentioned that it looked like coal, but even more said they thought it was charcoal. (Charcoal, I later pointed out to students, is entirely different; it's made by burning wood. Coal is a rock—fossilized plant matter found in veins or "seams" underground.)

> Coal is hidden in the curriculum, but also literally: We encounter coal every time we turn on a light, but—at least in urban areas—our students almost never see it.

I told students it was coal and recounted my difficulties in locating any to bring to class. I talked briefly about the importance of coal in the production of energy in the United States and the world and how it touches our lives every day, even if we don't think about it. I mentioned that when I'd moved to Oregon in 1977, I assumed that most of our power came from the mighty Columbia River—Woody Guthrie had written "Roll On Columbia" after visiting the power-generating Bonneville Dam just a half-hour drive up the river. But even here in water-rich Oregon, more than 40 percent of our electricity comes from coal, and the massive Boardman Plant burns three full railroad cars of coal every hour of every day. I wanted this introduction to emphasize coal's significance, but to be factual and neutral. So I didn't mention that coal produces twice the amount of climate-altering carbon dioxide per unit of energy as natural gas and a third more than oil—to say nothing of wind or solar—or that it is the country's (and the world's) single largest source of human-produced greenhouse gases, as well as airborne poisons like mercury. I didn't mention that mining and burning coal produces three times as much waste as all the municipal garbage in the country. I wanted students simply to "meet" coal before they thought about its social and environmental significance.

> I told them that to win the game would require some dexterity, some strategy, and, yes, some luck.

I found the game that I wanted to play with students in a remarkable book called *Coal Mountain Elementary*, compiled by Mark Nowak. This might be best described as a book-length prose collage, featuring pieces of testimony about the 2006 mining disaster in Sago, West Virginia; excerpts from newspaper accounts of coal mine accidents in China; and photos of Chinese mines by photojournalist Ian Teh. Lending an otherworldly feel, the book also includes pro-coal mining lesson plans for children developed by the American Coal Foundation (ACF), the nonprofit arm of the coal industry. The game I discovered in the book and then found detailed at www.teachcoal.org is a pedagogically clever activity on mining "coal" from chocolate chip cookies.

Playing a Game—with Two Minds

I sought to reproduce this unsettling juxtaposition between the grim reality of coal mining and ain't-it-great coal industry propaganda in class. I hoped students would play the industry-designed game and then be able to critique its premises as they learned a fuller story about coal mining. And Julie and Tim had just finished teaching about mining for natural gas via hydraulic fracturing—"fracking"—as described in the disturbing film *Gasland,* so the three of us felt that it would not be much of a stretch for students to see that there was much more to coal than presented in the cookie game.

I told students that before I explained the game I wanted to show them the prizes that I'd brought for them—a wide assortment of quality chocolate bars that I described one by one in delicious detail. "Of course, we don't have enough of these for everyone. Only for the winners." I told them that to win the game would require some dexterity, some strategy, and, yes, some luck.

I distributed the Cookie Mining Worksheet, and read aloud the ACF's description of the game: "The mining industry, like any other business, faces challenges to make itself profitable. To understand some of these challenges, you will attempt to conduct a profitable mining business in an experiment that requires you to mine the 'coal' chips from chocolate chip cookies." (Detailed instructions for the game are available at www.rethinkingschools.org/earth; the worksheet for the game is reproduced here on p. 207.)

I told students: "I'd like you to play the game with two minds. First, just get into the game and have fun. But second, as you're playing, think about what the game is meant to teach students. What values does it teach? What is missing from the game about what coal mining might actually mean in people's lives?"

At first glance, the game seems more complicated than it is. The basic idea is that students begin with a certain amount of money—for some reason, the ACF chose $19—and then must decide which "property" (meaning which type of chocolate chip cookie: Kentucky, $7; Pennsylvania, $5; or Montana, $3) and which kind of mining equipment (a paper clip, $6; a round toothpick, $4; or a flat toothpick, $2) they will purchase. Making these choices constitutes the strategy piece of the game. The dexterity component involves the speed and accuracy with which students can dig out the chips from the cookies—"labor" costs $1 a minute (students get five minutes to mine chips)—and how messy they are in

the mining process, because they will be charged for leaving cookie residue outside the original area of the cookie. (Each student begins by outlining his or her cookie on a sheet of graph paper.)

Here is how the ACF describes this "reclamation" process, which I read aloud before we began: "Coal companies are required by federal law to return the land they mine to its original, or an improved, condition. This process, known as reclamation, is a significant expense for the industry."

As mentioned, the game itself takes just five minutes to play, plus three minutes of post-mining reclamation. But it took a bit of time for students to choose which property and mining equipment they would purchase, and to review the Cookie Mining Worksheet, so everyone felt on top of the game before starting to dig out the chocolate chips—er, coal.

Tim, Julie, and I did this activity with three 9th-grade classes, each with about 30 students, in 90-minute block periods (including the initial coal

writing activity and follow-up writing on the game and discussion). Every single student was engaged in ferociously digging the chips out of these cookies as if there were no tomorrow, laughing and groaning. The air was punctuated with cries of "Man, these cookies are so hard," "Can I buy another toothpick?" "How am I ever going to put this cookie back together?"

Because the chocolate chips in each separate brand of cookie were of such varied sizes, afterward we distributed a "standard chip" to each student

> "How am I ever going to put this cookie back together?"

to measure chip production—students earned $2 for each chip mined. At the end of the game, we led students through their final tabulation: They began with $19; from this they subtracted the purchase of their property, their equipment, their labor, and their "reclamation" costs; and then added the gross profits from their chocolate chip mining to determine their final net profit. We awarded the delectable Toblerone, Equal Exchange, and Lindt chocolate bars to the students who ended up with the most profit—just like in the real world.

Students Analyze the Game

We wanted to give students a chance to step away from the frenetic competition and playfulness of the coal mining game, so we asked them to write briefly on two questions:

- What "works" about the ACF simulation? What is good about it as a teaching activity?
- What is missing from the simulation? What doesn't "work" about it? What is bad about it as a teaching activity?

I confess that I have a grudging admiration for the cleverness of this simulation. Who wouldn't want to design an activity that engages 100 percent of their students? And in their answers, students also expressed respect. As Henry said: "This is a good activity in that it teaches students the main costs that go to coal mining and especially reclamation. Also, it teaches students that it's difficult to profit from coal mining because the yield of a mine is not very predictable. It also teaches children that higher price has visible merits, as the paper clip never breaks, while toothpicks do."

Carmen added, "It's also more relatable because not many kids are interested in coal mines, but cookies are a shoo-in."

But students also saw the activity's limitations—or, one might say, manipulations. As Ally said, the activity "puts an image in students' heads that profit is the only thing to worry about. Also, having the cookie fit within the circle doesn't address how destroyed the cookie is and what effect it has on the entire paper (grease stains, crumbs)." Emily echoed this ecological critique: "It leaves out what effects the mining business has on the Earth. Kids need to know what burning coal can do to harm our environment. . . . Little kids

> As Ally said, the activity "puts an image in the students' heads that profit is the only thing to worry about."

could get the wrong impression about coal mining, since they'd think it's a good thing, since they had fun during the activity." As Alejandra summarized so eloquently: "This activity trivializes the effects of coal mining on the land, presenting destruction as the harmless hunting of chocolate. It sends the wrong message."

Students also noted the ACF's not-so-subtle linking of coal and chocolate: As Lili said: "It makes mining for coal look like a fun, happy cookie game. . . . Kids might also be attracted to the food aspect and think that coal is good." Finally, Carmen picked up on the way the game erases social class and "puts the jobs of two separate levels of the industry into one job." This was a brilliant point: In the game, each student plays both capitalist and worker, conflating the category of "labor." The activity entirely hides the fact that in the real world, one group profits off the work of another group.

Coal Mining: the Game vs. the Reality

Perhaps because this game came in the midst of a broader unit that Tim and Julie were teaching on climate change and environmental justice, almost all students were alert to the activity's failure to deal forthrightly with the consequences of coal mining on the Earth. However, I've used this game "cold" with groups of students who didn't have the environmental study context; participants still recognized the way the activity manipulates players to focus exclusively on profit and to disregard what happens to the cookie—the Earth.

With the end of the semester closing in on Tim and Julie, we were pressed for time, but we wanted to contrast the ACF game with at least some of the actual effects of U.S. coal mining. We decided to show pieces of the film *Burning the Future*, about mountaintop removal in southern West Virginia, and afterward to pair these with a choral reading of a collection of quotes about coal extraction and burning—some drawn from *Coal Mountain Elementary*, in homage to the book's idea of contrasting the coal industry's self-portrait with real-world consequences.

Before we began the film, we brought out the stool with the pieces of coal: "We want to give you your writing assignment now, so that you can take notes as you watch the film. Earlier, you wrote about this coal. After we finish the film and hear some passages about the effects of coal mining, we'd like you to write again about this coal. You can ask questions of the coal, write an interior monologue from the perspective of the coal, write from the point of view of the mountain where the coal was found, or write from one of the characters in the film about the coal. The idea is simply to see this coal more deeply and fully than we did the first time around." We mentioned that students might want to collect memorable passages from the film and complete the assignment as a "found poem." [See Julie Treick O'Neill's article on using found poems in "Our Dignity Can Defeat Anyone," *Rethinking Schools,* summer 2008.]

We showed students chapters 2 and 8 of *Burning the Future,* approximately 25 minutes' worth. Chapters 4 and 6, about the impact of mountaintop removal on water quality, are also excellent, but Julie and Tim thought those chapters might feel a bit repetitive because, as mentioned, they'd recently shown *Gasland* to study fracking and drinking water pollution. (A number of other films also would work well to alert students to the impact of coal mining, especially mountaintop removal, and also to the inspiring grassroots movement to protect communities and the environment; see "Coal at the Movies," p. 216.)

These segments of *Burning the Future* tell the story of Maria Gunnoe, the impact of mountaintop removal coal mining on her land and family, and efforts by members of her community to organize and draw attention to what's happening there. Gunnoe lives in Bob White, West Virginia. She explains that her family has lived in the area since the 1700s. "The land was kind of like the root of the family," she says, in a line that struck a number of students—and that they incorporated into their writing. In the film, Gunnoe walks the land near her home, pointing out how her family harvested ginseng, sassafras, and morel mushrooms, and tells of different ways her family used poke leaves and stems in cooking.

And then she tells of the arrival of mountaintop removal coal mining, the clearcutting of trees near her home, including a huge birch tree that her parents had carved their initials in when they were dating. The stripping of the land and filling in of the valleys left the land vulnerable to flooding. She describes her family's terror when the 2003 floods roared through her valley, carried off five acres of her land, and left her daughter traumatized. Gunnoe remembers:

> The mine company engineer come to my front yard the next day and told me that this was an act of God. He stood and looked me in the face and told me that this was an act of God. It just infuriated me. I mean, how dare he blame something like this on God? God didn't do this. God put what was here *before.* And it was beautiful, it was useful. And it was abundant with life; there were birds, and deers, and bears, and foxes, and coons, and grouses. It was—it was really abundant with life.

The coal industry uses the term "overburden" to refer to everything covering the coal—in other words, the land, teeming with life—that Gunnoe describes. In the film, a coal official explains that miners "take great pride as they mine the coal to make sure that we have a very small and gentle footprint on the scenic beauty of West Virginia. And specifically, what I'm referring to as the 'great reclamation' of the mining sites that our coal companies and our coal miners are taking great pains with. As they run the dozers to put the land back, it's almost an artistic activity to watch the way they sculpt the mountains in really a great manner." Meanwhile, we view images of this "art": bulldozers shoving rocks into gullies.

> **The coal industry uses the term "overburden" to refer to everything covering the coal—in other words, the land, teeming with life.**

I wouldn't say that the video ends on a hopeful note, but it does end on a defiant one, with Gunnoe and fellow activists visiting New York to testify before the U.N. Commission on Sustainable Development. We stopped the video as Gunnoe stands on Broadway in New York and cries out: "Do people realize that mountains in West Virginia are being leveled to keep this street lit up? Do you realize your connection? For the sake of the families in southern West Virginia, turn out the lights!"

We followed these excerpts of *Burning the Future* by distributing a collection of powerful brief readings about the impact of coal, and students read these aloud one by one. Some focused on coal mining and some on coal burning. (All the quotes we used with students are available at www.rethinkingschools.org/earth.) For example:

> My name is Martín Macías Jr. I'm 19 years old from Chicago. I live about two miles away from two of the biggest coal power plants in the Midwest region and the only two coal power plants in Chicago. [Coal is] responsible for about 50 deaths a year in my neighborhood, and it's responsible for toxic air, toxic soil. If you look at the demographics of these communities, it's mostly Latino working-class immigrants. It doesn't employ anyone from our neighborhood, and we don't get any energy.

"Black Cocaine"

We reiterated the writing assignment and asked students to suggest a few ways that they might approach "re-seeing" the coal sitting in front of them. Their writing was imaginative and, for the most part, heartfelt. Steve now described these little chunks of coal as "black cocaine, giving America its energy." David said: "Looking at this coal again is like looking at . . . a blood diamond. . . . I can picture a giant wrecking ball crushing through the walls of my very own house." And Alexandra lamented the "wilderness once without scars" that had turned into a "battlefield of destruction."

Several students saw themselves as the entire mountain. Henry lashed out at "those proud humans" who "make ridiculous claims that they can reclaim the land and return me to my original state. How dare they tell such an outright lie." Jessica-as-mountain wrote: "I used to be beautiful, covered in memories." Alejandra imagined herself as the Earth:

> I am the layer cake. I am the most ancient one of all and I have seen them all. I knew that man would destroy it if he could. . . .

In *Big Coal*, Jeff Goodell wrote that the coal industry's "goal is to keep us comfortable, not curious."

They hacked, hacked, hacked away, and black gems revealed themselves within my body. They carried pieces of myself to faraway big cities, burned me for light, for heat, for comfort. "Slow down," I cried. "Soon there won't be any of me left. I have forests and people to support. I am older than the oldest tree, older than your great-grandfather, older than time itself. I am the Earth. I am your mother.

William, on the other hand, was unmoved:

> The people from southern West Virginia are asking for all this sympathy. Stop using coal, blah, blah. . . . I know that not very many people live in southern West Virginia. A lot of people reap the benefits brought by clean coal. There are about 10,000,000 people in NYC and like 20,000 in southern West Virginia. Why should they all cater to the tree that three people care about? . . . Some streams were blocked, but it seems more is won than lost.

But William was an "outlier," as the expression goes. Some students may not have picked up on Maria Gunnoe's activism, but they certainly embraced her anger and her sense of loss. In a later discussion, Lili noted that she was particularly saddened by the fact that the destruction of land in West Virginia affected people who have passed down that land from generation to generation and who know it so intimately.

For Julie, Tim, and me, this was a quick curricular visit to the Land of Coal. But we'll be back. In *Big Coal*, Jeff Goodell wrote that the coal industry's "goal is to keep us comfortable, not curious." Our aim is just the opposite. As the great labor organizer in the coalfields, "the miners' angel" Mother Jones said, "My business is to comfort the afflicted and to afflict the comfortable." Pretty good words for teachers, too. ⊕

Bill Bigelow (bbpdx@aol.com) is curriculum editor of Rethinking Schools *magazine.*

Cookie Mining Worksheet

Name _____

Date _____

Costs

A. Land acquisition costs (price of cookie)
(Montana—$3; Pennsylvania—$5; Kentucky—$7)
Name of property _____

B. Equipment costs
Flat toothpick _____ x $2 = _____
Round toothpick _____ x $4 = _____
Paperclip _____ x $6 = _____

Total equipment costs $_____

C. Mining/excavation costs (chip removal)

Number of minutes _____ x $1 labor =   Total evacuation costs $_____

D. Reclamation
(Original number of squares covered before cookie was mined = _____)
Squares covered outside original outline after reclamation _____ x $1

Total reclamation costs $_____

Profit

E. Mining valuation
Number of whole chips mined _____ x $2
Gross profit $_____

Calculating Net Profit/Loss

Start-up funds
less total mining costs (A. + B. + C.)   -_____
less total reclamation costs (D.)   -_____
plus gross profit (E.)   +_____

Total Net Profit/Loss $_____

"I am a Mud River West Virginia Girl! More specifically, I am a Conley Branch Girl.

I loved the mountains that surrounded our little three-room house. It was as if the mountains were there to protect us.

The mountain to the east of our house was my absolute favorite. Amongst all of the trees that are indigenous to the area stood a huge pine tree. It jutted out far beyond the top of the forest as if to say, "I am here. I will protect and shelter you from harm."

There were sad times when I sought out the comfort of the mountain. I ran to my tree when my grandmother died.

I wish I could run there today, but the mining companies came after I left. Neither Conley nor Mud River will ever be the same.

Conley is now blocked off with a "No Trespassing" sign. The mountain at the turn into Conley is even gone. No trees. No wild flowers. No squirrels. Like a lot of places in the Appalachians, nothing is left except what the mining company did not want.

I pray that those of us who love this land are strong enough to stand up for the mountains that remain. They have provided strength, solace, protection, and even life, to us. It is now our turn to return the favor."

—Marlene Adkins Thames, iLoveMountains.org

WWW.MOUNTAINROADSHOW.COM

An Insult to the Moon

BY ERIK REECE

Putting my students in situations where they might learn and practice the art of real democracy has become a large part of my own teaching, and with this goal in mind I often take them to a place in eastern Kentucky called Robinson Forest. It is a brilliant remnant of the mixed mesophytic ecosystem, and it is home to the cleanest streams in the state. Yet only a short walk away from our base camp you can watch those streams die—literally turn lifeless—because of the mountaintop removal strip mining that is happening all around Robinson Forest.

A few years ago, I had one student (I'll call him Brian) who had only signed up for one of my classes because it fit his schedule. He was, in his own words, "a right-wing nut job," and he disagreed with virtually everything I said in class. But he was funny and respectful and I liked having him around. On our class trip to Robinson Forest, we all hiked up out of the forest to a fairly typical mountaintop removal site. The hard-packed dirt and rock was completely barren, save for a few non-native, scrubby grasses. To call this post-mined land a "moonscape," as many do, is an insult to the moon.

Brian was quiet as we walked, and then he asked, "When are they going to reclaim this land?"

"It has been reclaimed," I said. "They sprayed hydro-seed, so now this qualifies as wildlife habitat."

"This is it?"

"This is all the law requires."

Brian went quiet again, until finally he said, "This is awful."

Then he asked, "What do you think would happen if every University of Kentucky student came to see this?"

I pulled the old teacher trick and turned the question back on him: "What do you think would happen?"

Brian paused, and then said, "I think mountaintop removal would end." ⊕

Erik Reece teaches at University of Kentucky. Excerpted from Orion *magazine.*

Maria Gunnoe and Larry Gibson
on the EPA steps, Appalachia
Rising, September 2010.

"They Can Bury Me in These Hills, but I Ain't Leavin'"

The Story of Maria Gunnoe

BY JEFF GOODELL

Don Blankenship, head of Massey Energy, the largest coal producer in Appalachia and the fourth largest in the United States, was throwing a Christmas party in a town in coal country, in West Virginia.

Outside in the drizzling rain, Maria Gunnoe and her 10-year-old daughter waited in line to enter the party. Gunnoe was a thin, wiry 36-year-old woman with long, dark, wavy hair and a habit of speaking her mind. She lived up hollow in a little place called Bob White, a few miles south of Madison. Like most of the people at the event, she was from a coal mining family—her grandfather, father, and brothers all worked in the mines. Unlike nearly everyone else in line, Gunnoe was not a fan of Don Blankenship. In fact, she believed he was pretty much single-handedly responsible for destroying her beloved West Virginia.

Gunnoe came to her views the hard way. In the spring of 2003, Big Branch Creek, which ran only a few hundred feet from her house and was usually small enough to jump over, became a wall of black water roaring down out of the hollow. In the 50 years her family had lived in Bob White, nothing quite like that had ever happened before. Rocks the size of Volkswagens tumbled down the river. The force of the water yanked Rowdy, her rottweiler, right out of his collar and carried him off. Gunnoe dashed through waist-deep water to fetch her daughter at a neighbor's house, then carried her back through the rising current. She believed they would both drown. Somehow they made it through, and Gunnoe and her family spent the night huddled in her little house above the Big Branch, wondering if the water would wash them away.

Until that moment, Gunnoe had never quite grasped the consequences of the big new strip mines that had opened in the hills above her in 2001. She had heard the blasting and swerved out of the way when the coal trucks came barreling around the corner on one of the local roads. It was scary, but she'd dealt with it. Then the flooding began. In three years, Gunnoe was flooded six times. It was no mystery what was happening: As the mountains above her were disassembled, the rocks and debris were dumped into the headwaters of creeks and streams, creating what the coal industry innocuously calls "valley fills." When it rained, the naked mountains guttered the water into the hollows. The filled-in headwaters of the creeks only accelerated the momentum of the runoff during storms, often turning a small, docile-looking creek like the Big Branch into a raging torrent. This was not a problem particular to Bob White. More than 700 miles of streams had been filled in throughout Appalachia, changing the natural drainage patterns and making catastrophic flooding a springtime ritual in the southern coalfields.

Even more dangerous, Gunnoe realized, were the big slurry impoundment ponds that are often built at mining sites—huge man-made lakes designed to store the runoff from coal washing, which are often filled with sludge containing high concentrations of heavy metals such as lead, arsenic, and selenium. In heavy rains, the earthen dams that hold these impoundments back sometimes fail, sending tidal waves of black, polluted water down over the people living in the hollows below.

The floods woke Gunnoe up to what was happening around her. It wasn't just the blasting of the mountains and the floods in the hollows, she realized. It was the destruction of a whole way of life.

It was the fish that were gone from the streams, and the startling number of people she knew who had been diagnosed with cancer, and the kids with asthma, and the slurry impoundment ponds that leached chemicals into the drinking water. And most of all, it was the hopelessness and fear she saw all around her. Whenever she pulled up at a gas station in Boone County and saw a man with a particular look of sadness and desperation in his eyes, she wondered what coal company he worked for and whether more than 100 years of taking orders from the mine superintendents and coal barons had crushed something essential in West Virginia's soul.

Heaven No More

Not long after the Massey Energy Christmas Extravaganza, I visited Maria Gunnoe at her home in Bob White. Because she doubted I'd be able to find the place on my own, she met me in Madison, and I rode with her in a red Ford pickup for about 20 miles up into the steep hollow. More than once, she yanked the wheel to get out of the way of a loaded coal truck that swerved suspiciously close to the yellow centerline. "They all know my truck," she explained coolly. "Ever since I started talking in public about what the coal companies are doing, I've noticed that I'm not real popular with some folks around here."

Not long ago, she discovered that her truck's brake lines had been slashed. Another day, her family's dog was found dead at her son's bus stop.

We turned onto a smaller road, approaching the Pond Fork River. Before we got there, the road was blocked with impromptu barriers. Beyond were the remains of a washed-out bridge that had been partially destroyed during the big flood in 2003 and was deemed unsafe for vehicular traffic. The West Virginia highway department had still not gotten around to fixing it, which meant that for nearly two years, Gunnoe has not been able to drive up to her house. Instead, she had to park beside the barriers and walk over the river on narrow wooden planks.

When I got out of the truck, Gunnoe pointed to

> Not long ago, she discovered that her truck's brake lines had been slashed. Another day, her family's dog was found dead at her son's bus stop.

a small blue and white house about 100 yards away on the other side of the river. "That's my place up there," she said proudly. "I used to call it 'my little piece of heaven.'"

There was not much heavenly about it anymore. The house was on high ground, but it had the misfortune of being sited not far from where the Big Branch Creek tumbled out of the hollow to meet up with the Pond Fork. The hillside that the house was built on was rapidly being washed away. Besides wrecking the bridge, the flood had cut a gully about 20 feet deep and 70 feet wide in Gunnoe's front yard. It was a mean, nasty gash, still fresh looking, as if someone had tried to cleave the hollow in two with a dull knife. Gunnoe explained that since the flood had wrecked the bridge, she had to lug everything—groceries, furniture, Christmas presents, even water—over the Pond Fork and up the hill. But that she could handle it. What she could not handle was the peril of being washed out of her home.

We walked across the bridge and up toward her house. The Big Branch was just a trickle running out of the hollow. "The day of the big flood, we had 10 feet of water roaring through here," Gunnoe explained. "It just came down in a fury. I thought we were goners."

> "The first thing out of his mouth was 'We are not responsible for this. It was an act of God.'"

Gunnoe's paternal grandfather, Martin Luther Gunnoe, a full-blooded Cherokee, labored underground in nearby coal mines for 32 years. He earned about $18 a week, and after many years of struggle, he was able to save enough money to buy these 40 acres in Bob White. Gunnoe remembered helping her father and grandfather build the house. "I carried lumber, fetched the nails," she recalled. "We were proud of it. It was always real peaceful here."

The 2003 flood not only filled Gunnoe's barn with rocks and washed out her road; it also destroyed her septic system, ruined her well, and covered her garden with black sludge. Compared to others, however, she got off easy. All over the southern part of the state, homes were flooded, cars overturned, and lives destroyed. Seven people were killed, including a 6-year-old girl who was drowned when the car she was riding in was washed off a narrow bridge. And

that was after the 2002 floods, which killed six people and destroyed more than 200 homes. In 2003, West Virginia received $40 million in federal disaster relief aid, more than any other state in the country. No one blames the flooding entirely on the strip mines. In some areas, reckless timbering is also a factor. But poorly engineered mountaintop removal is the worst culprit. According to one EPA study, the runoff from these mine sites is sometimes three to five times higher than in undisturbed areas, which means that five inches of rain—about the amount that fell in southern West Virginia from June 13 to June 19, 2003—has the same effect as would 15 to 25 inches.

What bothered Gunnoe most was not losing her front yard to the floodwaters. It was the attitude of the engineer from the coal company who showed up the morning after. Gunnoe noticed him standing out in her yard and waded through the muck to see what he wanted. "The first thing out of his mouth was 'We are not responsible for this. It was an act of God.'"

Gunnoe, who takes her faith seriously, and who had been up all night listening to the water roar and praying that she and her two children would not be swept away, was furious. Not only was she angry at the lack of sympathy his remark displayed, but she was also upset by his passing the buck to God—who, in her opinion, would be outraged by what he saw the coal companies doing to West Virginia. "I didn't see God up there in those haul trucks, filling in the creeks," she shouted at him. "I didn't see God up there blasting the mountain with ANFO."

It took a while for the full tragedy of what had happened to sink in. But eventually Gunnoe realized that unless she could stop the mining, she would inevitably be forced to abandon her land. It was just a matter of time. This was reinforced a few months later when she was flooded again (not as severely). How long could she live in a place where every drop of rain struck terror in her heart? And it wasn't just the flooding. It was the fact that the place that was so much a part of her family life—where she knew every patch of ginseng and witch hazel, where she had spent the happiest moments of her childhood wandering in the woods—was being transformed into an industrial zone. The fish were long gone from the creeks, killed by the polluted runoff from the mines. The few black bears left in the area, aroused from hibernation by mining machinery, wandered around in an angry daze. The deer were gone, frightened off by the blasting. Soon she would be gone too.

Taking a Stand

So she decided to fight. She quickly learned that no one at the county offices and none of her state representatives wanted to hear about her troubles. "They all just danced around my questions, or promised to call me back and never did," Gunnoe recalled. "It took me awhile to figure out that they just didn't care. Or if they did call me back, they would talk to me about the importance of jobs. Jobs! When someone looks at me and says my job is more important than your life, they make an enemy of me."

Gunnoe began volunteering at Coal River Mountain Watch, a small group in nearby Whitesville that was headed by the late Julia (Judy) Bonds, a former Pizza Hut waitress who became one of the leading anti-mountaintop removal activists in the state. Gunnoe also attended rallies and public meetings with longtime activist Dianne Bady, the founder of the Ohio Valley Environmental Coalition, whom Gunnoe calls "one of the most courageous people I've ever met."

Bonds and Bady helped open Gunnoe's eyes to what was going on around her. She learned about the years of legal battles that environmentalists and local citizens had waged to slow mountaintop removal mining, and how the Bush administration had gone out of its way to subvert and delay those battles. Under the direction of Steven Griles, then-deputy secretary for the U.S. Department of the Interior and a former coal industry lobbyist, debris from mountaintop removal mining was reclassified from objectionable "waste" to legally acceptable "fill," despite the fact that the debris is known to leach acid and heavy metals into local streams. This one-word change had a huge impact in Appalachia, undercutting the legal challenges to existing mountaintop removal mining and clearing the way

for new mines. Gunnoe read one federal study that projected that over the next decade 2,200 square miles of land in Appalachia—an area larger than Rhode Island—would be impacted by mountaintop removal mining.

Gunnoe also learned about the dangers of coal slurry impoundments. In one instance, in 1972, the failure of a big slurry dam in Buffalo Creek, West Virginia, had sent a 20-foot-high wall of coal slurry into the hollow below, killing 125 people and leaving 4,000 homeless. Today there are about 135 slurry impoundments in West Virginia, some of them the size of a good-size lake and holding billions of gallons of black water. One of the largest slurry darns in the state, the Goals impoundment in Raleigh County, is less than a mile above an elementary school. If the dam were to give out, the children wouldn't have a chance.

> One of the largest slurry dams in the state, the Goals impoundment in Raleigh County, is less than a mile above an elementary school. If the dam were to give out, the children wouldn't have a chance.

Gunnoe soon realized that flooding is only the most visible and melodramatic danger from coal slurry impoundments. Perhaps more threatening to the long-term sustainability of West Virginia is the leakage of these impoundments into the drinking water. One of the regions where the health effects have been of most concern is Mingo County, where Massey's Blankenship went to high school and still lives today.

A few years ago, Dr. Diane Shafer, a busy orthopedic surgeon in Williamson, the Mingo County seat, noticed that a surprising number of patients in their 50s were afflicted with early onset dementia. In addition, she was hearing more and more complaints about kidney stones, thyroid problems, and gastrointestinal problems such as bellyaches and diarrhea. Incidents of cancer and birth defects seemed to be rising, too. She had no formal studies to back her up, but she had been practicing medicine in the Williamson area for more than 30 years, and she knew that many people who lived in the hills beyond the reach of the municipal water supply had problems with their water: Black water would sometimes pour out of their pipes, ruining their clothes and staining porcelain fixtures. Many people had to switch to plastic fixtures because steel ones would be eaten up in a year or two. The worst water problems were in the town of Rawl, near Massey's Sprouse Creek slurry impoundment pond, where millions of gallons of black, sludgy water is backed up. Were the health problems in the area related to the pollutants leaching into the water supply from the slurry pond? Dr. Shafer suspected they were.

> "The coal companies control everything down here," Dr. Shafer told me. "It's like the Wild West except there is no sheriff in town. They just do whatever they want and pretty much get away with it."

Dr. Shafer is the lone physician on the Mingo County Board of Health. Despite her urgings, she could get no one at an official level to take much interest in the water problems in the area. So at her recommendation, a group of concerned citizens contacted Ben Stout, a well-known professor of biology at Wheeling Jesuit University and an expert on the impact of coal mining on Appalachian streams, to study the water quality in the area. Stout tested the water in 15 local wells, most of them within a few miles of the Sprouse Creek impoundment and one just a short distance from Blankenship's home. Stout found that the wells were indeed contaminated with heavy metals, including lead, arsenic, beryllium, and selenium. In several cases, the levels exceeded federal drinking water standards by as much as 500 percent. Of the 15 wells tested, only five met federal standards. Stout says that the metals found in the water samples were consistent with the metals in the slurry pond and the most logical explanation for how those metals got into the Williamson drinking water was that the impoundment pond was leaking into the aquifer. He also pointed out that coal companies often dispose of excess coal slurry by injecting it directly into abandoned underground mines, where it can easily migrate into the drinking water.

Dr. Shafer is the first to admit that much more work needs to be done to prove that the health effects she has been seeing are caused by the metals in the coal slurry. But state and federal agencies have shown little interest in pursuing it. "The coal companies control everything down here," Dr. Shafer told me. "It's like the Wild West except there is no sheriff in town. They just do whatever they want and pretty much get away with it."

Worry About the Future

To Stout, the problems in Williamson are just a foreshadowing of what's to come in the rest of southern West Virginia. "We are taking one of the great freshwater supplies in the world, and we're screwing it up," Stout says. "You can't fill in a thousand miles of streams, then inject millions of pounds of toxic slurry into the aquifers, without it having an impact." He believes that in 20 years, a lack of potable water may make southern West Virginia uninhabitable. "It's not the mountains of West Virginia that I worry about," he says. "It's the people. Sometimes I think what's going on here is damn near genocide."

Sitting on the couch in her small, neatly kept house, Maria Gunnoe changed into her hiking boots, then disappeared into another room for a moment. When she returned, she was carrying a silver .32-caliber Colt pistol. "It was my grandfather's," she explained. "In case we run into bears." Or angry coal miners? "Unlikely," Gunnoe said matter-of-factly. "But you can't be too careful up there."

Gunnoe tucked the pistol under her belt, and we headed up the hollow, following the steep cut of the Big Branch. You could see more evidence of the floods—huge cracks and fissures cut by the rushing water, trees upended, their roots twisting up toward the sky. It was muddy and slick in places, but Gunnoe climbed with ease. She talked about how spring is the scariest time of year for her now. How big will the flood be this year? How fast will it come? She knows parents in Bob White who make their kids sleep in their clothes when it rains so they're ready to go when the waters hit.

As we hiked, I asked her the obvious question: Why not leave?

"If I leave, where am I going to go? This is my place, it is who I am. My memories are here, my life is here. I love this place. Who am I if I give all this up? Why should I give all this up? To make it easier for the coal company? Should I just admit that Don Blankenship has won, let him take over the state? My friends tell me I'm crazy, that this place is already ruined, so what's the point? But I'm not one to give in. My feeling is, you fight for what you've got, even if it's only worth a dime."

A half hour or so later, the hollow broadened. We passed several "No Trespassing" signs that marked the mine border. To Gunnoe, these signs meant less than nothing. We walked through a stand of pines and poplars, then abruptly confronted a huge wall of rock—a valley fill.

"Ain't it pretty?" Gunnoe said wickedly.

I had seen many valley fills before, but never quite from this angle. It was a wall of rock maybe 500 feet high, barren of trees, but with thin patches of grass growing here and there. Standing at the bottom, looking up, I thought it seemed both immense and fragile. It was, quite literally, the top of a nearby mountain that had been cut off and dumped in this narrow valley that had once been the headwaters of the Big Branch. Culverts had been installed to divert the water, but in the end there was nowhere for it to go but down the hollow. In the distance, we could hear the faint grind and roar of heavy machinery. Every few minutes, a haul truck would appear on the horizon at the top of the mountain, hauling a load of dirt and rock to another part of the mine. On one of the steps in the valley fill, the company had planted several fruit trees and carefully surrounded them with chicken wire to keep rabbits from gnawing at the bark. It was a touching gesture, but one that somehow made the larger devastation all around us seem all the more criminal. Gunnoe pointed out that one of the trees had already been smashed by flyrock from a mine blast.

Gunnoe makes this hike up to the mines several times a week. For her, it is a way of claiming ownership of the land, and of reminding the coal company that it's being watched. It's a risky trek—once she was almost hit by flyrock; another time she was confronted by a group of miners who made vague remarks about how dangerous it was for a woman to be walking alone in the woods. "The next day, I hiked up here with my shotgun and said, 'Who's first?' They left me alone after that," Gunnoe said.

> **Seeing Gunnoe standing in front of the valley fill reminded me of the famous picture of the student standing in front of the Chinese tank during the Tiananmen Square uprising.**

Seeing Gunnoe standing in front of the valley fill reminded me of the famous picture of the student standing in front of the Chinese tank during the Tiananmen Square uprising. In a way, this valley fill is Gunnoe's tank. Her life, her feelings for her land, her memories of exploring the woods as a kid and of helping her grandfather build his house—all that means less than nothing in the face of the industrial operation above us. She doesn't care. "They can bury me in these hills," Gunnoe said, facing the wall of rock, "but I ain't leavin." ⊕

Jeff Goodell is a contributing editor to Rolling Stone. *This article was excerpted from his book* Big Coal: The Dirty Secret Behind America's Energy Future.

See teaching ideas for this article, page 261.

ANTRIM CASKY/*THE LAST MOUNTAIN*

Dirty Business: "Clean Coal" and the Battle for Our Energy Future

By Peter Bull
(Center for Investigative Reporting, 2010)
88 min.

Dirty Business is a long film for classroom use, but it is also the most comprehensive look at global dependence on coal, and explores some promising alternatives. The film is built around the work of Jeff Goodell, who wrote the previous article on Maria Gunnoe, excerpted from the important book *Big Coal*. Goodell begins with the devastating impact of coal mining in Appalachia. He remembers when he first saw the impact of mountaintop removal mining: "It was like the first time you look into a slaughterhouse after you've spent a lifetime of eating hamburgers." The film travels to Mesquite, Nevada, where residents are fighting a coal-fired plant, and also to China to explore the health impact of coal there—an important piece of the story not included in any of the other films reviewed here. The film's strength is its exploration of alternatives to coal—wind, solar thermal, increased energy efficiency through recycling "waste

Coal at the Movies

BY BILL BIGELOW

heat"—which makes this a valuable resource for science as well as social studies classes. The treatment of carbon dioxide sequestration may confuse students—the film simultaneously suggests that sequestration is a terrible idea in North America but a good one in China. But, on the whole, *Dirty Business* is a worthwhile overview of a complicated issue.

Burning the Future: Coal in America

By David Novack
(The Video Project, 2008)
88 min.

Jim Hecker, of Trial Lawyers for Public Justice, offers this summary near the opening of *Burning the Future*: "What we're witnessing in Appalachia is probably the single most environmentally destructive activity in the United States today. Whole mountains are being chewed up and their waste is being dumped into nearby streams." *Burning the Future* shows us

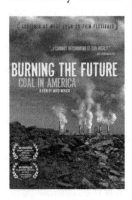 how mountaintop removal affects people living in the mountains of West Virginia without presenting them simply as victims. The film could be subtitled "The Birth of a Community Organizer": We watch Maria Gunnoe, one of the most compelling individuals in any of these films, as she turns her anger into activism. Chapter 4, which profiles the work of scientist Ben Stout, presents a wonderful example of science-for-society in action and could be used in biology or chemistry classes to show how "doing science" can make a real difference in people's lives. In chapter 6, coal country residents come to realize that their drinking water is killing them, and they commit themselves to do something about it. (See "Coal, Chocolate Chip Cookies, and Mountaintop Removal" on p. 200 for descriptions of chapters 2 and 8.) *Burning the Future* does not have much to say about climate change, and it waxes a bit nostalgic about "real" underground coal mining. Nonetheless, *Burning the Future* is a rich classroom resource that prompts students to think about where our electricity comes from. Each chapter is roughly 12 minutes long.

Deep Down: A Story from the Heart of Coal Country

By Jen Gilomen and Sally Rubin
(www.deepdownfilm.org, 2010)
57 min.

Deep Down offers an intimate look at what happens as neighbors are pitted against each other when a coal company proposes to strip-mine in the hills above Maytown, Kentucky. The film is built around Beverly May, who is determined to resist the coal company, and Terry Ratliff, who could sorely use the money the coal company is offering to lease some of his land for coal mining. The filmmakers present Ratliff's plight with sympathy, even as we cheer May's tireless efforts to save her community. At a hearing, May addresses the miners, whose livelihoods depend on continued mining: "I would like you to know that I work in a small clinic that takes care of people who are poor and who don't have insurance. I see every day many of your brothers. You are not my enemy. And I'm not yours. We are all victims of the same coal companies. It's just that you're on the top of the mountain and I'm down at the bottom. We are not enemies." As an antidote to cynicism, I wish every student in the country could meet the dedicated and compassionate Beverly May. *Deep Down* may be too slow, too "small" a story to hold some students' attention, but the film is a rare and remarkable teaching resource that shows the nitty-gritty process of organizing: the meetings, petitions, one-on-one conversations, phone calls, and demonstrations. The courage of Maytown residents is palpable. As one resident testifies late in the film: "Just imagine a society that is dependent on blowing up mountain after mountain after mountain. That there is a group of people that decided to stand up against it, that is exceptional."

The Last Mountain

By Bill Haney
thelastmountainmovie.com
(DADA Films, 2011)
95 min.

The Last Mountain refers to Coal River Mountain, slated for destruction by Massey Energy Co. through mountaintop removal. *The Last Mountain* exhibits and celebrates a tapestry of resistance to mountaintop removal mining—from the eloquent opposition of Robert F. (Bobby) Kennedy Jr., who "stars" in the film; to homegrown activists like Bo Webb, Ed Wiley, and Maria Gunnoe; to the courageous young people from around the country who commit civil disobedience to prevent this environmental devastation. Kennedy offers the outraged disbelief of the conscientious, engaged outsider: "I recently flew over the coalfields in the Appalachians. And I saw something that, if the American people could see it, there'd be a revolution in this country. We are cutting down the Appalachian Mountains, literally." At a weeklong energy teach-in, teachers at the public Sunnyside Environmental School in Portland chose to use *The Last Mountain* to introduce students to the ravages of mountaintop removal coal mining, but also to alert them to the imaginative activism in opposition.

The Mountaintop Removal Road Show

www.mountainroadshow.com

The Mountaintop Removal Road Show is a collection of slide shows and clips from videos that offer a critical look at the impact of mountaintop removal coal mining. Some of these feel like home movies and lack the production values that students expect. But what they lack in professional polish they make up for in authenticity. Most useful for the classroom is probably the 20-minute "Hidden Destruction of the Appalachian Mountains," a poignant overview of the impact of mountaintop removal mining on people and the environment. The slide show ends with some examples of resistance by mountain communities and their allies.

The 10-minute "McRoberts Residents Speak Out About TECO" is another piece that may be valuable to use in class. It offers startling images of the effects of mining-triggered flooding and the voices of residents who describe what it means for their lives.

The final segment, "Wake Up, Freak Out—Then Get a Grip," does not mention mountaintop removal or mining of any kind, but is a clever cartoon overview of the causes of and threats posed by global warming.

The Electricity Fairy

By Tom Hansell

(Appalshop, 2010)

52 min.

The title of this film comes from a former commissioner of the Kentucky Department for Natural Resources: "They reach out and flip the switch and the light comes on. Well, there's not a magic electricity fairy. That electricity comes from a power plant that feeds on coal." *The Electricity Fairy* describes a community's struggle against the construction of a coal-fired power plant in Wise County, Virginia, one of the poorest counties in the state. Because the film takes an expansive look at coal mining and the use of coal to fuel the rise of a consumer culture in the United States, it could be especially useful in U.S. history classes. We meet some imaginative activists, including Kathy Selvage of the Southern Appalachian Mountain Stewards, and student activist Marley Green, who delivers a powerful speech in opposition to the planned Dominion Virginia Power coal plant: "Coal is climate change. Coal is the mercury in our water, the asthma in our lungs, the soot in our air…Coal is the pusher that keeps America addicted, and the mountain destroyer that leaves the majesty of Appalachia a lifeless moonscape. We're trying out for a new world. Where bombs aren't dropped on our mountains, where streams aren't poisoned, and our society puts people before profit, and the health of communities before the convenience of suburbia." An important theme of the film is the seeming paradox that Appalachia is one of the most resource-rich areas of the country, yet also is home to some of the poorest people.

On Coal River

By Francine Cavanaugh and Adams Wood

(Downriver Media, 2011)

81 min.

In an emotional scene in *On Coal River,* Ed Wiley drops his granddaughter off at Marsh Fork Elementary School. As he drives away from the school, he says: "I tell you what, it's hard to let your child off at this place, knowing the dangers that's here. It's not right." Tears roll down his cheeks. The dangers at Marsh Fork are manifold: A 2.8 billion-gallon lake of toxic coal slurry sits above the school, held back by an earthen dam; coal dust from a nearby coal processing plant coats the playground and sidewalks around the school; the community's water is poisoned by mountaintop removal coal mining in the region and the "cleaning" of the coal in preparation for its shipment to coal-fired power plants; mountaintop removal explosions are nerve-jarring and put the earthen dam at continuous risk. But *On Coal River* is not just an exposition of the problems associated with mountaintop removal: We learn about the breadth of the problems by meeting the activists who tenaciously challenge the coal industry. One of the "stars" of *On Coal River* is Judy Bonds, the passionate organizer who became one of the most outspoken mountaintop removal opponents; she died in 2011 at age 58. The film is long and may be too slow for some classes. But in its attention to the details of one struggle in one small community, it tells a huge story. Late in the film, it dawns on one of the community activists, Bo Webb, that "we're on a mission to save the planet." It's no exaggeration. Struggles like the one described in *On Coal River* allow us to explore with students how local environmental justice work connects to the fight for planetary survival. ⊕

Mountaintop Removal Resistance

Numerous groups are fighting mountaintop removal coal mining in Appalachia. Some of these have valuable teaching resources at their sites:

Appalachian Mountain Advocates:
http://www.appalmad.org/

Coal River Mountain Watch:
www.crmw.net

I Love Mountains:
Ilovemountains.org

Kentuckians for the Commonwealth:
www.kftc.org

Ohio Valley Environmental Coalition:
www.ohvec.org

Southern Appalachian Mountain Stewards:
http://www.samsva.org

Children participate in a drawing competition as part of a "Say No to Coal in Manipur" campaign. Manipur, India.

Exporting Coal and Climate Change

BY BILL BIGELOW

If there were an award for environmental crimes, the proposal to export more than 100 million tons of coal to Asia every year would be a leading contender. Here's the plan: Strip-mine coal in Wyoming and Montana, put it on trains for more than a thousand miles to Oregon and Washington, then ship it to Asia to be burned for energy. According to the International Energy Agency, coal is the world's largest source of atmospheric carbon dioxide, the main greenhouse gas—now responsible for 45 percent of CO_2 emissions.

Despite coal's contribution to warming the planet, official school curriculum keeps students ignorant about coal. Textbooks mostly avoid the stuff. When it does make a rare appearance, coal is presented as a symbol of the bad old days, when small children toiled in mines. Prentice Hall's *United States History* includes only one mention of coal, on page 1,108, in a confused and brief passage about "the decline in American coal mining." In fact, although its use has gone down recently because of cheap natural gas, twice as much coal is mined in the

350.ORG

United States today than was 40 years ago. We burn about a billion tons a year, almost 15 pounds a day for every adult and child in the country. No doubt, Prentice Hall meant that there are fewer coal mining jobs in the country—which is true: The United States has about half the coal mining jobs it did 40 years ago, thanks to mechanized, union-busting, environment-wrecking strip mining.

According to a study by Sightline Institute, a Seattle-based nonprofit think tank, the annual carbon emissions from the coal exported from just two of the proposed three Northwest export terminals—in Longview and Bellingham, Washington—would exceed emissions from the dirty oil that would be carried by the proposed (and hotly contested) Keystone XL Pipeline from the tar sands in Canada. In brief: Coal exports are a really big deal.

I developed a "mixer" role play on the global impact of the proposed coal exports and taught it with Rethinking Schools editorial associate Adam Sanchez in a unit on climate change with one of his 9th-grade global studies classes at Madison High School. Madison is a public school of about 1,000 students in Portland, Oregon. Adam's class matches pretty accurately the overall demographics at Madison: roughly 40 percent white, 20 percent African American, 20 percent Latina/o, and 15 percent Asian American/Pacific Islander; about 70 percent free or reduced lunch.

Before we began the mixer activity, I held up a baggie of Powder River Basin coal and asked students to jot down some things they knew about coal. I was amused that the most common association was with Thomas the Train toy wooden railroad cars. I confess I spend many afternoons with my grandson, Xavier, on the living room rug playing with coal-powered engines like Thomas, Gordon, Percy, and Mighty Mac. I've wondered about the propaganda association between childhood play and Thomas the Train, and here it was, right in front of me. In fact, thanks to *Thomas & Friends*, students were surprised to learn that coal is no longer burned to power trains. Not so surprisingly, no 9th-grade student reported studying about coal in school.

Because I wanted to establish that coal is important, I shared statistics about the huge role coal plays in fueling the climate crisis—which students had been studying with Adam and me for a couple of weeks. I concluded with a quote from James Hansen, former director of NASA's Goddard Institute for Space Studies and perhaps the world's foremost climate scientist, who calls coal "the single greatest threat to civilization and all life on the planet" because of its enormous contribution to global warming.

Adam and I distributed a map showing the proposed route from the Powder River Basin in Wyoming and Montana that would bring more than 100 million tons of coal a year through the Columbia River Gorge, between Oregon and Washington. Of the three proposed projects, the two biggest are the ones in Longview and Bellingham, but the one furthest along in the permit process is a plan from the Australian company Ambre Energy to bring the coal by rail to Boardman in eastern Oregon, and then barge it down the Columbia River to Clatskanie, Oregon,

Thanks to *Thomas & Friends*, students were surprised to learn that coal is no longer burned to power trains.

where it would be put on ships for export to Asia. I didn't want to spend a lot of time lecturing, but I briefly shared a number of facts to emphasize that, love coal or hate it, if all three plans are approved, as many as 30 trains, each more than a mile long and carrying nothing but coal, would come through the gorge every single day—each pulled and pushed by four diesel engines.

The Coal Export Mixer

As we began the mixer, I explained to students that we were going to do an activity in which each of them would play the part of an individual whose life is connected to these potential coal exports—all actual people from around the world. (See the sample role at the end of this article.) In writing the roles, I tried whenever possible to draw exact language from the individuals I included. Some are in favor of coal exports, like Greg Boyce, CEO of Peabody Energy, the world's largest privately held coal company; Matthew Rose of the Burlington Northern Santa Fe Railroad, which makes more than 25 percent of its revenue from hauling coal; and Liu Guoyue, president of Huaneng Power International, which owns power companies in 18 Chinese provinces and is eager for Powder River coal. Others stand to lose, like Jasmine Zimmer-Stucky, an organizer with Columbia Riverkeeper in Hood River, Oregon, concerned about both water quality and climate change; and Manowara Uddin, originally from Bhola Island in Bangladesh, where the Padma River empties into the Bay of Bengal. Uddin was forced to migrate to a slum in Dhaka, thanks to the coal-powered climate chaos that every year produces more and more destructive storms.

We distributed a different role to each student, and asked everyone to read these more than once and to try to "become" these individuals. We suggested that they underline parts that seemed most important in determining their character's attitude about coal exports, then turn their roles over and list the three or four most important things about their character on the reverse side. We told them that when they circulated in the classroom meeting each other, we hoped that they would not read from their role, but would, as much as possible, know by heart the most important details about their individual. As students read and took notes, we distributed blank nametags for them to write their new "names." When students were ready, we distributed sheets with the questions that would guide their conversations with one another—for example, "Find someone who stands to gain from coal exports to China and Asia. Who is the person? How will he or she benefit if coal exports from the United States are approved?" And "Find someone who is taking action to try to stop the coal exports. Who is the person? What motivates this individual's work against coal exports?"

Before we asked students to get up, mingle, and find other individuals who could help answer one of the eight questions on the Coal Export Mixer Questions worksheet, we reminded them to speak in the "I" voice. They were attempting to become the individuals they were assigned, and we told them that an easy way to begin their conversations was simply to ask what connection the other person had to coal or coal exports.

We used to call this kind of activity a tea party until the far right hijacked the term. Mixer, scavenger hunt, block party, barbecue, or tea party—whatever name they go by, these activities are lively and efficient ways to learn lots of information. They also help students appreciate an issue's multiple perspectives and disproportionate impact.

After the frenetic 40 minutes or so of students circulating in the classroom talking to one another as executives, fisher people, ranchers, and activists, we asked students to turn over their mixer worksheets and write about what they noticed: patterns they recognized, new things they learned, people they met whose stories shocked them, or questions raised by the activity.

Some focused on what they'd learned: "I didn't know that fish could get mercury poisoning." "It surprised me that a cattle rancher in Australia and a cattle rancher from the United States could be going through the same thing and global warming is affecting both of them." "I learned that many, many people are affected by coal exports. I thought it was a small amount of people, but clearly I was wrong." "So many people are going to be affected by a few companies that are just looking for wealth." And Anna wrote succinctly and accurately: "I learned that coal poisons everything it touches."

> "I learned that coal poisons everything it touches."

"Why Don't People Take Action?"

Two questions recurred in students' writing: one about people's supposed passivity, the other wondering why this is happening. Elena asked: "Why don't people take action if it is affecting them harshly?" Many took aim at the coal companies, posing questions with typical teenage simplicity and directness. For example, Maria demanded: "Why haven't they stopped mining it if it has such a bad effect? . . . Why are we mining for more coal if it's the biggest contributor to global warming?" The question is so starkly obvious that we should raise it at every opportunity. And Anna offered this troubling question that came up regularly in our broader climate change unit: "How can adults doom our generation?"

Frankly, I was puzzled and dismayed by the "Why aren't people taking action?" questions. Many of the coal export roles *did* feature individuals who were taking action: Gail Small of the Northern Cheyenne, whose people had defied energy companies for generations; Andrea Rogers, mayor of Mosier, Oregon, whose city council went on record against the potential coal trains, and who continues to speak out publicly; Henry Smiskin, tribal council chair of the Yakama Nation, who submitted a brilliant and comprehensive critique of the coal exports grounded in treaty fishing rights (Smiskin wrote to the U.S. Army Corps of Engineers: "They tell us these destructive plans are part of the march of progress. But their plans for progress have left a wake of destruction that has nearly eradicated our indigenous culture"); Lowell Chandler of Missoula, Montana, an activist with the Blue Skies Campaign, which works to block coal through his community using a range of tactics, including civil disobedience; and Xin Hao, the waterkeeper of the Qiantang River in Hangzhou, China, who works with other waterkeepers throughout the world to protect the integrity of rivers.

However, students had focused not on these individuals' activism, which perhaps was too thinly described in the roles I'd written, and more on people's denunciations of the potential or actual impact of coal. As Marisol noted, "Why aren't they doing anything to stop it? Complaining won't do anything."

I've made efforts to address this in the coal export mixer's most recent iteration, but the students' comments served as a reminder that young people are accustomed to bad news about the environment. And, no doubt, the news is bad—worse than most of us realize—but that is not the full story. Imaginative activism for climate sanity and clean energy is global, and the coal mixer activity is an ideal vehicle to introduce students to these hopeful nodes of defiance. But subtlety in these roles was not a virtue.

A Critical Eye on Pro-Coal Ads

Adam and I wanted to give students an opportunity to put their new knowledge to use evaluating a pro-coal export advertisement that appeared on the op-ed page of a recent Sunday *Oregonian,* Oregon's most-read newspaper. The ad for this particular export proposal, Ambre Energy's Morrow Pacific Project, is headlined "A coal project that fits with Oregon's values." I have to hand it to them: It's clever—a blatant appeal to what could, with some justification, be considered Oregonians' "green chauvinism." Apparently, all the other states can have the dust and fumes shoved down their throats, but Ambre promises "to meet Oregon's high standards" and do business "the Oregon way."

We distributed a copy of the ad to each student. (See www.rethinkingschools.org/earth.) I began by asking students to read the ad critically. Adam and I encouraged students to treat reading as a conversation or perhaps an argument with the advertisement, which was written as an open letter from Clark A. Moseley, the CEO of the Morrow Pacific Project. We told students to mark up the ad, underline sections, write in the margins, circle problematic ideas, and—especially—to stay alert to the ad's "silences" about aspects of coal exports they had learned about in the mixer. When they finished marking up the ad, we asked them to turn it over and to write about which aspects of the ad seemed reasonable and which seemed wrong or incomplete.

Students took to heart our encouragement to read rebelliously.

Students took to heart our encouragement to read rebelliously. In fact, some students decided to "decorate" Moseley's photo on the ad. Decorations or not, all students found important things to comment on. Almost everyone noted that Ambre's plans to export coal would lead to jobs—although, earlier in class, we also discussed the vast discrepancy between the 1,000 "ongoing operation jobs" promised in Ambre's ad and the *Oregonian*'s estimate of 50 "permanent jobs." Andrea acknowledged simply that, "Creating jobs will help Oregon's economy." Adam and I could have done more to encourage students to reflect on ways that coal exports might hurt jobs—indirectly through climate change, or directly through a doubling of barge traffic on the Columbia River, endangering salmon fisheries and recreational uses of the river. And lots of students also credited Ambre with at least some awareness of the dangers of coal dust and the company's promises to cover the barges to "minimize or eliminate dust." Students acknowledged that this was a worthy gesture, even if they were skeptical. Marisol wrote: "I like that he thinks he can eliminate the coal dust."

The obvious silence in the ad was the impact of coal on the climate. As Elena shouted in bold dark ink on her ad: "Global Warming!!!" On the reverse of the ad, Elena wrote that the coal "is still gonna get burnt and still affect us." Others wrote: "What happens in Asia?" "But they will still be using the coal." "Missing the fact about global warming."

Students got that Ambre was engaged in a shell game, drawing our attention to the company's plans to deal with the danger of dust, all the while ignoring the much greater danger of what happens when those many millions of tons of coal arrive at their intended destination and are burned. Anna expressed contempt at Ambre's appeal to Oregonians' pride: "It is blatantly obvious that they are trying to tickle us into becoming more open to their plot."

The success of the activity reminded me how eager students are to read carefully and critically when they begin with background knowledge and are offered encouragement to question, to challenge.

Dirty Business

Adam and I wanted to close out our brief look at coal and the climate with a film that could show coal's impact in visual terms. Despite an assortment of classroom-friendly films that deal with coal and mountaintop removal mining in Appalachia (see "Coal at the Movies," p. 216), I am not aware of any film that looks at coal mining in the Powder River Basin, the source of coal exports to Asia and where more coal is mined than anywhere else in the United States—or one that focuses especially on coal's impact on the climate.

We settled, with some ambivalence, on *Dirty Business*. The film's strength is that, of all the coal films we're aware of, it takes the most expansive look—examining the devastating impact of mountaintop removal coal mining; discussing the connection between coal and climate change; spending time in China, the world's largest coal consumer; and evaluating supposed alternatives, from carbon sequestration to energy recycling to wind power. Journalist Jeff Goodell's narrative is compelling, and the film—we used the shorter, 60-minute DVD version—held students' attention.

However, students' written comments and our class discussion helped us face the limitations of *Dirty Business*. If the problem is dirty business, as the film suggests, then the solution must be clean business. The alternatives featured in the film are mostly technical, initiated by innovative entrepreneurs. The subtext: We'll find our way out of this mess created by dirty coal by allowing the genius of the capitalist market to discover that profit and ecological responsibility go hand in hand.

Students tended to focus on the technological fixes, not on what we need to do, the *organizing* needed to change course. Eduardo wrote, "There are a lot of better alternatives like the wind turbines. Those seem like a great idea." It is essential for students to recognize that, from an energy feasibility standpoint, one does not need to burn coal to keep the lights on. And the film's strength is highlighting conservation measures and non-coal alternatives. But we didn't want students to think of alternatives just in technical terms—solely as different sources of energy.

Grassroots activism is also an alternative, and organizing against the spread of coal has made a huge difference. Because of the anti-coal agitation in the Northwest, Ambre Energy announced it was delaying its permit request for its train-to-barge-to-ship coal export scheme, complaining that permitting agencies were not treating coal as just another commodity. Thanks to numerous community meetings and demonstrations, impassioned testimony by hundreds of people at hearings, and thousands of letters sent to newspapers, legislators, and regulators, Ambre's coal was not being treated as if it were any old commodity whose owner was simply seeking passage to new customers. And, thanks to anti-coal activism, three of the original six coal export proposals were scrapped.

Nor did we want students to believe that there could be an easy alliance between climate crisis solutions and the capitalist market. The stakes are too huge to leave the fate of the Earth to the mercy of the profit motive. The current carbon reserves of coal, oil, and gas companies are five times the amount that—if burned—would push the climate past a two degrees Celsius increase. Humanity has little hope of a livable future on Earth unless we challenge the right of private capital to pursue wealth as it sees fit. These stores of fossil fuels are worth an estimated *$27 trillion.* And corporate executives are falling all over themselves to search out even more coal, oil, and gas. As the activist author/journalist Naomi Klein has written, "With the fossil fuel industry, wrecking the planet is their business model. It's what they do."

There is absolutely nothing in *Dirty Business* that extols the fossil fuel industry, but the film does not offer an eyes-open critique of capitalism or celebrate grassroots activism. For our class, it was a curricular choice with mixed results.

We were finished with our focus on coal, but not the climate. In later activities, we continued to probe the roots of the crisis—especially through a "trial" role play to determine responsibility for global warming. And we imagined hopeful responses through an online research activity where students encountered activist organizations focused on climate issues and evaluated their strategies and tactics. A "teach-a-friend" assignment helped students share their knowledge with others. And students participated in a final "making-a-difference" letter-writing project. All of these activities aimed to encourage students to begin to see themselves as changemakers.

At the very least, I am confident that students will never again think about coal as the benevolent little hunks of rock that fuel Thomas the Train. In fact, in her letter to an alternative newsweekly, Maria concluded: "Think about it, I mean it's nice to have Thomas and his friends come by, but is it worth risking our environment, health, and home?" ⊕

Bill Bigelow (bbpdx@aol.com) is curriculum editor of Rethinking Schools *magazine.*

Sample Role

**Marianne Adlington, Cattle Rancher
Barham, New South Wales, Australia**

My world has dried up. Global warming. Climate change. Whatever you call it, it's destroyed my way of life. My husband, Malcolm, and I used to have 500 dairy cattle. Now we have only 70 left. We have nothing to feed them. Because of the worst drought in Australia's history, we are living in a desert. And we can't afford to buy grain, so slowly our cattle are starving. I can't stand lying in bed every night and hearing the cattle bellow from hunger.

Malcolm and I are 52 years old, and for 36 years—since we were just 16 years old—we've been in the dairy business. It's all we know. Malcolm told me the other night, "I have absolutely nothing to go on for." He was crying. Farmers around here have moved away. Some have even committed suicide. And, frankly, I'm worried about Malcolm.

Nature didn't cause the climate to change. People did—spewing too much carbon dioxide (a "greenhouse gas") into the atmosphere. Of all the fuels, coal damages the climate the most, and Australia continues to burn coal and even export it to Asia. One Australian company, Ambre Energy, even wants to dig up more U.S. coal in a place called the Powder River Basin in Wyoming and Montana, and export it to Asia—which will make the climate even hotter and drier. These people are murderers, plain and simple. Why not just hand my husband a shotgun and tell him to go blow his head off?

Fortunately, groups in Australia like Quit Coal are trying to put a stop to burning coal for energy—and to ban coal exports. It may be too late to help my husband and me, but it's time to stop the coal criminals. ⊕

All the roles for this role play are available at www.rethinkingschools.org/earth.

ERIK RUIN

This Much Mercury. . .

How the coal industry
poisoned your tuna sandwich

BY DASHKA SLATER

Rich Gelfond keeps his Oscar statue in a black cloth sack in the bottom drawer of his desk. He received it as CEO of the film-technology company Imax, for "the method of filming and exhibiting high-fidelity, large-format, wide-angle format motion pictures," although when I read the inscription aloud, he feigns surprise, as if he's forgotten how he came to own it. "Is that what it's for?" he muses. "An Oscar's kind of like potato chips—when you have one, you need more. Kind of like tuna sushi."

Tuna sushi—and the devastating repercussions of Gelfond's onetime passion for it—has been the topic of conversation for the past hour, and Gelfond smiles slyly and a bit ruefully at his joke. With his round face, long blond eyelashes, and startled blue eyes, he seems placid but with an underlying current of energy, like a guinea pig who just drank a latte. For years he was an avid tennis player who also loved running around the Central Park reservoir or near his home in eastern Long Island.

But about six years ago he began to feel oddly off-balance, as though he might fall at any moment. He tried running on grass instead of asphalt but finally had to give it up altogether. It was probably stress, he reasoned, opting to stick to tennis. It was only when he nearly fell over while trying to serve that he decided to see a doctor.

More symptoms came to light in the physician's office. Gelfond had a tremor in his hands and had trouble putting his fingers together. A neurologist, worried that the symptoms pointed to a brain tumor, ordered an immediate MRI. But there was no tumor, and as he underwent a battery of increasingly unpleasant tests and scans, Gelfond's symptoms worsened. His feet tingled (a condition called neuropathy), and his balance became so off-kilter that it made walking difficult. "If I was with someone, I would walk close to them so, if I fell, I could grab on," he recalls.

Then, six months into his illness, Gelfond's neurologist asked a seemingly random question: "Do you eat a lot of fish?"

"As a matter of fact, I do," Gelfond replied. It turns out that he had been eating seafood at two out of every three daily meals as part of his healthy lifestyle. What he didn't know was that some kinds of fish—particularly the tuna and swordfish he favored—are high in mercury, a potent nerve poison. A blood test revealed that his mercury level was 76 micrograms per liter (mcg/L), 13 times the EPA's recommended maximum of 5.8 mcg/L. It was so high that he got a call from the New York State Department of Public Health, asking whether he worked at a toxic waste site.

"I was just so frustrated that I was trying to do something good for my body and in fact I was poisoning myself," Gelfond tells me, leaning forward in his chair. "I had no awareness."

Neither did most of the people he talked to, including physicians. To them, mercury poisoning was something that happened to the mad hatters of the 19th century or to the victims of industrial waste in Minamata, Japan, in the 1950s. It didn't happen to 21st-century New York executives. Having found the source of his illness, Gelfond's neurologist had no idea how to treat it, and when Gelfond contacted other New York physicians, most told him that mercury couldn't possibly be causing his symptoms because adults aren't susceptible to mercury poisoning.

"There has been a tendency to say adults are resistant," says Michael Gochfeld, professor of environmental medicine at Rutgers University's Robert Wood Johnson Medical School, which has treated a number of people who have gotten mercury poisoning from fish. "We don't really understand why some adults are sensitive and others seem to be quite tolerant."

The main prescription for Gelfond's mercury poisoning was to stop ingesting it. Once he elimi-nated high-mercury fish from his diet, his levels began to drop; within six months he was down to 18 mcg/L. One year after his diagnosis, he was able to walk without assistance. Six months after that, he was back to playing tennis. Today he says he's about 60 percent recovered. He still has trouble running long distances, and his symptoms resurface when he's fatigued. Mercury has changed his life forever.

Risky Eating

A few weeks earlier I'd spent the evening at a salon in Billings, Montana, watching stylists give people what amounted to very tiny haircuts. The Sierra Club was offering free mercury testing there and at other places around the country, and about 40 people had gathered at the Sanctuary Salon to provide hair samples for analysis.

The room was perfumy with hair products and buzzing with blow-dryers. Young women sat down to have a few strands of hair clipped close to the scalp, murmuring the usual things women say in salons: "I hate my hair!" and "Can you get rid of the gray?"

They also wanted to understand whether their own eating habits put them at risk. Tierani Bursett, 27, asked whether she should be concerned about the walleye she catches while ice fishing (she should). A local newscaster, who didn't want to be named, said she was "addicted to sushi" and wondered if she should be worried (yes). The mother of a 4-month-old wanted to know if mercury passes through breast milk (it does), and an older man asked whether mercury is a concern for people over 60 (it is). Luzia Willis, one of the salon's manicurists, was feeling nervous about all the tuna she buys at Costco (with good reason). "Why is there extra mercury in the fish?" she asked. "What's causing it?"

The answer can be found all around her: at the Colstrip power plant east of Billings, which uses a railcar's worth of coal every five minutes; in the coal mining operations to the east and southeast; in the long chain of railcars that chugs through town each night full of black ore bound for boilers across

> "I was just so frustrated that I was trying to do something good for my body and in fact I was poisoning myself."

the country; and at the J. E. Corette power plant right in town. Although there is always going to be some mercury in the environment—it occurs naturally in the Earth's crust and can be released into the air during forest fires or volcanic eruptions—70 percent of what we're exposed to comes from human activities, and most of that comes from burning coal.

U.S. coal-fired power plants pump more than 48 tons of mercury into the air each year. The Martin Lake power plant in Tatum, Texas, spews 2,660 pounds per annum all on its own (it burns lignite, a particularly mercury-heavy form of coal). Compared with the vast amounts of mercury churning out of Asia, the U.S. contribution is fairly small—about 3 percent of the global total. Roughly a third of our emissions settle within our borders, poisoning lakes and waterways. The rest cycles through the atmosphere, with much of it eventually winding up in the world's oceans.

Inorganic mercury isn't easily assimilated into the human body, and if the mercury emitted by power plants stayed in that form, it probably wouldn't have made Gelfond and many others sick. But when inorganic mercury creeps into aquatic sediments and marshes (as well as mid-depths of oceans), bacteria convert it into methylmercury, an organic form that not only is easily assimilated but also accumulates in living tissue as it moves up the food chain: The bigger and older the fish, the more mercury in its meat. It takes only a tiny amount to do serious damage: One-seventieth of a teaspoon can pollute a 20-acre lake to the point where its fish are unsafe to eat. Thousands of tons a year settle in the world's oceans, where they bioaccumulate in carnivorous fish. Forty percent of human mercury exposure comes from a single source—Pacific tuna.

> **U.S. coal-fired power plants pump more than 48 tons of mercury into the air each year.**

Developing fetuses are particularly sensitive to the toxic effects of methylmercury. Two out of three large-scale studies have found that children born with it in their system have trouble with coordination, concentration, language, and memory—and continue to have the same deficits many years later.

Nancy Lanphear is a behavioral developmental pediatrician who works at a clinic in Vancouver for children with disabilities like autism or attention-deficit/hyperactivity disorder. Several years ago, a mother came into her clinic with a 4-and-a-half-year-old girl who had cerebral palsy as well as speech and motor delays. But what attracted Lanphear's attention was that the child was drooling.

"I'm looking at this 4-year-old and saying, 'This is mercury,'" Lanphear recalls, hypersalivation being a classic sign of mercury poisoning. The child's chart showed that a heavy metals screening at age 2 had found high mercury levels in both mother and child, as well as in the child's grandfather. The mother recalled being encouraged by her physician to eat fish during her pregnancy; she ate tuna or other seafood two to four times a week, sure that she was helping her baby's development.

"She knows that she's not to blame, that it was inadvertent, but there's still some grief there," Lanphear says of the mother. "It's not something that's going away, even though the child's mercury levels are now normal. The damage was done to the developing brain." Lanphear uses the story to remind obstetricians and pediatricians to be on the lookout for mercury poisoning in their patients.

The EPA estimates that at least 8 percent of U.S. women of childbearing age have blood mercury levels above 5.8 mcg/L. If you zero in on communities that regularly eat fish, the prevalence is much higher. In the Northeast, one out of every five women has a mercury level exceeding the EPA threshold. In New York City, it's one out of every four, and close to half of the city's Asian population has elevated mercury levels, as does two-thirds of the city's Chinese born outside the United States.

High-mercury pockets also exist on the West Coast. Between 2000 and 2001, San Francisco physician Jane Hightower tested 116 patients who said they frequently ate fish. She found elevated mercury levels among 89 percent of them, with half above 10 mcg/L. Many of these patients had reported nonspecific symptoms like headaches, nausea, depression, and trouble concentrating, and had been searching for an explanation for months or years.

Since that first survey, Hightower has treated hundreds of mercury-exposed people from all walks of life. Among her patients was then-5-year-old Sophie Chabon, the daughter of Pulitzer Prize-winning novelist Michael Chabon and Ayelet

Waldman, whose books include the best-seller *Bad Mother*. Sophie had been an early talker and walker, but then she seemed to hit a wall, suddenly unable to sound out words she used to know how to read and even forgetting how to tie her shoes. A blood test turned up mercury levels of 13 mcg/L. The culprit: twice-weekly tuna sandwiches.

As Sophie cut tuna out of her diet, her mercury levels dropped, and her stalled development surged ahead again. Now in high school, she has a passion for history, film, and French and shows no sign of any lasting effects from the mercury exposure. Still, Waldman fumes when she thinks about what might have happened if they hadn't caught the problem so early. "I blame our country for not [caring] about what we're spewing into the atmosphere," she says. "This is about coal, pure and simple. You wouldn't go and break your child's bones one by one, but we tolerate this kind of poison that's ruining their minds. It's insane."

Although Hightower's wealthy patients tend to eat sushi and expensive tuna, swordfish, and halibut, poor Americans eat canned light tuna—often subsidized by the federal Women, Infants, and Children nutrition program—and fish they hook themselves in local rivers, lakes, or bays. Immigrants are particularly likely to fish for food, often without understanding the risks of eating their catch. The average Latino angler, for instance, consumes twice as much mercury daily as the EPA considers safe, and a 2010 study of subsistence fishing in California found that some anglers were getting 10 times that dose. The same study found that anglers with children had a higher mercury intake than those without, probably because families with more mouths to feed rely more on food that can be caught rather than bought.

The boardwalk in Ocean City, New Jersey, is a 2.5-mile strip of salt air and stimulation, with arcades and carnival rides, pirate-themed mini golf, and fried clams. It was here that a young woman named Jaime Bowen stood in front of a microphone in June and nervously contemplated the crowd. A 31-year-old home healthcare worker with two children, Bowen had gone to a Sierra Club-sponsored

hair-testing event with an environmentally minded friend a month before, more as a lark than out of any real concern for her health. "It was kind of a joke going to get my hair clipped," she says. "Then, to get the results—it was a reality check."

Of the 36 people at the event who were willing to share their results, eight had elevated mercury levels. Bowen was one of them. Hers was 1.37 ppm—too low to cause health problems, but higher than the EPA considers safe for women of childbearing age. (Hair mercury levels are evaluated differently than blood mercury levels, but a hair level of 1.2 ppm is roughly equivalent to a blood level of 5.8 mcg/L.) Now she was concerned about her two children, who, after all, ate what she ate. "You hear, 'Don't break that thermometer.' You never hear about the fish," she says. "I made my kids tuna fish sandwiches the other day, and now I feel horrible. Tuna fish—it's just one of those things you wouldn't think to be scared of."

And so Bowen stood at the podium, gripping the paper that held her prepared remarks. She talked about fish and her fears about her children's safety, and about coal. To her surprise, she looked up to see that people up and down the boardwalk had stopped to listen. "I did want them to know," she says. "I'm just a regular person—I'm not doing anything different from those people."

Behind her, the ocean sparkled, sending salty breezes drifting over the boardwalk. A seagull circled, white and gray, its bright eyes scanning the scene below: the crowded boardwalk, a fish-filled sea, and, tucked in a bay just a little to the northwest, the lighthouse-shaped smokestack of the B. L. England generating station, producing 450 megawatts of electricity, powered by West Virginia coal. ●

Dashka Slater is a regular contributor to Sierra. *Her website is dashkaslater.com; she tweets @DashkaSlater. This article was funded by the Sierra Club's Beyond Coal campaign.*

 See teaching ideas for this article, page 262.

BURNING THE FUTURE: Oil

GORD HILL, KWAKWAKA'WAKW

I mages of oil-covered pelicans, sludge-soaked beaches, billowing black smoke, water on fire, and underwater plumes filled TV screens for 87 days. Reporters interviewed irate fishermen, their boats dry-docked and their lives upended. People across the United States were appalled and angry. So were my students.

The 2010 Gulf of Mexico oil spill coincided with my first year teaching modern world history at a high school in Portland, Oregon. The year was about to end, and I was just wrapping up a unit on World War II. U.S. intervention in Latin America was next—and time was running out. I hadn't taught students about energy policy or modern-day environmental issues, so when I woke up that Tuesday morning to the unfolding drama, I felt unprepared to help them make sense of the spill. The only

Environmental Crime on Trial

Students assess blame for the BP oil spill

BY BRADY BENNON

thing I had the time or energy to do was project live video streams of the underwater camera for a few minutes during daily warm-ups so we could follow BP's attempt to "kill the spill."

The BP disaster turned out to be the worst oil spill in U.S. history, with more than 200 million gallons of oil released into the ocean and more than 16,000 miles of coast contaminated. Some news stations compared the spill to the Exxon Valdez incident of 1989. I was an 8th grader at the time and I remember being stunned by the images of people in white jumpsuits spraying hot water as they tried to clean coastal rocks. I couldn't focus on my classes for a week. None of my teachers addressed it.

"It's awful and so, so, sad," Lisbeth wrote in her warm-up journal after watching the live feed one morning. "All those turtles and dolphins and birds." Other students echoed her sentiments.

Some, like Zach, focused their ire on the most obvious culprit, the company from whose well the poison gushed: "It'll probably take years for BP to clean that mess up."

> As the school year ended, I regretted that my minimal instruction on the oil spill had done little more than contribute to students' sense of sadness and helplessness.

I knew I had not done enough to deepen Zach or Lisbeth's understanding of the spill beyond the sound bites that flowed from pundits' mouths. Of course BP was culpable. But what about the regulators who green-light projects like these every year? What about consumers who purchase and drive gas-guzzling SUVs? How about the capitalist system that demands consumption and rewards greedy natural resource speculation and extraction? As the school year ended, I regretted that my minimal instruction on the oil spill had done little more than contribute to students' sense of sadness and helplessness.

I want my students to understand the historical roots of injustices and I want them to explore the social movements that have addressed those issues. But even the most social and environmental justice-oriented curriculum feels irrelevant if it does not connect to today's world. After all, historical roots grew into the tangled thorns and brambles we face today, and kids need the vision and skills to take on today's challenges.

The oil still spewed as the summer began, and I vowed to build real curriculum around this pressing issue. It wasn't too late to teach about oil spills. In fact, every year, dozens of oil spills pollute our oceans, rivers, and lands, as corporations' search for oil and fossil fuels increases. Learning about the Gulf oil spill can help students think about the relationships among people, government, business, communities, and the environment. Until people understand those relationships and their dysfunctions, we'll continue to see these horrific scenes play out year after year.

At the Oregon Writing Project's Summer Institute, I teamed up with Amy Lindahl, a biology teacher from a neighboring Portland high school. We decided that a trial role play would be the best way to help students explore the causes of the spill. We would turn our classrooms into courtrooms to help students grapple with this guiding question: Who or what is responsible for the devastation caused by the Gulf of Mexico oil spill? We designed the role play activity and wrote detailed "indictments" against BP, the U.S. government, the people of the United States, the people of the Gulf of Mexico, and global capitalism.

The next fall, I taught the role play as part of a mini-unit at the beginning of a schoolwide teach-in about the environment. I opened the lesson by showing images of oil-covered beaches, seabirds, mammals, and wetlands; and underwater footage of the oil billowing into the ocean. Without explaining the context, I asked students to describe what they saw and any feelings that arose.

Joan wrote: "In this slide show, I see oil, oil being pumped out of the ground, oil leaking into the water, covering and killing animals. Polluted beaches. Destroyed jobs, hobbies, and lives."

Kevin wrote: "A lot of fun activities are gone. The signs say 'the death of summer, the death of fishing and crabbing and playing on the beach.' What do they do now? What if this happened here in Portland?"

After the students read their responses to partners, I asked the students what they knew about the past summer's oil spill in the Gulf of Mexico. Students who had watched clips on TV explained the basic facts about the spill in deep water off the Lou-

isiana shore. Then I asked students: "Who or what was at fault?" All but four of my students said "BP" or "the oil company."

"Today and tomorrow," I told them, "this classroom will be transformed into a courtroom. In this role play, I will be the prosecutor and you will be the defendants. You are all charged with the crime of damaging the ecosystem and economy of the Gulf of Mexico. Before we start, it's important to know more about the damage the oil spill and the toxic chemicals sprayed on the oily water caused."

I organized students into six groups of five. I gave each member of the group a paragraph containing an excerpt from an article about the effects of the spill. They took turns reading the paragraphs out loud, and the group determined if it referred to damage done to human health, the economy, marine life, or natural scenery. For example, Ellie read an excerpt from Louisiana resident Margaret Curole's interview in Terry Tempest Williams' *Orion* article, "The Gulf Between Us."

> The [cleanup] workers are getting sick with contact dermatitis, respiratory infections, nausea, and god knows what else. The BP representatives say all it is is food poisoning

or dehydration. If it was just food poisoning or not enough water, why were the workers' clothes confiscated? As we say in these parts, Answer me dat!

Marco read a paragraph from the same article about the damage to marine life:

> We had a very small pod of sperm whales in the Gulf, nobody's seen 'em. Guys on the water say they died in the spill and their bodies were hacked up and taken away. . . . Dolphins are choking on the surface. Fish are swimming in circles, gasping.

From the article excerpts, students learned that two years after the spill, fish in the Gulf had open sores, mysterious black lesions, and parasitic infections; and that people who subsist on wild-caught seafood were ingesting dangerous chemicals. Commercial fishing dropped by 20 percent, and the tourism industry suffered massive economic hardship in the years following the spill. When marsh plants died from oil pollution, roots gave way and sediment could no longer withstand the tides that eroded miles of natural storm barriers into the ocean.

"You Are Charged…"

After students shared and classified the paragraphs, I explained that each group represented a possible plaintiff—an entity that might have some responsibility for the spill. I distributed indictments to each group—who represented BP, the U.S. government, the people of the United States, and global capitalism (see www.rethinkingschools.org/earth for the full text of the indictments). The first paragraph of each indictment is the same, and I read it aloud:

> You are charged with causing the oil spill in the Gulf of Mexico. Through carelessness and self-interest, you have set a disaster in motion that is killing this region. Because of your irresponsible actions, vast regions of the Gulf are turning into a dead zone; ecosystems and endangered species will be destroyed. Human communities, jobs, and economies will be devastated.

I explained that the groups would need to read their roles, identify information most damaging to their group, and build a defense. Each member of the group was required to present at least a portion of the case, so they would all be involved. I reminded them that sometimes the best defense is a good offense, so they should accuse at least one other group during their testimony. Any group accused during testimony could ask one rebuttal question of the defendants, so they should anticipate the types of questions they might be asked, and prepare to address them with evidence from their roles.

> **"You are charged with causing the oil spill in the Gulf of Mexico."**

The buzz of reading aloud filled the room. "You allowed companies to write their own rules. . . . Some of your agency employees are former oil executives," read Jasmine, representing the U.S. government. Her group let out a collective moan. "Oh, man," blurted Michael. "We might as well save the time and just put our group in jail."

"What are you complaining about?" countered Monica, from the BP group, sitting in a nearby cluster of desks. "We've definitely got the hardest role."

I pointed out to the students that the roles contained possible defenses that might help them figure out their arguments. I also handed groups a complete set of the roles in case they wanted to research other groups' vulnerabilities. I told them: "Lawyers always read and research carefully. Good lawyers turn that research into strong arguments." That was the encouragement they needed. Some groups divided up the research. Others used their cell phones to investigate more details about the spill online. All students were keenly engaged.

For the remainder of the period, students filled out the graphic organizers with defense arguments and accusations. They made lists of questions they could ask other groups if those groups attacked them. Finally, after dividing up the main points, students each wrote a paragraph for the opening statements. Students who did not finish completed it as homework.

The next day we began the trial. Our jury members—an instructional specialist and another teacher—entered the room carrying notebooks and wearing "Jury" signs. I've found that when students have authentic audience members, they often rise to the occasion with better performances and more careful arguments.

To kick off the trial, I re-read the common indictment and key passages of the first indictment, the one against the people of the United States:

> It's your job to learn all you can and push your elected representatives to regulate corporations. You could have joined with your neighbors to organize and pressure the government to reduce its use of fossil fuels. You could have demanded more fuel-efficient cars, but instead you drive and purchase gas-guzzlers.

One by one, members of that group made their case. Jared's analogy was effective: "Yeah, we're addicted to oil. We can't help it. Blaming us for that is like blaming a baby for being born to a meth-addicted mom. Of course we get addicted. The government and the capitalist system keep teaching us to stay addicted."

Juan added: "The president doesn't talk about the problems of oil consumption. Isn't it the government's job to educate the citizens?"

Because they were attacked, I allowed the members of the government group to respond with questions. Charlotte asked: "Why are you blaming us? Don't we live in a democracy? Aren't you, the people, in charge?"

Kate responded: "It's hard for people to make the right decisions when the government and BP don't tell us what's going on."

These exchanges were not always orderly, and they didn't resemble an actual court of law. At first, I was uncomfortable with this dynamic and frequently pounded my makeshift stapler gavel. Eventually I realized the sometimes-raucous discussion was leading to a deeper understanding of the issues. My students were arguing over resource consumption, democracy, capitalism, and corporate ethics. They were demanding clean energy and government oversight. They could barely remain in their seats as they grappled with these issues. We were going way beyond the simplistic "shame on BP" analysis from the year before.

One by one, the groups took the stand. The people of the Gulf of Mexico tried to explain how tough it is to say no to companies like BP when oil jobs help them stay afloat. Plus, weren't they helping keep the country from having to get oil from Iraq? Students representing capitalism claimed companies like BP were abusing the system. Doesn't the government create the system? Noah argued, "If people bought more green technology, the system would make more green technology." The government continued pointing to the trade-offs, including the need to have a strong economy and the need for the people of the United States to become more informed consumers. Shouldn't more people vote with their wallets?

The BP group used an "everything-but-the-kitchen-sink" approach. Why didn't the government enforce the law? Isn't the company just doing what companies do under capitalism—try to make money? Don't people want low prices?

After all the groups had their say, they had one minute to read closing remarks, summaries of their strongest arguments. Jamila, representing the U.S. government, explained why we should not let BP off the hook:

When we let companies like BP drill for oil, we trust them to follow safety rules and do a good job. BP is to blame because they betrayed our trust with carelessness and greed. Since 2007, they've had 760 major safety violations, while they mislead the American public with their "beyond petroleum" [branding]. They're not beyond petroleum and we shouldn't let them be beyond the law. We must make them pay.

In his follow-up to Jamila's statement, Mike showed his understanding of the connections at play:

Ladies and gentlemen of the court, we, British Petroleum, plead partially guilty. But the high demand for oil from the American people forces us to set up rigs in dangerous places. Doing so leads to safety violations, but the government never stopped us. We're a company and our only responsibility is to please our shareholders. That's a lot of pressure. Please go easy on us since everyone in this room brought on this tragedy.

A Broader View

I sent the jury into the hall to deliberate and asked students to step out of their roles as defendants and to step back into their roles as students. They were to consider the evidence presented in the trial and then write about whom they thought was ultimately responsible. I encouraged them to assign a percentage of the blame to each of the defendants and explain their thinking. While they wrote, I played Mos Def and Trombone Shorty's song about the oil spill, "It Ain't My Fault," and Johnny Cash's "Don't Go Near the Water" for background music.

> "Yeah, we're addicted to oil. We can't help it. Blaming us for that is like blaming a baby for being born to a meth-addicted mom."

Before the jury delivered its verdict, I called on several students to share their thoughts. Many students listed BP as the primary culprit, but most students also included the U.S. government and the people of the United States as sharing some of the blame. Justin read from his response:

BP is directly at fault for their carelessness, but so are the American people. We consume so much oil by driving big cars and buying food shipped long distances. Our demand causes BP to drill in hazardous and risky places. The government allows companies to do this even though they knew BP had such a poor record of performance.

Despite the feisty arguments made against people of the Gulf of Mexico, students consistently ranked them as the least at fault. This wasn't surprising, given the amount of time we spent learning about the health, economic, and environmental damage suffered in the region.

Capitalism consistently scored low on students' blame scale. Most students thought it was just too hard to convict a system for the crime. Kenyon wrote: "Capitalism is only about 5-10 percent guilty. It's a system controlled by people and the government. It's an idea. Yeah, it makes people greedy, but people don't blame the system for that until something goes wrong."

> Overall, the trial helped deepen students' understanding about the oil spill and it pushed them to see this disaster as a complex problem rather than as a simple story of corporate irresponsibility.

"Did anybody choose global capitalism as the most to blame?" I asked. Not a single student raised a hand. "Why not?"

Ellington responded: "Because people are responsible for doing something about the problems of the world, not the system. Capitalism is about profit, but people make those rules. Yeah, sometimes it's messed up, but people interpret rules for themselves, so people have to be the ones responsible."

Leah added: "Like if a driver crashes a car and kills a person, you can't blame the car. You can blame the driver, but not the car."

"But what if the metaphorical car was dangerous because it was made with shoddy materials because the car company wanted to maximize profits?" I asked. "What if the road didn't have proper stop-

lights because of cost concerns? Surely, systems and social structures that encourage reckless behavior own some blame for many of our world's problems."

With the end of class looming, I invited the teachers on the jury to announce their decision. They found the people of the Gulf of Mexico 10 percent at fault. The capitalist system received 15 percent guilt. The U.S. government was given 20 percent of the blame. The people of the United States were assigned 25 percent of the guilt. Finally, the jury announced that BP shoulders most of the blame, 30 percent, for taking risks and for deceiving the public.

Final Project

The next day, to conclude the mini-unit, I asked students to complete an action project that could display at the school's upcoming sustainability fair. The class split into two groups.

John had recently returned from a national global warming conference, Power Shift, where he heard green jobs activist Van Jones speak about the damage petroleum can wreak even if it gets safely to shore. For example, plastic bags and containers kill birds and take hundreds of years to decompose. John persuaded half the class to organize a campaign to support a local effort to ban plastic bags.

Students read and printed short articles about the issue. Then they created and printed postcards urging Portland's mayor to support the "Ban the Bag" initiative. At the fair, they handed out the articles and asked attendees to sign postcards. The other group of students decided to put together a poster with images from the Gulf oil spill. They chose quotes to strengthen the impact of the images and included poetry they had written earlier in the unit.

At the fair, my students took shifts staffing their displays and talking with the teachers, students, and community members. They were passionate and urgent. Many, like Karen, who is normally shy and withdrawn, beckoned strangers to their table to talk about the spill. Taking this lesson outside the classroom gave students a sense of agency and hope. I didn't want them to experience despair, because despair leads to hopelessness, which only reinforces the behaviors that got us to the oil spill in the first place.

Next Time

Overall, the trial helped deepen students' understanding about the oil spill and it pushed them to see this disaster as a complex problem rather than as a simple story of corporate irresponsibility. The trial format, and the assigning guilt, helped generate intellectual tension that hooked students into the role readings and discussion. It led to high engagement.

But the format and my facilitation were also limited. Students remained shallow in their analysis of capitalism's role in the spill, and they showed only a vague sense of how systems can perpetuate or even encourage harmful behaviors and decisions. However, Ellington's point about people being ultimately responsible for the rules of the game was an indication that the students felt they had a role to play. As Robert Reich said in a *Democracy Now!* interview: "The economy is not something out there, it is not kind of a state of nature. The economy is a set of rules. . . . And if our rules are generating outcomes that are unfair . . . we change the rules." These are ideas I wish I had pressed with my students. One of my aims was to get them thinking more systemically, but I was unprepared to facilitate a deeper discussion.

When I teach this in the future, I want to have students spend time after the trial reading articles that make more explicit connections among BP, the U.S. government, and global capitalism. I would have students draw concept maps or make metaphorical drawings to give them a chance to demonstrate their understanding of those connections.

And, unfortunately, I know I *will* be teaching this in the future. The impact of offshore drilling is only becoming more dire. And the regulatory response lags far behind the need.

For example, after a massive oil spill off Santa Barbara in 1969, Congress and the president drafted and signed into law several powerful modern environmental laws, including the Clean Water Act and the National Environmental Policy Act. In contrast, several years after the Gulf of Mexico oil spill, no major new policies have been passed, and people in the Gulf are still wrangling to get BP to pay for damages from the spill. In the real-life litigation, the defendants are limited to BP and the oil contractors who built and operated the wells. The people of the United States, the U.S. government, the people of the Gulf, and the global capitalist system are not on trial. But, at least for a while, my students got to see a more nuanced reality, complicated by the interconnections of corporations, consumers, government agencies, economic systems, and folks hungry for jobs. At least, for a day, my students took action on behalf of people, clean water, and healthy ecosystems. 🌐

Brady Bennon teaches history and literacy at Madison High School in Portland, Oregon. Students' names have been changed.

"We Know What's Goin' On"

Louisiana's Cajun shrimpers face dire threats

BY TERRY TEMPEST WILLIAMS

Kevin is working on his daughter's motor scooter, taking it apart in the middle of the sidewalk. I can't help but stare at the extravagantly colored tattoo on his back, a narrative needled and inked on flesh that depicts Godzilla standing on a shrimping boat battling other boats, with oil rigs looming in the background. He gets up, catches my eyes on his back, and shakes my hand. "It's a helluva good story if ya wanna hear about it."

Margaret and Kevin Curole are Cajun shrimpers. They have lived along the bayous in Galliano all their lives. Today, they are staying at their daughter's place in New Orleans, adjacent to a large cemetery. It's beyond humid and the searing heat leaves me drenched. Margaret has agreed to talk to us about the Gulf crisis as both a resident of the region and an activist who serves on the executive board of the Commercial Fishermen of America. She also serves as the North American coordinator of the World Forum of Fish Harvesters and Fish Workers, an NGO that works with the U.N.'s Food and Agriculture Organization to protect the rights of fishing communities around the world.

"It *is* a good story," she says, smiling at Kevin. She has a flower tattoo on her right breast showcased by her low-cut black T-shirt. "Let's get a couple of chairs and sit out back." Her dark, layered hair, shoulder length, accentuates her yellow-brown eyes. "Are you cool enough today?" she asks, smiling.

On May 16, 2010, Margaret Curole joined aerial artist John Quigley and sent three text messages, spelled out with human bodies on the beach in Grand Isle, Louisiana, to BP, the federal government, Congress, and other officials, calling for immediate action to address the economic and environmental devastation from the spill. Their message was simple and direct: *Never Again; Paradise Lost; WTF?!*

This last sentiment is where Margaret picks up with our conversation. "Did you see that there's another spill today, a barge hit ground off of Port Fourchon, not far from Grand Isle? That's in the Lafourche Parish where we're from." Margaret is referring to headlines in the *Daily Comet*: "New Oil Spill Sullies Locust Bayou Near Border of Terrebonne, St. Mary."

"About 500 gallons of light crude. It's the second spill this week in southeast Louisiana," she says. "It's endless and ongoing all over the world. I'm on my way tomorrow to a conference in Norway to talk about the state of fisheries and oil spills. Part of my job with the U.N."

Margaret tells me that her father was an oilman. In the 1950s, before she was born, her parents lived inside the British Petroleum compound in Saudi Arabia. "I was adopted. My birth mother was Cajun. I'm Cajun. The transaction was completed for the price of $500 and two new dresses for my mother. My parents are dead now, but I've lived in the same house in Galliano for 50 years."

"And your husband?" I ask.

"My husband has shrimped all his life, until the local fishing industry collapsed in 2000. Ask him about separating shrimp from a bucket for his grandmother when he was 3 years old. It's in his blood. He was fishing those waters as a kid. Loved it. Lived for it. We all did. It's how we raised our daughter. You know why he quit in 2000? 'Cuz he was feelin' violent—violent toward the government, violent for them not valuing an honest day's work. He just left what he loved and went and worked for oil. At least we were one of the ones who had options."

Margaret explains to us how the local shrimping industry has crashed in the bayous since 2000, due to America "dumping" Asian shrimp into the market. "Our shrimp aren't worth anything, certainly not worth all the effort that goes into harvesting them. My husband used to sell a pound of shrimp all cleaned up and put on a bucket of ice for $7. Then, after the Asian shrimp came in all covered with white blight and crowded out our own southern Louisiana shrimp, he'd get paid under a dollar. They treat our shrimp like trash. It's not just the money, it's our dignity. The ability to work hard is at the heart of Cajun culture."

"We are one generation removed from those speaking French, although Kevin still speaks the dialect. What you need to understand is that for us Cajun folk, fishing isn't a business, it's a way of life. It's something beautiful. We may be poor, but we never went hungry. We had shrimp, crabs, and coon oysters. We had a free and abundant food supply. In these parts, you either fish or you work in the oil fields. So if you take away the oil job, with the moratorium on deep-well drilling, and the fishing is gone, we're down to nothin.'"

Margaret's fast speaking clip slows down. "And then you've probably already heard about the dead zone in the Gulf of Mexico created by all the dumping of pesticides from farming—the nitrates from farms upriver?" She pauses. "My sense of hope is fading fast."

> "It's not just the money, it's our dignity. The ability to work hard is at the heart of Cajun culture."

She looks away and then her gaze becomes direct. "Don't believe 75 percent of what you hear about this blowout down here. Ask the people on the ground. People are not being allowed to talk. My husband has been working on the water for the past three months. Most of what is being done to clean up the oil is to make the American people think something is being done."

"So what's the story that isn't being told?" I ask.

"Two things: how much oil actually has gone into the sea and the amount of dispersants used to make it disappear," she says.

"The workers are getting sick with contact dermatitis, respiratory infections, nausea, and god knows what else. The BP representatives say all it is is food poisoning or dehydration. If it was just food poisoning or not enough water, why were the workers' clothes confiscated? As we say in these parts, Answer me dat!

"I never really got nervous until I got a call at 9:30 on a Sunday night from the BP claims office telling me to back off. But I'm speaking out. I kid my friends and family and say I'll leave bread crumbs. The other day, two guys from Homeland Security called to take me to lunch. I'm a chef. They tried to talk food with me, to cozy up and all, and one of them told me he was a pastry chef." Margaret shakes her head. "But I knew what they was up to, I'm not stupid. They just wanted to let me know I was bein' watched."

> "I don't feel like an American anymore," Margaret says. "I don't trust our government. I don't trust anybody in power."

"Here's the truth," Margaret says, now emotional. "Where are the animals? There's no too-da-loos, the little one-armed fiddler crabs. Ya don't hear birds. From Amelia to Alabama, Kevin never saw a fish jump, never heard a bird sing. This is their nestin' season. Those babies, they're not goin' nowhere. We had a very small pod of sperm whales in the Gulf, nobody's seen 'em. Guys on the water say they died in the spill and their bodies were hacked up and taken away. BP and our government don't want nobody to see the bodies of dead sea mammals. Dolphins are choking on the surface. Fish are swimming in circles, gasping. It's ugly, I'm tellin' you. And nobody's talkin' about it. You're not hearing nothin' about it. As far as the media is reportin', everythin's being cleaned up and it's not a problem. But you know what, unless I know where my fish is coming from, I'm eatin' nothin' from here."

Margaret and I sit in silence. I am suddenly aware of the shabbiness of the neighborhood, the cracking paint on the wooden slats, the weariness of the ivy in this dripping heat.

"I'm sorry," she says. "I haven't cried in a long time. I've been tough, I've been holding it all together, but it breaks me up." She looks at me with unwavering eyes, "Have you read 'Evangeline' by Longfellow?"

I can't speak.

"Read it. Read it again," Margaret says to me. "It's our story as exiles. If I wasn't speakin' out about this, I'd be havin' a nervous breakdown. I'll tell you another thing that nobody is talkin' about. At night, people sittin' outside on their porches see planes comin' into the marshes where they live, and these planes are sprayin' them with the dispersant. That's the truth. But hey, we're Cajuns, who cares about us?"

"I don't feel like an American anymore," Margaret says. "I don't trust our government. I don't trust anybody in power."

She leans forward in the heat as the pitch and fervor of frogs intensifies. "We might not be the most educated people schoolwise, but we know more about nature than any PhD. We know. We know what's goin' on."

Terry Tempest Williams is the author of Refuge: An Unnatural History of Family and Place, An Unspoken Hunger: Stories from the Field, Desert Quartet, Leap, Red: Passion and Patience in the Desert, The Open Space of Democracy, *and* Finding Beauty in a Broken World. *She also writes a column for* The Progressive *magazine.*

See teaching ideas for this article, page 262.

Which country is the No. 1 supplier of oil to the United States? Saudi Arabia? No. Iraq? No. Russia? No.

My geography students were surprised to learn it's my country—Canada. Most of that oil comes from the tar sands located in northern Alberta, in an area roughly the size of Florida. The Alberta tar sands are home to the world's second largest deposit of oil, after Saudi Arabia. They are also the source of controversy, seen by some as "Canada's greatest treasure" and others as "Canada's greatest shame."

As a Canadian teaching at an international school in New York City, I had long been planning to teach about the tar sands. Although many people are aware of the devastating effects of BP's oil spill in the Gulf of Mexico or Shell's human rights abuses in Nigeria, few know about the enormous environmental and social injustice caused by oil extraction just to the north. The perfect opportunity to teach about this issue arose in 2011 when the media began focusing attention on the proposed TransCanada Keystone XL Pipeline. If approved, this pipeline would bring as much as 700,000 barrels a day of crude oil from Alberta's tar sands to refineries in Texas.

Who would benefit most from the pipeline? Who would suffer? What would be the pipeline's long-term effects on the environment, economy, and on local communities? Should we support or resist increasing the capacity of the tar sands? These were the questions that I wanted my geography students (most of whom come from relatively privileged backgrounds) to consider.

To confront these questions, I wrote a role play where students take on the characters of six key stakeholders invited to an imaginary public hearing, chaired by then-U.S. Secretary of State Hillary Clinton (played by myself), to discuss whether or not the State Department and the president should approve the Keystone XL Pipeline.

I introduced the role play by providing students with some basic information and photos of the tar sands and the proposed pipeline. The handout I distributed also explained the political context:

> This project is unique in that it does not have to go through Congress. Because the Keystone XL Pipeline comes from Canada, it is a foreign project and foreign projects don't need approval from Congress. They need approval

NICHOLAS LAMPERT

Dirty Oil and Shovel-Ready Jobs

A role play on tar sands and the Keystone XL Pipeline

BY ABBY MAC PHAIL

from the State Department. So the State Department has to decide whether this pipeline is in the U.S. national interest. If they decide that it is, then the president has the last word.

Once my students were clear on the situation, I wrote the six different roles on the board and explained a little about each group before allowing students to write their names under the group they would like to represent. Groups were limited to five people, so once a group had five names, students had move to their second or third choice. I explained that the roles were based on information found on the organizations' websites and in media interviews given by their representatives. Here is a summary of each role. Full roles can be found at www.rethinkingschools.org/earth:

> "TransCanada is lying when it says that the pipeline will be safe."

TransCanada: Our company builds the infrastructure that transports energy throughout North America. As the proposed constructor of the pipeline, we have come to the public hearing to clear up any confusion surrounding the Keystone XL Pipeline. We want to assure the secretary of state and all those attending the hearing that the pipeline will be safe, will not cause damage to communities or the environment, and will provide thousands of jobs at a time when millions of Americans are unemployed.

American Petroleum Institute: API represents and speaks on behalf of 480 oil and natural gas companies. We support the Keystone XL Pipeline and call on the president to fulfill his promise to create thousands of jobs by approving this project immediately. Let's face it, the Canadian tar sands are already being developed, and the Keystone Pipeline will allow the United States to secure the energy we need from a friendly and reliable trading partner: Canada. Tar sands oil is a source of ethical oil—unlike Saudi Arabia, which exported 400 million barrels of oil to the United States last year. Why should we trade with

a country that does not allow its women to vote, drive, or even leave the house without a man? By getting our oil from Canada, we will stop funding the oppression of women.

Republican Party: We represent members of the Republican Party. We support the Keystone XL Pipeline because it is a shovel-ready, multibillion-dollar project that will create thousands of jobs. Extreme environmental groups are opposed to the pipeline because they say it will damage the environment and cause global warming. They are flat-out wrong. Pipelines are the safest way to transport oil. Many in our party believe that global warming is an unproven theory and that we need to stop wasting our time preparing for something that may not even be real, and start providing jobs for the American people.

Environmental Activists: We are opposed to the Keystone XL Pipeline and to the development of the Alberta tar sands. In fact, we were among the thousands of activists who protested outside the White House for two weeks in 2011 and later circled the White House to tell

the president to reject this pipeline. This pipeline will be "game over for the climate." We must dramatically reduce carbon emissions. We must decrease the capacity of the tar sands, not increase it. President Obama keeps saying that he wants to do something about climate change but Congress won't let him. Well, Congress has no say in this project, so here is the president's chance to prove that he is serious about protecting the environment and reducing climate change.

Indigenous Environmental Network: Our network works to protect the environment and build sustainable communities. We are opposed to the tar sands and to the Keystone XL Pipeline. Don't believe the oil companies when they say that "everything is fine" in northern Alberta or that tar sands oil is ethical. Tar sands oil is destroying our cultural heritage, ecosystems, and health. The First Nations communities surrounding the tar sands experience dangerously high rates of cancer. If you were to see with your own eyes how lives are being sacrificed for oil money, you, too, would be completely opposed to this project.

Bold Nebraska: We represent people from all walks of life and political parties who are opposed to the Keystone XL Pipeline. If constructed, it would cross through the Nebraska Sand Hills, a fragile ecosystem where a lot of our cattle are raised. It would kick farmers off land that their families have farmed for generations. The pipeline would cross the Ogallala Aquifer, our cleanest source of water, which we use not only for drinking, but also for Nebraska's main economic activity, agriculture. We are concerned about leaks. TransCanada is lying when it says that the pipeline will be safe. We know that there were 12 spills in 12 months on TransCanada's Keystone I Pipeline. We call on the president to follow through with his promises of healing the planet and reject this project.

Preparing for the Role Play

After students selected the organizations they would represent, I asked them to gather in their groups and spend the remaining class period preparing for the public hearing. I gave the groups their roles and asked them to first read silently before working to-

As students summarized arguments in their own words, I circulated among them, clearing up confusion and helping to build confidence in expressing their main points. I urged them to provide more details for arguments and, when appropriate, to bring in their knowledge of climate change and sustainability, two topics we had just covered. After they wrote their arguments, they introduced themselves to the other members of their group and explained why they were for or against the pipeline. By their third run-through, most were confident in their ability to articulate their positions.

Public Hearing: Should the President Approve the Pipeline?

The role play began the next day. We arranged the tables and chairs in a circle, and students found their places behind placards identifying the different groups. In my role as secretary of state, I welcomed all stakeholders to the meeting and assured them that both the State Department and the president were keen to hear their concerns and would take them into account when making the final decision on whether or not to approve the pipeline.

Each group made a brief introductory statement, presenting their group to the class and stating whether or not they were in favor of the pipeline. Once everyone was clear on the stakeholders present and their positions, I opened the floor to comments and questions. I asked for a show of hands from those who would like to begin and started a speakers' list on the board. At any time, students could raise their hand to be added to the list. We basically went in sign-up order, but I sometimes allowed groups to respond to comments directed at them.

Using the words of former NASA climate scientist James Hansen, Barry began the debate by stating his opposition to the pipeline, saying it would be "game over for the climate." He described how burning fossil fuels contributes to climate change and talked about the need to develop alternative forms of energy. He was backed up by fellow environmental activist Allan, who took on the role of environmentalist Bill McKibben. He described how he had

gether to create a list of their main arguments for or against the pipeline. Once they identified three to five arguments, I asked them to appoint one member of the group to be an expert on each of the arguments and to summarize that argument in their own words. The experts' task would be to explain their argument during the group's introductory speech, raise it with other groups during the role play, and be prepared to defend that position when it was challenged by another group.

For example, the TransCanada group identified several arguments in favor of the pipeline: safety, job creation, energy security, and TransCanada's policy of respecting local cultures and communities. The group decided that Jason would be the expert on the safety of the pipeline and would be responsible for defending this point whenever someone raised the issue of leaks or environmental damage. Henry's task was to emphasize the number of jobs the pipeline would create and accuse anti-pipeline groups of being "job killers." Homer would stress the energy security that the pipeline would provide to the United States, and Kelly would defend TransCanada's policy of respect for local cultures. Dividing the main points up in this way ensured that all students had a role and prevented the most vocal members of the group from dominating the debate.

created an organization called 350.org because 350 parts per million is the safe amount of carbon dioxide in the atmosphere.

This position was met with great opposition from the Republican Party. Timothy reminded his fellow Americans that there was "no proof at all that human activities, such as burning fossil fuels, cause global warming. We should not make any decisions on things that have not been proven." After this statement several more students raised their hands to be added to the speakers' list.

Next Molly from the American Petroleum Institute thanked Republicans for their support, saying she wished everyone had "as much respect for the oil and natural gas industry," based on the millions of jobs and millions of dollars the industry provides in the United States. She also talked about the 20,000 jobs that the pipeline would create.

Heather, from the Environmental Activists group, challenged Molly on the jobs number, pointing out that it seemed to change day by day. She added that if the president was really serious about creating jobs, he would invest in clean energy.

After a few more comments on the issue of jobs, the subject turned to the safety of the pipeline. Siena, from Bold Nebraska, questioned the route of the pipeline, through the Ogallala Aquifer and Nebraska Sand Hills. If the pipe leaked, drinking water and farming would be threatened. Jason, from TransCanada, assured the group there was no need to worry about leaks. Using a line straight from TransCanada's website, he stated that "the Keystone XL Pipeline will be the newest, strongest, and most advanced pipeline in operation in North America." He explained: "We will monitor our pipeline, using satellite technology, 24 hours a day, 365 days a year. If there is a leak, we will get to it right away."

Kara, from the American Petroleum Institute, followed: "Would you rather get your oil from Saudi Arabia where women are oppressed and can't even vote or drive cars? Or would you rather get your oil from our friendly neighbor Canada? This oil is ethical oil. Even Canada's minister of the environment agrees."

Next it was time for the Indigenous Environmental Network to reply. "There is no such thing as ethical oil," Peter began. "Let me tell you about how the tar sands affect the lives of the Indigenous peoples who live downstream. Our water has been polluted, our forests have been cut down, and we are dying of cancer because of this oil. Cancer rates increased 30 percent in Fort Chipewyan between 1996 and 2005. My town of 1,200 people has five cases of a very rare cancer whose rate is usually one per 100,000."

The role play continued for the remainder of our double period. When it was time to end, I thanked everyone for attending the public hearing, assured them I would bring all of their comments and concerns back to the president, and asked them to stay tuned in the following days for the decision.

In our next class, I asked students to take 10 minutes to list all the arguments for or against the pipeline, based on what they learned during the role play. Once they had written down as much as they could remember, I showed video clips of recent media interviews with many of the characters they had played. These included environmental activist Bill McKibben of 350. org, Canadian writer and activist

> "There is no such thing as ethical oil."

Naomi Klein, Cindy Schild of the American Petroleum Institute debating Jane Kleeb of Bold Nebraska on *Democracy Now!*, and Republicans Rick Perry, Mitt Romney, and Ted Poe. The final clip showed Canadian protesters and members of the Indigenous Environmental Network handing out goody bags of fake tar sands oil at the Durban Climate Summit. Hearing these ideas again, this time from the participants themselves, helped to cement them in my students' minds. They also found it amusing to see the characters they had played. Next, I told the students, I would ask them to share their own opinions on the pipeline.

Where Do You Stand?

To debrief the role play and encourage students to explore their personal opinions on the issue, I used a four-corners activity. I hung signs—"strongly agree," "agree," "strongly disagree," and "disagree"—in the four corners of the classroom. "I'm going to read some statements," I told them. "Listen carefully and go to the corner of the room that best represents your point of view. I'm going to call on volunteers to share opinions and to respond to the comments of others."

Then I read the statements one by one:

- People are suffering from lack of jobs in this country. We should build the pipeline and think about its consequences later.
- The pipeline is not sustainable.
- We should trust TransCanada when they say the pipeline will be safe and therefore we don't need to worry about leaks and damage to the environment.
- Tar sands oil is more ethical than Saudi Arabian oil.
- President Obama should approve the Keystone XL Pipeline.

I like this activity because it allows students to physically show where they stand on an issue and ensures that even the quieter students make their views known. It also allows for change. Sometimes, upon hearing classmates' comments, students changed their minds and silently walked to another corner of the room.

Letters to the President

As a follow-up assignment, I asked my students to write letters to President Obama, stating their opinions on the pipeline. Three students thought the pipeline should be built. Sheila, Jason, and Henry felt that letting the oil rest under the earth would be a waste and it should be used to support our economy. But the majority of students advised Obama not to approve the pipeline.

Their letters were their most sincere, thoughtful assignments so far. This, I believe, was due to the authenticity of the assignment, but also because the role play allowed them to empathize with those who suffer the social and environmental consequences of tar sands oil in a way that simply reading about it in the *New York Times* could not. Many of them told Obama about their geography class role play. Some used their characters' language of activism and critique:

> **The role play allowed them to empathize with those who suffer the social and environmental consequences of tar sands oil in a way that simply reading about it in the *New York Times* could not.**

Homer: In our role play, I played a TransCanada representative. When I read my role, I was in favor of the pipeline. But when I heard what the environmentalists and Indigenous people had to say about it, I changed my mind. I saw that the oil commercials are a lie.

Stacy: For the role play, I chose to be in the Republican Party group and realized how difficult it was for me to fight for something I do not believe in. Although I think the Republicans might be right that this will help the economy, this will only be for a short period of time. I believe the environmentalists when they say that this pipeline will be game over for the environment, and game over for the environment means game over for us.

Peter: The land surrounding the tar sands rightfully belongs to the Indigenous people. Increasing the tar sands will keep raising the rates of cancer in this area. The Indigenous people will be forced to move away from the land their families have lived on for generations or face cancer. Putting anyone in this situation is not ethical. Tar sands oil is not ethical oil.

My favorite moment came a few weeks after we completed the unit. In class, we had continued to monitor the ongoing demonstrations and activism around the proposal to complete the Keystone XL Pipeline. None of us knew the fate of the pipeline—and writing in the summer of 2014, we still don't—but it was clear to students that the protests had made this one of the key environmental issues of our time. People had made a difference. In a conversation about the impact of the activism around the Keystone XL Pipeline, one of my students, Heather, ended our class discussion. As she packed up her books Heather said: "The next time somebody says that protests or Occupy Wall Street is a waste of time, I'll tell them about the Keystone XL Pipeline."

Abby Mac Phail teaches high school humanities at an international school in New York City. All of the students' names have been changed. This role play was modeled on "Oil, Rainforests, and Indigenous Cultures," written by Bill Bigelow and published in Rethinking Globalization.

BURNING THE FUTURE:
Natural Gas and Fracking

R. BLACK

You can think of fracking as a hostage exchange program. A drill bit opens a hole a mile deep, turns sideways, and then, like a robotic mole, tunnels horizontally through the shale bedrock for another mile or more. The hole is lined with steel pipe and cement. To initiate the fracturing process, explosives are sent down it. Then, freshwater (millions of gallons per well) is injected under high pressure to further break up the shale and shoot acids, biocides, friction reducers, and sand grains deep into the cracks. Trapped for 400 million years, the gas is now free to flow through the propped-open fractures up to the surface, where it is condensed, compressed, and sent to market via a network of pipelines. The water remains behind.

Actually, only some of the frack water stays behind in the shale. The rest, now mixed with brine and radioactivity, shoots up to the surface with the gas. Finding a safe place to dispose of this toxic flowback is an unsolved problem. Sometimes, the waste from drilling is just dumped on the ground. That's illegal, but it happens. Sometimes the waste is dumped down other holes. In 2010, 200,000 gallons were poured down an abandoned well on the edge of Allegheny National Forest. Much of the flowback fluid is trucked to northeast Ohio, where it is forced, under pressure, into permeable rock via deep injection wells. This practice, the Ohio Department of Natural Resources has concluded, is the likely cause of the unusual swarm of earthquakes that shook northeast Ohio in 2011.

—from Sandra Steingraber, "The Fracking of Rachel Carson: Silent Spring's Lost Legacy, Told in 50 Parts," *Orion* magazine.

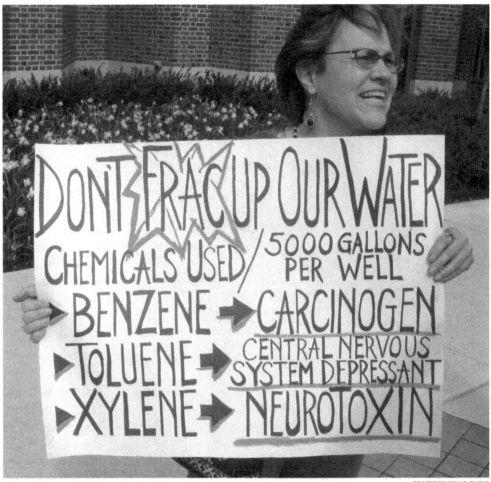

Teaching About Fracking

In the end, we're all downstream

BY JULIE TREICK O'NEILL

My cousin works in the natural gas fields of the Bighorn Basin, just outside of Worland, Wyoming.

It's a living-wage job, one of the few nonprofessional positions available in the area, and much easier than the work as a ranch hand she had been piecing together. If she decides to have children, their education will be fully funded. According to *Education Week*, Wyoming spends an average of $16,386 per student annually, the highest in the country. Thanks to the Hathaway Scholarship, funded since 2006 by federal mineral royalties, postsecondary study is virtually guaranteed to students willing to work hard and attend one of the state's community colleges or the University of Wyoming, my own alma mater.

Natural gas is king in Wyoming. It fueled the $437 million budget surplus for fiscal year 2011, when enough was pumped out of the tens of thousands of acres of natural gas fields to fuel one in three homes in the United States. Groups such as the American Gas Association, the Natural Gas Supply Association, and America's Natural Gas Alliance do their best to sell natural gas as an abundant, domestically available, and clean energy source. And most people seem to buy it. Maybe because they don't see the actual impact of production. Or perhaps they choose not to see because of the short-term benefits it provides in places like in my home state.

Today, I live and teach in Portland, Oregon, but I still consider Wyoming home. My recent visits to Wyoming are like snapshots—before and after images—of the changes brought about by natural gas production since the 1990s: When I was a child, we went to Jackson's slightly shabby main square to watch the stagecoach robbery re-enactment that was one of the highlights of the summer tourist season; now the shoot-out occurs in front of luxury stores.

Natural gas production in Wyoming is a lucrative and influential industry. All the perks— jobs, education spending, the budget surplus—are huge incentives for communities and the fossil fuel-friendly state legislature to play along, to stay silent. And many are willing. "The drilling is going to happen anyway," I've heard people in Wyoming say, "so the state might as well get the benefits."

There has been natural gas production in the state for almost a century, but gas, like all fossil fuels, is getting more difficult to find. As long as our country remains dependent on coal, oil, and gas, producers will utilize whichever techniques they can—including hydraulic fracturing—to meet the demand.

Gasland

I showed the film *Gasland* to my 9th-grade global studies classes at Lincoln High School as part of a larger unit on global climate change. I wanted students to look critically at our current energy system, which relies almost exclusively on fossil fuels and to begin to explore lower-carbon options—everything from specific alternative energy sources such as nuclear, wind, and solar power to market-based initiatives such as cap and trade, carbon offsets, and efficiency incentives. I wanted students to understand that every energy source has trade-offs. For example, harvesting the power of the sun has cloudy days, startup costs, storage issues, and, these days, the environmental impact of importing solar panels

from China, as well as the impact of manufacturing solar panels.

But I also wanted students to recognize that not all trade-offs are equal. Natural gas is a prime example, and *Gasland* was an engaging and shocking film for my students. Looking at hydraulic fracturing—"fracking"—was part of a broader exploration of what the energy industry is doing now that easy-to-find fossil fuels have run out. Fracking itself is a deadly serious issue, and the search for "extreme energy" also symbolizes our energy future, as long as our country is wedded to fossil fuel.

Before viewing the film, we watched a video clip titled *Facts About Natural Gas: the Natural Choice*, produced by a regional gas company in the Midwest. (This video is no longer available online, but any number of natural gas ads would work.) As they viewed the video clip, I asked students to tease out the claims supplied by the industry and write them as statements. For example: "natural gas is abundant," "natural gas is domestically available, ending our dependence on foreign sources of energy," "natural gas is clean."

> Fracking itself is a deadly serious issue, and the search for "extreme energy" also symbolizes our energy future, as long as our country is wedded to fossil fuel.

Students appeared to take all of these claims at face value. We had just completed in-depth studies of oil and coal, and the general consensus seemed to be that "clean-burning" natural gas had to be better than those.

"It makes you feel relieved," wrote Duncan, "that there are ways to be clean, it uplifted everyone." To my students, the message was clear and believable; after all, they'd seen many of the same kinds of ads from our local gas company, Northwest Natural: clean, safe, efficient, abundant.

"But how do we know these are facts?" I asked my students. "What more would you need to know to be certain of your answer? Add those questions to your list." I wanted to draw my students' attention to the claims made by the natural gas producers about the benefits of natural gas, so they could figure out for themselves what they needed to know in order to evaluate those claims.

"While you watch *Gasland*," I directed my students, "answer the questions listed on your paper. You may come up with more questions while you watch. We'll take time to discuss the claims after the film is finished."

Trading Money for Access

Gasland begins with a most basic trade-off, money for access. In 2006, Josh Fox, the filmmaker, was offered $100,000 for the rights to drill on his Pennsylvania family land, located on the Marcellus shale formation. This formation, "the Saudi Arabia of natural gas," stretches from Virginia into Ohio, Pennsylvania, and New York. Money for privately held natural gas seems simple enough. But there is a catch, a more insidious trade-off that the natural gas producers would like people to ignore. In 2005, with the help of my fellow Wyomingite, then-Vice President Dick Cheney, Congress created a loophole in an energy bill that exempts the natural gas industry from many of the environmental regulations that could hamper gas production: the Clean Air Act, the Clean Water Act, and the Safe Drinking Water Act. Now the trade-offs get much more serious, and this is what Fox details in his film: What exactly are you in for if you—or your state—lease land for natural gas production?

The film follows Fox as he investigates the industry and its practices—specifically fracking. Fox begins with his neighbors and moves across the country.

Fox was a hit with my students: He was real enough, cool enough, and smart enough to take on fracking, a process that—using up to 2 million gallons of water and 80,000 pounds of "proprietary" (secret) chemicals per well, injects fluid into formations like the Marcellus, causing enough pressure and disruption of the Earth's crust to release the natural gas trapped in the rock. Over and over during the film, students turned to me and to each other with looks of disbelief. Thomas, my perpetual critic, was enraged: "Where does that stuff go?" Minutes later, he finds out: 70 percent of fracking fluid stays in the ground, even though 65 chemical compounds used in fracking are known to be hazardous to humans, and those are just the ones we know about.

Story after story in the film exposed the truth about natural gas production: contaminated water wells; fracking disruptions that knock volatile gases such as methane free, creating "flammable" water taps in people's homes; the seeping of neurologically damaging fracking solvents into watersheds; toxic air pollution; mass deaths of fish and other water life in streams.

According to Fox:

Gas drilling and hydraulic fracturing is an inherently contaminating industrial process that injects millions of gallons of water, laced with toxic chemicals, at enormous pressure, to break apart rock and release gas from underground formations. Watersheds across the nation have been contaminated with plastics, carcinogens, neurotoxins and endocrine-disrupting chemicals—and with explosive natural gas. It causes massive land scarring, air pollution, a public health crisis, massive truck traffic, miles and miles of pipelines, blowouts, spills, accidents.

So, for example, Pinedale, close to the largest natural gas production site in Wyoming, spends more money on children than any other school district in the nation, plus providing bonus checks for teachers and new facilities. But it is one of the most dangerous places in the country to breathe the air, and the drinking water puts everyone at risk. What good is it, asks Fox, if the children aren't safe living there?

After we finished *Gasland,* students used their original facts/claims (from the natural gas ad), questions, and answers to write about natural gas and its trade-offs. Sydney summarized class opinion:

In the first commercial we watched, the company told us that natural gas is abundant, and judging by what I've seen in *Gasland,* that seems true. That's probably the only part that was. . . . The rest of the "facts" were created to lead the U.S. people into a false sense of security. . . . I know that retrieving natural gas is not efficient, it is not clean, and it is definitely not safe.

This is not exactly in line with the public relations message of the gas industry, and the industry has attacked the film repeatedly. A Google search for *Gasland* leads to all sorts of industry PR videos and disclaimers. But this search also takes you to Fox's website where he, along with environmental activists, educators, chemistry and biochemistry professors, an engineer, and a retired EPA official provide detailed information on "The Truth Behind *Gasland.*"

The film and these additional resources are an antidote to the smoke, mirrors, and very real money behind the industry. And, despite the devil's bargain the industry has pushed, all of Wyoming is not hoodwinked. Wonderful environmental groups such as the Powder River Basin Resource Council, based in my hometown of Sheridan, are actively trying to shape the energy and environmental future of the state. And people who live downwind or downstream certainly know what's going on as they struggle to get clean water or stay healthy despite the impact of fracking.

But in the end we're all downstream—and that was my purpose in showing this film. The health and environmental costs are too high, the trade-offs too costly, to see natural gas, or any fossil fuel, as a viable energy source. My students understood this. As Duncan said: "When we watched *Gasland* I believe everyone became scared and felt that there was no hope. But what is better? Knowing the real-life scary facts, or just to be ignorant and easily persuaded?" My students were now in a perfect place to begin to investigate alternatives—just where we need to be as a country. ⊕

Resources

Gasland. Directed by Josh Fox. Docurama, 2010. gaslandthemovie.com.

Julie Treick O'Neill teaches social studies at Lincoln High School in Portland, Oregon.

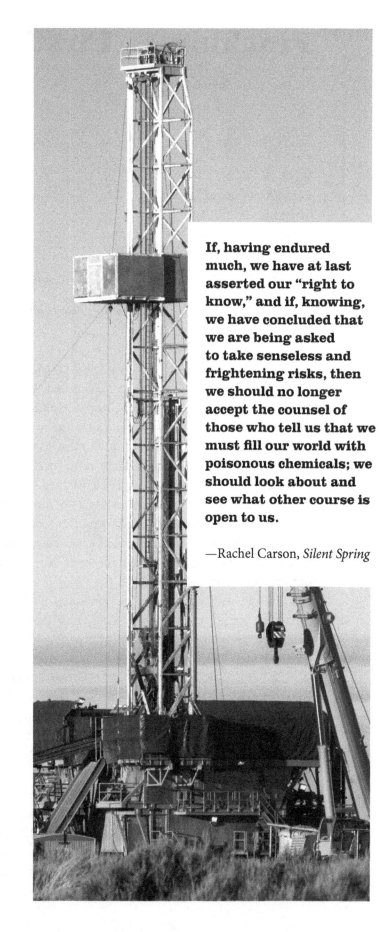

If, having endured much, we have at last asserted our "right to know," and if, knowing, we have concluded that we are being asked to take senseless and frightening risks, then we should no longer accept the counsel of those who tell us that we must fill our world with poisonous chemicals; we should look about and see what other course is open to us.

—Rachel Carson, *Silent Spring*

Fracking . . . Firsthand

A Letter to the Illinois General Assembly

May 28, 2013
Illinois General Assembly
Governor Pat Quinn
Attorney General Lisa Madigan
State House
Springfield, IL 62706

Dear Governor Quinn, Attorney General Madigan, and Members of the Illinois General Assembly,

We write today to urge you not to allow high-volume horizontal fracturing ("fracking") for oil and gas in Illinois. We, the undersigned residents of Pennsylvania, are among the many victims of fracking. Informed by extensive firsthand experience with the oil and gas industry and suffering from the impacts of fracking, we implore you with the greatest sincerity to protect the health and safety of the people of Illinois and swiftly enact a moratorium on fracking. We have learned the hard way that regulations—no matter how strict they sound on paper— do not provide adequate protection to human health or property, especially in tough economic times when the state agencies charged with enforcing the regulations are understaffed and underfunded. Also, regulations cannot prevent accidents, and this is an industry prone to accidents of an especially frightening nature and whose effects are not temporary.

The oil and gas industry promises that fracking is safe and that it will create jobs and bring your state riches, but Pennsylvania's experience in the past five years tells a very different story. In short, water contamination has been widespread; our air has been polluted; countless individuals and families have been sickened; farms have been devastated, cattle have died, and our pristine streams and rivers have turned up dead fish; only a fraction of the promised jobs and revenue for the state have come to fruition; and our communities have been transformed into toxic industrial zones with 24/7 noise, flares, thousands of trucks, and increased crime. What's more, the jobs have made many workers so sick that they can no longer work in the industry.

A week ago, the Scranton Times-Tribune revealed that oil and gas development from fracking damaged the water supplies of at least 161 Pennsylvania homes, farms, churches, and businesses between 2008 and the fall of 2012, as indicated by state Department of Environmental Protection (DEP) records. The Times-Tribune notes that this number is not comprehensive; an exhaustive analysis was made impossible by DEP's lack of transparency, poor record keeping, potentially inadequate testing procedures, and lack of cooperation with the investigation. Regardless, with around 4,000 wells drilled during that four-year time span, these 161 cases show how common and extensive water contamination is from fracking operations. These numbers are not surprising given the high rate of well casing failures. By the gas industry and the DEP's own data, well casing failure rate in Pennsylvania is 6.2 percent (rising to 8.9 percent in 2012). Failures occur when the layers of cement and steel that encase the well—providing a barrier between the toxic fracking fluid and freshwater aquifers—are damaged or become corroded. Even with the most careful workmanship cement can shrink, crumble, and crack as it ages.

Because the chemicals used in fracking operations are highly toxic, water contamination is a very serious problem. Although the industry blocks attempts to know what chemicals and combinations are used, we know that it is a cocktail whose ingredients are selected from a possible menu of around 600 chemicals. Those include many known carcinogens and endocrine disruptors. They include chemicals such as benzene, toluene, hydrochloric acid, and petroleum distillates. In addition to the chemicals used by the industry, the operation releases many hazardous materials from the shale itself, including radium, uranium and radon, arsenic, and mercury. Cows that have consumed water contaminated with used fracking fluid (flowback waste) have quickly died, and land where it has spilled has been scorched.

For us, fracking has been a public health disaster. Victims experience symptoms ranging from headaches, dizziness, burning eyes, sore throats, rashes, hair loss, severe nose bleeds, nausea, blood poisoning, liver damage, intestinal pain, neurological damage, cancers, and many more. Many fracking victims who have suffered these health symptoms sign legal agreements that force them to forfeit all rights to speak about what has happened to them in order to settle with multinational oil and gas corporations. Although many cases have been hidden from the public eye through these nondisclosure agreements, we have compiled a "List of the Harmed" that now well exceeds 1,000. Our efforts to create this lay registry of health problems is an attempt to compensate for the legally enforced silence of our medical community. After extensive lobbying by the oil and gas industry, the Pennsylvania Legislature passed Act 13, which, among other things, places a gag order on doctors who deal with victims of fracking and who wish information about the possible chemicals to which their patient may have been exposed.

The Southwest Pennsylvania Environmental Health Project—an initiative of medical experts—is working with Pennsylvanians affected by fracking and has concluded that health impacts are serious and that we still do not have enough scientific data to make an informed decision or to be able to claim that ANY regulations will protect public health.

One major, uncontrollable problem is hazardous air pollutants, which are emitted from wellheads themselves, as well as from flares, dehydration devices, compressor stations, and the thousands of diesel trucks that are needed to service each well. Silica dust—a known cause of lung cancer and silicosis—is also a problem in and around drilling and fracking operations. We live with the knowledge that our children are breathing in hazardous air, and are left to wonder what and how severe the ramifications will be in their future.

Our environment has been transformed seemingly overnight from beautiful countryside and farms into toxic, heavy industrial zones. Commutes that used to take 30 minutes now take two hours because of the truck traffic. Many of our schools and playgrounds are blanketed in carcinogenic silica dust. Towering flares light up the night sky, while health-damaging levels of noise penetrate our homes 24/7. Only a small fraction of the promised jobs and revenue have materialized, with most jobs going to out-of-state workers and most revenue accruing to only a few individuals. Meanwhile, the community has had to pay for road and bridge damage, increased accidents and need for more emergency workers, and we've had to live with increased crime rates.

In addition to the water contamination, air pollution, industrialized communities, increased crime rates, and ruined farms, we've also experienced countless spills, blowouts, and disasters. Communities have been evacuated because of explosions and uncontrolled leaks and fires.

As we have experienced the horrors of fracking firsthand for years, we have also carefully followed the industry in other parts of the country and watched the science that has emerged. We have followed what is happening in Illinois with great dismay. We are certain that your proposed regulations will not protect the health of Illinois residents, your farms, communities, environment, and everything that makes Illinois special. Please, do not make this mistake.

If you allow fracking to go forward as planned, you will bring to your state the same horrific experiences we have suffered in Pennsylvania. The oil and gas industry cannot and must not be trusted. We implore you to enact a moratorium in order to take the time to visit areas with fracking, bring scientists and medical experts into the process, and undertake an environmental and public health study. This is the only responsible course of action, and far too much is at risk to do otherwise. We would be glad to speak with you, and we invite you to our homes and communities to see fracking and its impacts firsthand.

Speaking on behalf of a broad network of communities, sincerely,

RON GULLA, Hickory, Pennsylvania
ADAM HEADLEY, Smithfield, Pennsylvania
DAVID HEADLEY, Smithfield, Pennsylvania
GRANT HEADLEY, Smithfield, Pennsylvania
LINDA HEADLEY, Smithfield, Pennsylvania
RAY KEMBLE, Dimock, Pennsylvania
JENNY LISAK, Punxsutawney, Pennsylvania
MATT MANNING, Montrose, Pennsylvania
TAMMY MANNING, Montrose, Pennsylvania
RANDY MOYER, Portage, Pennsylvania
VERA SCROGGINS, Silver Lake Township, Pennsylvania
CRAIG L. STEVENS, Silver Lake Township, Pennsylvania

See teaching ideas for this article, page 262.

The Environmental Protection Agency's (EPA's) Hydraulic Fracturing Public Informational Meeting was probably the strangest exhibition of performance art ever to grace the stage of the Broome County Forum Theatre in Binghamton, New York.

Over the course of two days, a panel of EPA officials heard 400 two-minute presentations by members of the public who had come to advise the agency, at its own invitation, on how it should design a scientific study. As ordered by Congress, this study will investigate the risks to drinking water posed by the Johnny-come-lately technology known as *high-volume slick water horizontal hydrofracturing,* which does to shale bedrock what mountaintop removal does to an Appalachian mountaintop: blows it up to get at a carbon-rich fossil fuel trapped inside.

In the case of fracking, the quarry is methane bubbles trapped inside impermeable layers of shale thousands of feet below the Earth's surface. To liberate the gas, millions of gallons of freshwater *(high-volume)* are mixed with sand and chemicals—some of which are carcinogens—and this slippery mixture *(slick water)* is forced, under immense pressure, into mile-long tunnels drilled sideways *(horizontal)* through bedrock. With the assistance of explosives, this poisonous solution shatters the shale *(hydrofracturing)* and releases a vaporous froth of petroleum, euphemistically known as natural gas, which floats up the borehole—along with brine, radioactive materials, and heavy metals.

So, last September in Binghamton, some 400 members of New York's citizenry signed up to express their particular views on the question of how one might go about studying the environmental impacts of this sort of energy extraction. The EPA panelists sat in chairs on the commodious stage of this tattered-but-grand former vaudeville house, while, one by one, each preregistered citizen advisor approached a podium in the orchestra pit and

Fracking Democracy

A two-day spectacle carved into two-minute chunks

BY SANDRA STEINGRABER

"Most of the state's fracking operations are set to take place in Pennsylvania's forests. To be precise, 64 percent of Pennsylvania gas wells are to be drilled in forested land, which includes state forests and natural areas. For each well pad sited in a forested area, an average of nine acres of habitat are destroyed, says the Nature Conservancy's Pennsylvania chapter (each well pad can accommodate up to six wells). The total direct and indirect impact is 30 acres of forest for each well pad. This does not include acreage lost to pipelines. On average, each well pad requires 1.65 miles of gathering pipelines, which carry the gas to a network of larger transporting pipelines."

—Sandra Steingraber, "The Fracking of Rachel Carson: Silent Spring's Lost Legacy, Told in 50 Parts," *Orion* magazine.

offered up opinions. After 120 seconds, the microphone turned off automatically, ending the presentation of a sometimes still-talking, still-gesticulating petitioner.

Then the next person on the roster was called to the mic. And then the next. And then the next. For four solid hours. And then the panelists took an intermission and came back for another four-hour round of two-minute testimonies. And then there was a second day of speeches.

For members of the audience, who could see only the back of the speaker as he or she addressed the onstage panel, the sole visual element was a giant digital timer projected onto a screen behind the panelists that ticked backward, second by second, from two minutes to zero, making the parade of speeches a cross between speed dating and a NASA countdown.

After my own 120 seconds of counsel—during which time I (rapidly) advised the EPA to consider revisiting its own prior investigation of PCBs in the Hudson River, at least some molecules of which seeped into the water through naturally occurring fissures and hairline cracks *(79 seconds; talk faster)* in the shale bedrock beneath General Electric's factory floor, migratory pathways not previously known or even thought possible—I had plenty of time to listen to the other presentations.

Because the EPA had signaled a possible willingness to expand the scope of its study to consider cumulative impacts, the pro-drilling contingent was on the defensive. One after the other, the self-identified "landowners"—which seemed to be code for "people who believe that the federal government should not get between a man and his gas lease"—urged the EPA to "restrict inquiry" and "resist the temptation" of more deliberation.

Back in the cheap seats, I practiced sympathy for this position. What would it be like, I asked myself, to view scientific inquiry as med-

FRACKING FOR NATURAL GAS CAN CAUSE EARTHQUAKES. FRACKING EXECS LOOK ON THE BRIGHT SIDE.

Shaken or stirred, sir?

Need you ask?

dlesome dithering? As someone who, in other circumstances, has argued that the time for action had arrived, I could almost understand the impatience of those who viewed fracking as a bold enterprise rather than complete lunacy.

But, soon, the repeated calls for expediency were followed by dismissive comments about water, and whatever empathy I might have felt for the opposition vanished. One man intoned rhapsodically, "Energy is life," and then added with a smirk, "Water is a resource." I thought that maybe I had heard it backwards, but then he repeated his assertion again, with even more sanctimony: "Energy is life; Water is a resource." It felt like a Monty Python drop-the-cow kind of moment, but, alas, no cows fell.

And then came the untruths. The millions of gallons of freshwater used by gas wells during fracking operations are exceeded, claimed one petitioner, by the leaks in the New York City water system. They are exceeded by the water used to irrigate golf courses claimed another. Huge amounts of water are wasted doing all kinds of things.

A geologist friend and I looked at each other in wonderment, and in my head, I began to imagine a 120-second rebuttal. It would go like this: Fracking constitutes consumptive water use, which is different from what happens to water when underground pipes leak and water re-enters the aquifer, or when irrigation leads to evaporation and cloud formation. When water is entombed in deep geological strata, a mile or more below the water table, it's permanently removed from the water cycle. As in, forever. It will never again ascend into the clouds, freeze into snowflakes, melt into rivulets, cascade over rocks, turn with the tide, soak into soil, rise through roots, or pour from your tap. It will never again become blood, tears, sweat, urine, milk, sap, nectar, yolk, honey, or the juice of a fruit. It will never again float a leaf boat, swell a bud, quench a thirst, fill a swamp, spill over an edge, slosh, dribble, spray, trickle, splash, drip, or glisten. Never again fog, mist, frost, ice, dew, or rain. It's gone. To conclude: Fracking turns freshwater into poison and makes the water disappear. That's something we've not done before on a large scale. And by the way, water is life. It's energy that's a resource.

An older man rose to speak. He announced he had a special presentation. And then he let 10 seconds of silence fill the theater while, before him, the monumental numbers projected on the screen blinked away.

After hours of ceaseless, rapid-fire speech, the sudden hush flowed through the overheated room like cool water. Someone giggled nervously. And then, finally, he spoke. That silence, he announced, represented the sounds of migratory birds. And tourists. And professors. And organic farmers. And thus with no words at all he reminded the audience of all the good members of our beloved community who would—if our land filled up with drill rigs, waste ponds, compressor stations, and diesel trucks—disappear, exit the cycle. As in, forever.

04. 03. 02. 01. Mute. And then he sat down. ⊕

> Fracking turns freshwater into poison and makes the water disappear. That's something we've not done before on a large scale.

Sandra Steingraber is an ecologist, biologist, author, and cancer survivor. She is the author of Living Downstream (Da Capo Press, 2010).

See teaching ideas for this article, page 263.

Life and Death in the Frack Zone

BY WALTER BRASCH

José Lara just wanted a job.

A company working in the natural gas fields needed a man to power wash wastewater tanks.

Clean off the debris. Make them shining again.

And so José Lara became a power washer for the Rain for Rent Co.

"The chemicals, the smell was so bad. Once I got out, I couldn't stop throwing up. I couldn't even talk," Lara said in his deposition, translated from Spanish.

The company that had hired him didn't provide him a respirator or protective clothing. That's not unusual in the natural gas fields.

José Lara did his job until he no longer could work.

At the age of 42, he died from pancreatic and liver cancer.

Accidents, injuries, and health problems are not all that unusual in the booming natural gas industry that uses horizontal hydraulic fracturing, better known as fracking, to invade the Earth in order to extract methane gas.

The process requires up to 21 million gallons of water for each frack, a problem in the drought-ravaged southwestern part of the United States. Of the 750 chemicals that can be used in the fracking process, more than 650 of them are toxic or carcinogens, according to a report filed with the U.S. House of Representatives in April 2011. Several public health studies reveal that homeowners living near fracked wells show higher levels of acute illnesses than homeowners living outside the "Sacrifice Zone," as the energy industry calls it.

In addition to toxic chemicals and high volumes of water, the energy industry uses silica sand in the mixture it sends at high pressure deep into the Earth to destroy the layers of rock. The National Institute for Occupational Health and Safety (NIOSH) issued a hazard alert about the effects of crystalline silica. According to NIOSH there are seven primary sources of exposure during the fracking process, all of which could contribute to workers getting silicosis, the result of silica entering lung tissue and causing inflammation and scarring. Excessive silica can also lead to kidney and autoimmune diseases, lung cancer, tuberculosis, and chronic obstructive pulmonary disease (COPD). In the alert, NIOSH pointed out that its studies revealed about 79 percent of all samples it took in five states exceeded acceptable health levels, with 31 percent of all samples exceeding acceptable health levels by 10 times. However, the hazard alert is only advisory; it carries no legal or regulatory authority.

In 2013 and 2014, trains carrying oil from the Bakken Shale through middle America to refineries in Philadelphia and the Gulf had more derailments and explosions than in the previous 40 years, according to the Pipeline and Hazardous Materials Safety Administration. In addition to the normal diesel emissions of trucks and trains, there have been numerous leaks, some of several thousand gallons, much of which spills onto roadways and into creeks, from highway accidents of tractor-trailer trucks carrying wastewater and other chemicals.

The process of fracking requires constant truck travel to and from the wells, as many as 200 trips per day per well. Each day, interstate carriers transport about 5 million gallons of hazardous materials. Not included among the daily 800,000 shipments are the shipments by intrastate carriers, which don't have to report their cargo deliveries to the Department of Transportation. "Millions of gallons of wastewater produced a day, buzzing down the road, and still nobody's really keeping track," Myron Arnowitt, the Pennsylvania state director for Clean Water Action, told AlterNet.

Drivers routinely work long weeks, have little time for rest, and hope they'll make enough to get that house they want for their families.

Desperate for Jobs

In the Great Recession, people have become desperate for any kind of job. And the natural gas industry has responded with high-paying jobs. Pennsylvania Gov. Tom Corbett is ecstatic that a side benefit of destroying the environment and public health is an improvement in the economy and more jobs—even if most of the workers in Pennsylvania now sport license plates from Texas and Oklahoma. However, his boast that Pennsylvania added more than 240,000 jobs is countered by the reality that the increase is more like only 17,500 new jobs created since 2007, according to the state's Fiscal Office.

The drivers, and most of the industry, are non-union or hired as independent contractors with no benefits. The billion-dollar corporations like it that way. It means there are no worker safety committees. No workplace regulations monitored by the workers. And if a worker complains about a safety or health violation, there's no grievance procedure. Hire them fast. Fire them faster.

No matter how much propaganda the industry spills out about its safety record and how it cares about its workers, the reality is that working for a company that fracks the Earth is about as risky as it gets for worker health and safety.

The Occupational Safety and Health Administration (OSHA) issued Rain for Rent nine violations for exposing José Lara to hydrogen sulfide and not adequately protecting him from the effects of the cyanide-like gas.

It no longer matters to José Lara.

The effects from fracking should matter to everyone else. ⊕

Walter Brasch is an award-winning journalist and professor emeritus of mass communications. His latest book is Fracking Pennsylvania, *an in-depth analysis of the effects of fracking upon public health, the environment, worker safety, animals, and agriculture.*

Divesting from Fossil Fuels

An open letter from frontline communities to student activists working to divest college and university pension funds from fossil fuels

Dear students,

We write to you from the front lines. Some of our communities have been fighting the fossil fuel empire for generations. Others have only recently joined this struggle. We send our support and gratitude for leading this fossil fuel divestment campaign. This is a mighty cause you are joining: challenging some of the biggest threats humans have ever seen and committing to what must become a global movement.

We support your mission to hold your universities accountable. Institutions of learning must challenge systems that endanger the future of younger generations. We believe that colleges and universities divesting from fossil fuels and reinvesting in clean energy will deliver a powerful political message. And yet, we—as frontline and Indigenous leaders—encourage you to dig deeper. We encourage you to understand your campaigns as part of a much longer struggle, one that has been going on for generations, for justice and health, and the environment.

The corporations you are targeting have pushed our people up against the edge of survival. We live in the land coal companies have stolen and destroyed. We live in the land the oil, fracking, and uranium industries have poisoned. As the climate crisis worsens, it is frontline and Indigenous communities who are hit hardest. When you demand that your colleges cut financial ties to ExxonMobil's or Peabody Coal's latest projects to pillage the Earth—it's our land and communities you're acting in solidarity with. Our work is deeply tied together.

Please join us. From the Indigenous peoples, to the coalfields, fracking wells, refineries, and communities facing all manners of extreme energy production. Fight the fossil fuel industry on campus, but not only on campus. Join us in our communities and our fights and bind your struggle to ours.

We welcome you to this movement with open arms. Together we can defeat the dirty energy industry and build a healthy, sustainable, and just world.

In solidarity,

ROBERT J. THOMPSON, REDOIL (Kaktovik, Alaska)
KIRBY SPANGLER, Castle Mountain Coalition (Palmer, Alaska)
VERONICA COPTIS, Center for Coalfield Justice (Greene County, Pennsylvania)
JANENE YAZZIE, Sixth World Solutions
CHIEF GARY HARRISON, Chickaloon Tribe, Alaska
DUSTIN WHITE, Ohio Valley Environmental Coalition, Southern West Virginia
IRIS MARIE BLOOM, Protecting Our Waters, Pennsylvania Marcellus Shale Region
BLAS ESPINOSA, Texas Environmental Justice Advocacy Services, Houston
VICTORIA CORONA, Texas Environmental Justice Advocacy Services, Houston
LIANA LOPEZ, Texas Environmental Justice Advocacy Services, Houston
THERESA DARDAR, Pointe-au-Chien Indian Tribe, Louisiana
MEAGAN DOCHUK, 1st Nations Aamjiwnaang
RON PLAIN, 1st Nations Aamjiwnaang
ELANDRIA WILLIAMS, Highlander Center, Knoxville, Tennessee

This letter is available at http://www.wearepowershift.org/blogs/open-letter-student-divestment-activists-front-line-communities-impacted-fossil-fuels

A Short History of the Three Ages of Carbon—*page 194*
By Michael T. Klare

Ask students to read, discuss, and write about "A Short History of the Three Ages of Carbon" in small groups. Here are some questions:

- Make a list of 10 things you did yesterday that used fossil fuels. What did you do that did not rely on fossil fuels?
  - List significant dates and statistics from the article.
  - What are the significant differences between each of the three "ages of carbon"?
- For each of the three ages of carbon, who do you see as the "winners" and "losers"? Do any groups show up consistently as winners or losers?
- Klare makes a prediction about the geopolitics of the third age of carbon. What's your response to his predictions?
- Klare makes several arguments about the future in the article's final section, "Surviving the Third Carbon Era." What's your response to his arguments and plan of action? What questions are you left with?
- Does Klare's article leave you more or less hopeful? Why?
- Review your list of 10 things that you did yesterday that depended on fossil fuels. What would need to happen for each of these to be dependent instead on renewable energy?

"They Can Bury Me in These Hills, but I Ain't Leavin'"—*page 210*
By Jeff Goodell

Write or discuss:

- List all the effects of mountaintop removal coal mining. If this practice is so detrimental to communities and to the environment, why does it continue?
- What role does the government play in mountaintop removal coal mining? Does it seem like the government takes sides? Whose side(s) does it take, and why?
- What would you find most difficult about Maria Gunnoe's experiences?
- What about Gunnoe do you admire?
- Dr. Ben Stout, professor of biology at Wheeling Jesuit University, said, "Sometimes I think what's going on here is damn near genocide." What is it about the effects of mountaintop removal that would provoke him to say this? Do you agree with his statement?
- Stop reading right after Dr. Stout's comment. Ask students to imagine for a moment that they are Maria Gunnoe. A reporter asks her, "Why don't you just leave?" Write Gunnoe's response. Afterward, compare what students wrote with how Gunnoe responded when reporter Jeff Goodell asked her this question.
- In order to help students appreciate the difficulties and dilemmas that Gunnoe faces, ask students to write on one of several topics. They should write these as personal narratives, not as essays. Encourage them to use dialogue and interior monologue.
  - Write about a time you stood up for something you thought was right.
  - Write about a time you defied someone with power.
  - Write about a place that is sacred to you and what makes it so sacred.
  - Write about a place you are familiar with that has been damaged in some way during your lifetime.

I Love Mountains (ilovemountains.org), a North Carolina group working to end mountaintop removal coal mining, created an extensive map of mountaintop removal sites within Google Earth. The "Appalachian Mountaintop Removal" feature can be found under "Global Awareness," located in the Layers box at the bottom left of the Google Earth screen. Once the feature is enabled, it outlines hundreds of mountaintop removal sites from Tennessee to West Virginia, giving students a scale of the destruction resulting from these mining practices. Within the feature, a users' guide developed by I Love Mountains offers access to videos, stories of the human communities affected by different mines, and a high-resolution "tour" of the creation of one mountaintop removal site in West Virginia.

This Much Mercury. . . *page 226*
By Dashka Slater

 Before students read "This Much Mercury. . ." tell the story of Sophie Chabon. "Sophie had been an early talker and walker, but then she seemed to hit a wall, suddenly unable to sound out words she used to know how to read and even forgetting how to tie her shoes." Ask students to offer some hunches about what happened to Sophie. (Note that this only works as an introduction to the article if students have not encountered the story of Sophie Chabon in the coal export role play.)

Ask students to use "This Much Mercury. . ." to put together a questionnaire to use with friends and family. Questions should focus on people's awareness of the potential impact of mercury in the seafood they eat.

Ask students:

- Who is to "blame" for Gelfond's mercury affliction? Is this his personal problem? Is it the seller of the sushi he regularly purchased? The fisherpeople who caught and sold the tuna? The coal companies for polluting the Earth with mercury? The U.S. government for permitting coal mining on public lands, coal burning, and coal exports?
- Who is to blame for Sophie Chabon's ailment? Do you agree with her mother: "I blame our country for not [caring] about what we're spewing into the atmosphere. . . . This is about coal, pure and simple. You wouldn't go and break your

child's bones one by one, but we tolerate this kind of poison that's ruining their minds. It's insane." What does it mean to blame "our country"? Are we all to blame equally?
- In what ways does this issue represent environmental racism? [The article points out, for example, that the average Latina/o fisherperson "consumes twice as much mercury daily as the EPA considers safe, while a 2010 study of subsistence fishing in California found that some anglers were getting 10 times that dose."]
- At the end of the article, Jaime Bowen, a 31-year-old home healthcare worker and mother of two, stands at a podium to speak about coal, fish, and children's safety. Write the opening of Bowen's speech.

"We Know What's Goin' On"—*page 238*
By Terry Tempest Williams

 Ask students to list what Margaret and Kevin Curole have lost.

Ask them what "reparations," if any, the Curoles should receive? Who should pay?

Pose this question:

- Terry Tempest Williams ends with a quote from Margaret Curole: "We know what's goin' on." How would you summarize "what's goin' on"? Even if the Curoles know what's goin' on, what should they do?

Fracking . . . Firsthand—*page 252*
By Pennsylvania residents

Ask students:

- Why do these citizens oppose fracking?
- Why do they think that regulations will not protect communities from the hazards of fracking?
- Why don't communities know the specific "cocktails" of chemicals used in fracking operations?
- How do you think the Illinois Legislature responded to this plea? Research the current status of fracking in Illinois and in your own state.

- What role does secrecy play in fracking?

Have students look up some of the chemicals and byproducts listed in the letter—for example, benzene, toluene, hydrochloric acid, petroleum distillates, radium, uranium and radon, arsenic, and mercury.

- How would the oil and gas industry likely respond to this critique. Find one example from a pro-fracking source that responds to one or more points raised in this letter.
- All the people who wrote this letter live in Pennsylvania, and yet the letter is addressed to leaders in Illinois. Why do people in Pennsylvania care about what goes on in Illinois?

Fracking Democracy—*page 255*
By Sandra Steingraber

Ask students:

- Why does Sandra Steingraber name this article "Fracking Democracy"?
- Assign students to write a two-minute testimony on fracking, or on how they would recommend 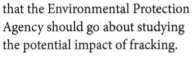 that the Environmental Protection Agency should go about studying the potential impact of fracking.

SUSAN SIMENSKY BIETILA

"It seems to me that our problem has a lot less to do with the mechanics of solar power than the politics of human power—specifically whether there can be a shift in who wields it, a shift away from corporations and toward communities, which in turn depends on whether or not the great many people who are getting a rotten deal under our current system can build a determined and diverse enough social force to change the balance of power. I have also come to understand...that the shift will require rethinking the very nature of humanity's power—our right to extract ever more without facing consequences, our capacity to bend complex natural systems to our will. This is a shift that challenges not only capitalism, but also the building blocks of materialism that preceded modern capitalism, a mentality some call 'extractivism.'"

—Naomi Klein, from *This Changes Everything: Capitalism vs. the Climate*

CHAPTER FIVE: Teaching in a Toxic World

ALEC DUNN

CARRIE NEUMAYER

"Underlying all of these problems of introducing contamination into our world
is the question of moral responsibility. . . . [T]he threat is infinitely greater to
the generations unborn; to those who have no voice in the decisions of today,
and that fact alone makes our responsibility a heavy one."

—Rachel Carson

INTRODUCTION
Teaching in a Toxic World

Toxic trespass. It's a term used by biologist, writer, activist, mother, and cancer survivor Sandra Steingraber. And it describes what this chapter is about.

In an interview with Bill Moyers, Steingraber defined it simply: Toxic trespass is "when chemicals without our consent enter our body..." And what could be a more intimate invasion? Invisible, uninvited, and often deadly.

Articles in this chapter describe how toxic trespass happens in far too many ways in our lives—here and around the world. It happens in the small particulate matter we breathe ("Science for the People"), the cosmetics we apply to our bodies ("Combating Nail Salon Toxics" and *The Story of Cosmetics*), the water we drink ("The Transparency of Water"), and the radioactive dust we inhale ("Uranium Mining, Native Resistance, and the Greener Path").

Toxic trespass happens to all of us—but, as we've shown throughout the book, not equally. Linda Christensen and Kevin Sullivan ("Reading Chilpancingo" and "Toxic Legacy on the Mexican Border") describe the damaging toxic stew that U.S.-owned factories pour into a poor Tijuana, Mexico barrio. Corporations repatriate enormous profits while children in Chilpancingo play in streams that run red then white then black in effluent emitted by maquiladoras on the mesa above. And those lightly regulated, poisonous nail salons? They're staffed almost entirely by immigrant women of color. And it was Navajo men who trudged home from uranium mines covered in radioactive dust with which they unknowingly poisoned their families—families that still today breathe contaminated dust from abandoned uranium mines and drink mining-tainted water.

Because of the intimate nature of the toxic trespass, students' first reaction may be self-protection or focusing on individual consumer choices, rather than collective action. This is an impulse corporations are all too happy to reinforce. For example, Elizabeth Royte describes the Keep America Beautiful campaign; the same corporations fighting pollution regulations and recycling requirements were the ones spending millions on advertisements like the so-called Crying Indian campaign that blamed pollution on litterbugs and thoughtless individuals. The chapter's opening article by Derrick Jensen, "Forget Shorter Showers," is a provocative one that directly challenges all of us to think beyond our individual actions. What sane person would suggest that composting would have ended slavery? Jensen asks. So why do some think that we can address the environmental crisis through personal lifestyle choices?

This chapter provides abundant evidence that people are taking collective action: Tony Marks-Block's students in Oakland investigating sources of airborne particulate matter and publicly sharing their results; Yukiko Kamea and other Fukushima refugees demanding an end to nuclear power in Japan; Lourdes Lujan and women's environmental justice activists organizing for a cleanup of a toxic landfill in Tijuana; and the indigenous activists Winona LaDuke describes who are building the "greener path."

In Bill Moyers' interview with Sandra Steingraber, she echoes the sensibility of this chapter:

> And so at this point in our history, it is the environmental crisis that is the great moral crisis of our age. And in that, I don't want to be a good German...I want to be one of the French resistance. One of the people who stand up and say, "This is not right. No matter how difficult this is to change, we're going to have to change it." ⊕

Forget Shorter Showers

Why individual action is not the problem—or the solution

BY DERRICK JENSEN

Would any sane person think dumpster diving would have stopped Hitler, or that composting would have ended slavery or brought about the eight-hour workday, or that chopping wood and carrying water would have gotten people out of Tsarist prisons, or that dancing naked around a fire would have helped put in place the Civil Rights Act of 1964 or the Voting Rights Act of 1965? Then why now, with all the world at stake, do so many people retreat into these entirely personal "solutions"?

Part of the problem is that we've been victims of a campaign of systematic misdirection. Consumer culture and the capitalist mindset have taught us to substitute acts of personal consumption (or enlightenment) for organized political resistance. Al Gore's film *An Inconvenient Truth* helped raise consciousness about global warming. But did you notice that all of the solutions presented had to do with personal consumption—changing lightbulbs, inflating tires, driving half as much—and had nothing to do with shifting power away from corporations, or stopping the growth economy that is destroying the planet? Even if every person in the United States did everything the movie suggested,

U.S. carbon emissions would fall by only 22 percent. Scientific consensus is that emissions must be reduced by at least 75 percent worldwide.

Or let's talk water. We so often hear that the world is running out of water. People are dying from lack of water. Rivers are dewatered from lack of water. Because of this we need to take shorter showers. See the disconnect? *Because I take showers, I'm responsible for drawing down aquifers?* Well, no. More than 90 percent of the water used by humans is used by agriculture and industry. The remaining 10 percent is split between municipalities and actual living breathing individual humans. Collectively, municipal golf courses use as much water as municipal human beings. People (both human people and fish people) aren't dying because the world is running out of water. They're dying because the water is being stolen.

Or let's talk energy. Kirkpatrick Sale summarized it well:

> For the past 15 years the story has been the same every year: individual consumption—residential, by private car, and so on—is never more than about a quarter of all consumption; the vast majority is commercial, industrial, corporate, by agribusiness and government [he forgot military]. So, even if we all took up cycling and woodstoves it would have a negligible impact on energy use, global warming, and atmospheric pollution.

Or let's talk waste. In 2005, per capita municipal waste production (basically everything that's put out at the curb) in the United States was about 1,660 pounds. Let's say you're a die-hard, simple-living activist, and you reduce this to zero. You recycle everything. You bring cloth bags shopping. You fix your toaster. Your toes poke out of old tennis shoes. You're not done yet, though. Since municipal waste includes not just residential waste, but also waste from government offices and businesses, you march to those offices, waste reduction pamphlets in hand, and convince them to cut down on their waste enough to eliminate your share of it. Uh, I've got some bad news. Municipal waste accounts for only 3 percent of total waste production in the United States.

I want to be clear. I'm not saying we shouldn't live simply. I live reasonably simply myself, but I don't pretend that not buying much (or not driving much, or not having kids) is a powerful political act, or that it's deeply revolutionary. It's not. Personal change doesn't equal social change.

So how, then, and especially with all the world at stake, have we come to accept these utterly insufficient responses? I think part of it is that we're in a double bind. A double bind is where you're given multiple options, but no matter what option you choose, you lose, and withdrawal is not an option. At this point, it should be pretty easy to recognize that every action involving the industrial economy is destructive (and we shouldn't pretend that solar photovoltaics, for example, exempt us from this: They still require mining and transportation infrastructures at every point in the production processes; the same can be said for every other so-called green technology). So if we choose option one—if we avidly participate in the industrial economy—we may in the short term think we win because we may accumulate wealth, the marker of "success" in this culture. But we lose, because in doing so we give up our empathy, our animal humanity. And we really lose because industrial civilization is killing the planet, which means everyone loses. If we choose the "alternative" option of living more simply, thus causing less harm, but still not stopping the industrial economy from killing the planet, we may in the short term think we win because we get to feel pure, and we didn't even have to give up all of our empathy (just enough to justify not stopping the horrors), but once again we really lose because industrial civilization is still killing the planet, which means everyone still loses. The third option, acting decisively to stop the industrial economy, is very scary for a number of reasons, including but not restricted to the fact that we'd lose some of the luxuries (like electricity) to which we've grown accustomed, and the fact that those in power might try to kill us if we seriously impede their ability to exploit the world—none of which alters the fact that it's a better option than a dead planet. Any option is a better option than a dead planet.

> **Why now, with all the world at stake, do so many people retreat into entirely personal "solutions"?**

Besides being ineffective at causing the sorts of changes necessary to stop this culture from killing the planet, there are at least four other problems with perceiving simple living as a political act (as op-

STEPHANIE MCMILLAN

posed to living simply because that's what you want to do). The first is that it's predicated on the flawed notion that humans inevitably harm their land base. Simple living as a political act consists solely of harm reduction, ignoring the fact that humans can help the Earth as well as harm it. We can rehabilitate streams, we can get rid of noxious invasives, we can remove dams, we can disrupt a political system tilted toward the rich as well as an extractive economic system, we can destroy the industrial economy that is destroying the real, physical world.

The second problem—and this is another big one—is that it incorrectly assigns blame to the individual (and most especially to individuals who are particularly powerless) instead of to those who actually wield power in this system and to the system itself. Kirkpatrick Sale again: "The whole individualist what-you-can-do-to-save-the-Earth guilt trip is a myth. We, as individuals, are not creating the crises, and we can't solve them."

The third problem is that it accepts capitalism's redefinition of us from citizens to consumers. By accepting this redefinition, we reduce our potential forms of resistance to consuming and not consuming. Citizens have a much wider range of available resistance tactics, including voting, not voting, running for office, pamphleting, boycotting, organizing, lobbying, protesting, and, when a government becomes destructive of life, liberty, and the pursuit of happiness, we have the right to alter or abolish it.

The fourth problem is that the endpoint of the logic behind simple living as a political act is suicide. If every act within an industrial economy is destructive, and if we want to stop this destruction, and if we are unwilling (or unable) to question (much less destroy) the intellectual, moral, economic, and physical infrastructures that cause every act within an industrial economy to be destructive, then we can easily come to believe that we will cause the least destruction possible if we are dead.

The good news is that there are other options. We can follow the examples of brave activists who lived through the difficult times I mentioned—Nazi Germany, Tsarist Russia, antebellum United States—who did far more than manifest a form of moral purity; they actively opposed the injustices that surrounded them. We can follow the example of those who remembered that the role of an activist is not to navigate systems of oppressive power with as much integrity as possible, but rather to confront and take down those systems. ⊕

Derrick Jensen is an author and environmentalist. His books include The Culture of Make Believe *and* Endgame. *This article first appeared in* Orion *magazine.*

See teaching ideas for this article, page 324.

> **"We, as individuals, are not creating the crises, and we can't solve them."**

The garbage landscape is littered with green-wash tactics, in which polluters pose as friends of the environment but spend more money advertising their green projects than on the projects themselves. One masterful example of corporate greenwash is the Keep America Beautiful campaign, which was founded by beverage companies and packaging executives in 1953 after magazine ads began promoting beverage cans as "throwaways" (one depicted carefree boaters slinging empties into a lake). Litter alongside roads, rivers, and farm fields had begun to accumulate, prompting Vermont to pass the nation's first bottle bill, which banned the sale of beer in nonrefillable containers. Beer companies didn't like that one bit. They lobbied hard against the law, and in four years it expired. (The state enacted a new bottle bill in 1972.)

In a stroke of marketing genius, Keep America Beautiful urged individuals to take responsibility for this waste, to "put litter in its place." In 1971, the organization sponsored one of the most successful public service announcements in history, a TV commercial in which a Native American, complete with braid and eagle feather, paddles down a pristine waterway until he reaches a teeming city. When he spots empty beverage cans swirling in the shallows, a tear rolls down his leathery face. Keep America Beautiful proudly called the "Crying Indian" spot an "iconic symbol of environmental responsibility." (Iron Eyes Cody, who played the Indian and claimed to be a Cherokee-Cree, was later outed as a Sicilian American named Espera DeCorti.)

But whose responsibility is the foul mess along the shore? The organization's underlying message is that individuals, not corporations who produce single-use containers, are responsible for trash, and that individuals must change their behavior, not manufacturers. Keep America Beautiful focuses on antilitter campaigns—which enlist millions of

Keep America Beautiful?

BY ELIZABETH ROYTE

volunteers a year to clean up beaches and road-sides—but it ignores the potential of recycling legislation and resists changes to packaging. Between November 1992 and July 1993, the American Plastics Council, a sponsor of Keep America Beautiful, spent $18 million on a national campaign to "Take Another Look at Plastic." The ads crowed that more than a billion pounds of plastic had been recycled in 1993, but they failed to mention that 15 billion pounds of virgin plastic were produced during that same eight-month period. According to a report by the Environmental Defense Fund, for every one-ton increase in plastic recycling between 1993 and 1996, there was a 14-ton increase in new plastic production.

> **But whose responsibility is the foul mess along the shore?**

For 20 years environmental groups, including the Sierra Club, the National Audubon Society, and the National Wildlife Federation, lent legitimacy to Keep America Beautiful by sitting on its advisory committee. Those relationships ended after a board meeting in July of 1976, when American Can Company chairman William F. May denounced bottle bill proponents as communists and called for a total mobilization against proposed bottle bills in four states. Today, Keep America Beautiful is funded by about 200 companies that manufacture and distribute aluminum cans, paper products, and plastic and glass containers, in addition to companies that landfill and incinerate all of the above. ⊕

Elizabeth Royte is the author of Garbage Land: On the Secret Trail of Trash *and* Bottlemania: How Water Went on Sale and Why We Bought It. *This article was excerpted from* Garbage Land *(Little, Brown and Company).*

See teaching ideas for this article, page 324.

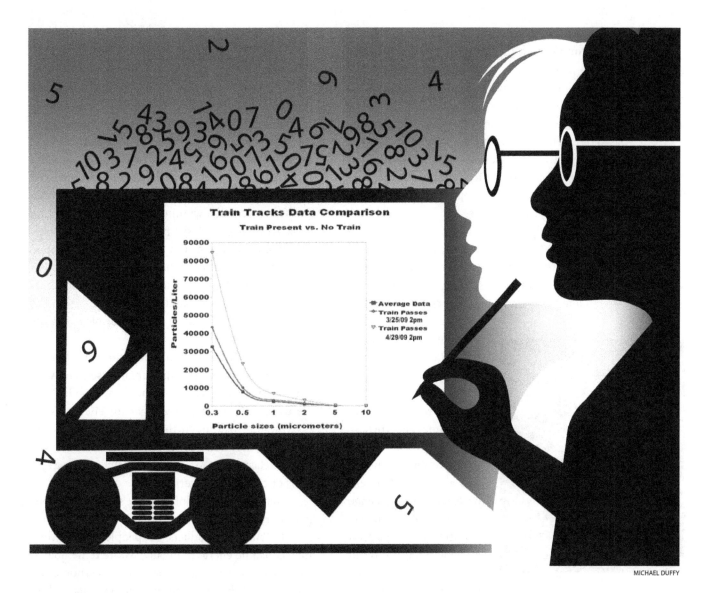

Train Tracks Data Comparison

Train Present vs. No Train

MICHAEL DUFFY

66 **W**ow—Tony, look at those numbers! They're like double the amount from before," a student in our after-school science research program exclaimed after a large train passed by. The students were using a machine to count particulate matter, microscopic particles primarily created through the combustion of various fossil fuels, along our air quality transect on Fruitvale Ave. in Oakland, California. What was once commonplace—a train speeding through the neighborhood—became a catalyst for students to understand the connections between energy and air quality.

Over a year, a small group of high school students risked their afternoons and summer to participate in a science program that was, as Maraya put it, "much different from science class." This was one of several after-school programs in Oakland

Science for the People

High school students investigate community air quality

BY TONY MARKS-BLOCK

and Richmond that I led as an instructor with the East Bay Academy for Young Scientists (EBAYS), a National Science Foundation project coordinated by the Lawrence Hall of Science at UC Berkeley. Students in our projects carry out local, community-based research. Given the proximity of many schools to freeways, train tracks, and industry, we decided to focus on energy production and its relation to air quality as a way to help develop science literacy and environmental justice leaders.

With the help of the high school staff, I recruited a group of 9th graders. Many of them were struggling academically and wanted the extra science credit offered for my after-school project. Although we explained that they'd be researching air quality, students later told me that they did not know what they were getting into and joined primarily for the extra credit.

When we started, environmental (in)justice was not a part of their vocabulary, but it was a part of their experience. My challenge was to help students connect their visceral understanding of racial injustice to a scientific process that could be used to develop deeper knowledge about community health and air pollution. If they could collect and analyze data on air pollutants, and then communicate their findings to others, they would be taking one step toward educating their community about the environment.

> My challenge was to help students connect their visceral understanding of racial injustice to a scientific process that could be used to develop deeper knowledge about community health and air pollution.

Test Tube Rockets

Before I introduced my students to air quality and particulate matter data collection, I led them through some energy-related activities structured to help them develop research skills and an understanding of the root problems underlying air quality. Unlike many science courses, where memorizing facts is the norm, our curriculum emphasizes the development of new knowledge through the manipulation and testing of various systems.

First, I introduced students to test tube rockets. I challenged them to find the combination of baking soda and vinegar that would propel a test tube sealed with a rubber plug the highest. It was a messy and competitive affair. In their excitement, students often forgot the proportions of each chemical used for a specific trial. I reminded students they could refine their systems better when they documented their measurements.

Once we had a winning combination of baking soda and vinegar, I asked: "Where does the energy come from that makes the rockets go?" I hoped someone could explain that the chemical reaction between the baking soda and vinegar produces carbon dioxide gas that builds up pressure within the tube and launches it from the rubber plug. But the room was silent.

So I asked them to think about their own bodies: "What sources of energy do humans use to move—whether with our legs or in our automobiles? Do we use baking soda and vinegar for energy? Why or why not?" Although they laughed at these silly questions, they had trouble answering them because in their classes, the world is rarely taught from a systems or comparative perspective.

I have found it helpful to ask students to compare human function with other systems, so I persisted: "What byproducts do people create when they consume energy?" Finally, Andre remembered from biology class that we breathe out carbon dioxide. This led the group to hypothesize that carbon dioxide may also be the gas produced when baking soda and vinegar are combined, as well as when fossil fuels combine with heat.

Of course, one experiment did not give students a full understanding of energy and the myriad ways humans generate energy to run their societies. I introduced them to electrical systems because of how integral—yet abstracted and hidden—they are in our daily lives. We experimented with moving electricity in different ways to produce different effects, from turning on a lightbulb to running a motor. I also introduced them to solar and wind as energy sources, and we compared them with a basic battery. Constructing circuits helped my students develop conceptual knowledge of an important system. The next question was "All our buildings and appliances depend on electricity. Where does it come from?"

Energy Use and Air Quality

Most electricity, of course, is generated through the burning of fossil fuels. To make the connection to air quality, I introduced students to common fuels that generate electricity and, appropriately, we burned them. Before the thrill of actually setting gasoline, diesel, ethanol, and wood on fire, I asked students to predict which fuels would generate the least particulate matter. I also introduced them to our Fluke 983, a machine that pumps ambient air and counts the number of particles of 0.3, 0.5, 1, 2, 5, and 10 micrometers with laser technology. The smallest particles are the most dangerous because they can travel farther into the body and can accumulate in the lungs and bloodstream. This machine is powerful for students because it produces data within seconds. They can easily operate the Fluke and, while it pumps a liter of air, they can watch it count particles in real time. When the students compared the quantitative data from the Fluke and the qualitative data from their visual observations of the fuels we had just burned, they were excited to see that their observations had a strong correlation to the Fluke data. (Note: unfortunately, the Fluke costs about $5,000. For a lesson using inexpensive materials, see http://enviromysteries.thinkport.org/breakingthemold/lessonplans/indoorair.asp)

My students were astonished to learn how many microscopic particles are in the air, and instantly had many questions and comments: "Where do these particles come from?" "Can I bring this to my house?" "Will there be more particles in the bathroom?" "How much does this thing cost?"

My students began to generate their own research questions. I encouraged them to develop a method for answering their questions: "How would you figure out which room has more particles?" "If you only tested the air once, would that be enough proof that a particular room always had higher levels of particulate matter?" My prompting resulted in a high-energy discussion. Pedro asked what particulate matter is made of. Although we did not have time that day to delve into an answer, the discussion certainly helped my teaching. I began to prepare future lessons to help them learn what they wanted to know.

In order for students to understand the health impacts of particulate matter, I brought in a guest speaker to discuss how particulate matter affects the

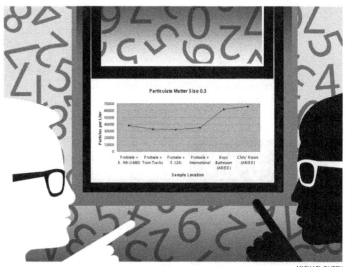

MICHAEL DUFFY

body's respiratory and circulatory systems, and how it can cause and exacerbate asthma, lung disease, and heart disease. We also did background research on health statistics for Alameda County. The table of numbers with columns on race, gender, hospitalizations, and deaths intimidated the students. They had done little data analysis in school, so asking specific questions such as "Which race has the highest rate of asthma in Alameda County?" frustrated them. They weren't eager to do activities that were more "like school" and less "hands on."

It wasn't just the math that made my students reluctant to get into this. Across the board, African Americans and Latinos in the county have higher levels of disease. For many of my students this fact felt like another obstacle; and for some it generated cynicism. When I asked why they thought asthma affected African Americans and Latinos at a higher rate than whites in Alameda County, Raúl said it was because their neighborhoods are closer to environmental pollutants, but Cecilia said "all Mexicans do is smoke weed," reflecting how uncomfortable many of the students were with the discussion. Like many youth of color, they were wrestling with the perceptions and stereotypes of their communities in U.S.

> The smallest particles are the most dangerous because they can travel farther into the body and can accumulate in the lungs and bloodstream.

society. All in all, I did not receive the outpouring of student-initiated questions as I had when I introduced the Fluke. Instead of students leading the discussion with their questions, I struggled to move the discussion away from self-deprecation.

I wish I had more examples of communities with data showing that the reduction of specific environmental pollutants lowered disease. Analyzing that data would be a more fulfilling experience for us all. In the end, these lessons did help students cement connections among energy, air quality, and health, despite the difficulties students had fully engaging in the material. When asked how energy use impacts their community, Huong summed it up:

> "Which race has the highest rate of asthma in Alameda County?"

The way you make energy . . . you need fossil fuels to convert to electricity, and all of that burning causes a lot of smoke and stuff in the air . . . and it just gives people breathing problems . . . and asthma.

Taking Research Out to the Neighborhood

Drawing on this new background knowledge of energy, air pollution, and scientific research, we began to design an experiment with the Fluke particle counter to test student hypotheses. I brought out some local maps and asked, "If we want to find out where in the neighborhood there is more particulate matter, where should we collect samples?" Students knew that automobiles and diesel trucks emit a lot of particulates and suggested we collect samples in the parking garage. We tried it several times. Then I steered students back to my original question and asked if just testing in the parking garage would give us an understanding of particulate matter in the community as a whole.

Eventually, we decided to sample for particulate matter at numerous locations in the neighborhood. I asked if they knew of any other areas with lots of vehicular traffic, and they immediately pointed to the nearby I-880 freeway. "What do you think?" I asked them. "Will particulate matter increase or decrease as we get farther from the free-

way?" They hypothesized that particulate matter would be higher near the freeway and would decrease with distance.

We began collecting samples at each intersection between the freeway and International Blvd. along Fruitvale Ave. every Wednesday afternoon after school between March and May. This was sometimes tedious, but students developed an understanding of why it was important to collect samples over and over again: Air quality changes each day, and they wanted to understand particulate matter patterns in the area over time. One week, when they found higher levels of particulate matter at the freeway, I asked them which variables impacted the particulate matter count. Students responded with a list they now understood: "Temperature, humidity, wind speed, and traffic levels."

I asked students if they had ideas for a research method that would yield the most comprehensive results. Veronica responded that an on-site monitor, always collecting data, would give us a better understanding. Other students added that if the machine could also detect the chemical composition of the particles, we could know whether any of the chemicals in the air were at higher levels, and potentially determine the source of the particles.

These wonderful responses raised such great political issues, I had to follow up with further questions: "Why isn't this monitoring occurring if there are such high levels of disease and illness in this community? Who should be leading those monitoring efforts?" My students hadn't really thought about that, and just assumed "other scientists" should be leading the effort.

"Why do you think other scientists aren't doing this kind of research?" I asked.

"Scientists don't really care about Mexicans and Fruitvale," Francisco said.

"That might be true. Do you think that scientists and the government agencies that regulate air pollution even have the resources to carry out this research?"

"Ha! Of course not," several students responded. From their experience with their education and the quality of life in their neighborhood, the dots were connected. If the government put so few resources into other social services, why would there be resources for health or the environment? Through discussions like these, I tried to integrate new vocabulary, including environmental racism and injustice, to define what my students had already explained.

More Research Questions

Through our data collection along the transect, students became interested in using the Fluke to perform other tests. For example, they saw how often trains passed by and they wanted to see the effect on particulate matter. So they began to note when the trains arrived, and positioned themselves near the tracks to take samples. "Particulate matter sizes increased significantly, most likely because of the trains' diesel fuel exhaust combined with the accumulation of dust particles swept up from the ground as the train passes with great speed," they concluded.

They also began to collect data inside one of their school classrooms and bathrooms to test indoor air quality, since that is where they (unfortunately) spend most of their day. Particulate matter counts for the smallest particles (particulate matter size 0.3 micrometers) turned out to be much higher in the school restrooms and classrooms than outdoors. At first, this confused students, since they had hypothesized that levels would be lower indoors. One student proposed that it was an anomaly—maybe someone had just sprayed cleaner in the bathroom? But the higher levels of particulate matter persisted each week, and when we finally averaged our data at all sample locations, the indoor locations still were significantly higher for particulate matter 0.3 compared with outdoor locations.

Because we had not done background research on indoor air quality, I brought in the Environmental Protection Agency's (EPA) fact sheet to help us devise hypotheses for the higher levels. We used the EPA information to narrow down the probable sources of high indoor particulate matter counts to furniture or paint off-gassing, cleaning fluids (as had been previously predicted), and/or poor ventilation.

To continue to push students to use scientific methods, I asked them to devise a method to test one of the potential causes of poor indoor air quality. Delia thought we should just open the classroom door to see if air quality improved when outside air came into the classroom. Sure enough, particulate matter counts went down when the door was opened to the outdoor courtyard. As the students

explained later on their research poster: "Opening the door let outside air flow into the room, and we found a sevenfold decrease in levels of all particulate matter particle sizes. However, soon after the particulate matter levels decreased upon opening the door, we began to smell barbecue from a nearby restaurant, and particulate matter counts in the room immediately increased again."

But why were these small particulates accumulating in the first place? I proposed that the building's ventilation system, which filtered air into the building, was malfunctioning or not designed to filter out such small particles. This was an air quality issue we could do something about.

Turning Data into Analysis

The first step was to consolidate and analyze our data. Coincidentally, at just this time, the students had the opportunity to create a scientific poster of their work for the upcoming American Geophysical Union conference in San Francisco. This task was not easy. I led them through the most basic data analysis step by step, as they had never used spreadsheet software to organize quantitative data. After they had input all of their data, I asked, "How can we see if a particular point has more particles than others over time, and not just on one day?" I hoped students would recognize that they could use their existing knowledge of averages and apply it to their data, but I got a lot of blank stares. For many of the students, the data was a morass of seemingly impenetrable information. I showed them how to use functions to average particulate matter at each location, and then to create graphs to compare the averages of each sample location. Then we worked on everyone being able to explain the graphs.

Although this was a frustrating process for all of us, I believed the experience would help them with their mathematical skills in the future. Due to time constraints, I couldn't teach them the statistics that

> They also began to collect data inside one of their school classrooms and bathrooms to test indoor air quality, since that is where they (unfortunately) spend most of their day.

MICHAEL DUFFY

approaches to articulating their understanding to an audience that had initially felt intimidating. The conference also showed professional scientists the work that needs to be done in under-resourced communities and it demonstrated the capabilities of students who are severely underrepresented in the field. This intersection of cultures challenged both our students and the scientists, who were blown away by our students' rigorous work.

Turning Analysis into Activism

After the conference, we returned to the question of how to address the high levels of particulate matter indoors. We began a discussion about who had the information and/or the power to do something about the air quality problems they had identified at the school. The students identified the principal as the place to start. The principal sent them to the building manager, who immediately referred them to the building engineer.

Unfortunately, the engineer dismissed the students' concerns. He told them that the ventilation system functioned fine, and that the level of student foot traffic must be contributing to the high particle counts. When we asked to see the filtration and ventilation system to learn how it functioned, the engineer successfully gave us the runaround and stood us up numerous times.

In the meantime, students began to notice that the vents in their rooms had accumulated layers of dust that were impeding air from leaving the room. I suggested the principal could apply the necessary pressure to do something about the classroom vents. One day several months into our efforts, students noticed that management had finally sent in staff to clean off the layers of dust that had accumulated. As soon as the vents had been cleaned, students said, "Let's test our rooms now!" Sure enough, counts were consistently lower over the subsequent weeks.

Although cleaning the vents did not address the root cause of particulate matter, the students were successful at addressing the problem. They had gained confidence, and I began to connect them to other organizations, the media, and higher-ups in

could make their arguments stronger, but we did articulate their basic findings.

Their next task was transforming their data and graphs into a slide show and a poster presentation. I asked them to arrange their graphs and write a discussion section that would review their findings. When graphs lacked titles and labels, or their scaling made it difficult to read, I asked students if they thought others would be able to read the graph, or even know what it represented: "What unit does the y-axis represent? I only see that there is 70,000 of something at Fruitvale and International—is it 70,000 gum wrappers, cars, flies?" Keeping things humorous prevented total meltdowns and helped us get through a process that was entirely new to them. Once students made their graphs and wrote paragraphs explaining their purpose, methodology, and conclusions, we worked on oral communication skills. It was important to make sure everyone could discuss all aspects of their project, even though each student had focused on a particular element of the slide show and poster.

At the San Francisco conference, many different scientists approached the poster simultaneously, so each student had to engage with "real" scientists and explain their project. The conference pushed students to think about questions they had not answered in their poster, and to develop new the Oakland Unified School District. Interestingly, the district had recently formed an indoor air quality committee and the nurse leading the effort was pleased to hear about my students' work. They were invited to speak to and eventually join the committee. Their experience gave them a level of authority and expertise that others in the district did not have, and they were asked to begin testing the air in other schools. I hope to see my students train other students to carry out the research within each of their schools. As Eric said:

> This can be the beginning, I don't think we are going to finish this all the way to the end, so that everybody in the community knows about particulate matter, but we can help with that, and start it, and in the future, others can continue the work.

These words are evidence of the community perspective my students have gained. They recognize that their work was done as a team, not as individuals, and the more people involved, the more they can accomplish.

In the beginning, my students were shy and lacked confidence, but their experiences with many different audiences helped them become excellent advocates for reducing air pollution and creating alternatives to dirty fuels. They found that scientists, teachers, and peers respect their work. Evelina said:

> I felt good showing all the scientists my work; we told them about all the particles at all of the sample locations. I felt professional. With science I can tell my community what's bad for you, and if you use [fossil fuels] too much that your kids could get diseases or asthma. I can tell my family about what's in the air. My house is close to the freeway, too. Doing more presentations and research will benefit the community.

I think this sense of themselves as scientists who "own" their research and who can see their positive role in developing solutions to health problems had a profound effect on their development and esteem.

The students have many ideas for future research. They want to collect data on specific pollutants, like carbon monoxide and nitrogen dioxide, as well as do similar testing in other parts of the East Bay. They also want to share their findings with more audiences.

Creating space for our students to collaborate on community actions is an important step toward making their research applicable and relevant to their lives. Eric said he "would invest more money into solar panels and renewable energy on houses and would use science to learn how to build cheaper solar panels." Do we have the infrastructure or resources to support those goals? If that is what justice looks like to him, how do we support him in developing this vision?

All students who are aware of the issues need outlets and organizations they can work with to build alternatives for their community. To really integrate environmental justice from classrooms to community, teachers need to participate in community movements so we can get our students involved. The school only took action to clean the vents after many people pushed to get them cleaned. If it takes that much effort to get some vents cleaned, we'll need forces beyond a few staff members in our school districts demanding healthy air. As teachers, activists, and organizers, our role is to work with students to develop campaigns that challenge the routing of trucks and freeways in our communities, as well as encourage alternative modes of transportation and production, so we can move to cleaner sources of energy. If we want our students to become leaders and environmental justice activists, then we need to exercise some leadership as well. ⊕

> **"With science I can tell my community what's bad for you, and if you use [fossil fuels] too much that your kids could get diseases or asthma."**

Tony Marks-Block is program coordinator for the East Bay Academy for Young Scientists. Mills College Educational Talent Search was a collaborating partner on this project. Students' names have been changed.

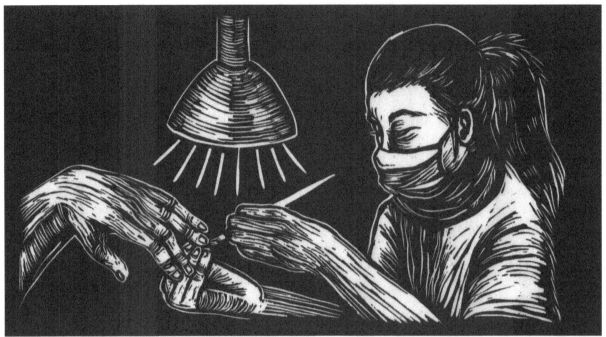

MEREDITH STERN

Combating Nail Salon Toxics

BY PAULINE BARTOLONE

Oakland, California-based Alisha Tran remembers the first time she suffered strange symptoms while working as a manicurist.

One day I was working with my client, and I feel my face numb. I feel a numbness on my finger. . . . I cannot close [my hand] and I cannot open it. . . . I sweat too, [I] sweat a lot. . . . And I was talking with client and I said, "call ambulance."

The doctor at the emergency room told her she was anemic and sent her home. Two weeks later, Tran had another episode on the job. She was taken to the emergency room again and saw the same doctor.

"And then he asked me, 'What kind of job you working?' And I said, 'I am working for nail salon.' And then he said, 'I think you should quit your job.'"

The nail salon industry is booming nationwide. The number of salons in California has more than tripled in the past two decades, according to the California Healthy Nail Salon Collaborative, whose report also notes that an overwhelming majority of the manicurists are women of color—59 to 80 percent are Vietnamese—and of reproductive age.

Every nail salon carries products loaded with chemical compounds with hard-to-pronounce names, including the commonly occurring "toxic trio"—dibutyl phthalate, toluene, and formaldehyde. Animal studies of dibutyl phthalate have shown reproductive and developmental effects. Formaldehyde is classified as a known carcinogen. And toluene has been shown to depress the nervous system.

Unfortunately, not many studies have been done on workers exposed to the "toxic trio" every day. Which is why Dr. Thu Quach, an epidemiologist with the Cancer Prevention Institute of California (CPIC), decided to study the prevalence of breast cancer among nail salon workers in California.

When the CPIC matched the names on the California cancer registry with names of licensed manicurists in the state, they did not find any alarming trends. Quach believes that is because the data was limited. The workforce, for the most part, is young and has not been in the industry for more than 10 years. And cancer takes a long time to develop.

Regardless of the findings, health advocates see cancer as just one of many health outcomes to be concerned about. Dr. Quach sees the cancer study as a first step in monitoring the workforce.

"While we're waiting for the research evidence—which often takes so long when it comes to epidemiology and population health finds—we need to phase out these chemicals now rather than waiting for research to show the dead bodies," she says.

The Push for Cosmetic Regulation

The Food and Drug Administration currently regulates cosmetics, which includes nail products. But Jamie Silberberger, director of programs and policy at Women's Voices for the Earth (WVE), believes that "regulation" may be an overstatement.

"The federal law that governs the cosmetics industry is two-and-a-half pages long and it hasn't been updated in 70 years," says Silberberger. "Because the law is so weak, companies can use ingredients that are known to cause cancer—or cause reproductive harm—and it's perfectly legal to do that."

Groups like WVE and the California Healthy Nail Salon Collaborative want more research, more government regulation, and market pressure to force manufacturers to change the chemical ingredients in cosmetics.

According to Silberberger, a cosmetics industry review panel exists, but it has only reviewed 11 percent of more than 12,000 cosmetic chemicals for safety. What's more, she says, the law requiring disclosure of ingredients on cosmetic retail products does not apply to items used in salons.

"Nail salon products are not required to be fully labeled," Silberberger explains. "If you buy a bottle of nail polish at a retail store, you'll see at the bottom a full list of ingredients. And that's required by federal law. But with salon products, there's a loophole and no requirement for them to be labeled. So nail salon workers don't know what's in them."

However, groups like the National Healthy Nail & Beauty Salon Alliance—of which WVE and the California Healthy Nail Salon Collaborative are a part—advocate for more worker protections, not just chemical policy reform. They want changes in the permissible exposure limits set by the Occupational Safety and Health Administration (OSHA).

"The [OSHA] standards were created in the 1960s for an industrial setting," explains Silberberger. "The intention was to protect against acute exposures. But these permissible exposure limits don't take into consideration the effects of a combination of multiple chemicals over the long term, or the chronic health effects of exposure, such as asthma, cancer, or reproductive harm." And many nail salons are poorly ventilated, she adds, which increases the exposure to harmful chemicals.

> "We need to phase out these chemicals now rather than waiting for research to show the dead bodies."

U.S. legislation around chemicals in cosmetics and around toxic products in general contrasts sharply with the laws in the European Union. According to the Campaign for Safe Cosmetics, the FDA has only banned or restricted 11 chemicals found in cosmetics, whereas the European Union has banned 1,000.

Silberberger calls the European Union's laws around toxic products "precautionary"—an approach that the United States does not subscribe to. But there is change on the horizon.

If signed into law, the Safe Cosmetics Act would ban the use of known carcinogens, genetic mutagens, and reproductive toxins, and require pre-market assessment for safety of ingredients. It also would close the loophole that allows salons to purchase and use unlabeled products.

The state of California passed a Safe Cosmetics Act in 2005, which provides for light cosmetic safety regulation, but groups are still waiting for more comprehensive national reform.

Advocates for Salon Workers Speak Up

In the lobby of Asian Health Services in Oakland, 61-year-old Lam Le is sitting in a fold-up chair. She is accompanied by her interpreter for the day, My Tong, an associate with the California Healthy Nail Salon Collaborative. After surviving the Vietnam War and living in refugee camps, Le came to the United States in the late 1980s.

> The calls for environmental justice were many—too many for everyone's voices to be heard.

"She thought that being a manicurist was pretty OK for her," says Tong, interpreting for Le. "She can make a living. But she knows that the chemicals probably won't be too good for her health."

Le, who worked in the industry for 12 years, has been diagnosed with a thyroid condition and has suffered from asthma and skin rashes. She had breast cancer too, and beat it twice. Le was never told directly by a doctor that her ailments were related to her work in the salons, but she says that she is "scared" of the businesses.

"Just smelling those chemicals makes her want to run away!" laugh Tong and Le.

For all her hardships, Le is quick to smile and speak up. She is at Asian Health Services not as a patient, but as an advocate. The California Healthy Nail Salon Collaborative trained Le in patient leadership, and she is waiting for other nail salon workers to arrive before attending a special EPA hearing on environmental justice.

"She wants to ask Congresswoman [Barbara] Lee and [EPA] Administrator Jackson to improve the conditions for the nail salon workers," explains Tong.

At the Oakland Federal Building, Le and Tong prepared her statement for Barbara Lee as they waited for the EPA town hall meeting to begin. Le had to convert her two-page testimony into a single question on a small index card, which was read out to the representatives by Phaedra Ellis-Lamkins of Green for All.

Sitting beside Le was Connie Nguyen, a nail salon worker for 17 years who suffered from asthma. She wanted help identifying formaldehyde-free disinfectants that might be better for her breathing.

The calls for environmental justice were many—too many for everyone's voices to be heard. Among the voices that did not get heard that day by the EPA representatives was Le's and she was clearly disappointed, although she managed to force a smile.

Patients as Educators

On a Wednesday morning at the clinic in Oakland, about 20 Vietnamese-speaking patients gather around a conference table. They are patient leaders being educated about a wide range of health care topics so that they can go out and discuss them with the rest of the community.

"Recently, we [have been] talking a lot about nail salons," says Alisha Tran, an ex-manicurist and nail salon owner who now leads the Patient Leadership Council at Asian Health Services. "Most of the time, I talk about toxics."

Tran talks to patients about the need for ventilation, wearing masks, taking a break, and using green products in the workplace. She also talks about the "toxic trio" found in nail salon products and their strong link to negative health impacts and other harmful chemicals. She hopes the information will eventually reach the Vietnamese-speaking workers in nail salons.

"I need their spirit to spread the news," she says.

Le is also present at that Patient Leadership Council meeting, wearing a blazer suit, earrings, and a big smile. And although she speaks through interpreter My Tong, her voice and her words carry power.

"I want to talk about this issue so that other people won't have the same problem," she says. "Other people may be quiet, but not me. I want to speak out." ⬤

This story was adapted from a radio feature Pauline Bartolone produced for Making Contact, *www.radioproject. org. The journal* Race, Poverty & the Environment *co-sponsored the original report using the journalism crowdfunding site www.spot.us.*

 *See teaching ideas for this article, page 324.*

David didn't come to school for two days. He came back with a note from his mom: "David was absent because his asthma was really bad."

Years ago, I worked with a girl named Diana, who seemed to be in a perpetual fog. I checked her individualized education program, talked with our special education teacher, and learned that she *was* in a fog: She had lead poisoning and a learning disability, and she had an extremely difficult time understanding, processing, and recalling information.

That same year, two of my students were diabetic and others were overweight and out of shape. They, too, were part of a national trend.

But what does this have to do with life in my classroom? Diana has a hard time retaining and processing information. Martin, an outgoing, athletic boy, has to take it easy in gym class and on the playground because exercise can trigger an asthma episode. Overweight 9-year-old Shayla can't run halfway around the school without stopping to walk. María has to have a special nurse come to school every day to give her insulin and monitor her diet. So what should I do?

Teaching About Toxins

BY KELLEY DAWSON SALAS

Of course, as a teacher it is my job to make sure individual children are cared for and to be sensitive to their medical needs. But asthma, lead poisoning, obesity, and diabetes are more than individual health problems. They are public health problems that plague my city of Milwaukee (and all others in the United States) and that disproportionately affect poor people, people of color, and other populations that are concentrated in urban areas.

Because these issues affect many of my students, I decided to engage my 4th-grade class in a study of public health problems. I also required them to choose projects on health topics for our annual science exhibition.

> I wanted them to understand that asthma is a public health issue and a disease that disproportionately affects poor, urban people of color.

With this unit, I wanted my students to learn about asthma, lead poisoning, obesity, and diabetes. I wanted them to link their personal experiences with these illnesses with new information, and to understand that these diseases plague some communities more than others because of environmental and economic injustice. I wanted them to understand that the illnesses we were studying disproportionately affect poor communities and people of color. And I hoped we could think through some actions that young people could take to combat the injustice.

We began by studying some basic facts about asthma and lead poisoning: What causes the illnesses? What are the symptoms? What can we do to protect ourselves from the things that can trigger or cause these illnesses?

I used materials on asthma that I got from a nurse at a local health clinic called the Sixteenth Street Health Clinic and from the American Lung Association website. The materials included several handouts that the clinic shared with families and asthmatic children. The fliers contrasted healthy lungs with lungs during an asthma attack. Other pamphlets described asthma triggers and symptoms signaling the onset of an asthma attack. All the materials were child friendly with illustrations and they used language that was age appropriate for my 4th graders. Best of all, the materials were in Spanish, the target language of my two-way bilingual classroom.

We also read the book *De modo que tú también tienes asma! (So You've Got Asthma Too!)*, which explains and illustrates what happens to the body when lungs are breathing normally and then again during an asthma attack.

The class worked together to make a poster of asthma triggers—things that bring on asthma attacks in asthmatics. We took the triggers from the health clinic's flier: cigarette smoke, dust, strong odors, mold, humidity, pets, abrupt climate change, cold and dry air, air pollution/ozone, cockroaches, pollen, food allergies like peanuts, colds, exercise, being overweight, strong emotions and stress, and the night.

Each child created a 3x5 card with a picture of a trigger and wrote a short paragraph that read something like "Dust could be an asthma trigger. Kids with asthma should stay away from dust." (In reality, people with asthma must learn to recognize and avoid only those triggers that provoke attacks for them personally, but I wanted the kids to have an awareness of the wide range of situations that can cause attacks in asthmatics.)

After learning about what triggers asthma attacks, we studied what an asthmatic should do during an attack. We read and discussed a flier from the American Lung Association's *Open Airways for Schools* curriculum that explains the basic steps an asthmatic should follow during an episode.

Obviously this information is useful for children with asthma, but for my students without asthma, it is useful for them to know how to respond when a classmate or relative has an asthma episode. It is especially important for children to know how to behave toward an asthmatic child. Children sometimes treat asthmatic kids differently—especially if exercise is a trigger and a child is not able to play or participate in physical education at the same level as other children. Kids need to be able to ask questions and get answers about asthma, and they also need to know that we expect them to treat asthmatic classmates with respect and compassion.

Public Health Implications

So far we had studied the causes and symptoms of asthma, and learned how asthmatic children can avoid or respond to asthma episodes. But I needed my students to go beyond that. I wanted them to understand that asthma is a public health issue and a disease that disproportionately affects poor, urban people of color.

We began to discuss the numbers and percentages of people with asthma. I shared with them the statistic that, according to the American Lung Association, about 10 percent of children in the United States have asthma. Because percent is not an easy concept for 4th graders to understand, I asked the students to take a blank grid of 100 squares and color in 10 of the squares. Under that grid we wrote: "Ten of every 100 kids in the United States have asthma." Then I gave them another copy of the same grid and told them that in big cities in the United States, 14 percent of children have asthma. They colored in 14 squares and wrote underneath: "In big cities in the United States, 14 of every 100 kids have asthma." (Washington, D.C., leads the nation in cases of childhood asthma with 18 percent.)

We discussed possible reasons why more children in cities have asthma. One student mentioned that cities have more factories and another mentioned that cities have more cars and traffic. We talked about older houses and schools like ours where there is a lot of dust in the vents and the air is not as clean. I asked why they thought the asthma rate is lower in the suburbs and rural areas. Their responses were limited: They guessed it was because there were fewer factories and cars, the houses were newer, and the air was cleaner.

The students' knowledge of asthma as an illness that targets specific groups was still superficial. I asked them: If we do a research project in our class, what percent of kids do you think will have asthma? They hypothesized that a lot of kids would have asthma, but did not state their hypotheses in terms of percentages, and they did not link their hypotheses to the national or urban percentages we had discussed. Instead, their hypotheses were anecdotal: "I think a lot of the kids in our class will have asthma because a lot of my friends have asthma."

We did the study with both groups of students I worked with. In one group, two of 25 kids (8 percent) had asthma. In the other group, six of 25 kids (24 percent) had asthma.

Studying Lead Poisoning

We moved on to study lead poisoning. I explained that lead poisoning is another sickness that affects people much more in big cities, especially in older houses in poorer urban areas. As a resource, I used a packet I downloaded from the National Institute of Environmental Health Sciences website called "Aventuras del club de los detectives del plomo"—"Adventures of the Lead Detectives Club."

Another Spanish-language, kid-friendly resource, this interactive website walks students through the dangers of lead, discusses the causes and symptoms of lead poisoning, and shows actions kids can take to protect themselves and their younger siblings from lead poisoning. It asks student visitors to respond to questions, click on lead poisoning hazards, play board games, and do word searches. (http://kids.niehs.nih.gov/explore/pollute/lead.htm).

After students learned what causes lead poisoning and how they can protect themselves and their younger siblings from lead hazards, I reminded them that lead poisoning is much more of a problem in big cities than it is in suburban or rural areas. We talked about why this is true. We discussed how most of the homes in the neighborhoods we live in are older homes, and I explained that paint used to be made with lead, but it is not made with lead anymore. (Lead paint was banned in the United States in 1978, and leaded gasoline was banned in 1986.) I told the kids that newer homes in suburban areas are less likely to have lead paint.

> "In big cities in the United States, 14 of every 100 kids have asthma."

I did not share statistics with the students on lead poisoning, because we moved on quickly to their science projects. But looking back, I think I should have spent more time working on percentages and helping kids to see and understand the drastic inequalities in how different populations experience lead poisoning. I could envision having students color a map of the country and a map of Milwaukee that was divided into neighborhoods to show where the highest concentration of lead poisoning cases is.

Science Projects

For their science projects, I asked students to investigate something related to children's health. I offered the two illnesses we'd studied, asthma and lead poisoning, and I also offered the general area of nutrition, diabetes, and obesity. (Although we had not done any previous study of these issues, several students in my class expressed an interest in doing projects related to diet, obesity, and diabetes.)

The class brainstormed several different questions that could be investigated for a science experiment. The process of arriving at questions that are appropriate for a 4th-grade science experiment is grueling: 4th-grade students typically don't know how to write a question that lends itself to a science experiment, and they need a lot of guidance. But I find that starting with the students' own questions and reworking them one by one into appropriate questions is the best approach—rather than just providing students with a list of questions and asking them to choose.

Questions we came up with included: How many 2nd graders take medication for asthma? What provokes asthma attacks for 4th graders who have asthma? How clean is the air in our classroom? Which classroom has the cleanest air, Room 33 or Room 32? How many 4th graders have lead paint in their homes? How many 1st graders know what lead poisoning is before and after a lesson on lead poisoning by 4th graders? What is the difference between the diet of a nondiabetic student and a diabetic student? What is the difference between the diet of a person who eats a lot of fast food and the diet of a person who eats a lot of home-cooked food?

> I would like students to understand the connection between air pollution and asthma. And I want to help them think of creative ways to speak to their community about the importance of improving air quality to improve children's health.

Working in pairs, students had to choose a question, write a hypothesis, write a procedure, conduct the experiment, make observations, gather data, display their data, and write a conclusion. Then they created displays for our science exhibition. The process took several weeks and a lot of guidance. Students needed support with the content and the steps of the scientific process; and they also needed assistance with Spanish-language arts skills. I worked with students in partners and in focus groups (the asthma group, the lead poisoning group, and the nutrition group).

Along the way, we made several discoveries about public health in our school: Marcela, Lorenzo, and Diana found out how many 2nd-grade students had asthma. Anita and Cristina learned that of 15 1st graders surveyed, not one knew what lead poisoning was—but they all claimed to understand what it was after Anita and Cristina's lesson. Rosa and Michelle found out what triggered 3rd graders' asthma, and Roberto and Jason learned that many 4th graders didn't know whether their house had been tested for lead paint or not. Rosario and Julia found out that there are more calories and fat in a fast-food diet than in a home-cooked diet (what a surprise), and Raquel and Sara unfortunately found out that the air in our classroom was not very clean. (Their experiment involved placing a Tupperware lid smeared with Vaseline in the room for several days, then examining it under a microscope to see what kinds of particles had settled on it—not pretty, but revealing.)

As a result of this learning, we made some changes in our classroom. We tried to keep things cleaner so the air would be cleaner. We got rid of our carpet after we learned it could harbor dust and mites that can trigger asthma episodes. I became more aware about pets and animals in school, and helped asthmatic students stay away from animals that were brought into the school. I began to sweep more often behind bookshelves (always a struggle) and tried to make sure my return air vent wasn't blocked.

All these changes, though, were only at the classroom level. What about all of the sources of air pollution in our community that seem to be beyond our control: cars, buses, power plants, and factories, to name a few? In the future when I teach about asthma, I would like students to understand the connection between air pollution and asthma. And I want to help them think of creative ways to speak to their community about the importance of improving air quality to improve children's health.

There are several other aspects of my teaching I hope to improve upon. To start with, I need better materials and more resources. I need to engage students in more meaningful lessons that will do a better job of helping them understand how race, poverty, and health intersect, so they will come away with a clearer understanding of how these illnesses afflict poor, urban communities.

I also need to do a more thorough job of teaching the content about what these illnesses are, their symptoms, and how to prevent them. I realized this when four of my students asked me to proofread a coloring book on lead poisoning that they were preparing for 1st graders. One page showed a dead girl lying in front of a house with peeling paint. The text read "Lead paint can make you sick and even kill you." When I talked about it with the girls I realized that although we had covered the effects of lead poisoning (brain damage, flu-like symptoms, learning disabilities), they had not retained the information—and they were ready to spread misinformation to the 1st graders. We went back to the website and re-read the part about what lead poisoning does to the body, and they corrected that page of their book before teaching their lesson to the 1st graders.

The next time I teach this subject, I want to include community outreach and require students to educate their own families as a major part of their work. I can envision students producing a pamphlet explaining what causes asthma, why asthma is more of a problem for people living in cities, why that's unfair, and what we could do to change it. Likewise, students could create a similar pamphlet about lead poisoning and include it as one piece of a lead poisoning prevention kit. The kit could also include a coloring book to teach kids how to avoid lead poisoning, information for parents on how to have their homes tested for lead paint, and possibly even a lead test kit. We could distribute the pamphlets and kits to 4th-grade parents or make them available to interested parents at all-school events. Students could also prepare an education/outreach packet on nutrition and obesity that explains what child obesity is, why it affects so many children in our country, and lists some things kids and families can do to prevent it.

Studying toxins and health hazards in our own communities offers children an opportunity to use science, math, reading, and writing to understand—and hopefully have some impact on—serious problems in our neighborhoods. Right now, too many students are sick: asthmatic, lead poisoned, learning disabled, diabetic, obese. Their illnesses are not their fault. The environment they live in is sick—and that is not their fault either. Someone has to make changes in our community and in our environment, so that they can be healthy. Why shouldn't they help make those changes? ⊕

Kelley Dawson Salas taught 4th grade at La Escuela Fratney in Milwaukee. She is communications and publications director at the Milwaukee Teachers' Education Association. All students' names have been changed.

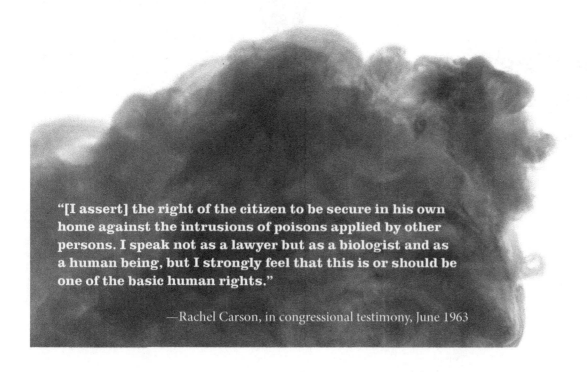

"[I assert] the right of the citizen to be secure in his own home against the intrusions of poisons applied by other persons. I speak not as a lawyer but as a biologist and as a human being, but I strongly feel that this is or should be one of the basic human rights."

—Rachel Carson, in congressional testimony, June 1963

Reading Chilpancingo

BY LINDA CHRISTENSEN

When I met Lourdes Lujan and saw her contaminated river, I knew I had to teach about Chilpancingo, a neighborhood in Tijuana, Mexico, where corporations' toxic presence and women's organizing against the monster in their backyards make for David and Goliath teaching lessons.

When our Rethinking Schools-Global Exchange tour first entered Colonia Chilpancingo and descended downhill to the place where the waste from hilltop factories bubbled into Río Alamar, the stench of burning rubber and untreated garbage filled the air. Lujan, a local activist, walked down from her home perched above the river. "When I was growing up, I played and fished in this river," she said. "My family had picnics at that park." She pointed to a sandy, trash-filled triangle at the edge of the river filled with discarded tires, pop cans, plastic bottles, baby diapers, and a lone picnic table. Her arms are pocked with rashes.

"Now, when the stream starts to flow, and it isn't even the rainy season, the kids play in the puddles. You know how kids are. They play in the water, and get blisters all over their feet."

Lujan is part of a collective that works with women in their local communities and across the U.S. border to battle the giant corporations and governments that have made her home a nightmare. But the horror story unleashed by Metales y Derivados, a U.S.-owned battery recycling company whose owner abandoned thousands of tons of hazardous chemicals that contaminated Colonia Chilpancingo, didn't stop with rashes and blisters. Lujan described children born without brain stems, children whose parents slept with them at night, fearful they would drown in their own blood from spontaneous nosebleeds, *maquila* workers who suffered miscarriages and had babies with birth defects, and neighbors with abnormally high rates of cancer. She pointed to the buildings on the ridge above Chilpancingo and explained how Santa Ana winds blow contaminated waste down into her village, how the rains sluice down the side of the mesa and pool in the grade school at the bottom of the hill.

I watched two boys walking on a muddy path, heading to their homes constructed from wooden pallets and tarps, homes without running water or sewage connections. The men, women, and children who seek shelter along this toxic riverbed have traveled from southern Mexico, hoping to find work on the *maquila*-saturated hilltop and a toehold into a better life. We asked Lujan if they knew about the contamination. She shrugged. It's hard to get people to care about potential harm when they worry every day about getting enough food for their children.

Later, we learned about elevated lead counts, the high price of lead tests compared to *maquila* wages, and the long fight that Lujan and her group have waged to educate and organize their neighbors about the factories' toxic waste. We learned about the lawsuits they've sponsored to make Jose Kahn, the San Diego-based owner of Metales y Derivados, and the U.S. and Mexican governments clean up the poisons surrounding Chilpancingo.

Back Home

To bring the lesson back to my junior students at Grant High School in Portland, Oregon, I start with

a photograph of the hill opposite Chilpancingo that includes a shot of the makeshift housing for *maquila* workers and a polluted stream. (Available at www.rethinkingschools.org/earth) Eventually, they will read an article about the toxic waste, but I use the photograph to stir their interest.

My students are a delightful mix of colors, attitudes, and ambitions. What unites them is that they didn't choose to be in any of the honors, AP, or advanced English classes that Grant High School offers. Many struggle with reading and writing; about a third spend at least one period of the day in special education classes or they have a caseworker; others have sophisticated literacy skills, but have disengaged from school for a variety of reasons. Any lesson I bring to my class must include reading and writing strategies for this diverse group. I've discovered through our lessons on race, class, and Hurricane Katrina that they care about the world beyond their cell phones. I attempt to build units that teach literacy skills embedded in larger world issues, and I try to find places where they can learn to read critically, but that also give them examples of how people have worked together to confront oppression. Chilpancingo fits my criteria.

I project the photograph of Chilpancingo. I tell students: "Reading a picture is like reading a text. You read on a number of levels. I want you to read the picture first. Just make a list of everything you see in the photo. Don't make any judgments about what's there. Make a list." My students verbalize while they work. Perhaps not everything they think comes out of their mouths, but they are not a quiet group: "Shacks," one student says.

> She pointed to a sandy, trash-filled triangle at the edge of the river filled with discarded tires, pop cans, plastic bottles, baby diapers, and a lone picnic table. Her arms are pocked with rashes.

"Wait. Wait. You're making a judgment," I tell them. "Just write what you see. Someone give me an example." Ann Truax, a Portland ESL teacher, taught me to begin reading lessons with visual texts to draw students in, but also to give English language learners pictures in their heads as they encounter new vocabulary. I've discovered this strategy works with all students as a pre-reading strategy.

The students list details they notice in the picture: Water. Bottles. Cans. Dead tree. Shadow. Smoke. Blue tarps. Water. Gray skies. Truck. Houses. Rust. Their lists grow long. They crowd around the overhead to get a better look. Later, I see that Charlie has written "Sedges a small weed-like plant that grows by water."

Students share their lists with a partner. Then we go around the class and everyone adds items from their lists. When Dontay calls out "Dirty water," I stop the shout-out.

"Wait. That's a judgment. Where's your evidence? How can you tell that the water is dirty?"

Dontay looks back at the picture. "See the bottles and paper in the water? That's what makes it dirty."

"OK. Now when you do that in reading, it's called an inference. You gather up information, then you make a judgment about it. For example, if you read about a man who slapped his child in the grocery store, what kind of inference would you make about him?"

> **"Where did the garbage come from? Why doesn't someone clean it up?"**

"Child abuser," Alley shouts.

I admit: I am shamelessly didactic, but too many of my students have bought into the idea that they can't read. My job is to show them that they are "reading" all the time. I want to name their reading strategies, so when they read word texts they can remember they know how to do this.

"Now, I want you to list your questions." When no one writes, I ask, "Who has a question?"

Josh starts: "The five W's. Where is the picture taken? What is in the picture? When was it taken? Why was it taken? Who took the picture?"

"What's in the river?" Katie asks.

"Is there piss in the river?" Vernell jokes to get a laugh.

"Now, on your own." I use this question technique prior to introducing the reading because often my poor readers read the surface of words. They skim over the paragraphs, forgetting what they've read as soon as their eyes pass the words. They even talk about their "comprehension problems." I use activities like this one to teach them how to read with questions in their heads to provide hooks to slow

them down, but also to capture images, facts, and ideas that slide by when they read without purpose.

"When we share out, I want you to write down other people's questions. When you read the article about this place, I want you to read to answer all of these questions." I've found that it's not enough to practice strategies without discussing them with my students. When I make my reasons for using the strategies transparent, they understand what I'm doing, so they can transfer the process. At parent conferences, I was pleased when the parent of one of my mainstreamed special education students explained how her son was transferring the strategies he'd learned in my class to his history class. He showed her how he highlighted and wrote marginal notes and told her that it helped him retain the information.

After students review their questions with a partner, we share in the large group again. I write the questions on the overhead as students call them out. It's clear they've started caring more about what's happening in the picture. Their cute, glib remarks are gone. "Who lives there? Is this humans' destruction or nature's? Where did the garbage come from? Why doesn't someone clean it up? What caused this mess? Where is the water coming from? Are there any dead bodies? Is this a dumpsite? Do people live there? Why is that tree dead? What is in that mound? Is this a wetland? Is that a neighborhood behind the trees and shrubs? Do people drink the water? What are they trying to show with the picture? Why don't the people move to a better place?"

Making Connections

I move to the next pre-reading strategy—connecting new information to previous knowledge. Skilled readers and learners do this automatically, but struggling readers don't. "Now, I want you to make connections between the picture and your lives. Does this remind you of any place you've been? Movies? News programs? Something you've read? Who can make a connection?"

No one answers. "Does it look like a spot on the way to the Oregon coast?" Still no answer. I continue: "When I showed this picture to a group of teachers, one of them said that it looked like a field where he and his friends hung out in high school. It kind of reminds me of Beggar's Tick, the site where my daughter Gretchen took water and soil samples when she was in Mr. Street's environmental science class."

"It looks like the off-ramp to the dump," Ethan says.

"Great. The reason you want to make connections between the picture and your life is that when you make a connection, your brain finds a way to remember it. This is the same thing you do when you read. But it's also to think about the similarities in their lives and ours—as well as the differences."

Calais raises her hand, "This reminds me of Tijuana when Heather and I went there last summer with our church group to build a playground."

"OK, now write your connections."

Thomas writes that it reminds him of the ads about adopting poor children. "They always show them surrounded by trash." Other students write: a relative's backyard, New Orleans after the hurricane. Charlie writes that it reminds him of the "crap Tim Robbins had to crawl through at the end of *Shawshank Redemption*." Brittany writes, "Driving through LA in September."

"Now, it's time to read about this place. All year, we've been keeping two ideas in front of us as we read: injustice and hope. When we read about Hurricane Katrina, we examined where there was injustice, and where we found hope. When we read *Thousand Pieces of Gold*, we talked about the injustice that Lalu faced, and where she found hope. Now, I'm going to give you an article that was published in the *Washington Post*; I want you to read to identify both the injustice and the hope in this situation. I also want you to use the same strategies we used to 'read' the picture while you are reading the article. Keep those questions in front of you as you read."

The class moves into their work groups. In this class of diverse abilities, I strategically place strong and weak readers together. I hang group members' names on the sides of the room where I've clustered their desks. I distribute highlighters and copies of Kevin Sullivan's article "A Toxic Legacy on the Mexican Border," p. 294, which details the environmental devasta-

Lourdes Lujan

CALIFORNIA NEWSREEL

tion. It is a difficult read for many of my students.

"Feel free to read this out loud in your group. Highlight places in the article that answer your questions. Write new questions that the article raises for you in the margins. Find the justice and the hope." Students' outrage is immediate. I can hear Ethan cursing in his corner overlooking the soccer field. I hear Ryan: "Damn, man, this is cold." The bell rings before we can discuss the article.

Adding Details

The next day I draw a circle on the overhead; I write "Chilpancingo" in the middle and ask students to return to their articles and notes. "Let's review the reading from yesterday. What key pieces of information should I add to the 'map' of the article?"

Hillary says, "You need a section for birth defects."

Russell adds, "You need a part for effects of the toxic waste on the people who live there."

"How about a section on what happened to the environment?"

The overhead fills up as students tell me where to add details. Josh points to the overhead, "By birth defects you want to add stillbirths, born without lower body and without skulls."

Katie adds, "By the environment, you need to add toxic waste; Metales; Jose Kahn, the owner of Metales."

I had worried that students wouldn't understand the article, but without even looking back at their notes, they were shouting out answers. When they did return to the article, they looked for specific numbers—the amount of waste left behind, the cost of the cleanup.

"Now, tell me what new questions came up for you in your reading? What do you still want to know that this article didn't answer?" I wind the overhead to a new sheet as students call out their questions. "You write them down too," I say, "because we will read more articles about Chilpancingo."

I list their questions on the overhead: What happened to the children? Did Jose Kahn ever clean up the waste he left behind? Are there still birth defects? Who helped clean it up? What happened to Carmen's new baby? Their questions are specific. Pointed. They want to know what happened to these people.

> Their questions are specific. Pointed. They want to know what happened to these people.

"I want you to get back into your groups. Each person in your group is going to read a different article that gives more information. In order to answer your questions, you will each need to read carefully, take notes, and then share your information with your group."

I distribute color-coded articles written from different perspectives about Chilpancingo. (These articles are online at www.rethinkingschools.org/earth.) For example, I give one student in each group Mariana Martínez's article "Empowerment Brings Change." This article describes how Factor X, a Tijuana women's group that provides education to *maquila* workers, teaches women how to advocate for themselves and their communities. As Martínez explains:

> Little by little people are learning to speak out, and become an agent of change. An example of the people taking charge was the case against Metales y Derivados, a company owned by New Frontier Company in San Diego. After many complaints from the community to the Mexican environmental authorities, after gathering over 500 signatures and organizing protests, the case was finally brought to the attention of the Environ-

mental Cooperation Committee, which was established as part of the Free Trade Treaty [NAFTA], who . . . established that the chemicals that this company manages are of "grave danger to human health" and that "Mexican authorities had failed to enforce their own environmental laws."

Sadly, Factor X no longer exists, but the impact of its work lingers.

I give the article "Environmental Health and Toxic Waste" to my struggling readers because it contains pictures of Lourdes Lujan and Magdalena Cerda, an activist from San Diego, as well as pictures of the Metales plant. It has shorter passages, but it provides some relevant information about how the toxic waste generated in Tijuana doesn't stay there:

> The air and water are shared. The runoff from the Industrial City flows into a stream in Colonia Chilpancingo. . . . The pollution produced in Tijuana equally affects the people of San Diego. In addition, the capital that comes to Tijuana is American capital, which for the people here produces only a little money and a lot of pollution. People in Mexico need this work desperately, but it doesn't allow them to live in dignity or comfort. This type of injustice is not tolerable, and this is why we work together.

I realize that I didn't clarify to students that the second round of readings would contain overlapping information with the first. I also didn't point out the different dates on the readings, so they were not as alert to the changes within each article as I hoped they would be.

"When you finish your reading, share your information with your group. Add this new information to your map. Then write a paragraph summarizing what you learned about injustices in Chilpancingo and where you find hope."

Most express both outrage and hope about the environmental destruction wreaked on this community. Charlie's anger at Jose Kahn is echoed by his classmates, but he does find hope in the community's organizing:

> I believe what Jose Kahn did was horrible and the United States should make him pay for the

cleanup. The community around the pollution is paying instead. There is a lot of hope in that community, along with a lot of strength. They have come together and formed programs to educate themselves on how to deal with the problem. Meanwhile, the U.S. needs to do the right thing and make Jose Kahn pay.

Brittany blames Jose Kahn as well, but also finds hope in the community's resistance:

It was definitely wrong for Jose Kahn to just leave all that toxic waste in Chilpancingo. It was killing all of the residents. Not immediately, but slowly and painfully. . . . Jose didn't have to pay because he crossed the border. But what is cool is that the residents are being informed and informing each other about how to protect themselves and deal with the toxins. There is a five-year project to clean up Chilpancingo.

My students didn't travel to Mexico with me. They didn't stand on the banks of Río Alamar, smell the acrid odor of a town drowning in toxins, see the rash on Lujan's arms. But sitting in a classroom near the banks of the Columbia River, they learned how to step into a picture and connect with a community on the other side of the border and question why it's OK for a U.S. corporation to leave toxic waste behind. And they discovered how women organizing in local communities can tackle giants—and win. ⊕

Linda Christensen is the director of the Oregon Writing Project at Lewis & Clark College. She is a Rethinking Schools *editor and is author of* Reading, Writing, and Rising Up: Teaching About Social Justice and the Power of the Written Word *and* Teaching for Joy and Justice.

A City of Factories

An excellent film chronicling women's community organizing in Tijuana is *Maquilapolis.* In *Rethinking Schools,* Julie Treick O'Neill writes:

Push, assemble, remove, push, assemble, remove. A line of women are dressed alike in blue smocks that indicate their respective positions in one of Tijuana, Mexico's 4,000 factories. They are the manufacturing "machines" corporations so desire in the global economy. Silently, they push, assemble, remove, push, assemble, remove.

But as the film continues, the power of *Maquilapolis (City of Factories)* is evident—the women come alive, sharing their dilemmas, resistance, and hope. The film follows two former maquiladora workers, Lourdes Lujan and Carmen Duran, as they take on the multinational corporations harming their community and infringing on workers' rights. The women are *promotoras,* members of a social justice group organized to educate and empower the thousands of Tijuana maquiladora workers.

As the film demonstrates, the same corporations that mistreat workers also mistreat the environment. *Maquilapolis* presents the outrages described in "Reading Chilpancingo" and "A Toxic Legacy on the Mexican Border," But the film shows people fighting back, with imagination and determination. ⊕

A Toxic Legacy on the Mexican Border

BY KEVIN SULLIVAN

Andrea's monster lives up here. It breathes lead dust that coats her windows and her baby toys. It sweats rivers of arsenic and cadmium and antimony that seep into her water and the soil where her children play. It squats on a hilltop above her home, horrible and poisonous.

"There it is," says Andrea Pedro Aguilar, breathing heavily from the hike up the hill.

She is standing in front of her monster, the derelict remains of a lead smelter that everyone here calls Metales. For more than a decade, an American-owned company, Metales y Derivados, took in thousands of U.S. car and boat batteries, cracked them open to extract their lead, melted it into bricks and shipped the bricks back to the United States.

Mexico shut the plant in 1994 and the next year its owner, a U.S. citizen named Jose Kahn, crossed the border back into San Diego. Mexican arrest warrants were outstanding, charging him with gross environmental pollution. He still lives in a comfortable neighborhood of San Diego.

According to the Mexican government, he left behind up to 8,500 tons of toxins from battery guts that lie strewn over three acres, in open piles, rusted barrels, and rotted bales. Every time the wind blows or the rain falls, more of the toxins end up in Colonia Chilpancingo, a workers' village of 10,000 people directly below the plant.

According to Mexican environmental officials and the U.S. Environmental Protection Agency, the toxic dump here exemplifies how much of the border area is a no-man's-land, a place where international companies have polluted the environment.

When the Metales furnaces were still burning in 1990, a Mexican university study found levels of lead more than 3,000 times higher than U.S. standards and levels of cadmium more than 1,000 times higher in a stream that runs through the community and eventually flows north over the border into the United States. A 1999 study by the enforcement division of Mexico's environment ministry found lead concentrations in the soil near the plant 50 times higher than the limit set by Mexican law. That report called the Metales site a "major health risk."

A cleanup of the site could cost $6 million or more. In 2002, the state of Baja California and Kahn filed a joint loan request for $800,000 from the North American Development Bank, which was created as part of the 1994 North American Free Trade Agreement (NAFTA). A bank official said the unusual request—coming amid rising demands from residents for a cleanup—is being reviewed. He said one concern is that the loan might not cover the cost of the cleanup.

In the meantime, the toxins bake in the sun and blow in the wind. The pollution keeps flowing into Chilpancingo, from Metales and from some of the other 130 factories, known as maquiladoras, in the huge industrial park where it sits.

"Danger, hazardous waste" is stenciled on the concrete wall that partially surrounds Metales. But the place is still a favorite for dare-taking kids who scoot through holes in a fence into the forbidden site.

Reached by telephone, Kahn, who is in his late 80s, said, "We are negotiating a loan to clean up the place. I really can't tell you anything more than that." He declined further comment on Friday. In an interview published in the *San Diego Union-Tribune*, Kahn said the loan request shows that he is serious about cleaning up Metales: "We all want a solution. No one wants to walk about without a cleanup."

It hasn't rained in Chilpancingo for nearly two months, but dirty water still runs down the middle of Andrea's street. It starts in a gaping drainage pipe that emerges from beneath the industrial park that emits a milky white flow of God-knows-what that flows downhill to Andrea's neighborhood. Factories there are required by law to treat their own hazardous waste, but state environmental officials say many still dump illegally.

"I don't know what they were thinking," says Andrea, who had two feet of acrid, filthy water in her living room when heavy rains caused flooding last year. "People live down there."

Neighbors like her kids. Lupita is 4 and Ivan is 6. They ride scooters in their living room and watch *Monsters Inc.* and *Rugrats* for hours on end. Andrea thinks it's safer for them to be inside even though her little lead-testing kits have turned up elevated levels of the toxin on her dishes and on the sill of her kitchen window. Outside, the fruit trees and grass that her mother planted 20 years ago have all died.

Just before Christmas, 20 Chilpancingo children under the age of 6 were tested for lead. Officials from the Environmental Health Coalition, a San Diego-based organization, said that all the results showed significant and potentially dangerous levels of lead in their bloodstreams. Lupita's blood had the highest level, 9 micrograms of lead per deciliter, just under the level of 10 micrograms per deciliter classified as elevated for children by the U.S. Centers for Disease Control and Prevention. Lead, especially in children, can damage organs and severely retard mental development, and studies suggest it may cause cancer and birth defects.

Officials at the CDC said arsenic, cadmium, antimony and other byproducts of the smelting process are carcinogens. CDC officials also said that exposure to those metals can cause skin rashes, nosebleeds, and hair loss.

Lupita's hair slips out by the brush-full every day, and she has suffered spontaneous bleeding in her nose and throat for the past couple of years. It got so bad in November that Andrea and her husband slept with Lupita, out of fear that she might drown in her own blood. Andrea says they did not know what was causing her problems. Then her lead test came back positive.

Now Andrea stands before the monster, a mile south of the U.S. border, shaking her head in disgust.

Wenceslao Martinez, a physician, runs a health clinic a few blocks from Andrea's house. He says he constantly sees patients with suspicious diseases, from chronic rashes to cancers to fatal birth defects.

"For a colonia of only 10,000 people, what we see here is very strange," he said. "There is definitely a link to the maquiladoras. But it's hard to prove. So who gets the blame? Nobody."

He treated Margarita Jaimes' 3-year-old son,

Serafin Vidrio, who turned up one day last July with swelling in his neck and eyes. He was diagnosed with acute leukemia on Aug. 6. He died on Aug. 24.

Margarita, like the others, is frustrated no one has spent the money to study whether the illnesses around these factories are linked to the toxins they have dumped. As she talks, her daughter, Eva Paulette, 6, sits on her lap. She has been having nosebleeds. Her hair is falling out in clumps. The doctors cannot explain it.

Carmen Garcia used to walk to work every day past Andrea's house, past the open piles of sludge at Metales to a factory where she assembled stereo speakers. When she became pregnant two years ago, she knew her factory was not the best environment, because in the previous two years three of her co-workers had delivered stillborn babies.

> The enforcement division of Mexico's environment ministry said that "with alarming regularity" foreign-owned factories are being abandoned, with their hazardous waste left behind.

Then on Nov. 3, 2000, Carmen delivered Miguel Angel, who suffered from anencephaly, a fatal defect in which babies are born with little or no brain or skull. Miguel Angel's empty skull was open wide like a tulip. He survived for two months.

"It's like a trap here," Carmen said. She's pregnant again. "I'm so scared."

A CDC spokesperson estimated that anencephaly occurs in two to four of every 10,000 births in the United States; hydrocephaly, a related disorder, occurs in about six of 10,000 births.

The state of Baja California, which includes Tijuana, began conducting its first major study of those two birth defects in 2001. Moises Rodriguez Lomeli, the state's chief epidemiologist, said the study was launched after state officials realized the rate of those birth defects in the state was abnormally high. In one two-block area of Chilpancingo, residents count eight babies born with those two defects in recent years.

Andrea and other community leaders, working with the Environmental Health Coalition, filed a complaint about Metales with the Commission for Environmental Cooperation, NAFTA's environmental watchdog agency. The commission issued a report in 2002 noting that "exposure to these heavy metals can severely harm human health" and called the site's cleanup "urgent."

In the commission's 154-page report, the enforcement division of Mexico's environment ministry said that "with alarming regularity" foreign-owned factories are being abandoned, with their hazardous waste left behind. It also said the EPA viewed the Metales situation as "exemplifying a critical public policy issue in the border region: the use of the border as a shield against enforcement." The Mexican government has been reluctant to clean up foreign-made messes, and when the foreigners return home they are beyond the reach of Mexico's laws.

Black smoke is rising from a burning car behind Andrea's house. She's standing in her fenced-in yard, where she rents out a couple of small shacks to make a little money. The woman who lives in one of them gave birth a few months ago to a baby missing most of its lower body. The previous tenant in the same house woke up with her neck swollen like a bullfrog's. "Two months later, she was dead," Andrea says. "Nobody ever knew why."

Andrea and other women in the community, with help from the Environmental Health Coalition, are now trying to educate residents about the hazards around them. They pass out lead-testing kits and arrange blood tests for children. They write to government officials and hold all-night vigils outside their offices. They marched on Kahn's office in San Diego, holding up signs with such messages as "Jose Kahn: You forgot something in Tijuana."

A couple of unhurried firemen arrive to begin hosing down the burning car. Andrea, who is pregnant again, says she dreams of the day when all the toxic pollution is gone, when Chilpancingo is clean and healthy, filled with flowers and trees, the way she remembered it as a girl.

Then she closes her eyes against the thick, black smoke. ⊕

Reprinted from the Washington Post.

"**W**ho actually owns the water?" One of my students asked me as I introduced a mini-unit on water to my 5th graders.

"Well, who do you think does?" I responded.

"The government!"

"Nobody!"

"Bill Gates!"

"The fish!" were among the chorus of replies.

I wrote the question and responses on a wall chart that listed what the kids already knew about water.

The students already knew many basics: About 70 percent of the Earth is covered with water; most of it is salt water; it exists in three different forms; most of our bodies are composed of water; and more.

Their questions included: Who contaminates water? What country has the most water? How much does water cost? Does the chlorine put in the water in the state's largest water park hurt animals?

I told students they could add more questions throughout the day and I would type and categorize them for us to discuss the next day.

I had decided to teach a water unit for several reasons. I wanted the "data and statistics" unit in the math curriculum to be engaging and socially relevant; the school science night loomed a couple weeks ahead; and I had learned some fascinating things about water on a recent trip to the Tijuana border area during a Rethinking Schools "From the World to Our Classrooms" curriculum tour. I had been particularly amazed at how much water cost in the impoverished community of Chilpancingo, just outside of Tijuana. (See "Reading Chilpancingo," p. 288.)

RICARDO LEVINS MORALES

Measuring Water with Justice

BY BOB PETERSON

I had more specific goals as well. I wanted students to learn the importance and power of data in understanding significant scientific phenomena and problems while also getting practice representing that data and communicating it to others. And I wanted students to learn more about water—which many of us take for granted—and the central role water plays in our lives.

After school, I categorized the students' questions, pulled relevant books from our school library, and downloaded images from the internet to make a short PowerPoint presentation. I included the students' questions, images of people from Africa and Asia collecting water, and the aftereffects of oil spills.

The next day we reviewed a couple of key math concepts and skills. I had each student draw and label a circle graph of the percentage of the Earth's surface covered by water. In the process we reviewed equivalent fractions and percent. We looked at a chart of a person's average water usage in the United States and calculated how much water a person would use over the course of a week and year.

Then I showed them the PowerPoint presentation and some books I planned on sharing with them. We concluded by sharing the students' questions from the day before, which I had divided into four categories:

> **I had more specific goals as well. I wanted students to learn the importance and power of data in understanding significant scientific phenomena.**

- Water basics—fresh and salt water; how much do we use?
- Who can get clean water?
- How much does water cost?
- Oil spills.

Images of oil-soaked birds and seals generated the most interest and I realized that if had let the students indicate their preferred areas of study, all would have chosen oil spills. I decided to postpone the choosing, and instead read to them the picture book *Prince William,* by Gloria and Ted Rand, a fictionalized account of the rescue of a baby seal during an oil spill.

For homework that night I asked students to record every time they used water and estimate how much water they consumed.

That evening I decided to divide the oil spills category into two: oil spills of the world and the Exxon Valdez oil spill. [This activity was taught before the BP oil disaster in the Gulf. —eds.] That would allow more students to focus on oil spills and their consequences. I also decided to have all the students do a science/language arts activity on oil spills to tap their interest.

The following day I explained in more detail the five topics, the questions they would try to answer, and some suggested activities. I asked students to list their top three choices on a piece of paper.

Writing on Oil Spills

I read aloud and we discussed another picture book, *Washing the Willow Tree Loon,* by Jacqueline Briggs Martin. The book describes how a bird is rescued from an oil spill, cleaned, and returned to the wild. I also gave each student a two-page fact sheet on water and pollution from the National Oceanic and Atmospheric Administration (NOAA) website. (See end of article for URL.)

After reading the materials we brainstormed all the different animals that might have been affected by the spill and what aspects of their lives would have changed. I had students choose one animal from the list and asked them to write from the animal's perspective. Before writing, we talked about what might be in an animal's "diary"—how it first saw the spill, how the oil affected its food supply, living situation, members of its family, etc. The students worked on their animal diaries during writing workshop and continued them for homework. The next day they worked with their peers to write second drafts.

Over the next couple of days all the students completed their diaries. Eventually, the students glued their final drafts to construction paper, added either a drawing or an internet photo of an oil-soaked animal, and displayed them on the bulletin board. The writing was uneven. Some students used lots of details about how the animals might have reacted to spills, while a few had the animals die immediately.

Aliza wrote from a duck's perspective:

I was swimming until blackness filled the

water. A beautiful day it was, until it started to get very cold. I thought I was a Popsicle, frozen, couldn't move, could hardly breathe. I tried to clean myself but the taste was unbelievable. So disgusting like black coal, thick, hard to swallow. I tried to escape, I tried to fly but I couldn't. I was trapped. That whole day I thought I was going to die.

Another student, Margarita, described the thoughts of an otter:

Then some black ooze came toward us. It was smelly, thick, slimy, and smelled gross. It covered all of me except my head. It was like glue, I couldn't move. The black ooze was everywhere. I thought what is this? Will I die, are these the last days of my life? Then suddenly this person took me to a place that had a sign saying "Rescue Center for Animals." Of course I didn't know what it meant. I was wondering where I was. I couldn't see any water.

When we were inside, they put me into water and it was so relaxing. They cleaned me off and it felt great. I heard them say, "We got the oil off and she looks great." I wondered what oil was. One thing was for sure I was cleaned off. I was saved. I was so content these were not the last days of my life. I was the happiest otter ever.

The essays were basic, but they encouraged students to look at human-made phenomena from the perspective of other species. I think in this country we take too many things for granted—like the availability of water—and we rarely stop to think about the impact of our actions on nonhuman species.

Water Basics

I posted the research groups at the beginning of the day so students could finish complaining about whom they were or were not grouped with before math time started. At the beginning of the math period I reviewed my expectations for group work, especially the requirement that all students need to be engaged and that students need to stay in their assigned groups. I also presented a list of activities I expected each group to complete during the next four math/science periods. I reminded students to show their math work, to use their own words in presenting information, and to put their final work on a tri-fold poster display board.

The "water basics" groups worked on graphically showing the role of water in everyday life. They graphed the percentage of water covering the Earth's surface, the amount of water in a human being, and the amount of water that a typical person from the United States uses every day. One of the group members exclaimed, "Water is everywhere. Inside you and you use it all the time. It's almost like we are fish. Well, not really."

BOB PETERSON

Students in the group working on who has access to clean water were among the most startled. Reading the book *For Every Child: The U.N. Convention on Rights of the Child in Words and Pictures* and a selection from the Universal Declaration of Human Rights, they found that access to clean water is considered a human right by various international conventions. They were surprised to learn that despite this, more than 750 million people don't have access to clean water.

The group spent considerable time printing photos from the internet depicting how different people get water. I insisted that they write captions describing the pictures and identifying their locations. Two of the group members wrote a dialogue poem between two girls—one with access to clean water and one without—contrasting their different lives.

At the end of the activity, the group wrote "Clean Water Is a Right!" in huge letters on their poster board.

Water Access and Cost

The group researching water costs had a hard time tracking down data because of differences in how water costs are reported and rates that vary on the amount of water use. Students ended up comparing the costs of water in Milwaukee; Chilpancingo, Mexico; and Cochabamba, Bolivia.

Two 11-Year-Old Girls

We are 11-year-old girls.

I get my water in my house.

I go to the sink in my kitchen
and turn on the faucet.

When I turn on the faucet
I can drink hot or cold water right away.

I can take a shower or bath
in my house for as long as I want.

My mom puts our dirty clothes in
the washing machine and the
washing machine does all the work.

I am happy to have clean water
in my house.

We are 11-year-old girls.

I get my water from the river.

*I have to walk a mile to the river
and scoop it up with a bucket.*

*I have to boil my water because it
is contaminated.*

I have to take a bath in the river.

*I have to wash my clothes directly
in the river and I have to do all the
work myself.*

*I wish I had clean water
in my house.*

—Alejandra Guevara and Brenda Villanueva

The data that the group used were varied and interesting. The water costs from Milwaukee were based on my water bill. Some students were surprised to learn that we pay for water at all. The information on water prices from the community of Chilpancingo I had collected as part of the "From the World to Our Classrooms" curriculum tour I had taken. We found water costs in Cochabamba, Bolivia, on the web. We even found scans of actual water bills for individuals, showing a spike in prices once Cochabamba officials privatized the water, after contracting with Bechtel, the giant transnational corporation.

The students made a chart that had information on the cost of water per gallon and per 100 gallons, the minimum wage in the area, and percentage of daily minimum wage to pay for 100 gallons of water. The contrasts were startling. They found that "in Milwaukee people paid $1.18 for 748 gallons of water. One hundred gallons of water cost only 15 cents. At the time, the minimum wage in Mil-

waukee was $5.15 per hour and $41.20 per day. The percentage of the daily minimum wage to pay for 100 gallons of water was about one-third of 1 percent." In Chilpancingo, students found, the people paid 15 pesos, about $1.50 for 55 gallons of water that a truck delivered to their homes, which lacked running water. So 100 gallons cost a bit less than $3.00. The minimum wage in the area was $5 per day. They concluded that the percentage of the daily minimum wage to pay for 100 gallons of water was more than 50 percent.

The group filled a plastic milk jug with water and labeled how much it would cost in Milwaukee and Chilpancingo.

The group didn't make a direct comparison for Bolivia because the pricing varied depending on the number of cubic meters. "The cost went high because some new people—Bechtel Corporation— bought the water company," they concluded. "They doubled the prices or in some cases raised them by

50 percent. After lots of protests in which people got killed, the government took back the water company from Bechtel."

Ivory, one of the group members, wrote:

> In Chilpancingo, Mexico, people pay 3 cents for a gallon of water. That's about 20 times more than what people pay in Milwaukee. In other words, we can get 20 gallons for what they pay for one, and people in Milwaukee make more money in the first place.

> In conclusion, we think everybody should pay the same prices for water as in Milwaukee. It's not fair that in Chilpancingo people have to pay so much money for so little water, especially because it takes up about half of their salary. For the amount of money that people in Tijuana pay for 44 gallons we in Milwaukee could get 1,000 gallons. We learned that water is expensive in some places and in some places it is cheap. In Bolivia the people protested to keep the price of water lower. Next time I drink a cup of water I am going think about how much it is going to cost in other places.

Oil Spills and Exxon Valdez

The students studying oil spills graphed the number of oil spills, starting in 1970, and marked the location of the 50 largest ones on a large world map. They also made circle graphs that showed the causes behind large oil spills (more than 100 tons) and smaller ones, noting that "the larger ones usually are caused by collisions (28 percent) or grounding (34 percent), while the smaller spills usually happened during loading and discharging." They were surprised that the Exxon Valdez oil spill, which we had all read about, was not the largest spill ever.

The group looking at the Exxon Valdez had the most information of any group—in the form of children's books and websites. They wrote a brief history of the spill, drew a map of the affected area, and in-

BOB PETERSON

cluded photos of oil-soaked sea animals.

To show how much 11 million was—the number of gallons of oil spilled in Prince William Sound by the Exxon Valdez—students filled a one-gallon plastic milk jug with water dyed with food coloring to make it look like oil. Then they set it next to a math sheet with 5,000 dots that the class had used in a previous math assignment. The group calculated that it would take 2,200 pages of paper, each with 5,000 dots on it to equal the amount of oil spilled by the Exxon Valdez. "Just imagine all that dirty stuff going out into the water," said one student after finishing the calculations.

> **The group calculated that it would take 2,200 pages of paper, each with 5,000 dots on it to equal the amount of oil spilled by the Exxon Valdez.**

The group reflected on the research:

> We learned that 11 million gallons of oil spilled when the Exxon Valdez ran aground trying to avoid icebergs off the coast of Alaska in Prince William Sound. The water was polluted and it killed millions of animals like sea birds, otters, seals, whales, eagles, and millions of salmon and others. The carcasses (the dead bodies) of more than 35,000 birds and

1,000 sea otters were found after the oil spill but since most carcasses sink, this probably is only a fraction of all those who were killed. The best estimates are 250,000 seabirds, 2,800 sea otters, 300 harbor seals, 250 bald eagles, up to 22 killer whales, and billions of salmon and herring eggs.

Science Night and Reflections

Each group made an oral presentation to the rest of the class. The presentations were somewhat anticlimactic because they were laden with data and the students hadn't carefully thought through the best way to present the material. The two most engaging parts of the presentation were the dialogue poem and the students' new knowledge about the large numbers of people who do not have access to clean running water in their homes. I challenged students to not use the faucets in their homes for the entire next day.

The final connection students made was when they looked at the pictures of the street demonstrations in Bolivia, where people protested the privatization of their water supply: "That's just like in the Civil Rights Movement," said one student. "The women's fight to vote movement, too," added another. "Geesh, just for clean water," added a third. "That's amazing."

The afternoon before science night, the students set up their displays close to the bulletin board that had the animal diaries. When parents arrived, they read the diaries as well as the information on their display boards. Several expressed surprise at the information students had gathered, including the amount of oil spilled throughout the world and the number of people without access to clean water. A number of the parents had grown up in rural Mexico where they had to spend lots of time fetching clean water, so they told students they heartily agreed that water should be a basic right for all.

One indication that the students valued the project is that they all wanted to take the final project home. And they often asked me if they could start working on their water projects. The project suffered, however, from some of the same problems group projects often suffer from: In a few groups, a couple of people did more than their share of the work, which led to some feelings of resentment. I tried to address this through the individual writing assignments and by assuring students that I was observing who was doing what work. Next time, I will have students reflect in writing on their roles as group members as well as sharing what they learned overall. I also plan to give individual feedback to each student.

I also realize that we never answered the crucial question: Who actually owns the water? Or a related question that we uncovered while researching oil spills: Why is there such a thirst for oil?

That's one of the benefits of teaching. There's always next year. ⬡

Bob Peterson taught 5th grade at La Escuela Fratney in Milwaukee and is an editor of Rethinking Schools. *He is president of the Milwaukee Teachers' Education Association.*

WEBSITES

http://response.restoration.noaa.gov/index.html. Office of Response and Restoration, National Ocean Service, National Oceanic and Atmospheric Administration. Includes age-appropriate background information and descriptions of science experiments.

www.unicef.org. Do a search for "water" on the site and find many useful articles.

MARTHA MERSON

I f you are an environmental organizer, like Selene, your classroom is a conference room, the community garden, or a church parking lot. Your students are everyone—from toddlers to the elderly; they come with a variety of levels of formal education. Your goal is to increase environmental justice, community well-being, and individuals' health.

If you are a math educator/curriculum writer with an interest in data, like Martha, you teach in adult ed and K–12 classrooms, libraries, and living rooms—anywhere you can sneak in math. The teaching starts with a provocative statistic or a document with unfathomable numbers. Your students are often math averse. Their motivation could be to earn a high school diploma or simply to learn more. Many have plans to put learning to use in their communities, churches, and families. Your goal is to encourage adults and youth to take a new look at numbers, to ask questions.

Statistics for Action (SfA) brought us together. The organization provides organizers and community members with tools and resources for under-

Transparency of Water

A workshop on math, water, and justice

BY SELENE GONZALEZ-CARILLO AND MARTHA MERSON

standing and using scientific data in communities affected by environmental contamination. During the project, we led an SfA-inspired workshop in Spanish designed to probe participants' distrust of tap water and arm them with skills and knowledge to take on water quality/delivery issues.

Do You Drink Chicago's Tap Water?

At 9 a.m., volunteers recruited for a six-week leadership program filed in for their fourth session. Among the participants were Carmela, a single mother, college student, and intern at Little Village Environmental Justice Organization; Elena, who is studying for a commercial trucker's license; and two middle school students, Luna and Maria, who brought along their pet rooster. The eight participants ranged in age from 10 to 60. Their project facilitator, Norma, also a neighborhood resident, has a long history of social justice work.

> We treated this opportunity with great respect, conscious that it takes courage to learn the basics in public.

We began a taste test of Mountain Spring bottled water, filtered water, and tap water. Selene instructed the participants to sample water from each of three pitchers. They examined the water in their cups, swished it around, and swallowed.

"This is good because I'm dehydrated," said Carmela.

Sandra added: "I'm a dummy; they taste the same to me."

"It's all the same water, I get it." Carmela added, "Is it?"

Although divided in their preferences, the participants were surprised by how difficult it was to discriminate between bottled and tap water.

Martha then asked the attendees: "Do you buy bottled water?"

We wanted to understand why and to what extent the participants paid for bottled water. We hoped they would recognize some of their reasons in *The Story of Bottled Water*" (storyofstuff.org), which we had queued up.

"I buy bottled water because it's cleaner," said Yoana.

Carmela related what happened when her girlfriend offered her water:

I'm like, "What are you doing?" because that's tap water—forget about it. [But] the tap water tasted better and I would tell my mama, "Drink water from the tap, it tastes very good," and she would say to me "No! Are you crazy? It has chemicals."

We didn't judge participants' choices. Distrust of tap water can run deep in communities of color. Little Village, the Chicago neighborhood where the participants live, is home to more than 90,000 residents, nearly half of whom are immigrants, according to the University of Chicago. Many come from places with a history of serious water issues. A 2011 study published in *Pediatrics & Adolescent Medicine* found that nonwhite parents were more likely to offer only bottled water to their children, citing reasons of taste and safety.

In 2008 the Environmental Working Group published results of their investigation of 10 popular brands of bottled water in nine states and found a total of "38 chemical pollutants . . . with an average of eight contaminants in each brand," including industrial solvents and fertilizer residue. Consequently, the reliance on bottled water in communities of color is a serious concern. In addition to the ecological and economic issues related to bottled water use, these communities are disproportionately at risk for health consequences associated with bottled water.

Selling the Public on Bottled Water

Selene introduced the group to the video, which illustrates the trend away from tap water and toward bottled water, situating this shift in a larger context of manufactured demand. She asked the group: "Twenty years ago, people in the United States rarely drank bottled water. So how did so many people's opinions change?"

Annie Leonard, who narrates *The Story of Bottled Water*, throws out intriguing facts. In a common sense way, she explains how bottled water is a problematic product along its whole life cycle: It takes petroleum to make the bottles, and we know how devastating oil spills can be; it takes fuel to move the bottles around and that pollutes the air; and then the bottles end up in trash cans or by the side of the road. Of those that make it to recycling centers, only a few are actually recycled. Leonard's message is simple: "We're trashing the planet; we're trashing each other; and we're not even having fun." Although ads may promise glamour and bottled water manufacturers say their product meets consumer demand, in taste tests across the nation, people chose tap water over bottled water. The animation of this point, a stick figure spitting out bottled water, adds comic relief to the disturbing story Leonard spins.

Two minutes into the video, we paused to let this fact sink in: Bottled water costs 2,000 times more than tap water. At 3:46, we heard and captured this: One-third of bottled water sold in the United States is filtered tap water. At 4:42 we paused to record: 80 percent of plastic water bottles go to landfills.

Leonard goes on to explain why the trend toward buying bottled water took off by defining *manufactured demand* as the force that drives the production of goods. "In order to grow you have to sell stuff." When soft drink companies feared a drop-off in business, they fabricated the need for bottled water by inducing fear, making bottled water look seductive, and misleading the consumer. Marketed as a beverage, bottled water companies avoid the rigorous testing the Environmental Protection Agency requires of municipal water systems.

In the final minutes of the film, Annie sends out the battle cry: "Take back the tap!" She makes clear that viewers can take action and make a difference to save our right to clean water. After viewing the video, we asked: Was there anything surprising?

Sandra spoke up: "I did not think that the bottles would be [piled up] . . . just like that, mountains and mountains [of bottles]. . . . I used to consider myself 'damned' because I could not buy this type of water. A friend would tell me 'You're drinking dirty water from the tap!' But tap water isn't bad; they're just charging us double."

That was an important point. We wanted participants to create sound bites drawing on one of the three statistics and their experiences, imagining that they were talking to a neighbor who hadn't seen the film. Selene planned to elicit an example, stating a fact with a fraction, a percent, and a ratio.

Selene: "Instead of a third, what's the percentage?"

Elena: "*¿Un tercio?*"

Carmela: "Oh, we learned this in school—1.3."

Elena: "9 percent."

Carmela: "How did you get that?"

Elena: "Cause you divide three times the decimal."

Selene: "What do you know about a third? Is it more than a half?"

Elena: "It's more."

Juan: "It's less."

When Elena said a third was more than half, we realized she was unclear about more than the fraction-percent conversion. Because others were struggling too, we opted to slow down, to play with different ways to show these fractions. We treated this opportunity with great

> Bottled water is overpriced, and bottled water companies are generating big bucks from consumers' willingness to pay the price.

respect, conscious that it takes courage to learn the basics in public.

Selene: "If this is a complete bottle of 100 percent, half would be 50 percent and a third would be a little less, because it is three parts that make it complete, and so it has to be of three equal parts—one, two, three."

We tore open a 24-pack of bottled water brought as a prop, inviting the group to show us one-half. They quickly separated the 24-pack into two groups: 12 in each. We wrote "12 is half of 24" on the board and asked them to explain how one-half is like 12/24. The school-age participants smiled knowingly. We all agreed that another way to say one-half is one out of every two.

"One-third of the bottled water sold in the United States is from the tap." We challenged the group to show us with the 24-pack. Now they could see that "one out of two" is more than "one out of three." Half the 24-pack was 12; one-third was only eight. A participant demonstrated one-fourth to emphasize the relationship between the change in the denominator (increase) and the number of bottles set apart (decrease).

> We want people to make informed decisions based on data rather than on suspicion or misinformation or even faith in us.

We spent a little time on the second statistic, especially because Sandra had mentioned that bottled water costs double. Her statement communicated the gist of the situation—consumers are paying dearly for water—but it massively understated the amount. The film draws an analogy between paying for bottled water and a consumer paying $10,000 for a hamburger. To explore the comparison between the cost of bottled water and tap water, the group listed current prices and calculated price tags 2,000 times higher. For one gallon of gasoline—$8,000!

Selene summarized: "Would one spend that much money on gas? But that is what we are spending in reality on water."

Murmurs and head shaking indicated the group members were impressed. Bottled water is overpriced, and bottled water companies are generating big bucks from consumers' willingness to pay the price.

The Transparency of Water

Ultimately, we want people to make informed decisions based on data rather than on suspicion or misinformation or even faith in us. In the beginning of the video, Leonard reminds viewers that "in many ways bottled water is less regulated than tap." The Food and Drug Administration (FDA), which regulates bottled water in the United States, "does not have the ability to require the submission to the agency of results from the testing conducted by and on behalf of bottled water manufacturers, and . . . does not have specific authority to mandate the use of certified laboratories," according to testimony given by the FDA to the U.S. Department of Health and Human Services.

We felt it was vital to draw this distinction. Although people might have legitimate concerns about what lurks in tap water, at least it is a matter of public record. People voice all kinds of fears about the tap water, but they can read water test results, answer some of their own questions, and make specific demands of their government. Having information about our drinking water allows people to take action. This is the case in California, where residents are fighting to be the first state to set a limit on the carcinogen hexavalent chromium.

To this end, we examined Chicago's water report, which we distributed to participants. Municipalities have to report to residents annually, but few people know that or have ever read them. We asked participants: "What do you know? What do you want to know?"

When Norma admitted to feeling overwhelmed, Selene explained: "Parameters, what they tested for, go down the side. See anything familiar?"

All: "Chlorine, fluoride, copper, lead."

Juan: "The water has all of this?"

Carmela: "What! There's cyanide in our water?"

Selene: "Well, let's see. These are all the things they test for. In the next column, we have numbers from the Illinois Environmental Protection Agency. What do you think? Why would they put numbers in a water chart?"

Yoana: "To communicate?"

Selene: "Yes, they are communicating that 200 micrograms of cyanide per liter of water is the maximum safe level. If we see a higher number there—like 250—it's for sure a known risk. What else?"

Elena: "Some are blank."

Sandra: "Because they are trying to figure out if it is good or bad?"

We explained that the federal guidelines for maximum contaminant levels in drinking water regulate about 100 of the chemicals in use. Municipalities test for other chemicals, but under the Illinois EPA regulations, there is no set limit for comparison, so the cell is blank.

Next we pointed out the columns listing the level of contaminants in the "raw" lake water. After treatment, sodium levels go down, while chlorine levels increase. Participants paired up and picked one parameter, tracing how the level changed before and after treatment.

Carmela: "Cyanide is allowed in drinking water up to 200 micrograms per liter and it shows that it came in at . . . wait, when there is a "less than" sign. . . . What does that mean?"

Selene draws "<" on the board.

Carmela: "The alligator eats the bigger number!" Everyone laughs. "So less than five."

Selene: "Exactly. That's how I remember. I used to draw little teeth. What the alligator is eating tells us that the actual amount is less than whatever number it's eating."

The participants nodded and Carmela finished her assessment: The amounts for cyanide before and after treatment were less than five. Less than five what? The column specifying units indicates less than five micrograms per liter.

The time went so quickly. We had to put off an exploration of units as well as the critical notion of "safe" levels. Identifying (or explaining or determining) "safe" levels is a frustrating endeavor. For carcinogens such as arsenic, there is no such thing as a safe level, only what has been deemed "acceptable" risk. Generally standards are set to protect safety and are set based on known risks, yet many agree that standards are inadequate, because researchers know little about synergistic effects, the likely accumulated health impacts from exposure to a variety

of chemicals across a life span. Over time, researchers may compile a body of evidence that shows negative health effects at lower levels than previously thought, but it can take years for regulations to catch up.

Complex as it is to make sense of levels and standards, clean water activists believe politicians will succumb to pressure from industry, as they have in the past, if communities are silent on these issues.

Responses and Reflections

With limited time for exploration, it is tempting to avoid messy data sets and questions that have no easy answers—for example, "Is < 5 µg/L of cyanide

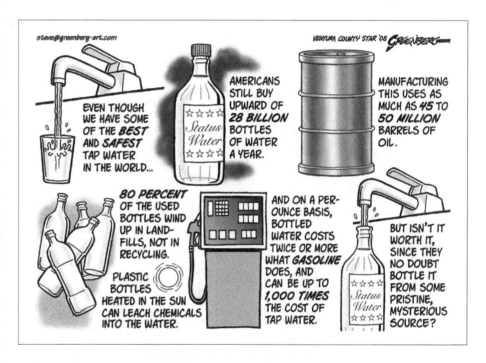

safe?" Yet we wonder whose purposes are served if we take a pass. Resources from SfA include activities and data sets that make teaching the math of environmental data a bit easier, and hints for facilitators and participants to take control of the math (sfa.terc.edu).

Although the Common Core standards set forth a prescribed sequence for math learning, our experience shows that local, relevant data spark engagement at all levels. Regardless of their past success or failure with math, participants grasped concepts like reporting limits, explored persistent misunderstandings like the relative size of a half

"I was curious about where the plastic bottles that I put in recycling bins go. I found out that shiploads were being sent to India. So, I went there. I'll never forget riding over a hill outside Madras where I came face to face with a mountain of plastic bottles from California. Real recycling would turn these bottles back into bottles. But that wasn't what was happening here. Instead these bottles were slated to be downcycled, which means turning them into lower quality products that would just be chucked later. The parts that couldn't be downcycled were thrown away there; shipped all the way to India just to be dumped in someone else's backyard."

—Annie Leonard, *The Story of Bottled Water*
Storyofbottledwater.org

THE STORY OF
STUFF
WITH ANNIE LEONARD

and a third, and coordinated information from rows and columns to identify contaminant levels. Young participants made connections to school learning.

Taking Action

We considered this workshop a success. As math educators, we were pleased that the people stayed with the math instead of counting themselves out. The group got started on the path toward digging into the numbers. As environmentalists we observed participants make connections between purchasing choices and environmental consequences like mountains and mountains of trash.

On a short evaluation form, participants told us what they valued learning:

- I can drink water from the sink.
- I'll know not to buy bottled water and save money.
- Cost of H2O (bottled) 2,000 x more!

Conversation continued as people left. As she walked out the door, Yoana said, "They are trying to confuse the people; people should have access to water without paying so much."

Months later, the participants of this workshop joined a protest against the mayor's plan to privatize Chicago's water system. They had the background to contribute to discussion about who is in the best position to monitor and safely deliver drinking water.

Clean drinking water is in a precarious position, threatened by fracking, pesticide runoff, and, in Chicago, privatization. Even in cities where the water routinely gets high marks, residents can't afford to be complacent. Access to clean water, as Annie Leonard says, is our birthright; but in a capitalist economy, nearly everything is for sale and the water—and water system—we all rely on and need to live requires vigilant protection.

Many of us memorize and spout statistics that affirm our beliefs without so much as a glance at data sets like water quality reports. As educators and organizers, we advocate interpreting data to describe conditions. We advocate shedding light on how to calculate, quantify, and explain the costs of injustice. Workshops like this one set the stage for broader involvement in developing and understanding these statistics. They are part of a respectful approach to convincing people who've grown accustomed to buying water that they are shouldering an unjust financial and health burden. Intuitively, many know the situation is inequitable, but facing the data can ignite a sense of urgency. Then the hard work of identifying steps for collective action begins. ⬤

Selene Gonzalez-Carrillo has worked as the open space coordinator for Little Village Environmental Justice Organization and on outreach for Statistics for Action. Martha Merson is the project director for Statistics for Action, based at Technical Education Research Centers. She is a co-author of the EMPower curriculum series for nontraditional students.

At the end of biology class, Ana, a sophomore, asked if she could talk with me in private.

"Ms. Lindahl, do you know who is a good cancer doctor in Portland?"

I was stabbed with sadness—we were in the midst of our cancer unit. I've included a study of cancer every time I teach mitosis. It helps students see the powerful and personal implications of errors in cell division. But, without fail, painful stories arise during and after class.

I have spent the last six years teaching high school science in Southeast Portland. We are located in the heart of the largest immigrant area in our city, and at least 80 percent of our students qualify for free or reduced lunch. Every day I witness how poverty and language barriers create serious challenges in my students' lives.

Ana came to me with the desperate hope that I was an expert on cancer treatment. I'm not. I listened to her story. Her father had just been diagnosed with advanced cancer. They were looking everywhere to find treatment and they lacked insurance. Because her parents spoke little English, Ana was trying to negotiate a system that could confuse and intimidate the most privileged adult. She was frustrated and frightened. All I could give her was sympathy and general suggestions. I was unable to provide any real comfort. Mixed with my grief for her was anger that she was in this position and that, as a society, we couldn't do better.

Ana's story was one among many. So many students have raised their hands on the first day of the unit to tell the class about a family member who has survived, is suffering, or has died from cancer. One year, a father came to parent conferences ravaged by the late stages of lung cancer. His son had shared his father's diagnosis with his classmates one day during the cancer unit. Not long after, I found out that his younger brother had also been diagnosed with cancer and was facing a series of surgeries.

KATHERINE STREETER

Facing Cancer
Social justice in biology class

BY AMY LINDAHL

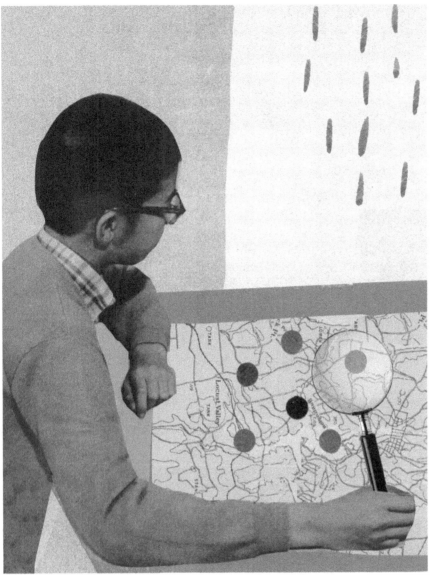
KATHERINE STREETER

After these discussions, the students learned current theories about how and why cancer develops. The unit ended with each student taking on the role of an oncologist. I assigned them each a patient, and they created brochures that explained how a particular chemotherapy drug disrupted cell division and could stop that patient's cancer from spreading. Some years I gave students the additional challenge of performing in front of their peers; they acted in a short scene, meeting with their patients and teaching them about how their chemotherapy drug worked.

In many ways, this curriculum was innovative and effective. My students saw the real-life implications of cellular biology. They were involved in interactive role plays and they delved into writing about how chemotherapy works. Despite all that seemed to be going right, I became troubled.

I recognized that my curriculum never directly addressed the sadness and fear these lessons evoked in my students. I worried that I had used a topic because I knew it would generate an emotional response and draw them in, but that I wasn't directly acknowledging their fears, their pain, and their difficult questions:

"Why do so many people in my family get cancer?"

"What happens if we can't get insurance?"

"What if my whole family is eating the stuff you said we shouldn't be eating?"

I came to see that teaching about cancer needed to be about more than the specifics of mitosis and mutations. My students needed to hear the difficult and troubling story of this disease: We live in a world that puts some of us at more risk than others. I needed to develop a new curriculum that would teach the biology of cancer while also exploring its complex origins and outcomes. Finally, I needed to arm my students with the tools and knowledge to

As the years passed, I became more and more convinced that cancer rates, severity, and survival in this community were worse than in the white, upper-middle-class community I was raised in. As I began to look at research around cancer rates and mortality, my fears were confirmed. Race and class play a key role in who gets a cancer diagnosis and who lives to tell about it.

My unit had always started with students reading fictionalized stories of a diverse range of people who develop cancer. As a class, we collected data on risk factors and the types of cancer their characters were diagnosed with. Students often pointed out that, in our scenarios, people with less access to health care were more likely to die from the disease.

face this disease and to consider the kind of social changes necessary to address both its causes and effects. They needed a curriculum that could help them save their own lives.

Beyond Mitosis

So this year I spent time with cancer statistics. A lot of time. I combed through resources for information on cancer disparities and worked with other teachers to imagine what my lessons could look like. I faced a daunting task. I needed my students to understand that cancer is a disease of societal inequity, genetic predisposition, and personal choice—albeit choices rooted in the nature of our society. My lessons needed to take a hard and direct look at the uncomfortable questions I had sidestepped. And the curriculum had to guide students to a place of hope and activism.

I decided we would start off with the biology of cancer and the role play, but after that, students would analyze demographic disparities in rates of cancer diagnosis, treatment, and outcome. Then, I would assign students a specific community and they would delve deeply into data that showed which cancers were most common in that community. They would read about the reasons behind these disparities. I knew I would expose my students to troubling data, so we needed to spend significant time thinking about how we might decrease these cancer inequities. Student teams would brainstorm ways to decrease disparities, design action campaigns, and present to one another. Finally, they would advocate for a community in real life.

Despite all my planning, I was scared to start, worried that I was launching into something too difficult and too depressing, fearful about effectively guiding discussions about race and class. Despite my misgivings, I knew it was time to take a leap.

When the day arrived, I started with a lesson on the deeper causes underlying common cancer risk factors. Many cancer risk factors, such as smoking and diet, are often thought to be issues of personal choice. However, these simplistic judgments quickly break down on deeper analysis. I wanted to get my students thinking about the role of risk factors in different stages of the disease. Some, like pesticide exposure, can increase cancer incidence, while others, like lack of cancer screening, can affect mortality.

To prepare students for this type of analysis, I assigned small teams several risk factors (for example, eating fast food or living near a mine) and they placed them into risk categories: behavior, genetics, workplace safety, environmental safety, and access to health care. They discussed whether each factor fit in one category or multiple categories. They also considered if the risk factor affected cancer incidence, mortality, or both. As groups reached consensus on risk category assignment, I had each team present their ideas to the rest of the class. I asked them to pick the risk factor that required the most thinking and discussion. Some comments from that day:

> Race and class play a key role in who gets a cancer diagnosis and who lives to tell about it.

> I don't think it is right to say eating healthy food is only about choice and behavior. What if you don't have a lot of money? What if you live somewhere with only fast-food restaurants? I think that maybe it is environmental safety, too.

> If you live near a mine and chemicals get into your water . . . that could increase your incidence. But we thought it could also increase your mortality because you get your cancer treatment, but then you go right back home to the environment that made you sick.

Who Is at Risk?

On our next day in class, it was time to move beyond speculation and spend time with real data. I gave each student a National Cancer Institute (NCI) graph on how incidence and mortality vary by race/ethnicity and by cancer type. As they examined it, I had them write about patterns and disparities. I asked them to think about potential problems with sorting people by race and ethnicity. One example students noted was that multiracial people are not identified at all.

After students shared their ideas in small groups, I gathered everyone together for a discussion. We went over norms: Keep an open mind, build on each other's ideas, respect air time, use the text, and speak to each other. As I do before every Socratic seminar, I reminded my students that we were going to work together to explore complex ideas, and that all real data are open to multiple interpretations and further questioning. Careful listening and taking time to think before talking would help them reach a deeper understanding.

I asked them to share what they noticed about disparities, to talk about reasons they think these disparities might exist, and to consider the strengths and weaknesses of breaking down cancer statistics in this way. Students started by identifying patterns they saw in the graph, and we started to explore risk factors that might be at work.

I was taken aback, again and again, by their careful and nuanced thinking. "I noticed that black women are less likely than white women to get breast cancer. But they have higher mortality. What if that is because of access to health care? If African Americans have higher poverty rates, maybe they aren't getting screened or treated in time?" or "Lots of Asian Americans still have a different diet because of their culture. In my family we eat a lot more fish because we eat Japanese food. Maybe that changes our risk."

> But, as I look back and face those questions more honestly, I see that our students need and deserve a new curriculum. As science teachers, we must guide and support them as they grapple with the difficult questions our lessons inspire.

One student questioned thinking about cancer through the lens of race: "How can a race like 'white' or 'black' mean anything? Not all people of a race have the same genetics. I don't think it is right to make this graph. I think it is racist."

I pointed out how looking at statistics by race can help us see the effects of racism. But later I realized I need to do more careful teaching on this point. Because a vulnerability to some cancers (breast and other gynecological cancers) can be genetically transmitted, it's easy to confuse race with genetics. Race is socially constructed—it's an institutional and historical category with no basis in biology. So when we're looking at racial statistics on cancer,

we're looking at the structural impact of racism on communities.

Cancer and Social Justice

It was time for the students to dig deeper into how cancer affects specific communities. I assigned each small team a different group (for example, African Americans, rural poor, the elderly, Asian Americans/Pacific Islanders). These groups, which obviously overlap, have been identified by the Intercultural Cancer Council (ICC) as facing significant cancer disparities. I gave each student group a packet of data and fact sheets specific to that community. Each packet included ICC fact pages on cancer disparities facing the group and relevant NCI data and graphs. For example, the group researching Latina/os saw that cervical cancer incidence in this community is somewhat higher than average, but that cervical cancer mortality is extremely high. They read about reasons for this: late diagnosis, language barriers between doctors and patients, fear of screening.

We spent one class period identifying the most worrying disparities and likely causes: Was family income and lack of access to health care a driving force? Could workplace or living conditions account for these disparities? As they discussed their ideas, I circulated around the room and talked with teams.

Many students quickly identified risky behaviors as a cause for disparities. When I heard this, I pushed each group to move beyond the simplistic view that cancer is a result of unhealthy personal choices. "Which causes are beyond the control of that community?" I asked. "How might corporate and government policies, pollution, and racism play a role?"

Finally, students brainstormed actions that would decrease cancer rates in their community. "If you had the power to help this community in the best ways possible, what could be done to lower cancer rates and increase survival?" I asked. "These solutions will likely look different in different communities. Imagine the many small solutions that could eliminate the big disparities you've been seeing."

Ideas for solutions, small and large, began to pour out of my students. Some saw the need for education: "Could we help Asian kids understand how important it is not to smoke?" Others saw how targeted cancer screening could help: "How do we increase rates of pap smears among Latina women?" And many students identified the need for making screening and treatment more affordable: "How do we pass a law to bring health care to everyone?" Working in their small teams, my students were charged with planning a step-by-step action campaign for their community. What disparity were they most disturbed by? What were its causes? What were some possible solutions? What stakeholders would need to be involved in this effort? Each team created a poster that proposed a way to narrow the health gap in their community.

A team researching the rural poor produced a particularly thoughtful action campaign. On reading that rural poor women have elevated rates of breast cancer (both incidence and mortality), they determined that lack of access to health care was a major barrier for this group of women, especially in Appalachia. They read that many of these women live far from medical facilities, work long hours, and report that visits to the doctor are prohibitively difficult and expensive. The team determined that the best way to help these women would be to bring free mobile breast cancer screening clinics to work sites.

Other teams focused on education and legislation. One proposed K–12 education programs on healthy eating and several groups presented on the need for universal health care legislation. Other groups designed billboards and storyboards for public service announcements. On our presentation day, students shared their campaign ideas with classmates and ended the day writing about the approaches they thought would have the biggest and best effect.

With the first semester drawing to a close, we needed to move on to our genetics unit, but I organized biweekly lunch meetings for the students who wanted to take action. One student created a refrigerator checklist she gave to all of her family members. It provided concrete steps for protecting against skin cancer, a common diagnosis in her family. Two other students began exploring the need for medical interpreters in Oregon. They contacted local immigrant rights organizations to discuss the need for new legislation in our state.

I won't deny that this project scared me. As my students began asking probing questions about injustice, racism, and poverty, I was often over-whelmed and unsure of how to answer. I was asked, more than once, why we weren't doing our usual labs and science content. As each day of new curriculum drew to a close, I was painfully aware of my missteps and fumbles. Still, despite my misgivings, I ended the unit with the growing certainty that I was heading in the right direction. When the bell rang at the end of each period, students lingered and wanted to talk. I saw more hope than sadness. When my students evaluated the unit, they wrote about their desire for action, on both personal and larger scales. They were glad we spent time with real data. They felt well informed and equipped to fight cancer inequities.

I realize other science teachers may balk at the idea of trying to tackle so many social issues in a biology class. Believe me, I had the same misgivings. Each previous year, I had been lulled by the little voice that insisted, "Isn't teaching the health inequities in our society somebody else's responsibility? Aren't social studies teachers supposed to do that?" Overwhelmed with the science standards and skills I am responsible for teaching, I had made my excuses and walked away from the hardest questions my students ask. But, as I look back and face those questions more honestly, I see that our students need and deserve a new curriculum. As science teachers, we must guide and support them as they grapple with the difficult questions our lessons inspire. Our students are ready to look at the prejudice and inequities in our society straight on. They are waiting to be taught how to demand a better and healthier world. Let's help them do it. ⊕

> They needed a curriculum that could help them save their own lives.

Amy Lindahl teaches science at Grant High School in Portland, Oregon. Previously, she taught at BizTech High School, Marshall Campus, in Portland, Oregon.

Outrageous Hope

GARY PACE

I come to the Bodega Headlands often to make offerings and prayers. Offerings to the land—the wild Northern California Coast—and a prayer of gratitude to people who 50 years ago had the vision to block construction of a nuclear plant on this spot straddling the San Andreas Fault. In essence, a prayer to what didn't happen.

The "Hole in the Head" is a small pond tucked behind dramatic cliffs dropping into the ocean, close enough for me to hear sea lions and glimpse migrating whales. This crater was the initial excavation for the foundation; now it only hints at the disaster that could have unfolded. During the 1906 earthquake, this coastline moved 15 feet. In the event of another quake, the largest nuclear reactor of its era would have been perched on the edge of the fault.

What outrageous hope led people to believe they could halt construction of a nuclear site once

MEG McCOLLISTER

building had begun? Who could have predicted the unlikely outcome of PG&E's plan as outlined in this understated park sign?

> *Proximity of the San Andreas Fault would have significantly increased risks to marine, tidal, and atmospheric environments. Citizens and scientists collected signatures, filed lawsuits, wrote letters, and appeared at hearings ranging from Sonoma County to Washington, D.C. The project was finally abandoned in 1964 after eight years of controversy and citizen action.*

If it had opened as planned, the reactor would have outlived its utility by now. I would be standing before the shuttered structures of cooling towers, looking out over the yard littered with barrels of spent fuel rods, a lonely witness to the radioactivity,

which would persist for scores of human lifetimes. The electricity produced would have long since passed through local residents' lightbulbs and dishwashers, with new sources of power being tapped for contemporary needs.

My thoughts drift to those folks who worked for such a reasonable outcome. As the sea breeze blows across my face, I arrange flowers in honor of the legacy these activists left behind, an ephemeral monument to something that never came to pass. ⊕

Gary Pace is a family practice doctor living in Northern California with his wife and daughter.

 See teaching ideas for this article, page 325.

"We Want to Stop It Now"

Fukushima's nuclear refugees

In March of 2011, in the wake of a tsunami, the Fukushima Daiichi nuclear reactors in Japan experienced meltdowns. Anti-nuclear activist Dr. Helen Caldicott wrote:

The world was warned of the dangers of nuclear accidents 25 years ago, when Chernobyl exploded and lofted radioactive poisons into the atmosphere. Those poisons "rained out," creating hot spots over the Northern Hemisphere. Research by scientists in Eastern Europe, collected and published by the New York Academy of Sciences, estimates that 40 percent of the European land mass is now contaminated with cesium 137 and other radioactive poisons that will concentrate in food for hundreds to thousands of years.

Board, Stefan Possony, suggested. The phrase "Atoms for Peace" was popularized by President Dwight Eisenhower in the early 1950s.

Nuclear power and nuclear weapons are one and the same technology. A 1,000-megawatt nuclear reactor generates 600 pounds or so of plutonium per year: An atomic bomb requires a fraction of that amount for fuel, and plutonium remains radioactive for 250,000 years. Therefore every country with a nuclear power plant also has a bomb factory with unlimited potential. The nuclear power industry sets an unforgivable precedent by exporting nuclear technology—bomb factories—to dozens of non-nuclear nations.

The only time one country intentionally used nuclear weapons against another nation during wartime was when the United States bombed Hiroshima on Aug. 6, 1945 and then three days later at Nagasaki. In addition to the tens of thousands killed instantly, the bombs left almost 400,000 homeless.

Today's nuclear refugees are also Japanese. One of these, Yukiko Kameya, was interviewed in 2014 by Mike Burke of the radio/TV show *Democracy Now!* She was attending a demonstration outside the official residence of Prime Minister Shinzo Abe. Kameya describes what her life is like, now that she cannot return home:

Yukiko Kameya: My name is Yukiko Kameya. I'm from Futaba, which was 2.1 kilometers from Fukushima Daiichi nuclear facility. Right now, we have evacuated, and we are living in temporary housing in Tokyo in a space provided by the government. It's close here, so I'm coming here every Friday to demonstrate against the nuclear power facilities. When we fled Futaba, we had nothing. We lost everything. We couldn't bring anything from our house. We didn't have a toothbrush. We didn't have a blanket. We didn't have towels. We had nothing. It was truly hell, and we thought it would be much better to die. But now we are here,

"When we fled Futaba, we had nothing. We lost everything."

Wide areas of Asia—from Turkey to China—the United Arab Emirates, North Africa, and North America are also contaminated. Nearly 200 million people remain exposed.

In her article, "After Fukushima: Enough Is Enough," Dr. Caldicott traces the history of nuclear power and its relationship to nuclear weaponry:

The concept of nuclear electricity was conceived in the early 1950s as a way to make the public more comfortable with the U.S. development of nuclear weapons. "The atomic bomb will be accepted far more readily if at the same time atomic energy is being used for constructive ends," a consultant to the Defense Department Psychological Strategy

Mike Burke: Do you think you'll ever be able to return to your home?

Yukiko Kameya: The Futaba where I lived is not livable, and the government says so. So I know we are never going back in my entire life. But for the Fukushima prefecture, it is still not safe. The radiation level is still very high, so I don't think it's safe, and I don't think we'll go back there.

Mike Burke: And what is your assessment of how the government has handled the crisis?

Yukiko Kameya: We expected our government to do a better job when this accident happened, but they don't really do what you want. They ignore all the problems we are having. There are many young people between 15 and 19 in Fukushima who are in high school, who have died suddenly. For example, this morning, I saw online a story that a 17-year-old died from leukemia. In the morning, when his mother came to wake him up, he was found dead in bed. Everyone says this was caused by the radiation level from the nuclear accident, but our government never recognized it. And there are many children, 59 children, with thyroid cancer. They will never recognize it as being caused by the radiation.

Mike Burke: We're standing right now outside the prime minister's official residency. What is your message to the prime minister?

Yukiko Kameya: We told the prime minister many times, every week here, that we are against the reopening of the nuclear facilities. But it doesn't seem that he gets it. He just does whatever he wants to do anyway. In Futaba, when we had a meeting with Tepco [Tokyo Electric Power Company], we were told the facility is very, very safe. Now we know it is not safe at all. And we've been telling the prime minister it's not safe, so do not restart, but it doesn't seem like it's getting through to him. My real feeling is that I want to go back to Futaba-machi in Fukushima, but I know we can't go back. And I dream of it every day. And I know we cannot go back, so I don't want anyone in the world to feel this way. We want to stop it now. ⊕

and we can't really give up. We want to fight for this cause. When I fled from Futaba, I couldn't even talk. I couldn't even have friends over or anything. But my people encouraged me to be standing here right now. Without the people's help, I couldn't be here. That's why I appreciate them, and I want to join them in their call.

Mike Burke: And what is your message to the Japanese government and the world about nuclear power?

Yukiko Kameya: I don't want anyone in the world to experience what we have experienced. We have houses in Futaba, but there's nothing there. Everything is robbed. All the furniture was broken. I can't really go back there. We know it. We don't want anyone in the world to be in the situation we are in. There are 59 children with thyroid problems, and there are hundreds more on the way. The real problem in Fukushima is children cannot go out and play. They have to stay inside, and this is not the way children should grow up. I don't want anyone in the world, or Japan, to experience this type of situation for children, so I want to stop nuclear facilities now, and I don't want them to be continued.

> **"I know we can't go back. And I dream of it every day. And I know we cannot go back, so I don't want anyone in the world to feel this way."**

 See teaching ideas for this article, page 325.

There were almost 400,000 survivors of the nuclear explosions at Hiroshima and Nagasaki in 1945—called Hibakusha, *literally translated as "A-bomb received persons." In 1967, two* Hibakusha *women, Shizuko Takagi and Kazue Miura formed the Women's Section of the Osaka Association of A-bomb Victims. One part of their work was memoir writing.*

"Kazue, Alive!"

Hiroshima's nuclear refugees

BY KAZUE MIURA

When I finished school in 1941, I began to work as an operator in the Central Telephone Exchange. Dec. 8 of that year was an especially busy day, on which Japan declared war against the Allies. As the war grew in intensity, the Telephone Exchange was staffed almost entirely by women, mobilized high school girls among them.

The Telephone Exchange was located within 500 meters of the hypocenter of the A-bomb explosion, and I was one of the few people who survived in this innermost zone. I was hurled to the floor by the fierce blast, and felt warm blood spurting from my nose and mouth. After a momentary silence, the shrill voices of my workmates rose to a mournful chorus. At one of the second floor exits, I found a girl who was thrown through a window and whose face, full of glass, was bleeding profusely. I held her in my arms, and led her out of the building.

Beautiful Hiroshima was now a wasteland of debris. I desperately wanted to make my way home, 400 meters from the hypocenter, but the heat from the burning houses was too intense. I decided instead to go with my companion to her home in the north of the city. She was in such fear that she would not part her hands from mine, even for a moment. Stunned, expressionless, monsterlike people, young and old, cried out for their mothers and begged for water. When I tried to comfort children, words would not come, only tears.

> "Why did you give birth to me, Mom? You are a bomb victim, so you should not have brought me into the world."

The next morning I was able to return to the place where my house had stood so sturdily. It looked as if the house had been melted and coagulated. My father, mother, little brother, and sister were nowhere to be seen, and I learned later that they had all perished during or soon after the bombing. There was nothing I could do but write "Kazue, Alive!" on the wall of the water tank, now completely dry. I walked back and forth between what had been my home and my place of work, ignorant of the terrible effects of residual radioactivity, looking desperately for my family and friends.

I was suffering intensely from diarrhea. I got weaker and thinner and felt like a ghost. At the hospital doctors were sure I would die from the terrible, mysterious symptoms that had already claimed so many lives. But through the kindness of a family friend, who took me to his quiet home by the sea and fed me fresh fish and oranges, I miraculously began to recover.

In November, although I was still weak, I went to my two sisters in Osaka, carrying the ashes of our parents in an urn wrapped in cloth. My sisters nursed me back to health. They introduced me to a good man, and we married in 1948.

My first baby was stillborn, as was the case with one of my older sisters who was A-bombed in Hiroshima. I was hesitant to have another, but we wanted children very badly. In 1950 I gave birth to a boy, and in 1953 I had a girl. My daughter Maki is troubled by anemia and low blood pressure.

As she grew older, Maki noticed that newspapers in the summer featured stories of the bombings and deaths of survivors. She came to hate all reminders of the bombing because of the pain it had caused me and her fear that I, too, would succumb. When she was 14, she looked me in the face reproachfully and asked, "Why did you give birth to me, Mom? You are a bomb victim, so you should not have brought me into the world."

I had long anticipated that question, but no amount of emotional preparation could have softened the blow of those words. I told her that I had thought a lot before giving birth to her and didn't know whether she might get a bad disease, not wanting to mention leukemia. "And what would you do if it happened to me?" she asked. What could I answer her? In painful honesty I told her that there was nothing we could do about it. That was the saddest and most heartbreaking moment of my life.

In late 1976 I began to suffer from symptoms of anemia, and a gynecological examination revealed myoma of the uterus [a fibro-muscular tumor]. Now I had joined the ranks of the seriously ill *Hibakusha*, many of whom had been operated on for uterine cancer, myoma, or cystoma [ovarian cyst]. To my happiness, the Women's Section published my life story, *Survival at 500 Meters in Hiroshima*, in December 1979, and I hope that it may serve to prevent any other human being from experiencing the horror of nuclear war. ●

(Kazue Miura died of stomach cancer on April 25, 1980.)

Reprinted from The Progressive *magazine, August 1981.*

See teaching ideas for this article, page 326.

Milton Tso, president of the Cameron Chapter of the Navajo Nation, uses a geiger counter to measure radiation on a sweat lodge built with soil from an abandoned uranium mine in Cameron, Arizona.

In a Diné creation story, the people were given a choice of two yellow powders. They chose the yellow dust of corn pollen, and were instructed to leave the other yellow powder—uranium—in the soil and never to dig it up. If it were taken from the ground, they were told, a great evil would come.

The evil came. More than 1,000 uranium mines gouged the earth in the *Diné Bikeyah,* the land of the Navajo, during a 30-year period beginning in the 1950s. It was the lethal nature of uranium mining that led the industry to the isolated lands of Native America. By the mid-1970s, there were 380 uranium leases on native land and only four on public or acquired lands. At that time, the industry and government were fully aware of the health impacts of uranium mining on workers, their families, and the land upon which their descendants would come to live. Unfortunately, few Navajo uranium miners were told of the risks. In the 1960s, the Department of Labor even provided the Kerr-McGee Corporation with support for hiring Navajo uranium miners, who were paid $1.62 an hour to work underground in the mine shafts with little or no ventilation.

All told, more than 3,000 Navajos worked in uranium mines, often walking home in ore-covered clothes. The consequences were devastating. Thousands of uranium miners and their relatives lost

Uranium Mining, Native Resistance, and the Greener Path

The impact of uranium mining on indigenous communities

BY WINONA LADUKE

their lives as a result of radioactive contamination. Many families are still seeking compensation. The Navajo Nation is still struggling to address the impact of abandoned uranium mines on the reservation, as well as the long-term health effects on both the miners and their communities, many of which suffer astronomical rates of cancer and birth defects.

As a college student, I worked for Navajo organizations, trying to inform their people about the uranium mining industry and the large corporations—Exxon, Mobil, United Nuclear—that proposed to mine their lands. It was a humbling experience, seeing some of the richest corporations in the world faced by courageous peoples who fought for the two things that mattered to them more than money: their land and their identity. The Navajo people joined with many others across the country who felt that there was a much better way to make energy. In the end, the people did prevail—new mining proposals evaporated as tribal resistance and legal and administrative battles merged with economic forces. Eventually, contracts for uranium were canceled by utilities, which no longer sought to build unpopular nuclear power plants.

> Now I feel like I am having very bad déjà vu—only this time nuclear power is seen as the answer to global climate destabilization.

Now I feel like I am having very bad déjà vu—only this time nuclear power is seen as the answer to global climate destabilization. In 2005, the Navajo Nation passed a moratorium on uranium mining in its territory and traditional lands, which was followed by similar moratoria on Hopi and Havasupai lands, where mines are proposed adjacent to the Grand Canyon. "It is unconscionable to me that the federal government would consider allowing uranium mining to be restarted anywhere near the Navajo Nation when we are still suffering from previous mining activities," Joe Shirley Jr., Navajo Nation president, explained at a congressional hearing on opening uranium mines in the Grand Canyon area. To the north, the Lakota organization *Owe Aku* (Bring Back the Way) is an intervener in a Nuclear Regulatory Commission hearing to allow the Canadian corporation Cameco to expand its Crow Butte uranium mine, just over the Nebraska border from the reservation.

Mining in Aboriginal Lands

I recently traveled to Australia, the country with the largest known uranium reserves in the world. In my Sydney hotel room the television broadcaster summarized Australia's economic strategy: "We dig it up, and they buy it." The mining industry, in a world bent upon combusting and consumption, looks to be very healthy. Australia's uranium mines include the Beverley Mine, which is in the territory of the Kuyani and Adnyamathanha peoples. Olympic Dam (operated by BHP Billiton—the largest mining corporation in the world) is the country's second-largest uranium operation and is in the traditional territory of Aboriginal people as well. In fact, most major mining operations in Australia are within Aboriginal territory. These are some ancient civilizations—resilient in the face of a deep history of genocide and destruction, which continued well into the 20th century. Aboriginal people did not even get the right to vote until 1967. Due to their relative isolation in the outback, many of these tribes have had few interactions with outsiders. That is, until recently.

Kakadu is the longtime home to the Aboriginal Mirrar people, as well as a recent intruder: British-based Rio Tinto. In the 1970s, Kakadu's Alligator River System became the focal point of Europe's uranium demands. Built right in the center of the Mirrar homeland, the Ranger Uranium Mine is one of the largest uranium mines in the world. But the Ranger mine is also in the center of Kakadu National Park, one of just 25 UNESCO World Heritage sites in the world designated on the basis of both cultural and ecological significance. Kakadu includes more than 190 major Aboriginal rock art and sacred sites.

The Ranger Uranium Mine opened in the early 1980s, after much protest from the Mirrar people, who made it clear that they opposed the mine. Rio Tinto has assured Australians, UNESCO, and the Aboriginal owners that it is operating under "world's best practices" of uranium mining, a term some would argue is an oxymoron. Meanwhile, radioactive groundwater contamination is reported to be spreading through the park. A 2004 incident allowed a number of workers to drink, ingest, and shower in heavily contaminated water, with a large amount spilling out of the site itself. And in 2006, Cyclone Monica delivered extreme rainfall, causing the radioactive containment ponds to fill. The company responded by lifting tailings dams, redirecting runoff into streams, and using the contaminated water for irrigation.

In 1999, Jacqui Katona, a Djok Aboriginal woman, and Yvonne Margarula, a Mirrar woman, won the Goldman Environmental Prize for their struggle to oppose development at Jabiluka, another mine proposed for Kakadu National Park. Yvonne explained that an agreement to open the mine "was arranged by pushing people, and does not accurately reflect the wishes of the Aboriginal people who own that country." In 2005, after a long and heated battle, the Mirrar people fought off the proposal to open a uranium mine at Jabiluka. But now, with demand for uranium on the rise, the threat is once again looming on the horizon.

With some 16 percent of Australian land controlled by Aboriginal people and with many of the mine sites in the Aboriginal heartland, the upcoming pressure on communities to buckle to the largest mining companies in the world will be daunting. Coinciding with the proposed ramp-up of the nuclear industry is the negotiation of land settlements for a number of these Aboriginal first nations. If history is any indicator, many of these land rights settlements will mirror what happened in Alaska, where the Alaskan Native Claims Settlement Act—promoted by oil companies that deemed it necessary to negotiate some agreements between themselves and Aboriginal people—established Alaskan Native Corporations, which today create a complex set of divided loyalties and communities. This is perhaps best illustrated by the case of the Gwich'in people, who find themselves not only opposing oil companies that want to drill in the Arctic National Wildlife Refuge, but also Alaskan Native Corporations, whose income has derived from the exploitation of the land and its resources.

There is another prophecy that is relevant to this story, though. Ojibwe legends speak of a time when our people will have a choice between two paths: One path is well worn and scorched, but the second path is not well traveled and it is green.

There is an alternate economic future for Indigenous Peoples, and it too is green. In order to stabilize carbon emissions in the United States, the country will need to produce around 185,000 megawatts of clean new power over the next decade, which could mean up to 400,000 domestic manufacturing jobs. The Intertribal Council on Utility Policy

estimates that tribal wind resources alone represent 200,000 megawatts of power potential. In fact, Native American nations are some of the windiest places in the country.

The Rosebud Lakota put up the first large native-owned windmill in 2003, a 750-kilowatt turbine right in the middle of the reservation. The Turtle Mountain Ojibwe just erected a 660-kilowatt wind turbine; 10 more megawatts are planned for Rosebud; and the White Earth Anishinaabeg have several projects under way in Minnesota. Proposals for up to 800 megawatts of power for northern Plains states are being put forth by the Intertribal Council on Utility Policy. There's also a 50-megawatt project on lands held by the Campo and Viejas bands of Kumeyaay people in Southern California, and a 500-megawatt project in which the Umatilla Tribe of Oregon is a partner. Boston-based Citizens Energy is working with a number of tribal communities in the United States and Canada to bring green power from the reserves to the grid.

In the United States, native communities have an opportunity to lead the way to a green future. We have a chance to create a just energy economy in the most wasteful and most destructive country in the world. We need help, though. Ensuring that climate-change legislation does not reboot the nuclear industry will be a critical part of supporting native struggles to choose the green path over the scorched one. ⊕

Winona LaDuke is an Anishinaabekwe (Ojibwe) enrolled member of the Mississippi Band Anishinaabeg who lives and works on the White Earth Reservations, and is the mother of three children. She is also the executive director of Honor the Earth and serves as co-chair of the Indigenous Women's Network. She is the author of six books, including The Militarization of Indian Country, Recovering the Sacred: the Power of Naming and Claiming, *and* All our Relations: Native Struggles for Land and Life.

This article was first published in Orion *magazine.*

See teaching ideas for this article, page 326.

Forget Shorter Showers—*page 268*
By Derrick Jensen

Ask students to discuss or write on:

- Derrick Jensen writes, "Consumer culture and the capitalist mindset have taught us to substitute acts of personal consumption (or enlightenment) for organized political resistance." Can you put that in your own words? What examples can you think of to illustrate Jensen's point?
- Why does Jensen criticize measures like recycling, using more energy-efficient lightbulbs, or driving less as a response to the environmental crisis?
- Who does Jensen think is to blame for environmental crises?
- "Personal change does not equal social change," Jensen writes. Why not?
- Jensen quotes Kirkpatrick Sale, who says, "We, as individuals, are not creating the crises, and we can't solve them." If *we* can't solve them, who can?
- If we cannot consume—or "unconsume"—our way to a better world, what does Jensen suggest that we do?

Presuming students have done other activities from *A People's Curriculum for the Earth*, ask them to relate Jensen's article to other lessons—for example, the "Thingamabob Simulation," "The Indigenous People's Summit on Climate Change," or "The Mystery of the Three Scary Numbers."

Keep America Beautiful?—*page 271*
By Elizabeth Royte

Show students the iconic "Keep America Beautiful" ad, which is often referred to as the "Crying

Indian" ad (https://www.youtube.com/watch?v=j7OHG7tHrNM). In pairs or small groups, ask students to brainstorm as many ideas as they can about the commercial. They might consider: When was this commercial made? Who made it? Why did they make it? What did the producers of the ad want viewers to think? What did they want them to do or not do? Who do students think paid for the ad?

Have students read "Keep America Beautiful?" and write on any of their predictions that were borne out. Which of their questions were answered?

Ask students: Why would it be in corporations' interest to support the "Keep America Beautiful" campaign?

Ask students to imagine that they have as much money for an ad campaign as the "Keep America Beautiful" producers had. (The full TV ad is a minute long.) Tell students their task is to design an ad or ad campaign with the aim of teaching viewers something fundamental about an aspect of the environmental crisis. What do they want people to understand, to know, to do? What images will they use? Depending on time and resources, students might perform or film their ads.

Combating Nail Salon Toxics—*page 280*
By Pauline Bartolone

Ask students the following questions:

- According to Jamie Silberberger with Women's Voices for the Earth, only 11 percent of more than 12,000 chemicals used in cosmetics have been studied. Why would there be so little study and regulation about chemicals that are used regularly and so intimately?
- There is less regulation of nail products for workers

than for consumers. Why would this be the case?

- "According to the Campaign for Safe Cosmetics, the FDA has only banned or restricted 11 chemicals found in cosmetics, whereas the European Union has banned 1,000." What explanations might there be for this difference?
- Why do women work in nail salons if they know that the chemicals there are toxic and might hurt their health?
- What rights should workers have to know about the chemicals they use at the workplace on a regular basis?
- Lam Le prepared a two-page testimony to give at a congressional hearing with Rep. Barbara Lee. She had to reduce her testimony to one question. If you had only one question to address to a member of Congress at a hearing, what would it be?
- This article originally appeared in the magazine *Race, Poverty & the Environment.* Are toxic nail salon products a racial issue or an issue of poverty?

Bring bottles of nail polish and nail polish remover to class—or other cosmetic products. Ask students to choose several of the chemicals found listed on the ingredients and research their health effects. It's worth pointing out to students that the government does not require manufacturers to list all the chemicals used in cosmetics on the container.

Watch the amusing yet troubling *The Story of Cosmetics,* narrated by Annie Leonard (http://storyofstuff.org/movies/story-of-cosmetics). Leonard provides a helpful annotated script here: http://safecosmetics.org/downloads/SoCos_footnoted_script.pdf.

Outrageous Hope—*page 314*
By Gary Pace

As he explains in this short article, Gary Pace comes to the "Hole in the Head" to pay tribute to "something that never came to pass"— the building of a massive nuclear power plant near the San Andreas Fault on the California coast. Much of working for a better world depends on maintaining the "outrageous hope"

that our efforts will make a positive difference, despite the enormous wealth and power of vested interests. Brainstorm with students other things that they hope will never come to pass. Examples might include the Keystone XL Pipeline, coal export facilities in the Pacific Northwest, fracking facilities, mountaintop removal strip mines, etc. You might also brainstorm possible effects of global warming that students hope will not come to pass: submerged islands or cities, farmland turned to deserts, forests ravaged by wildfires, species extinction, etc. Ask students to imagine that—as in Gary Pace's article—the efforts and "outrageous hope" of activists prevented these outcomes. Ask them to choose one of these outcomes they wish to avoid and write a sign commemorating this activism, as the understated park sign does at the "Hole in the Head."

"We Want to Stop It Now"
Fukushima's nuclear refugees—*page 316*

Ask students to write on or discuss:

- According to Dr. Helen Caldicott, what is the connection between nuclear weapons and nuclear power?
- Why will nuclear power always make nuclear war a threat, according to Dr. Caldicott?
- If the plutonium generated by a nuclear reactor stays radioactive for 250,000 years, how could such a long-lasting poison be stored safely? How could one communicate to future generations that this material is hazardous?
- Nuclear power is often presented as a "cleaner" alternative to burning fossil fuels because it does not create carbon dioxide pollution in its generation of electricity. Do you agree or disagree that nuclear power is the cleaner alternative? What more would you need to know to be more certain?
- Imagine that you are Yukiko Kameya. What would you find most painful or difficult about your situation?
- What gives Kameya hope?
- How did the Fukushima accident especially affect children?
- Kameya says that the Japanese government

ignores all the problems that the Fukushima refugees are having. Why would the government not be more responsive?

Ask students to write a conversation between Kameya and Kazue Miura in the following reading on "Hiroshima's Nuclear Refugees" about their different—and similar—experiences. Students might complete this as a dialogue poem.

Tell students: Imagine that you are Yukiko Kameya. Come up with a slogan to express your grievances against the government, and draw the poster that you will carry during the demonstration at the prime minister's residence.

"Kazue, Alive!"
Hiroshima's nuclear refugees—*page 319*
By Kazue Miura

Ask students to discuss or write on:

- How was Kazue Miura's *Hibakusha* experience different from a man's experience because she was a woman?
- What can you find that is hopeful or positive in this story?
- What allowed Miura to continue, to not give up and decide to take her own life?
- What makes radiation pollution especially horrible?

Ask students to imagine and write the conversation between Kazue and Maki when Maki says to her, "Why did you give birth to me, Mom? You are a bomb victim, so you should not have brought me into the world." This could also be completed as a dialogue poem.

Ask students to use language in Yukiko Kameya's interview and Miura's memoir to create a "found poem"—snipping and assembling evocative lines in the fashion of a poetic collage.

Uranium Mining, Native Resistance, and the Greener Path—*page 321*
By Winona LaDuke

Ask students to discuss or write on:

- Why is it that so much uranium mining has happened—and is happening—on the lands of Indigenous Peoples?
- LaDuke mentions the 3,000 Navajos who worked  in uranium mines and would walk home in ore-covered clothes, putting not just themselves but also their families at risk. Why were workers not warned of the dangers of radiation?
- Why would the federal government today consider allowing uranium mining near the Navajo Nation?
- What does Winona LaDuke mean when she writes that some people argue that the "world's best practices" uranium mining is an oxymoron?
- Is what LaDuke describes in the article an example of environmental racism? Why or why not?

Show students the excellent segment from the March 14, 2014, *Democracy Now!*, "'A Slow Genocide of the People': Uranium Mining Leaves Toxic Nuclear Legacy on Indigenous Land," which illustrates many of the themes in LaDuke's article, and features clips from the film *The Return of Navajo Boy.* www.democracynow.org/2014/3/14/a_slow_genocide_of_the_people.

Valuable Films on Our Toxic Nuclear Legacy

(All are available from the Video Project, videoproject.com.)

Nuclear Savage: The Islands of Secret Project 4.1
By Adam Jonas Horowitz
(2012)
60 and 87 min.

This disturbing film helps students grasp how U.S. nuclear testing in the Marshall Islands, beginning in 1946, "terrorized and traumatized" people there, in the words of a Marshallese government official. It's hard to overstate the racism and depravity of U.S. officials who intentionally treated Marshallese—especially those from the island of Rongelap—as human guinea pigs. In 1956, Merril Eisenbud, director of the U.S. Atomic Energy Agency's health and safety laboratory, described the government's plans for sending Marshallese back to Rongelap, just three years after the largest nuclear test in history: "That island is by far the most contaminated place on Earth and it will be very interesting to get a measure of human uptake when people live in a contaminated environment." Eisenbud added, "While it is true that these people do not live the way Westerners do, civilized people, it is nevertheless also true that these people are more like us than the mice." *Nuclear Savage* not only chronicles the experimentation on the Marshallese but also introduces us to individuals who continue to work for justice. It's a film that needs to be a staple in U.S. and modern world history curricula.

The Atomic States of America
By Don Argott and Sheena M. Joyce
(2012)
70 and 90 min. versions

These days, nuclear power is being marketed as "clean energy"—free of all those nasty greenhouse gases generated by burning coal, oil, and natural gas. In *The Atomic States of America*, President Obama announces that, thanks to $8 billion in loan guarantees, the United States is about to break ground on the first new nuclear plant in 30 years. This is the best audiovisual overview of nuclear power that we've seen—clear, engaging, moving, story-rich. Based on Kelly McMasters' *Welcome to Shirley: A Memoir from an Atomic Town*, *The Atomic States of America* raises profound concerns about our nuclear future. It deserves to be widely viewed, in school and out.

Into Eternity
By Michael Madsen
(2009)
58 min. classroom version (75 min. version included on DVD)

Into Eternity, a film about Onkalo, the "permanent" nuclear waste facility in Finland, asks how it is possible to bury poisons and keep them safe for 100,000 years. This haunting film may be too slow for some students, but it raises key questions about our responsibility to future generations that should be asked in high school science, global studies, economics, and government classes.

ERIC DROOKER

"In nature nothing exists alone."
    —Rachel Carson, *Silent Spring*

CHAPTER SIX:
Food, Farming, and the Earth

MEREDITH STERN

"Peasant-led agroecology is the real solution to global hunger. Not only do peasant farmers feed communities, they also cool the planet and protect Mother Nature. Unlike agribusiness, peasants do not treat food as a commodity for speculation profiting out of hunger. They do not patent nature for profit, keeping it out of the hands of the common man and woman. They share their knowledge and seeds, so everyone can have food to eat... La Vía Campesina reminds society and governments that if we really want to put an end to hunger, then we must accept the central role of the peasants, and support them to feed humanity."

—La Vía Campesina

Food, Farming, and the Earth

We start the book with the motto "everything is connected," and we finish on the same note. Food embodies many of the ecological problems and social injustices highlighted throughout the book. And similarly, it calls out for activism that recognizes the interconnectedness of these issues. The global industrial food system holds an inherent contradiction: It is a major source of global warming pollution, and at the same time it is threatened by increasing climate chaos. This same food system currently leaves close to 1 billion people hungry, not for a lack of food production or "overpopulation"—as many textbooks tell students—but because the global market privileges the profits of multinational corporations over the human right to food.

Although Christian Parenti's article, "Reading the World in a Loaf of Bread," shows up in the first chapter of this book, it occupies a symbolic place in this chapter as well: The Arab Spring revolts several years ago were sparked by record food prices, largely caused by the worst droughts in a century. Global climate change, global capitalism, and the constant struggle of people to maintain power and control over their lives (especially food) are all intricately woven together—and it's only by turning a blind eye to increasing global inequalities that we continue to treat these in the curriculum as different subjects.

Food embodies the connections among the problems we face, and it also represents the connections linking the solutions we must work toward. The La Vía Campesina role play in this chapter, based on the global social movement of 200 million peasant farmers, offers one of the best examples of how a deep response to the food crisis is also, necessarily, a deep response to the connected crises of the climate, global inequality, forced migration, and public health, among others. It's immensely hopeful to see how the farmers of La Vía Campesina, situated around the world, are fighting against the corporate food system that has wreaked havoc across the global countryside. The farmers are creating "food sovereignty" through local seed banks, co-operative farm schools, mass land occupations, and direct pressure on national and international food policy makers.

Finally, examples like La Vía Campesina could be lost on students unless we help them recognize the power they possess to effect change in the current food system. In "Food Secrets," Michi Thacker walks her students through the seemingly simple exercise of researching where their food comes from—only to discover that food corporations seem happier to take our money than our questions. This secrecy protects the corporations from public scrutiny: Once we start to inquire about the source of our food, we are likely to ask more questions about how it is grown, by whom, and under what conditions. And the inevitable conclusion, in the words of the farmer, poet, and philosopher Wendell Berry, is that "how we eat determines, to a considerable extent, how the world is used." ⊕

STEPHANIE McMILLAN

ORGANIC

PROCESSED

FORTIFIED

CONVENIENT

IMPROVED!

MICHAEL DUFFY

Food Secrets

Students explore the secretive journey from farm to table

BY MICHI THACKER

ood plays an important role in the curriculum at our school, an alternative public elementary school in Olympia, Washington, which enrolls just under 300 students. We have a large organic garden and greenhouse, and during the school year, our students dig, plant, weed, water, compost, harvest, and cook fresh produce. All classrooms use the garden to help teach measuring, graphing, and weighing. And we all work together to plan and prepare for a yearly harvest festival.

Years ago, our school piloted a new lunch program for our school district. After parents and students approached our district's food services supervisor, the district started working with local farmers to offer organic produce and non-meat protein choices. It made sense to begin with our school, as students were already used to eating fresh organic produce from our garden. After the changes succeeded at our school, the district expanded the program and contained costs by eliminating sugary desserts and other processed foods. The number of children choosing fresh fruits and vegetables went up by 50 percent, along with the number of children buying lunch at school. The district now supports eight regional farms.

I had been inspired by reading two provocative essays, "The Oil We Eat," by Richard Manning, and "Lily's Chickens," by Barbara Kingsolver, and decided to take our learning about food a little deeper. Our class began investigating food and where it really comes from. In a multiage classroom of 3rd and 4th graders, my goal was to increase children's understanding of where our food comes from by looking at individual food products and their ingredients to find out how far they traveled from farm to factory to store shelves. In the process we would look at food costs, the use of fertilizers and pesticides, who was harvesting and processing the foods, and ultimately, the benefits and costs of local vs. non-local, and organic vs. non-organic foods.

We are privileged to have our own school garden and to have access to a variety of locally grown, organic produce in our lunches; but I wanted my students to have a deeper understanding of why we garden organically at our school, and why we would choose to make these changes in our lunch program. Because many families cannot afford or do not choose to buy local and organic food, I wanted children to be able to explore the issues without feeling guilt or shame that their families were not making the same choices.

I began our study with the question "Where does our food come from?" Children's responses were predictable: "The store," "farms," "my garden," "cows," "a factory." After our initial brainstorm, I said I would like to explore the question a little more, and that we would take a field trip to a local grocery store to begin our research. I had chosen a store where they would find a diverse selection of foods, including some organic and natural foods. I invited children to think of a food they would like to research. They could choose any food on the shelf;

my only criterion was that the list be diverse (no repeats), with a variety of brands. Children raised their hands to share ideas as I wrote them on a large piece of newsprint. The list included well-known packaged brand-name items like Kraft macaroni and cheese, Gatorade, Nancy's yogurt, Pace salsa, and Diane's tortillas, as well as various organic and non-organic fresh fruits and vegetables.

Before our field trip, I asked the children what questions we would need to ask to find out exactly where our food items had come from, and we generated a list on another piece of newsprint. Using a combination of their questions and mine, I put together a data-gathering sheet and a "Food Product Research" sheet.

The Field Trip

On the day of the field trip, I explained to the students and adults who joined us: "Your task is to find as much contact information about your product as possible. Record everything you can find on the label or package, and be sure to pay attention to accuracy when you copy down information." I distributed clipboards, pencils, and data-gathering sheets. Because several parents joined us, I was able to assign a small group to each adult. The groups enthusiastically traveled together down the aisles, hunting for their items, pulling them from the shelves, helping each other inspect the boxes and packages for the critical information, and diligently recording it.

> Because many families cannot afford or do not choose to buy local and organic food, I wanted children to be able to explore the issues without feeling guilt or shame that their families were not making the same choices.

They listed brand names, parent companies if applicable, distributors, and any other potential contact information they could get from the labeling on their food items—phone numbers, addresses, web or email addresses. A few students needed to ask the store manager for additional help to find information about farms and distributors for the fresh produce. (Fortunately, the manager was welcoming and friendly, considering that I did not have the forethought to alert him ahead of time.)

After gathering the initial contact information, we returned to school and I assigned a research project. I explained to the class that the goal was to find out where our food had come from and how far it had traveled from farms to factories to store. I gave students a list of questions about their chosen products to ask when they contacted the products' manufacturers. (I didn't expect each student would be able to find answers to all of the questions, but I wanted them to ask them all so they would gather as much information as possible.) They asked the following questions:

- Where was your product made?
- What are the ingredients of your product?
- Where did each of these ingredients come from?
- Who harvested the fruits and vegetables that were used in your product?
- Where were they harvested?
- What company shipped the different ingredients to the factory and the final product to the shelves?

It was clear that this research project was going to require too much assistance to be completed at school, so I assigned the project as homework. I included a note to parents on the homework sheet: "Parents—please work on this research together with your child! Even if your child is capable of making the phone calls, they may come across people who do not take their questions seriously. Also, if you do not have internet access at home, please let me know so that we can arrange for some help at school."

I gave students two weeks to complete their research to make sure parents could find time to work with them to finish it. (Phoning can be complicated with time zones and children in school the majority of the day.) I had already introduced this project to parents at our first parent meeting, so most of them had had opportunities to ask ques-

> **As children began to bring in and share their information, we were perplexed and stunned to find that most of the companies refused to give us the information we were looking for.**

tions, and meeting notes went home to those who didn't attend. I knew this was a lot to ask of parents, but parent involvement is embedded in our school's philosophy. Because parents choose the food children eat, I hoped that parents and children working side by side might encourage some interesting discussions at home. What I didn't know was how truly difficult the assignment would be.

We started mapping our information right away with the initial information that children gathered at the grocery store. We had two large maps on the wall of our classroom, one of the United States and the other of the world. Each child wrote the name of his or her food item on a Post-It note. I helped children find and mark the places on the map where the product companies were located. Each time a student brought in new information about the source of an ingredient or location of a farm, distributor, or other company, we would add a Post-It to one of the maps (we used the world map for anything from outside the United States). I explained that we would be looking at how far our food had traveled and thinking about the resources used in transporting foods.

I also made a large chart to go on the wall with each child's name, the name of his/her product, and several columns where they could record the answers to our research questions.

While the students were working at home on their research, we took a field trip to Common Ground, a local organic farm that practices community supported agriculture (CSA), a system in which participants buy a share of the farm's produce and receive a weekly box of seasonal vegetables and herbs. (Some CSA farms also provide meat, cheese, flowers, or fruits to their members.) The owners of the farm asked children what they knew about organic farming, explained their system to students, and took them on a tour. Many of the students were familiar with CSAs because their families had bought shares. The farmers explained why they farmed organically and why buying and eating locally benefits small farmers. Then students formed a "bucket brigade" line and heartily joined in the effort to load boxes of produce for delivery.

Serendipitously, at home I discovered a short piece titled "Why it's a good idea to eat certified organic foods" printed on the inside of a box of Nature's Path cereal. I made copies and assigned children to work in small groups, reading and discussing the text. From this and other readings we learned about nutrients in the soil, erosion, and some of the consequences of short- and long-term use of pesticides and fertilizers.

Secretive Corporations

As children began to bring in and share their information, we were perplexed and stunned to find that most of the companies refused to give us the information we were looking for. Kraft Foods responded to one child by email, "As much as we'd like to help you, the information you are seeking is considered confidential. We hope you can understand our position. We'd be glad to share with you a brief history of one of our signature brands."

Children continued to hear back from other companies (Gatorade, Post, Frito-Lay, Campbell's) that the information was "classified information," "confidential," or "not available." Sometimes a company gave the name or location of a distributor, but said it was "not allowed" to tell where the ingredients came from. Several corporations never responded to emails or phone calls. Children were able to track each product back to its parent company, but often that was the company whose information was printed on the box or package, and usually it wasn't the place where the product was manufactured. Information about individual ingredients—names of farms where they were grown and harvested, names of factories, and so on—were "unavailable." Sometimes companies gave locations, but no names.

Each time we gathered together to share information, I asked children why it might be that some companies wouldn't share basic information about what we were eating. Students were, for the most part, stumped. Eventually one of the children had an answer: The concern was that the information might be given to a competitor. The implication was that we were spies! I asked the students why passing on information about sources of food might be a problem. They speculated that maybe there was something about the food that wasn't good for them. One child suggested that companies might not want us to know if they use genetically modified organisms (GMOs). I had some ideas of my own about the reasons for this secrecy, but I wanted children to sit with the question. The issue was keeping them en-

gaged and they were curious about other children's findings.

Not every company refused to give us information. Most of the companies that willingly shared (and were the friendliest and most welcoming to children) were small businesses, regional farmers, and/or growers of organic produce. Some sent us information packets. Nancy Van Brasch Hamren (of Nancy's yogurt) from Springfield Creamery in Eugene, Oregon, told her student-researcher the name of the farm their milk came from and where all of the other ingredients of their yogurt came from. We learned that their vanilla came from Madagascar, farther than any other food item we were able to track.

After all the students' information had been recorded on our chart, I pointed out which companies were smaller farms and businesses and asked why they thought those businesses were more willing to share information. Some suggested that they had nothing to hide, or that that the food might be healthier. I explained that small businesses were owned by a single person or a few people who worked closer to or even directly with planters and harvesters, rather than large corporations whose CEOs were far from the workers in the fields and whose focus was primarily on profit. That didn't necessarily guarantee that the food was healthier, I explained, but at least the owners were more likely to be informed about the conditions and details of the farm, food, and work environment. The owners of companies that produced organic foods often expressed a commitment to the health and environmental benefits of the food they produced.

On our map, the closest Post-It was in Oregon. None of our products had come from Washington state; many had come from the Midwest and the East Coast, and a few from overseas. Although it was clear from our research that the food we eat travels long distances, we had been unsuccessful, for the most part, in tracing foods all the way back to farms. Still, even without all of the information, we were able to look at the map together and see the great distances many of the products had traveled. I estimated with students the mileage traveled for a few of the items. I asked them what other energy costs might be involved; for example, what kinds of things are done to preserve food. Students came up with the costs of processing, packaging, and refrigerating foods. I added that the farther a food item travels, the more gas is required, which contributes to air pollution and global warming, and the longer the food items need refrigeration. We talked about how processing and packaging foods helps preserve them, but has additional environmental costs because of the process itself, as well as the plastic and paper used for packaging.

Visitors

At about this time, a delegation of representatives from the U.S. Department of Agriculture visited our school to learn about our lunch program. Our principal had asked me if I would be willing to have them join us for a lesson and I agreed. With the visitors present, I decided to conclude our study by reflecting as a group on the reasons for the changes in the district's lunch program—why we choose organic and locally grown foods.

I posted two large pieces of newsprint on the board. One had the question "Why buy local?" written on one side and "Why import foods?" on the other. The second had "Why buy organic?" on one side and "Why buy non-organic?" on the other. I asked children to reflect on their learning and address the questions one at a time. I recorded their thoughtful responses on the newsprint. We stopped occasionally to clarify or question the student answers. Student responses to "Why buy local?" included the following:

- Trucks don't have to drive as far, so it costs less to transport and causes less pollution.
- Food is fresher and less likely to rot.
- Buying locally supports local farmers in our community.
- Local foods cost less because they don't have to be shipped as far. (We explored this and talked about why this may or may not be true. For example, workers get paid differently depending on where the food is coming from, but imported food has farther to travel.)
- It is less likely to need pesticides because it isn't stored for as long.
- It tastes better because it is fresher.

And here were some of their responses to "Why import foods?"

- Some things don't grow well here. (We listed some examples: mangos, coconuts, vanilla, avocados.)
- Some things aren't available here year-round.
- Maybe it's cheaper? (We revisited the comments above.)

To the question "Why buy organic?" students responded:

- No pesticides are used; pesticides aren't good for you. (When I asked why, a number of students responded with the following points.)
- Pesticides kill good bugs too.
- Pesticides kill everything in the soil: microorganisms, worms, nutrients, roots.
- Pesticides are harmful to the people who harvest the food.
- Pesticides pollute water.
- Pesticides can lead to more erosion in the soil.

Another student said, "Organic food tastes better." I asked if this was a fact or an opinion, and one student responded that it would taste better because it is fresher. Another responded that she knew because she eats organic food. Another child added that eating organic food is better for biodiversity because pesticides are not killing off different kinds of insects.

When I asked, "Why buy non-organic?" students agreed that non-organic food is often cheaper. Because I wanted children to be clear that I didn't expect all of their parents to start buying organic food, I added that I don't always choose to buy organic for that reason—and that most people can't afford to always buy organic. But I also shared that buying things in larger, bulk quantities sometimes makes organic food cheaper and creates less waste. Other students noted that non-organic food is easier to find, not all foods are available organically, and that food might look better if it didn't have "bugs or bug holes." It's easier to ship, someone added, and doesn't rot as fast, and GMOs naturally resist some pests.

I concluded our brainstorm discussion by summarizing students' comments, adding that if our criteria for buying food were to be to live and eat more sustainably, it seemed clear that eating organic and closer to home made more sense.

Then children redirected their focus to our visitors from the USDA. I asked students to share about their research projects. Several spoke about how they had been unable to find the information they were looking for. I asked the delegation why we had such difficulty getting information about the food we eat. One of the representatives responded that many companies have confidentiality agreements with their providers, processors, and distributors. Each company that produces a product contracts individually with farms, factories, and distributors to get the lowest price possible. Those prices vary from company to company (or farm to farm) and sharing information might put one company at an advantage. Ultimately, the goal for each company is to maximize profits. One USDA representative confirmed that competition was the motivator behind the secrecy.

This is where I left the secrecy issue with students, but I'm not fully satisfied. As public interest and awareness around food sourcing and food safety continues to grow, corporate resistance to transparency has also been on the rise. In 2013, corporations poured millions of dollars into fighting a statewide initiative in Washington that would have required labeling foods with genetically modified ingredients. The initiative would not have banned GMOs, only required that their presence be announced. Corporations' resistance to consumers' right to know the content of our food underscores their desire to monopolize every aspect of production and distribution of food. And secrecy is an inherent part of this control.

In the end, we had to leave many blank spaces on our chart, and in many places we entered "not available," "classified," "confidential," or "no response." Throughout our discussions I had asked the class how they felt about not being able to get specific information about the food they eat, and they had repeatedly responded with a sense of confusion and betrayal. What felt like important information to us was determined to be none of our business.

Looking Back

This was not an easy unit to teach. If I were to attempt it again, I would simplify the investigation. Students might track only one or two main ingredients (produce, grain, or meat) in each item. I would also ask children to document attempts at communicating with farmers and corporations, and the results of each contact. I might collaborate with an older class of students with more computer experience, to minimize the need for parent help.

Ultimately, I think the investigation would be a better match for middle and high school students who could take full responsibility for the assignment, and, working over a period of weeks, document each attempt at contacting farmers, producers, and distributors.

A year after this research project, I went back to some of these students to ask what they remembered from the study. All of them remembered how difficult it had been to get information, even if they themselves had been successful. In a comment that was indicative of how many of us felt, one student said, "I felt like Kraft was telling me to buzz off. It made me wonder if they're not telling me something; are they hiding something from me? It made me be a little more aware of what's going into my food— check the contents, the ingredients, think more about it."

Although we were all disappointed we couldn't get some of the information we sought, this lesson provided some powerful learning and lots of food for thought. The children gained a deeper understanding about the purpose behind our new lunch program, and some families reported changes in the choices they made at home. In an emergent unit like this, where I was learning along with my students, the point was not to have all of the answers, but to be able to wonder with children about how the world of information operates, to teach children to ask the questions, and ultimately, to help them act on what they have learned. My long-term hope was that children would continue to think critically about their own food choices and continue to ask questions about the business of food. ⊕

> Corporations' resistance to consumers' right to know the content of our food underscores their desire to monopolize every aspect of production and distribution of food.

Michi Thacker taught at Lincoln Elementary School in Olympia, Washington, for 23 years. She is currently teaching in the Master in Teaching program at The Evergreen State College.

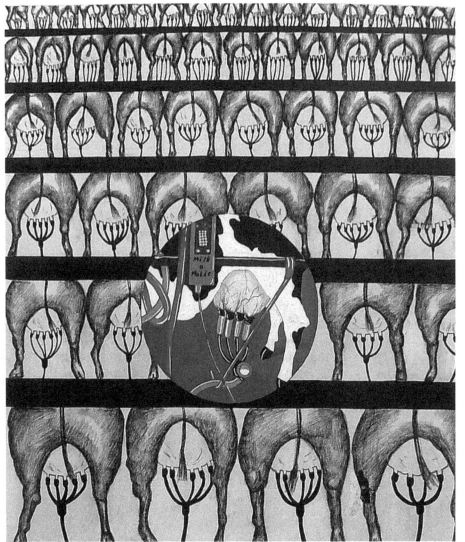

SUE COE: *MACHINE COW* ©1990 SUE COE, COURTESY GALERIE ST. ETIENNE, NEW YORK

Got Milk, Got Patents, Got Profits?

Students probe the corporate-controlled food system and patents on genetically modified seeds

BY TIM SWINEHART

ot milk? Want strong bones? Drink milk. Want healthy teeth? Drink milk. Want big muscles? Drink milk.

"The glass of milk looks nice and cold and refreshing. If I had a warm, homemade chocolate chip cookie, it would make my day. They go perfect together."

Ari and Colin could have been writing radio spots for the Oregon Dairyman's Association, but instead they were writing about the glass of milk I had set out moments earlier in the middle of the classroom. My instructions to the students were simple: "Describe the glass of milk sitting before you. What does it make you think of? Does it bring back memories? Do you have any questions about the milk? An ode to milk?"

From the front row, Carl said, "Mmmmm...I'm thirsty. Can I drink it?"

"Why don't you wait until the end of the period and then I'll check back with you on that, Carl," I responded.

We had spent the last couple weeks discussing the politics of food in my untracked 11th-grade global studies classes. And while students—mostly working class and European American—were beginning to show signs of an increased awareness about the implications of their own food choices, I wanted to find an issue that they would be sure to relate to on a personal level. One of my goals in designing a unit about food was to give students the opportunity to make some intimate connections between the social and cultural politics of globalization and the choices we make as individual consumers and as a society as a whole. A central organizing theme of the unit was choice, which we examined from multiple perspectives: How much choice do you have about the food that you eat? Do these choices matter? Does knowledge about the source/history of our food affect our ability to make true choices about our food? How does corporate control of the global food supply affect our choices and the choices of people around the world?

I wanted to encourage my students to continue asking critical questions about the social and environmental issues surrounding food, even outside the confines of the classroom. I wanted to develop a lesson that would stick with them when they grabbed their afternoon snack or sat down for their next meal, something they might even feel compelled to tell their friends or family about.

Milk turned out to have the sort of appeal I was looking for. For almost all my students, milk embodies a sort of wholesome, pure "goodness," an image propped up by millions of dollars of advertising targeted especially at children. My students had been barraged with the message that "milk does a body good" for most of their lives. Parents, teachers, celebrities, and cafeteria workers had encouraged them to drink milk as a healthy part of their diets. But I believe that my students—along with the vast majority of the U.S, public—haven't been getting the whole story about milk. I wanted to introduce them to the idea that corporate interests, oftentimes at odds with their own personal health, hide behind the image of purity and health.

Growth Hormones and Milk

I wanted to help my students reexamine the images that milk evoked by presenting them with some unsettling information about the artificial growth hormone rBGH, originally developed by Monsanto. Recombinant Bovine Growth Hormone (rBGH—also known as Bovine somatotropin, bST, or rBST) is a genetically engineered version of the growth hormone naturally produced by cows, and was approved by the federal Food and Drug Administration (FDA) in 1993 for the purpose of increasing a cow's milk production by an estimated 5 to 15 percent. Monsanto began marketing rBGH, under the trade name Posilac, and promoted the hormone as a way "for dairy farmers to produce more milk with fewer cows, thereby providing dairy farmers with additional economic security." But with an increased risk of health problems for cows stressed from producing milk at unnaturally enhanced levels—including more udder infections and reproductive problems—critics argue that the only true

economic security resulting from the sale of Posilac is the huge corporate profits made from the product. In 2008, Eli Lilly and Company purchased the Posilac business and patent from Monsanto for more than $300 million.

The human health risks posed by rBGH-treated milk have been an issue of intense controversy since rBGH was introduced more than a decade ago. Monsanto and the FDA say that milk and meat from cows supplemented with rBGH are safe. On the other hand, a number of peer-reviewed studies, most notably those of University of Illinois School of Public Health Professor Samuel Epstein, MD, have shown that rBGH-treated milk contains higher than normal levels of Insulin-Like Growth Factor 1 (IGF-1). Although IGF-1 is a naturally occurring hormone-protein in cows and humans, when increased above normal levels it has been linked to an increased risk of breast, prostate, and colon cancers. Monsanto itself, in 1993, admitted that rBGH milk often contains higher levels of IGF-1. The uncertainty surrounding these health risks has led citizens and governments in Canada, the European Union, Australia, New Zealand, Israel, Argentina, and Japan to ban rBGH.

The continued use of rBGH in the United States points to the political influence of large corporations on the FDA's regulatory process. When, in 1994, concerned dairy retailers responded to the introduction of rBGH with labels indicating untreated milk as "rBGH free," the FDA argued that there was no "significant" difference between rBGH-treated milk and ordinary

> For almost all my students, milk embodies a sort of wholesome, pure "goodness," an image propped up by millions of dollars of advertising targeted especially at children.

milk and warned retailers that such labels were illegal. The FDA has since changed its position and now allows producers to label rBGH-free milk. Paul Kingsnorth, writing in *The Ecologist* magazine, offered one explanation for the FDA's protection of rBGH: "The FDA official responsible for developing this labeling policy was one Michael R. Taylor. Before moving to the FDA, he was a partner in the law firm that represented Monsanto as it applied for FDA approval for Posilac. He has since moved back to work for Monsanto." Not an isolated incident, this example illustrates what critics refer to as the "revolving door" between U.S. biotechnology corporations and the government agencies responsible for regulating biotech products and the safety of the nation's food.

The story of rBGH encapsulates many of the worst elements of today's corporate-controlled industrial food system. Despite the illusion of choice created by the thousands of items available at the supermarket, consumers have little knowledge about where food comes from and how it is produced. By uncovering the story behind rBGH, I hoped students would begin asking questions about the ways corporate consolidation and control of the world's food supply have drastically limited the real choices and knowledge we have as food consumers.

To familiarize ourselves with Monsanto's point of view, we spent a day in the computer lab exploring the corporation's website (www.monsanto.com). I asked students to look for arguments made in favor of biotechnology and genetically modified foods: Why do companies argue that these technologies are important? What benefits do they offer to humans and the environment? On Monsanto's website, some students were impressed with descriptions of a genetically engineered soybean designed to reduce trans fats in processed food, and others mentioned drought-resistant crops that required less water.

Drew, however, was skeptical of the language Monsanto used to describe its research and products. "Why don't they ever use the terms 'genetically modified' or 'genetically engineered' and always use 'biotechnology product' instead? I find it ironic that Monsanto's 'pledge' is to uphold integrity in all that they do, even though genetically modified foods threaten the integrity of people and the environment."

The Corporation

Carl's request to drink the milk we had used as a writing prompt made a nice segue into showing students a short clip about rBGH from the documentary film *The Corporation* (from 29:15 to 32:30 on the DVD). As we viewed the clip, which includes powerful images of cows with swollen udders and compelling testimony from Dr. Samuel Epstein that links rBGH to cancer, students reacted. "Is that a real cow?" "Gross!" "Is that in our milk?" and "That's messed up, dude!" came from various corners of the room. But while sick cows and potential cancer risks are important, I was hoping to impress upon students how the risks of rBGH have been ignored and hidden from the public by Monsanto and by those who license its use at the FDA.

I showed the clip from *The Corporation* as a pre-reading strategy for Kingsnorth's article "Bovine Growth Hormones." The article is technical and can be a difficult read for some students, so I hoped to encourage their interest and give students a purpose for reading before I passed it out. I asked students to list questions or concerns as I paused the DVD. I was encouraged by their curiosity: "Do hormones get into the milk and how do they affect us?" "Is there pus in our milk?" "Is milk truly healthy for us?" "Why is rBGH necessary, if we already have too much milk?" "If they knew that the drug made cows sick, why do they still use it?" "What can we do about it?"

Then I passed out highlighters and told students to choose five questions from our list and to read "Bovine Growth Hormones" with those five questions in mind, highlighting important information as they came across it. The article is comprehensive, and students found answers to the majority of their questions, including everyone's favorite: "Is there pus in our milk?" Truth be told, all milk, including organic milk, has small amounts of somatic cells or "pus" in it, but the FDA has strict quality standards for the somatic cell count (SCC) above which milk may not be sold to consumers. What students learn from the article—and what the warning label accompanying all Posilac reads—is that cows treated with rBGH are more likely to produce milk with increased SCCs due to the heightened risk for udder infections

Drawing from the information from the website, film, and article, I wanted to give students another chance to respond to the glass of milk still sitting in the center of the room. I asked them each to draw a line under their initial descriptions and to write a second response: "What additional thoughts

> "Gross!" "Is that in our milk?"

or questions do you have about this glass of milk?"

Ari had initially extolled the many health virtues of milk but now seemed equally concerned about possible health risks: "Apparently, I get calcium, pus, and an increased risk of uterine, breast, and various kinds of cancers. Now, when I look at that glass half full of milk, I see cancer in a glass with a thin layer of pus as a topping. Now I don't think I can look at milk in the same way."

Ari's comment brings up a legitimate concern that by teaching students about rBGH I am scaring them away from milk and toward less attractive alternatives, including soda. Such risks were a constant concern while teaching students about the myriad problems associated with industrially produced foods. After learning about the health and environmental risks of pesticides, herbicides, hormones, and genetically modified food, I had more than one student ask in exasperation: "But Mr. Swinehart, what can I eat?"

We are fortunate in Portland, Oregon, to have a vibrant local food system that makes healthy, safe, and affordable food available to many. Several Portland-area dairies, including Sunshine, Alpenrose, and the nation's second largest producer of natural chunk cheese, Tillamook, have all committed to producing only rBGH-free milk products. Because these are not organic dairies, their rBGH-free milk tends to be less expensive and a more reasonable alternative for students than certified "organic" milk. Dairies in many other parts of the country have made similar pledges Being able to recommend these local dairies not only presented students with a viable alternative to giving up milk completely, but also gave them a chance to apply their knowledge of controversial rBGH labeling during the next trip to the grocery store.

Compared to Ari, Eron wasn't too worried about rBGH's health risks, but he did express a willingness to rethink his decisions as a consumer: "I still love milk and will drink it, but maybe I will make a change and buy organic milk instead so that I don't get all of the health risks. It seems this might benefit me the most and I will be happy about the choices I made." Of course, many students will choose to continue drinking milk regardless of where it comes from or what it has in it, but their knowledge of rBGH and the corporate politics behind unlabeled milk cartons makes this a considerably more informed choice than most U.S. consumers have.

Eron's comment also raised one of my primary concerns in trying to teach students about the global politics of food. I was confident going into the unit that students would react strongly to issues surrounding the health of animals and their own personal health, but my goals for the unit were larger than this. I was encouraged to see Eron thinking about the effects of rBGH on his own personal health, but I also wanted students to make broader connections to ways the corporate control of the food system takes knowledge and power out of the hands of small food producers and consumers around the world. Do some countries and corporations benefit more from a global industrial food system than others? Do the environmental costs of this same food system pose a substantially greater risk for the world's poor, who still depend on a direct connection to the Earth for their means of sustenance?

Patents on Life?

Since students' comments during the milk lesson seemed to focus on personal choices, I realized that we needed to broaden our focus from the politics of health surrounding rBGH to include an exploration of how a global food system, increasingly controlled by a few multinational agribusiness corporations, affects lives and cultures around the world. I wanted students to look at how corporations are changing the nature of food. Through the science of genetic engineering, biotechnology companies are experimenting with the biological foundations of what is arguably the world's most important life form: the seed. Biotech companies tend to downplay the revolutionary nature of this new science by suggesting that humans have influenced plant genetics, through selective breeding and hybridization, since the dawn of agriculture.

But because genetic engineering allows for the DNA of one organism, including animal and virus DNA, to be placed in a completely unrelated plant species, it crosses natural barriers that were never breached by traditional plant breeding. Without adequate testing or knowledge of long-term consequences, genetically modified (GM) crops are now grown around the world, posing what many argue is a serious threat to global food security. Through the natural and highly uncontrollable process of cross-pollination, GM crops have the potential to contaminate the genetic code of the traditional crops that have provided people with food for thousands of years.

> I wanted students to look at how corporations are changing the nature of food.

It is not, however, just the seed itself that is changed through the process of genetic engineering, but the very idea of the seed is transformed. By altering the DNA of traditional seeds, biotech companies claim the new seed as an "invention" and secure their right to ownership through the legal system of patents. Global production of biotech crops and the number of corporate-owned patents on seed have increased dramatically over the last two decades. Monsanto alone owns more than 11,000 seed patents.

To help students grapple with the international politics of seed patenting and GM foods, I designed a role play that would encourage them to confront the often unequal effects of the global food system and the global economy in which it operates. I set up the role play as a special meeting of the World Trade Organization (WTO), the primary governing body for international trade law. I asked students to debate how GM foods should be regulated internationally by taking on the following roles: farmers from India, U.S. Trade representatives, European Union commissioners, U.S. consumers, Greenpeace, and Monsanto. I asked them to reconsider WTO rules that set U.S. patent law as the de facto international standard for determining who has "ownership" of certain foods. In the introduction to the role play handout, I explained:

You are delegates to a special summit of the World Trade Organization (WTO). This meeting has been called to debate genetic engineering and patenting of foods. Due to worldwide resistance to genetically modified (GM) foods and the patenting of seeds, the WTO has been forced to reconsider its position on patents and the rights of multinational corporations to trade GM foods and seeds. . . . Your task for this summit is to determine to what extent GM foods deserve regulation, who should be responsible for any regulations that are necessary, and what these rules should look like.

(Note: the roles and full role play description are available here: www.rethinkingschools.org/earth)

This "special" meeting included voices that would never be heard at the actual, much-more-exclusive meetings of the WTO, but I wanted students to make their decisions in the role play based on a fuller representation of global perspectives.

To encourage students to consider the issues at stake in the role play, I asked them to write interior monologues—statements where they imagined details about family, background, hopes, dreams, and fears, all from the perspective of their roles. I wanted to give them opportunities to create personal connections to the characters they would embody during the role play, while also engaging with the critical issues surrounding GM foods and seed patenting.

Julia's monologue from the perspective of an Indian farmer was particularly insightful:

I don't have the heart to tell my mother about TRIPS (Trade-Related aspects of Intellectual Property Rights), because I don't think her body could handle the stress. TRIPS is an agreement of the World Trade Organization, an organization I could not have cared less about until a few years ago. TRIPS requires member countries to protect patents on all kinds of life. This means that if someone was to put a patent on the type of rice that I am growing, I would be unable to grow and sell my crop without a payment to the patent holder. In addition, I wouldn't be able to save my seeds from one year to another—something every generation in my family has done as far back as anyone can remember....By saving our seed, we become acquainted with every plant on our field. I know that some of the seeds that I have stored away date back to my father's time. When I plant my saved seed, I plant not only rice, but also my heritage.

Of course, not all my students displayed such a sophisticated understanding of something as complex as international patent law. Looking back, I may have taken on a little too much with the content of the role play. Many students struggled to understand how the specific concerns of their characters should translate to recommendations at the WTO meeting. There were times when I felt ill prepared to answer students' questions about the international debate surrounding genetically modified foods or the current status of WTO trade laws. But when it came time to discuss the issues at our meeting, I felt encouraged by the students' ability to not only articulate the perspective of their own roles, but to ask questions of one another that showed a solid grasp of the concerns represented around the room.

Will, speaking as the U.S. trade representative, said:

It's our belief that the companies that create GM foods are the most capable of testing them for safety. Companies like Monsanto spend millions of dollars each year on research, so they have an expertise that an international testing body wouldn't. And as far as saying that people may have allergic reactions to GM foods—well, we just don't feel that this is a sufficient reason for banning them completely. I mean, look at how many people are allergic to peanuts, but we don't ban peanut butter, do we?

Amber chimed in as the Monsanto representative:

Yeah, if you think about it, it's in our interest to produce safe foods. I mean, we want people to keep eating them, right? And I'd like to remind you that the FDA fully approves all of the GMOs that are used in food in the United States.

Colin, representing Greenpeace, said:

But isn't it true that there are some GMOs that are not approved for use in food for humans? Mix-ups occur. How can we be sure what we are eating? If GM foods aren't labeled, how can consumers protect themselves?

And Julia, as an Indian farmer, said:

It's not just allergies that we're worried about. There are countries in Africa that have refused GM food from the United States because they are afraid that it will mix with native crops and contaminate them. Farmers from my country are worried about the same thing. You tell us that these things are safe, but you're the same people that made Agent Orange into a pesticide to use on food. How can we trust you?

Although we finished the role play with some ideas for how it could be improved next time, the discussion showed me that my students were leaving with an understanding of the politics of food. They had gained knowledge about the way the issues of GM foods and patenting play out on a global scale, privileging a few powerful agribusiness corporations at the expense of the world's food consumers and small, local farmers.

After discussion, the class decided to follow the "precautionary principle," which guides policy in many European nations, and institute a worldwide moratorium on GM foods until they could be proven safe, and to require labeling of any GM foods that were approved for consumption. Furthermore, the summit voted to take away the right of any person or corporation to patent food.

> "You tell us that these things are safe, but you're the same people that made Agent Orange."

Of course, in the real world, the voices of traditional Indian farmers are not heard in the same conference room as those representing the world's largest corporations. Furthermore, the WTO is not likely to institute a ban on GMOs or radically reform patent laws any time in the near future. In this respect, the role play did not

MEREDITH STERN

A Storage Place of Culture

The seed, for the farmer is not merely the source of future plants and food; it is the storage place of culture and history. Free exchange of seed among farmers has been the basis of maintaining biodiversity as well as food security; it involves exchanges of ideas and knowledge, of culture and heritage...The new intellectual-property-rights regimes allow corporations to usurp the knowledge of the seed and monopolize it by claiming it as their private property... To secure patents on life forms and living resources, corporations must claim seeds and plants to be their "inventions" and hence their property. Corporations like Cargill and Monsanto see nature's web of life and cycles of renewal as "theft" of their property.

—Vandana Shiva, *Stolen Harvest*

result in practical solutions to the problems facing farmers and consumers of food around the world. I was concerned that this might do a disservice to students. But after spending close to a month studying the crises of our global food system, I believe that I would have done a greater disservice by not prompting them to envision a more equitable and sustainable food economy.

When we started the unit several weeks earlier, most students were unable to see beyond how the choices we make about food affect anything other than personal health. The milk lesson was intended as a hook to reach students through their concerns about personal health with the hope of transforming this concern into a broader appreciation for our fundamental right to know and control where our food comes from and how it is produced. The current state of the industrial food economy, as Julia wrote in her final paper, "results in a public denied of their right to knowledge and proper choices about their food." Changing this economy will require the sort of resistance embodied in the role play by the farmers of India and the advocacy of groups like Greenpeace.

One of my greatest hopes in teaching students about food is to foster an understanding of the important role food plays in today's global economy and the even more important role it will play in creating more local, democratic, and sustainable economies of the future. ◉

Tim Swinehart (timswinehart@gmail.com) was a student teacher at Franklin High School in Portland, Oregon, when he taught this unit. He currently teaches at Lincoln High School in Portland. In 2002, Swinehart and his wife, Emily Lethenstrom, founded the Flagstaff Community Supported Agriculture (CSA) project in Arizona.

In a mild late-winter afternoon, 5th graders at Verde Elementary School in North Richmond, California, squat on soggy ground, poking beans into the dirt with thin sticks. They move on to carrots, marveling at the tiny seeds that get stuck on the palms of their hands. Fava beans, bright yellow and orange calendula, and a whole pharmacy of herbs are flourishing in the garden's rock-rimmed plots.

In one recent year Verde Partnership Garden produced close to 1,000 pounds of vegetables. The students set up a farmers market in front of the school every two weeks. Parents were so eager to buy that they sent orders in with their children, said garden co-coordinator Bienvenida Mesa. North Richmond, like many depressed communities across the nation, has more than its share of liquor stores, but no stores that sell decent, much less organic, produce.

Verde Garden gives much to the school as well as to the community. Teachers work it into lesson plans. Students come for recess or respite and don't want to leave. Mesa and the other garden coordinator, Cassie Scott, have to gently herd them back inside.

The garden started in 1995, after a group of Laotian immigrants simply began digging there. Hmong and Mien women who went to an ESL class at Verde decided they needed a garden and began hand-tilling the rubbish-filled vacant lot next to the school. Scott worked as a play therapist at Verde at the time.

"I saw a large number of women with hoes working there and within a short time they'd dug up a vacant piece of land next to the football field. And I was inspired," says Scott.

Modern Diggers

Like the "Diggers" who began farming on Saint George's Hill in the spring of 1649 and inspired a brief but radical chapter in British history [see p. 18], the Laotian women's act called into question the

MEREDITH STERN

Greening for All

The right to access healthy food

BY MARCY REIN AND CLIFTON ROSS

meaning of "public property." They exercised their right to land by putting hoe to dirt, and joined Richmond's deeply rooted and lively gardening tradition.

Other gardeners here have dug on vacant land, but asked permission first. Today, some are finding they need to claim their rights first in the city planning process. Urban agriculture, with its potential to build food security along with community well-being, has hit the public policy agenda. Funds are becoming available, and questions of access and inclusion are raising their heads.

In the early 1900s, the North Richmond area around Verde School was called "Cabbage Patch" for its dozens of truck farms. Victory gardens bloomed

> The Laotian women's act called into question the meaning of "public property." They exercised their right to land by putting hoe to dirt, and joined Richmond's deeply rooted and lively gardening tradition.

Very Human Needs

I want you…to think about change very personally, in the way that people, for example, have changed in Detroit. In the 1970s and '80s, all you could see were vacant lots. Abandoned houses. Rot. Blight.

Then some African American women who had lived in the South saw these vacant lots as places where you could grow food to meet a basic need. And they didn't see it only in terms of belly hunger. They saw urban kids growing up without a sense of process, without a sense of time. And they thought urban agriculture would be a means for cultural change in young people. That's how the Urban Agricultural Movement developed—out of that reality and the very human needs of people.

—Grace Lee Boggs, author of
The Next American Revolution: Sustainable Activism for the Twenty-First Century

around Richmond during World War II, when waves of rural migrants swelled the city's population to around 100,000. Okies and Arkies and African Americans from the South worked the shipyards and defense plants, and grew their fruits and vegetables.

"Sure, we had gardens all over the city," says longtime community and environmental activist Lillie Mae Jones. "People had to have gardens so they could survive." Like today's urban gardeners, they used techniques such as sheet mulching, composting, and companion planting to raise more and healthier crops.

Today Richmond has about a dozen garden projects, including an urban agriculture class at Richmond High School, several school gardens—at least two of which may be casualties of the school district's financial woes—and "Lots of Crops."

North Richmond has upwards of 40 vacant lots that invite illegal dumping and become eyesores and health hazards, according to Saleem Bey of North Richmond Green, which runs Lots of Crops. So far, the group has gotten permission from owners of 10 of those lots to use them. It plans to build raised beds, employ young people from the community as gardeners, and distribute the produce in the neighborhood—free or at low cost.

"We consider these lots public lands," Bey says. "We walk by them every day. We think that if you have a need and the land is available, the greater good supercedes ownership. While they are not being used, we have the right to use the land in our community."

Food Self-Sufficiency

Richmond has been known for crime and violence, poverty, and pollution since the postwar years when industry fled and malls destroyed downtown. "But Richmond could be the most food self-sufficient urban community in the U.S.," says Park Guthrie, who works with the nonprofit Urban Tilth and the 5% Local Coalition, which wants to see 5 percent of west Contra Costa County's food grown locally. "We have a climate that lets us grow year-round, a number of agrarian traditions, lots of open space and public officials who are interested," says Guthrie.

City planners see the newly minted Richmond Greenway as a prime site for urban gardens. The

Greenway runs a mile up an old railroad right-of-way, past cyclone fences and boarded-up windows, broken-down factories, and neat bungalow homes. Near the west end of the trail, the Lincoln School Garden and Lincoln Community Garden make use of the Greenway's wide swath of open space.

At the east end, Khmu immigrants—who like the Hmong and Mien come from the hill country in Laos—raise eggplants and peppers, cilantro, cucumbers, squash, and beans. Father Don McKinnon and Sr. Micaela O'Connor of the Catholic Church's Khmu Pastoral Mission helped them work through the city's bureaucracy to get permission to use the land—a process that seemed odd to the Khmu.

"In my country, when you want to garden, you just find some place, cut down the trees and plant," says Kham Sousamphan. "We don't have fences."

Avoiding Gentrification

Gardens and open space can be attractive amenities that boost property values, according to a 2006 study done by New York University. Property values went up as much as 9.4 percent over five years in neighborhoods next to well-kept community gardens in the Bronx—and low-income areas benefited most from the gardens.

This means Richmond will need to move carefully to be sure the Greenway doesn't turn into "a Roman road to gentrification," says Doria Robinson from the 5% Local Coalition. The Romans first built roads and made other "public improvements" in areas they conquered or hoped to conquer.

"Communities like mine deserve greenways and open space," Robinson says, "but when we're moving forward with greening we can't get myopic. We have to be clear that we're working for the people who live here, not doing things that will end up moving them out and moving in people with more means."

A real commitment to "greening for all" will involve patient work in the neighborhoods to draw people in and be sure they have a meaningful voice in land use planning. History suggests another path as well, the one taken by the Hmong and Mien women at Verde School. From the 1400s to the mid-1600s, Europe's landed gentry gradually "enclosed" or privatized public lands that had been used for grazing and farming. People who depend-

ed on those lands for survival launched struggles to reclaim them. The Diggers chose direct action against the gentrification of their time: they began to dig and plant, hoping to overturn society with the turning of the soil.

Marcy Rein is a writer, editor, and organizer. She has engaged with a range of social movements and organizational forms over the last 40 years, including publication collectives, labor unions, and community organizations. Clifton Ross is a translator, filmmaker, and writer. His book of poetry, Translations from Silence, *was the recipient of PEN Oakland's 2010 Josephine Miles Award for Literary Excellence. They co-edited* Until the Rulers Obey: Voices from Latin American Social Movements.*

See teaching ideas for this article, page 386.

A Grassroots Revolution

This is a grassroots revolution that is changing the way we grow and distribute food. It is creating a new agricultural system that builds community rather than destroys it. It is connecting small farmers directly to consumers, and preserving their livelihoods. It is bringing fresh food into urban communities like the one where I work in inner-city Milwaukee. It believes in the idea that everyone should have access to healthy, affordable food.

This revolution began simply as a movement, with a handful of organic farmers who decided decades ago to reject the industrial model of agriculture. I call it a "revolution" now because we have reached a critical mass. More and more people are becoming involved in growing their own food, buying from farmers markets and rising up against the industrial methods of agriculture. This includes black people and white people, rich and poor, young and old. We all realize that our industrial food system has made us sick.

—Will Allen, director of Growing Power,
an urban farming project in Milwaukee

King Corn

Teaching the food crisis

BY TIM SWINEHART

"**A**ll right, folks, we're going to start today by breaking the rule about eating in class. I'd like you to take all of your snacks, drinks, gum, mints, and any other food in your backpacks and pile it up on your desk. If you don't have any food with you today, borrow something from a neighbor. Start reading through the ingredients and looking for corn. How much corn do you think we have in the room today? Any guesses?"

I have the good fortune of planning my 9th-grade global studies curriculum with my colleague Julie O'Neill. We decided to start class with this "food-from-your-backpack" activity as an introduction to the documentary *King Corn*. Recent advertising campaigns against high-fructose corn syrup have helped to make some students more aware of how pervasively that ingredient is used in processed foods, but we were hoping to show just how "creative" industrial food producers have been at incorporating America's largest crop into the snack foods that students eat every day.

We projected a list of commonly included ingredients that are likely to be made from corn (including such unlikely suspects as caramel, dextrose, sorbitol, food starch, and xanthan gum), and then asked students to guess what percentage of their foods were made of corn. Although the accuracy of their estimates was questionable (ranging from zero to 90 percent), the exercise gave students a way to immediately connect to the film, which opens with the two filmmakers undergoing isotopic hair analysis to see how much corn they are made of. When Curt Ellis and Ian Cheney are told by Stephen Macko, University of Virginia professor of environmental science, that more than 50 percent of the carbon atoms in their bodies are composed of corn, our students seemed to share the shock that Ellis and Cheney display in the film.

Our goals in showing the film extended beyond raising awareness about the foods that our students eat. Our food curriculum focuses on "choice": the extent to which we do or don't have true choices about the food we eat, and what that means for people and the environment. In part, we want to confront the narrative of the all-powerful consumer that looms so large in U.S. culture: the idea that we are all in control of our diets, health, and happiness through the decisions we make in the marketplace. Without stripping power and agency from students, we want to show that most of us don't have as much choice as we perceive about the food that we eat. This is especially important when we talk with students about solutions to the social and environmental problems we discuss over the course of our food unit—their natural first reaction is often to focus on making better individual food choices.

In fact, for the past six years, I've created units on the politics of food for almost every social studies class I've taught—from world history to economics—and I've started each unit by asking students to think and write about the question of choice. In the discussions that follow, we've bounced from the inevitable critiques of school food to differing perspectives on how families shop, cook, and eat at home.

As a whole, these discussions are some of the most lively and impassioned that we have all year, which is not surprising, given the highly personal nature of food. But our introductory conversations rarely branch into the realm of the political. Perhaps this should not come as a surprise, given the lack of any real national discussion about how today's industrial food system—with its reliance on fossil fuels and single-minded focus on increased production—is inextricably linked to the crises we face with regard to issues like climate change, energy, and health care.

King Corn

I asked students to return their food to their backpacks, and it was time to watch the movie. Before the film began, I told students: "As you watch *King Corn*, I want you to take some notes." The expected chorus of groans erupted across the classroom. "But when I take notes, it makes it hard to pay attention to the movie," Sam argued compellingly. In fact, Julie and I wanted to help our freshman students give the film their full attention and had decided to ask them to take graphic notes: "When something comes up in the film that captures your attention, really makes you think or ask questions, make a drawing in your notes to represent the idea. If you want to jot down a few words to go along with the drawing, that's fine, but the goal is simply to help you remember what you are thinking while you watch the film. You'll have the chance later to expand an issue from your notes into a metaphorical drawing that represents one of the messages you think we should take away from the film."

> Without stripping power and agency from students, we want to show that most of us don't have as much choice as we perceive about the food that we eat.

This method of note-taking has been a good compromise with our highly visual and kinesthetic freshman students: It allows them to stay focused on the film while also giving them a surprisingly effective way to recall important points once we've finished.

An Acre of Corn

Before watching *King Corn* for the first time in 2007, I hadn't given much thought to the role of corn in my own diet—let alone our entire food system. The story told in the film, of two friends who move to Iowa after college to grow an acre of corn, forever changed my understanding of how this single crop has come to dominate our industrial food system. *King Corn* tells a "cradle-to-grave" story of their one-acre corn crop; this turns out to be a much more complicated project than Ellis and Cheney likely imagined to begin with. As Cheney remarks in the film:

> It was already clear that, when the time came to say goodbye to the corn from our acre, we would never know exactly where it would end up. After the crop is delivered to the elevator, following corn into the food system becomes a game of probability. Of the 10,000 pounds of corn our acre is likely to produce, 32 percent will be either exported or turned into ethanol, in neither case ending up in our food. Or in our hair. But 490 pounds will become sweeteners, like high-fructose corn syrup. And more than half our crop, a full 5,500 pounds, will be fed to animals to become meat.

As they follow their acre of corn, the filmmakers are forced to confront the environmental and social consequences of a food system reliant on cheap, industrially produced corn. I like to pause the film in a number of places to give students the chance to process these issues and add their own commentary. One example is when Curt sticks his hand through an "observational" hole that has been cut into the side of a living cow's stomach—a cow being used to research the health effects of a diet composed of so much corn. The gross factor is pretty high for most students during this scene, but I think it's important to harness their initial interest (and disgust) and to push for a broader understanding of what this individual sick cow represents.

After stopping the film, I asked: "If you can put aside for a second how totally disgusting that is, I'm curious if anyone has a thought about why it's so common in this country to feed cows a diet that actually makes them sick?"

Sammy eventually hit the nail on the head: "Because it's how they can make the most money. Corn is cheap and it makes the cows get fat quick. That way you can make a bigger profit than the ranchers who only feed their cows grass."

The use of antibiotics in concentrated animal feeding operations (CAFOs) was another theme in the film that captured students' attention. Each year, corn-fed livestock in these industrial feedlots require more antibiotics to stay "healthy" (read: keep them alive) than are given to the entire human population of the United States. This provided a rich point in the film to stop and discuss. Students' notes and drawings often depicted sick cattle getting antibiotics and producing massive amounts of waste and pollution on the CAFO in Colorado that Ellis and Cheney visit. Claire added a little humor to her notes about CAFOs by adding a caption above her cow that read "No mooooooooore corn please!"

When Julie and I compared the graphic notes from our different classes, a few themes showed up again and again, giving us a sense of what our students were taking away from the film. Not surprisingly, many students included pictures of the isotopic hair analysis from the beginning of the film. We also saw various depictions of Ellis and Cheney driving massive tractors to apply fertilizer and herbicide to their crop. Evan's tractor drawing captured a great deal with surprising simplicity: Above a tractor pulling a tank of ammonia fertilizer, he drew dollar symbols to indicate that the use of synthetic fertilizers had led to higher and higher yields, but that the fertilizer also meant higher costs for farmers. Behind the tractor he drew a dead bird in the field to depict the ecological consequences of large-scale industrial agriculture, which he described as "collateral damage."

Why Agricultural Policy Matters

The film also shows the government policies that have encouraged the massive expansion of corn production in the United States over the last several decades—and that federal agricultural subsidies benefit a few large corporate producers more than the farmers and rural communities. Toward the end of the film, Ellis and Cheney visit former Secretary of Agriculture Earl Butz in his retirement home and question him about policy changes he presided over during his time in the Nixon administration. As a fierce advocate of the so-called "free market," Butz helped to dismantle New Deal supply management policies, effectively "deregulating" the corn market. Under Butz, the USDA abolished long-standing New Deal programs that paid farmers to keep land fallow during times of overproduction, which had the effect of stabilizing prices for farmers, as well as easing the demands made on the land. In contrast, Butz repeatedly told the nation's farmers to "get big, or get out," and encouraged them to "plant fence row to fence row," promising that excess production could be sold through trade in foreign markets. In the interview, Ellis and Cheney ask Butz to respond to the criticism that these policies produce more food than we need:

> Well, it's the basis of our affluence now, the fact that we spend less on food. It's America's best-kept secret. We feed ourselves with approximately 16 or 17 percent of our take-home pay. That's marvelous; that's a very small chunk to feed ourselves. And that includes all the meals we eat at restaurants, all the fancy doodads we get in our food system. I don't see much room for improvement there, which means we'll spend our surplus cash on something else.

I stopped the film here to ask students to respond to Butz's claims. On the surface, they are difficult arguments to find fault with—most students agreed that cheaper food is a good thing. Although it's true that on average people in the United States spend less of their income on food today than their parents' and grandparents' generations, it's also true that the agricultural policy changes pushed by Butz resulted in much bigger gifts to a few big food corporations—as well as food that is more problematic from the standpoint of human health, the global economy, and the environment.

Like so much commentary that views low U.S. consumer prices as the highest good, Butz's remarks ignore the impact on communities and cultures around the world. For example, Butz's oceans of cheap corn have drowned farm communities throughout Mexico for whom corn is not a simple commodity, but represents survival and a cultural touchstone.

This is where *King Corn* comes up a little short: the filmmakers do not explain the heavy influence companies like Cargill and Archer Daniels Midland had on Butz-era agricultural policies—and how those companies profited marvelously from the corn economy they helped to create. This silence in the film makes follow-up readings, including Tom

Philpott's "The Butz Stops Here: A Reflection on the Lasting Legacy of 1970s USDA Secretary Earl Butz," particularly important.

Julie and I created a jigsaw reading activity that allowed students to choose from a selection of readings that go into more depth on a few of the issues raised in the film (see Resources). We gave students time to begin the readings in class, and asked them to highlight important passages and add notes in the margins when they came across sections that connected to the film or raised new points or questions. We also asked them to revisit the question that had introduced our food unit a week earlier: "How much choice do you have about the food that you eat?"

The Politics of Food

When our students returned to class the next day, we organized small groups of students who had looked at dissimilar readings and asked them to compare notes: "Each of you should share at least two or three important quotes or film connections from your readings. Then, think as a group about how you might revise your answers to our question about choice after watching the film and reading these articles." After students had exhausted their quotes and thoughts in small group discussions, we transitioned to a larger class discussion, where students shared favorite quotes and important themes from the small group discussions.

Will shared a quote in class from "Unhappy Meals" about school lunch programs: "At a time when weight-related illnesses in children are escalating, schools are serving kids the very foods that lead to obesity, diabetes, and heart disease." In the discussion notes he later handed in, he reflected on the quote:

> "At a time when weight-related illnesses in children are escalating, schools are serving kids the very foods that lead to obesity, diabetes, and heart disease."

> Now that I think about the food situation we have in America, I don't think I had as much choice as I thought I had. But for kids in elementary schools, especially poorer children, they really don't have a lot of options because

STEPHANIE McMILLAN

Eating Oil

After cars, the food system uses more fossil fuel than any other sector of the economy—19 percent. And while the experts disagree about the exact amount, the way we feed ourselves contributes more greenhouse gases to the atmosphere than anything else we do—as much as 37 percent, according to one study.

Whenever farmers clear land for crops and till the soil, large quantities of carbon are released into the air. But the 20th-century industrialization of agriculture has increased the amount of greenhouse gases emitted by the food system by an order of magnitude; chemical fertilizers (made from natural gas), pesticides (made from petroleum), farm machinery, modern food processing and packaging and transportation have together transformed a system that in 1940 produced 2.3 calories of food energy for every calorie of fossil fuel energy it used into one that now takes 10 calories of fossil fuel energy to produce a single calorie of modern supermarket food. Put another way, when we eat from the industrial food system, we are eating oil and spewing greenhouse gases. This state of affairs appears all the more absurd when you recall that every calorie we eat is ultimately the product of photosynthesis—a process based on making food energy from sunshine. There is hope and possibility in that simple fact.

—From Michael Pollan, "Farmer in Chief,"
New York Times, October 9, 2009

they can't afford to bring a homemade lunch to school every day. They have to eat the processed, high-fat foods that are served to them during lunch.

A number of students were captivated by Michael Pollan's argument in "Farmer in Chief." Pollan says that our industrial food system facilitates the overconsumption of high-fat, high-sugar foods that are linked to heart disease and diabetes; requires intensive and unsustainable fossil fuel inputs; and produces as much carbon dioxide as any other sector of the economy. As a result, we will need to fundamentally reform our food system if we hope to ever make meaningful progress in our attempts to address health care, energy, and climate change. In her notes, Hannah offered these thoughts in response to Pollan's argument:

> I never thought the food I eat could have anything to do with health care, energy independence, or climate change, but I guess it does. I usually try to pay attention to what I eat, because I want to be healthy, but this gives me more reasons to make good choices.

Hannah's statement raises an important issue I've learned to tease out of the class discussions we have about *King Corn*. It concerns the role each of us plays as individual agents of change within our food system, as compared to the massive ability of the government to effect change within the same system. Here in Portland, we pride ourselves on being a hotbed of local food activism—the locavore concept has become a way of life for many people across the region. Farmers markets and community-supported agriculture projects are not only ways to eat well, but also strategies for engaging in one's community, strengthening local economies, and nurturing stronger bonds to the people and land that produce our food. For some, these strategies are a form of resistance against a corporate-industrial food system.

But there is an inherent risk in focusing on local strategies for creating change, especially if we fail to connect our local efforts to broader political and economic transformation. I've come to realize that I neglected to give kids the tools to understand and think critically about the necessary role of government food policy in creating a more just, sustainable food system. By starting food units with reflection questions about personal choices, I encourage students to make important real-life connections to the curriculum. But if the emphasis remains on the power of consumer choice, students come away thinking that if we could all shop at farmers markets, buy organic produce, and support local food producers, the power of personal actions would transform our entire food system.

In so many ways, *King Corn* tells the story of how government food policy has entirely remade the food landscape in the United States over the last 40 years. From the massive expansion of the number of acres of corn grown across the country, to the ever-increasing ways that corn is incorporated into the food production process, to the industrial feedlots that produce most U.S. meat, *King Corn* illustrates how our food system is not only the product of corporate greed, but also of government policy that was intended to produce many of the results we see today.

It's important for students to understand this, to recognize that the history helps explain our current reality. This is especially true in an age when we face so many crises—from diabetes to climate change—that will require not just changes in personal lifestyle, but deliberate and focused policy from our government as well. As Alejandra wrote at the end of our unit:

> All this cheap corn made the heavily processed, unhealthy outputs like burgers (corn-fed beef!) and soda (high-fructose corn syrup!) cheap, too. And so, indirectly, the government did begin to subsidize Happy Meals instead of healthy ones. (And no, that is not my line.) The government, the farmers, and the people—generally poor people who had very few options when it came to food and were forced to buy what was cheap—all suffered, and only the processed food corporations came out on top.
>
> Federal agricultural policy does matter. Many people blame it for the obesity epidemic, and rightly so. I am sure many Iowa corn farmers (and Mexican corn farmers out of work) blame it for the loss of their livelihood over the years, and the evidence is on their

side. And now that we know that all these laws and policies do matter, it makes every one of us *that* much more enlightened and *that* much more capable of raising a racket. Why is it important to know the history of federal agricultural policy? Because just knowing is a window into how the system works. And once we understand how it works, we can start working to manipulate it for positive change. Now we know what to do.

Seeing how the system works, and how we got here, is essential to creating a better food system—one that provides food as a human right, not simply as a commodity to be bought and sold. By seeing the bigger picture, I hope that my students can help to "raise the racket" necessary to re-craft food policies that serve the interests of people and the planet over the interest of food corporations. ⊕

> **"Just knowing is a window into how the system works. And once we understand how it works, we can start working to manipulate it for positive change."**

Tim Swinehart (timswinehart@gmail.com) teaches at Lincoln High School in Portland, Oregon.

Resources:

Independent Lens, *King Corn*, pbs.org/independentlens/kingcorn/index.html.

Michael Pollan, "Farmer in Chief," michaelpollan.com/articles-archive/farmer-in-chief. An abbreviated version of this *New York Times* commentary is available at the Center for Ecoliteracy website, ecoliteracy.org/essays/farmer-chief.

Tom Philpott, "The Butz Stops Here: A Reflection on the Lasting Legacy of 1970s USDA Secretary Earl Butz," grist.org/article/the-butz-stops-here.

Barry Yeoman, "Unhappy Meals," motherjones.com/politics/2003/01/unhappy-meals.

"We Have the Right..."

A Youth Food Bill of Rights

Before a recent Youth Leadership Summit, organizer Maya Salsedo, winner of a 2012 Brower Youth Award, said, "As youth, we are on the front lines of our broken food system—that's what motivates us to make real change. Like all movements for justice in our country's history, we believe youth are the catalyst for systemic change."

Salsedo participated in an annual gathering sponsored by Rooted in Community (RIC), which describes itself as "a national grassroots network of youth and adults working together toward community resilience through urban and rural agriculture, community gardening, food security, culinary training, social enterprise, and environmental justice work."

One of the products of this work is the "Youth Food Bill of Rights." The most recent version of this document, included at its website, www.youthfoodbillofrights.com, includes a manifesto of 17 rights. Here, we reproduce seven of these.

—the editors

Youth Food Bill of Rights

In order to reshape our broken food system, we the youth have come together to name our rights.

1. We have the right to nutritional education. We demand government funding to educate and inform youth and parents about nutrition.
    a. Education on things such as seasonal eating, organic farming, sustainability, and diet-related illness should be provided so that people can make better informed decisions.
    b. We recommend that schools recognize youth-led fitness programs as tools for success.

2. We have the right to healthy food at school. We the youth demand more healthy food choices in our schools, and in schools all over the world. We want vending machines out of schools unless they have healthy choices. We need healthier school lunches that are implemented by schools with the ingredients decided on by the youth. We demand composting in schools and in our neighborhoods.

3. We have the right to poison-free food. We the youth absolutely don't want any chemical pesticides in our food!

4. We have the right to beverages and foods that don't harm us. We the youth demand a ban on high-fructose corn syrup and other additives, and preservatives that are a detriment to our communities' health. This must be implemented by our government, and governments around the world.

5. We have the right to local food. We demand food to be grown and consumed by region to cut the use of fossil fuels and reduce the globalization of our food system.

6. We have the right to cultivate unused land. We demand that a policy be enacted allowing for unused land to be made available for communities to farm and garden organically and sustainably.

7. We have the right to leadership education. We the youth demand that there be more school assemblies to inform and empower more youth with the knowledge of food justice. The continuation of the movement for food justice, food sovereignty, and cultivation of future leaders is necessary for feeding our youth, our nation, and our world.

See teaching ideas for this article, page 386.

"**W**ear green on St. Patrick's Day or get pinched." That pretty much sums up the Irish American "curriculum" I learned when I was in school. Yes, I recall a nod to the so-called Potato Famine, but it was mentioned only in passing.

Sadly, today's high school textbooks continue to largely ignore the famine, despite the fact that it was responsible for unimaginable suffering and the deaths of more than a million Irish peasants, and that it triggered the greatest wave of Irish immigration in U.S. history. Nor do textbooks make any attempt to help students link famines past and present.

Yet there is no shortage of material that can bring these dramatic events to life in the classroom. One resource is Irish-born poet Nigel Gray's "When the Hunger Was Upon Us," which includes this wrenching passage:

Our scarecrow children,
lay like sacks of sticks
scattered on a sprinkling of straw
and let go of life without a
        murmur;
or died at the roadside,
their mouths stained green
from chewing grass,
their bodies, no more than
        parchment-covered bones,
half eaten by rats and starving
        dogs.
Though they were leaf-light,
we were too weak to bury them
beneath even a handful of stones.

By contrast, Holt McDougal's U.S. history textbook *The Americans* devotes a flat two sentences to "The Great Potato Famine." Prentice Hall's *America: Pathways to the Present* fails to offer a single quote from the time. The text calls the famine a "horrible disaster," as if it were a natural calamity like an earthquake. And in an awful single paragraph, Houghton Mifflin's *The Enduring Vision: A History of the American People* blames the "ravages of famine" simply on "a blight." The only contemporaneous quote comes, inappropriately, from a landlord, who describes the surviving tenants as "famished and ghastly skeletons." Uniformly, social studies text-

Hunger on Trial

An activity on the Irish "Potato Famine" and its meaning today

BY BILL BIGELOW

books fail to allow the Irish to speak for themselves, to narrate their own horror.

These timid slivers of knowledge exemplify much of what is wrong with today's curricular reliance on corporate-produced textbooks.

First, does anyone really think that students will remember anything from the books' dull and lifeless paragraphs? Today's textbooks contain no stories of actual people. We meet no one, learn nothing of anyone's life, encounter no injustice, no resistance. This is a curriculum bound for boredom. As someone who spent almost 30 years teaching high school social studies, I can testify that students will be unlikely to seek to learn more about events so emptied of drama, emotion, and humanity.

Nor do these texts raise any critical questions for students to consider. For example, it's important for students to learn that the crop failure in Ireland affected only the potato—during the worst famine years, other food production was robust. Michael Pollan notes in *The Botany of Desire*, "Ireland's was surely the biggest experiment in monoculture ever attempted and surely the most convincing proof of its folly." But if only this one variety of potato, the Lumper, failed, and other crops thrived, why did people starve?

Thomas Gallagher points out in *Paddy's Lament* that during the first winter of famine, 1846–47, as perhaps 400,000 Irish peasants starved, landlords exported 17 million pounds sterling worth of

grain, cattle, pigs, flour, eggs, and poultry—food that could have prevented those deaths. Throughout the famine, as Gallagher notes, there was an abundance of food produced in Ireland, yet the landlords exported it to markets abroad.

The school curriculum could and should ask students to reflect on the contradiction of starvation amidst plenty, on the ethics of food exports amidst famine. And it should ask why these patterns persist into our own time.

More than a century and a half after the "Great Famine," we live with similar, perhaps even more glaring contradictions. Raj Patel opens his book, *Stuffed and Starved: Markets, Power, and the Hidden Battle for the World's Food System*: "Today, when we produce more food than ever before, more than one in 10 people on Earth are hungry."

Patel's book sets out to account for "the rot at the core of the modern food system." This is a curricular journey that our students should also be on—reflecting on patterns of poverty, power, and inequality that stretch from 19th-century Ireland to 21st-century Africa, India, Appalachia, and Oakland. Together we should be exploring what happens when food and land are regarded purely as commodities in a global system of profit.

The Irish Famine Role Play

I designed a trial role play to highlight the "crime" of famine and to encourage students to reflect on responsibility for that crime. The Irish Potato Famine lends itself to this teaching strategy, because students plainly recognize the enormity of the famine, but the causes for the Irish suffering are not self-evident and require more consideration. I wrote five detailed "indictments": British landlords, Irish tenant farmers, the Anglican Church, the British government, and "Political Economy"—the system of colonial capitalism. (Excerpts from these appear on p. 359.)

This last role requires some explanation for students, because unlike the others, it is not a specific group of humans: It is a system of ownership, production, and distribution. But asking students to think systemically—reflecting on how the "rules of the game" reward and punish particular behaviors— is a key aim of my global studies curriculum. I don't want to dehumanize responsibility for injustice, but I want students to look beneath the surface to try to account for why people make the choices they do, and not to rely on glib explanations like "greed."

Each group was charged with the same crime, but for different reasons:

> You are charged with the murder of one and a half million Irish peasants who died in the famine years of 1846 and 1847. These were needless deaths. Even without the potato, there was more than enough food produced in Ireland during those years to feed everyone in the country and still have plenty left over. The action—or lack of action—taken by your group led to untold misery. You are to blame.

Role Play Set Up

I used this role play with my 11th-grade global studies classes at Franklin High School in Portland, Oregon, as part of a broader unit on colonialism and the history of global inequality. One reason I appended a short unit on British colonialism in Ireland to this unit is because I wanted my classes of largely working-class European American students to see that colonial exploitation affected "white" people, too— although white deserves quotation marks because the British constructed the Irish as a separate race from themselves [see Noel Ignatiev's book, *How the Irish Became White*].

I opened the unit by bringing to class a potato—an extraordinary food contribution from Native America to the rest of the world. I told students that the word potato comes originally from the Taíno word, *batata*, for what we know today as the sweet potato. The Incans in South America cultivated more than 3,000 varieties of potato, ingeniously working out a way to freeze-dry potatoes to make storage and transportation easier; Incan freeze-dried potatoes could be stored for up to five years. An acre planted in potatoes produces twice as many calories as an acre planted in wheat, requires less labor to tend, is less prone to damage from storms,

and produces less tooth decay. Its introduction to Europe led to a population boom.

I explained to students that potatoes would be at the center of a role play that examined British colonialism in Ireland. Because there were so many poor people in Ireland—seven out of every eight people on the island—and because they had so little land, the Irish poor relied on the potato. But beginning in 1845, they began to notice the arrival of a blight. It turned the potatoes black, gooey, and bad smelling. But the blight afflicted only the potato crops. There was no drought, and three-quarters of Ireland's cultivable land was planted in crops other than potatoes, so there was no need for anyone to go hungry.

> **The aim of the trial is not to lead students *Perry Mason*-like to some definitive guilty party for the Irish Potato Famine—although I do want them to recognize that the potato is not guilty and nature is not guilty.**

Before we began the trial, students read excerpts from chapter five of *Paddy's Lament,* watched parts of the PBS video *The Irish in America,* and listened to Sinead O'Connor's haunting "Skibbereen," a mother's expression of grief offered to her son about why she left Ireland, with verses like:

> Oh son I loved my native land, with energy and pride
> 'Til a blight came over on my prats, my sheep and cattle died,
> The rent and taxes were so high, I could not them redeem,
> And that's the cruel reason why I left old Skibbereen.
> Oh, it's well I do remember, that bleak December day,
> The landlord and the sheriff came, to drive us all away
> They set my roof on fire, with their cursed English spleen
> And that's another reason why I left old Skibbereen.

I wanted to humanize the effects of the famine, but at the outset I tried not to offer any material that

Potato Famine Memorial, Dublin

explored its causes. Thus, I saved Sinead O'Connor's hip-hop-influenced song "Famine," from *Universal Mother,* until the conclusion of the role play.

After introducing students to some of these voices of the famine, I reviewed the charges that confronted the class and then divided students into five groups representing the defendants and a sixth for the jury. Each indictment "role" detailed the specific charges against the group but also indicated the outlines of a possible defense. I distributed packets of all the roles to each group. Students were free to read only the role for their group or, if they wanted more evidence, they could read the entire packet. Their responsibility was to fashion a defense to their group's indictment. They could plead guilty, but in their defense, they had to accuse at least one other group. The jury received all the roles and was responsible for preparing at least three pointed questions for each of the defendant groups.

After preparing their defenses, students sat in a large circle for the trial, each defendant group sitting together. I played the prosecutor and passionately delivered the charges against the first defendant group. Immediately following presentation of the charges against a group, members of that group defended themselves and accused other groups. The jury questioned the group on the hot seat, and then

I opened the floor to other groups who could also question the defendant. We repeated the process until all groups were charged, had defended themselves, and had been questioned.

The aim of the trial is not to lead students *Perry Mason*-like to some definitive guilty party for the Irish Potato Famine—although I do want them to recognize that the potato is not guilty and nature is not guilty. I'm hoping instead to nurture a pattern of questioning: Who holds a society's wealth and power? What determines how resources are used? What human-created institutions and behavior are at the root of suffering? Too often, the curriculum promotes the notion that we all share a "national interest," as if our students and the CEO of Exxon lived in the same family. But countries are not families, as Howard Zinn said, and suggesting they are miseducates students. Societies are stratified, especially by race and class (and by immigration status). "Ireland" didn't starve; the poor of Ireland starved—needlessly.

Helping students think clearly about the past equips them to think clearly about the present. As in Ireland 160 years ago, hunger is rampant today. And like 160 years ago, the cause has little to do with genuine scarcity. If one aim of the curriculum is to help students imagine solutions to social problems, then such imagination needs to begin with an eyes-open analysis of the root causes of those problems. ⬤

Bill Bigelow (bbpdx@aol.com) is curriculum editor of Rethinking Schools *magazine.*

Excerpts from Famine Trial Indictments

(Note: Full roles are available at www.rethinkingschools.org/earth)

British Government

In the first winter of the famine, while 400,000 Irish starved to death, English landlords continued to export food from Ireland to England. The British government could have outlawed the export of food while people starved. However, the government allowed the landlords to export 17 million pounds sterling worth of grain, cattle, pigs, flour, eggs, and poultry—food that could have fed 12 million people, twice the number of Irish tenant farmers dependent on potatoes.

Anglican Church

In some ways it could be said that the Anglican Church is the most to blame for the starvation of the Irish. Supposedly, you are God's representative on Earth. The Bible says to love your brother as yourself. Yet how did you respond in those two fateful years, 1846 and 1847? You knew that the English landlords forced the Irish to live on the worst land—land that was good only for growing potatoes. The landlords used the best Irish land to get rich, by growing wheat, barley, and oats; by raising pigs; and by grazing cattle and sheep. Did the landlords keep any of this food in Ireland, even when people began to starve? No. They exported it to England. The government enforced the will of the landlords with military might. And how did the Anglican Church respond? Did you protest? Did you tell the landlords that they were too greedy, that they were not doing their Christian duty? No. You did nothing.

"Political Economy"—The System of Colonial Capitalism

There really are no evil people here. Sure, people did evil things, but it was the capitalist market that was mostly to blame. Yes, the British landlords exported lots and lots of food while the Irish starved. But why? Because they were devils? No. They did it because that's what people do in a capitalist economy: They sell their produce where they can get the best price. The poor in Ireland had little money. So the landlords sent their wheat, barley, oats, cattle, sheep, and pigs to England. Not bad people, a bad system: you, capitalism.

British Landlords

You are directly responsible for the terrible famine resulting from the potato blight. You owned the land that the Irish peasants worked. When the potato crop failed, you had a choice: You could either allow your tenants to stop paying rent temporarily, and allow them to eat the crops grown on other parts of your land, or you could force them to pay rent even if they would starve as a result. You chose this latter course, which resulted in so much starvation, disease, and death.

Irish Tenant Farmers

It's true that the British landlords, backed up by the British government, turned the potato blight into a famine that killed more than a million Irish. However, what did you do to stop the crimes committed by the British? By not organizing massive resistance to the British, you are also to blame. And you knew that the solution was simple. There was a saying in Ireland at the time: "Sure, this land is full of barley, wheat, and oats. The English have only to distribute it." What a foolish hope. You had to take it from them. You must have known that they wouldn't just give it to you. The most that the Irish did in the way of "redistributing the wealth" was to steal a few sheep.

The Irish Famine Trial—Instructions

1. I review the indictment (included on each role sheet) with the class, and count students off into six groups—five defendant groups and one jury. Each indictment role details the charges against the group but also indicates the outlines of a possible defense.

2. To the defendant groups, I distribute packets of all the roles. Students are free to read the role only for their group or, if they want more evidence on each group, they can read the entire packet. I explain that these might be useful as they considered which other groups have more guilt than theirs. Students' responsibility is to fashion a defense to the indictment for their group. They can plead guilty, but in their defense, they have to accuse at least one other group.

3. To the jury, I distribute packets of all the defendant roles. Jurors are responsible for reading each of the roles and developing at least three probing questions for each group. I also distribute blank placards and markers for students to write and display the name of their group.

4. I circulate throughout the class, talking with different groups and helping the defendant groups with their defense. I encourage each defendant group to prepare its defense/presentation in a way that involves as many students as possible. (In role plays, I offer more points per group if the group is able to involve every member in a substantial way in their presentations—apologies to Alfie Kohn, author of *Punished by Rewards* and the leading opponent of classroom bribery schemes.) Students needn't write out their defense presentations word for word, but sometimes this process helps make sure that the presentations are well argued.

5. After students prepare their defense presentations, we form a large circle, with students from each group sitting together, placards displayed.

6. I introduce the jury and indicate that I play both judge and prosecutor. (If a second adult is available, it's best to have a separate judge to keep order—and if this individual is unknown to students, so much the better, as it lends an air of seriousness to the proceedings. But this is not absolutely vital, and more often than not, I don't have another adult involved in the trial proceedings.) I review the process: the prosecutor will lay out the charges against one group; that group will defend itself and, if they like, accuse others; the jury will raise questions of the group that just defended itself; and any group specifically charged or attacked by the defendant group will be able to raise questions or make counteraccusations.

7. As the prosecutor, I try to be an equal-opportunity accuser—arguing the charges with even-handed vigor against each group. Sometimes the jury raises sharp, difficult questions of each defendant group. But sometimes a jury is too soft, in my opinion, and I supplement their questions by raising difficult ones of my own. However, the best exchanges are always student-to-student, so I encourage other students in the class to also raise questions of the defendant group.

8. After each group has been prosecuted and defended itself, I ask the jury to retire to the hallway and deliberate about who or what they think was "guilty" for the Irish famine. Meanwhile, I ask other students to step away from their roles and to write on this question, and I suggest that they assign percentages of blame—adding up to 100 percent—to each defendant group. They are, of course, free to argue that any group is not guilty. (I used to require students to use percentages of guilt in their write-ups, but in a different trial role play, one of my students criticized this requirement as it inhibited her from discussing how factors were interrelated.)

9. The jurors return, announce their verdict(s), and we discuss based on the jurors' findings and their own written judgment about the "guilt" for the famine.

Why so much hunger? What can we do about it? To answer these questions we must unlearn much of what we have been taught. Only by freeing ourselves from the grip of widely held myths can we grasp the roots of hunger and see what we can do to end it.

Myth 1: There is not enough food to go around.
Reality: Abundance, not scarcity, best describes the world's food supply. Enough food is produced to provide every human being with *2,800 calories a day*. In volume, it comes to three to four pounds for each of us; and that's after feeding a third of the world's grain to livestock and wasting a quarter of the calories. The problem is that many people are too poor to buy readily available food. Even most "hungry countries" have enough food for all of their people right now. In fact, many are net exporters of food and other agricultural products.

Myth 2: Nature is to blame for famine.
Reality: It's too easy to blame nature. Human-made forces are making people increasingly vulnerable to nature's vagaries. Food is always available for those who can afford it—starvation during hard times hits only the poorest. Millions live on the brink of disaster in South Asia, Africa, and elsewhere, because they are deprived of land by a powerful few, trapped in the grip of debt, or miserably paid. Natural events rarely explain deaths; they are simply the final push over the brink. Human institutions and policies determine who eats and who starves during hard times. Likewise, in the United States many homeless die from the cold every winter, yet ultimate responsibility doesn't lie with the weather. The real culprit is the belief that a market economy must be driven by the narrow rule of what brings highest profit to the largest wealth holders. Their economic power then shapes public policies to their interests, denying fair opportunities to all.

Myth 3: There are too many people.
Reality: Birthrates are falling rapidly worldwide. Although population growth remains a serious concern in sub-Saharan Africa, nowhere does population density explain hunger. For every Bangladesh, a densely populated and hungry country, we find a Mozambique, Honduras, or a Bolivia, not densely populated, where abundant food resources coexist

GREENBELT MOVEMENT

10 Myths About Hunger

BY FRANCES MOORE LAPPÉ AND JOSEPH COLLINS

with hunger. Or we find a country like the Netherlands, where very little land per person has not prevented it from eliminating hunger and becoming a net exporter of food. Rapid population growth is not the root cause of hunger. Like hunger itself, it results from underlying inequities that deprive people, especially poor women, of economic opportunity and security. Rapid population growth and hunger are endemic to societies where land ownership, jobs, education, health care, and old age security are beyond the reach of most people. Those Third World societies with dramatic and rapid reductions of population growth rates—Sri Lanka, Costa Rica, Cuba, and the Indian state of Kerala—prove that the lives of the poor, especially poor women, must improve before they can choose to have fewer children.

Myth 4: We will have to make sacrifices to the environment in order to feed more people.

Reality: A trade-off between our environment and the world's need for food is not inevitable. Efforts to feed the hungry are not causing the environmental crisis. Large corporations are mainly responsible for deforestation—creating and profiting from so-called developed-country consumer demand for meat, tropical hardwoods, and exotic or out-of-season food items. Most pesticides used in the Third World are applied to export crops, playing little role in feeding the hungry, while in the United States they are used to give a blemish-free cosmetic appearance to fruits and vegetables, with no improvement in nutritional value.

Alternatives exist now and many more are possible. The success of environmentally sound agricultural alternatives from the United States to Africa proves that safe, ecological practices can be as or more productive than environmentally destructive ones. In fact, with these benefits in mind, policymakers in Kerala, India, have required all its growers to farm organically by 2020.

Myth 5: The Green Revolution is the answer.

Reality: Thanks to the new seeds and production advances of the "Green Revolution," millions of tons more grain a year are being harvested. But focusing narrowly on increasing production will not alleviate hunger because it does not change the tightly concentrated distribution of economic power that determines who can buy the additional food. In fact, it contributes to economic concentration by making farmers dependent on costly chemical fertilizers and pesticides. That's why in several of the biggest Green Revolution successes—India, Mexico, and the Philippines—grain production and in some cases, exports, have climbed, while hunger has persisted and the long-term productive capacity of the soil is degraded. The prospect of a "New Green Revolution" based on biotechnology (genetically modified seeds) threatens to make inequality worse.

Myth 6: We need large farms.

Reality: Large landowners who control most of the best land often leave much of it idle. Unjust farming systems leave farmland in the hands of the most inefficient producers. Small farmers in the Global South typically achieve at least four to five times greater output per acre, in part because they work their land more intensively and use integrated, and often more sustainable, production systems. Without secure tenure, the many millions of tenant farmers in the Third World have little incentive to invest in land improvements, to rotate crops, or to leave land fallow for the sake of long-term soil fertility. This undermines future food production. On the other hand, redistribution of land can favor production. Historically, comprehensive land reforms have markedly increased production in countries as diverse as Japan, Taiwan, Mexico, and Chile. A World Bank study of northeast Brazil estimates that redistributing farmland into smaller holdings would raise output an astonishing 80 percent.

Myth 7: The free market can end hunger.

Reality: A "market-is-good, government-is-bad" formula can never help address the causes of hunger. Such a stance implies that a society can opt for one or the other, when in fact every economy on Earth combines the market and government in allocating resources and distributing goods. The market's efficiencies can only work to eliminate hunger, however, when purchasing power is widely dispersed.

Therefore, all those who believe in the usefulness of the market and the necessity of ending hunger must concentrate on promoting not the market, but the consumers. In this task, government has a vital role to play in countering the tendency toward economic concentration through genuine tax, credit, and land reforms to disperse purchasing power toward the poor. More privatization and deregulation are definitely not the answer.

Myth 8: Free trade is the answer.

Reality: Promoting more global trade in food has failed to alleviate hunger. In many Third World countries, exports have boomed while hunger has continued without change or has actually worsened. For example, while soybean exports boomed in Paraguay—mainly destined to feed livestock in Asia and Europe—hunger and poverty have only worsened, as over 100,000 small-scale farm families have been pushed off the land and wound up in the ranks of the urban jobless. Where the majority of people have been made too poor to buy the food grown on their own country's soil, those who control productive resources will, not surprisingly, orient their production to more profitable markets abroad. Export crop production squeezes out basic food production. So-called free trade treaties like the North American Free Trade Agreement (NAFTA) and those of the World Trade Organization (WTO) pit working people in different countries against each other in a "race to the bottom." People in different countries are forced to compete against each other for who will work for less, without adequate health coverage or minimum environmental standards.

Myth 9: More U.S. aid will help the hungry.

Reality: Most U.S. aid works directly against the hungry by reinforcing, not changing, the power status quo. Where governments answer only to elites, our aid not only fails to reach hungry people, it shores up the very forces working against them. Our aid is used to impose free trade and free market policies, to promote exports at the expense of food production, and to provide the arms that repressive governments use to stay in power. Even emergency, or humanitarian aid, which makes up only a small percentage of the total, often ends up enriching U.S. grain and shipping companies while failing to reach the hungry, and it can dangerously undercut local food production in the recipient countries. Food aid for emergencies therefore should be short term and procured as locally to the emergency as possible, so it would be far more timely and in many cases provide support to nearby small farmers, who are often poor.

Myth 10: We benefit from their poverty.

Reality: The biggest threat to the well-being of the vast majority of North Americans is not the advancement but the continued deprivation of the hungry. Low wages—both abroad and in inner cities at home—may mean cheaper bananas, shirts, computers, and fast food, but in other ways we pay heavily for hunger and poverty. Poverty in the Third World jeopardizes U.S. jobs, wages, and working conditions as corporations seek cheaper labor abroad. In a global economy, what North American workers have achieved in employment, wage levels, working conditions and environmental regulations can be protected only when working people in every country are freed from economic desperation.

The growing numbers of working poor are those who have part- or full-time low-wage jobs yet cannot afford adequate nutrition or housing for their families. Educating ourselves about the common interests most North Americans share with the poor in the Third World and at home allows us to be compassionate without sliding into pity. By working to clear the way for the poor to free themselves from economic oppression, we free ourselves as well. 🌐

Frances Moore Lappé is the author of 18 books, including the 1971 Diet for a Small Planet. *Her most recent is* EcoMind: Changing the Way We Think, to Create the World We Want. *With her daughter Anna Lappé, she directs the Small Planet Institute in Cambridge, Massachusetts, and Oakland, California.*

Joseph Collins in 1975 co-founded with Frances Moore Lappé the Institute for Food and Development Policy. His books include (with Lappé) Food First: Beyond the Myth of Scarcity *and* Aid as Obstacle; *among his other books are* No Free Lunch, Chile's Free-Market Miracle: A Second Look, How to Live Your Dream of Volunteering Overseas, *and* AIDS in the Context of Development. *Since 1990 he has been a consultant in Africa, Asia, and Latin America to the United Nations and various international nonprofits.*

 See teaching ideas for this article, page 387.

Selected References:

Note: for a full list of references, see Lappé, Frances Moore, Joseph Collins, and Peter Rosset. 1998. *World Hunger: Twelve Myths.* New York: Grove Press. A new edition, *World Hunger: 10 Myths*, will be published in 2015 in which the authors analyze these issues in light of 21st-century realities.

Myth 1:
1. Data calculated from FAOSTAT. 2014. Rome: United

Nations Food and Agriculture Organization.
2. Otterdijk, Robert Van and Meybeck, Alexandre. 2011. *Global Food Losses and Food Waste: Extent, Causes and Prevention.* Rome: Food and Agriculture Organization of the United Nations. http://www.fao.org/docrep/014/mb060e/mb060e.pdf.

Myth 2:
1. Sen, Amartya. 1981. *Poverty and Famines.* Oxford: Clarendon Press.
2. Timberlake, Lloyd and Anders Wijkman. 1984. *Natural Disasters: Acts of God or Acts of Man?* London and Washington, D.C.: Earthscan.

Myth 3:
1. United Nations Children's Fund. 2013. *Improving Child Nutrition: The achievable imperative for global progress.* New York. http://www.unicef.org/media/files/nutrition_report_2013.pdf.
2. The World Bank. Data: Population density (people per sq. km of land area). http://data.worldbank.org/indicator/EN.POP.DNST.
3. Data calculated from FAOSTAT. 2014. Rome: United Nations Food and Agriculture Organization.

Myth 4:
1. International Assessment of Agricultural Knowledge, Science and Technology for Development. 2009. Global Report: Agriculture at a Crossroads. Washington, D.C. http://www.unep.org/dewa/agassessment/reports/IAASTD/EN/Agriculture%20at%20a%20Crossroads_Global%20Report%20(English).pdf.
2. Pretty, Jules. 2006. *Agroecological Approaches to Agricultural Development:* Version 1. RIMISP Background Paper for the World Development Report 2007. https://openknowledge.worldbank.org/bitstream/handle/10986/9044/WDR2008_0031.pdf?sequence=1.
3. Thottathil, Sapna E. 2015. *India's Organic Farming Revolution: What It Means for Our Global Food System.* Iowa City: University of Iowa Press.

Myth 5:
1. Shiva, Vandana. 1991. *The Violence of the Green Revolution: Third World Agriculture, Ecology, and Politics.* Penang, Malaysia: Third World Network.
2. Tiwana, N.S. et al. 2007. *State of Environment Punjab-2007.* Chandigarh: Punjab State Council for Science & Technology. http://www.moef.nic.in/soer/state/SoE%20report%20of%20Punjab.pdf.
3. Fagan, John, Antioniou, Michael, and Robinson, Claire. 2014. *GMO Myths and Truths.* Second edition. London: Open Earth Source. http://earthopensource.org/files/pdfs/GMO_Myths_and_Truths/GMO-Myths-and-Truths-edition2.pdf.

Myth 6:
1. Sobhan, Rehman. 1993. *Agrarian Reform and Social Transformation.* London: Zed.
2. Thiesenhusen, William C. 1995. *Broken Promises: Agrarian Reform and the Latin American Campesino.* Boulder, C.O.: Westview Press.
3. Griffin, Keith, Khan, Azizur, and Amy Ickowitz. 2002. Poverty and the distribution of land. *Journal of Agrarian Change* 2(3):279-330.

Myth 7:
1. Collins, Joseph and John Lear. 1995. *Chile's Free Market Miracle: A Second Look.* Oakland: Food First Books.
2. Bello, Walden. 1994. *Dark Victory: The United States, Structural Adjustment and Global Poverty.* London: Pluto Press, Food First Books, and Transnational Institute.

Myth 8:
1. (World agricultural commodity export data) FAOSTAT data. 2005. Rome: United Nations Food and Agriculture Organization.
2. Goldsmith, Edward and Jerry Mander, eds. 1996. *The Case Against the Global Economy: And for a Turn Toward the Local.* San Francisco: Sierra Club Books.
3. Putana, Alicia. 2012. Mexican Agriculture and NAFTA: A 20-Year Balance Sheet. *Review of Agrarian Studies* 2(1):1-43. http://ras.org.in/mexican_agriculture_and_nafta.
4. Oxfam. 2014. Smallholders at Risk. Oxfam Briefing Paper #180. http://www.oxfam.org/sites/www.oxfam.org/files/bp180-smallholders-at-risk-land-food-latin-america-230414-en.pdf.

Myth 9:
1. OEDC. 2003-4. Total DAC Aid at a Glance. www.oecd.org/dataoecd/17/39/23664717.gif
2. Lappé, Frances Moore, Schurman, Rachel, and Kevin Danaher. 1987. *Betraying the National Interest: How U.S. Foreign Aid Threatens Global Security by Undermining the Political and Economic Stability of the Third World.* New York: Grove Press.

Myth 10:
1. Sklar, Holly. 1995. *Chaos or Community? Seeking Solutions, Not Scapegoats.* Boston: South End Press.
2. Lowe, Eugene. 2002. A *Status Report on Hunger and Homelessness in America's Cities: A 25-City Survey,* December 2002. Washington, D.C.: United States Conference of Mayors.
3. Bonanno, Alessandro, Busch, Lawrence, Friedland, William et al., eds. 1997. Food Security and Agricultural Trade under NAFTA. Minneapolis: Institute for Agriculture and Trade Policy.

Institute for Food and Development Policy Backgrounder Summer 2006, Vol.12, No. 2, updated in August 2014.

BEC YOUNG

Two Food Systems

Let's at last recognize that there are two food systems, one industrial and one of small landholders, or peasants if you prefer. The peasant system is not only here for good, it's arguably more efficient than the industrial model. According to the ETC Group, a research and advocacy organization based in Ottawa, the industrial food chain uses 70 percent of agricultural resources to provide 30 percent of the world's food, whereas what ETC calls "the peasant food web" produces the remaining 70 percent using only 30 percent of the resources.

Yes, it is true that high-yielding varieties of any major commercial monoculture crop will produce more per acre than peasant-bred varieties of the same crop. But by diversifying crops, mixing plants and animals, planting trees—which provide not only fruit but also shelter for birds, shade, fertility through nutrient recycling, and more—small landholders can produce more food (and more kinds of food) with fewer resources and lower transportation costs (which means a lower carbon footprint), while providing greater food security, maintaining greater biodiversity, and even better withstanding the effects of climate change.

—Excerpted from Mark Bittman, "How to Feed the World," *New York Times*, Oct. 14, 2013

Food, Farming, and Justice

A role play on
La Vía Campesina

**BY BILL BIGELOW, CHRIS BUEHLER,
JULIE TREICK O'NEILL,
AND TIM SWINEHART**

La Vía Campesina is arguably the largest social movement in the world. As of 2014, it comprised 164 local and national organizations in 73 countries, and represented about 200 million farmers. And yet, try finding a mention of La Vía Campesina in mainstream textbooks.

One of the reasons textbooks fail to include discussions of La Vía Campesina—and other social movements, for that matter—is because the publishers presume that change comes from the top. Textbooks teach students to look to Great Individuals, governments, corporations, multilateral organizations, the United Nations. According to the official stories offered in textbooks, power flows downhill, from the commanding heights. Another source of La Vía Campesina's invisibility is textbooks' core narrative—that humanity is in the midst of a pageant of progress, powered by science, technology, and capitalism. Peasants, by contrast, represent backwardness, pockets of ancient history waiting around in the countryside for the beneficent arrival of the modern world.

This role play on La Vía Campesina challenges both these notions. Not only does La Vía Campesina's work call into question a future of corporate-dominated globalized markets and homogenized bigness, it has dramatically changed the global conversation about agriculture, food, and hunger as well. This role play allows for these vital conversations to occur in our classrooms. When they represent members of La Vía Campesina, students grapple with core issues that many in the industrialized West choose to ignore, or have never thought to question.

We designed this role play to teach some of the facts of La Vía Campesina, but in a way that invites students to "become" La Vía Campesina activists throughout the world, discovering differences, common circumstances, and shared objectives—and defining together what they mean by "food sovereignty," the animating principle of the organization. Through role play, students take on personas representing six of the many constituent organizations that make up La Vía Campesina: the Basque peasants union (EHNE); Brazil's Landless Workers Movement (MST); Haiti's Group of 4 and the Dessalines Brigade; the National Peasants Union of Mozambique; the Korean Women's Peasant Association; and the Tamil Nadu Women's Collective of southern India. Members of each group learn about one another in a mixer activity and meet some of the actual La Vía Campesina activists by viewing a film produced by the organization. They encounter an entirely different strategy to feed the world, one articulated by the world's wealthiest G7 countries. And finally, students propose and organize for alternatives to the G7 agricultural vision.

The role play helps students see that how we farm determines how we live. Food is about nutrition, no doubt, but it's also about community, public health, the climate, economic justice, jobs, and nature.

One final note: We taught this role play with 12 different classes—mostly 9th graders, but some 11th and 12th graders. (We've also used it in workshops with teachers and prospective teachers.) We revised and improvised as we went along. Thus, in the suggested procedure that follows, we occasionally offer different choices for how to proceed. We're confident that everything we suggest here "works," but we try to explain a bit about alternative ways to approach the role play. Special thanks to Matthew Plies, who taught the role play with his students and shared important insights with us.

Materials needed:

- "Facts on Food, Farming, and Hunger"—enough for every student in class.
- Construction paper for making name placards, one for each of the six groups.
- Colored markers.
- Copies of each of the six La Vía Campesina roles—enough so that each student in a group gets a role sheet.
- "La Vía Campesina Mixer" questions—enough for every student in class.
- Recommended: Nametags for each student for the La Vía Campesina Mixer activity.
- Copies of "A Proposal to Feed the World"—enough for every student in class.
- Copies of "La Vía Campesina's Response to the G7's 'Proposal to Feed the World'"—enough for every student in class.
- Materials for students to make metaphorical drawings—colored pencils, markers, or crayons; blank 8.5- x 14-inch paper.

Suggested procedure:

1. Begin by telling students that they're going to be doing a role play on the largest social movement in the world. Ask for students' thoughts on what it might be. If they have not heard of La Vía Campesina, mention that it includes more than 160 organizations with a combined membership of 200 million people in more than 70 countries. If students have not heard of it, you might ask them why not. Why would a social movement representing so many small farmers all across the world not be more well known? If they have an assigned textbook, ask students to look in the book's index to see whether there is an entry for La Vía Campesina.

2. Tell students the role play turns on the question of how we are going to feed the world, given that the United Nations estimates that close to a billion people are chronically hungry. Distribute the student handout, "Facts on Food, Farming, and Hunger." Point out the

assignment to students: to read the facts and to come up with three things that strike them as interesting, surprising, or new; and to generate at least three questions that the facts leave them with. Ask students to complete this in pairs or threes, as it is best if students have a chance to consider the questions with partners.

3. In the full class, ask students to share thoughts about the "Facts" reading. Our students were especially struck by the contradiction that the world produces enough food to feed everyone on Earth one and a half times over. One student wrote simply: "If we produce so much food, why are children dying?" Another wrote: "Why can't we all share if we have enough for everyone and more?" Students were also puzzled by the huge amount of hunger in India even though India is the world's second largest grain producer. Next to these paired facts on the handout, one student wrote in the margins of his paper "WOT?" If students have done the Irish Famine Trial (see p. 355) they have encountered the stain of hunger amidst plenty—as the mostly absentee landlords in Ireland exported vast amounts of food during even the most deadly years of the famine.

    As for questions, some students were puzzled by farmers in Haiti burning huge amounts of Monsanto-donated seeds, even in the wake of the devastating earthquake. One student wrote, "I get that it's a social movement, but would you rather starve or stand up to GMOs?" Another asked, "Why don't we hear about La Vía Campesina in schools?" This is an activity that we hoped would alert students to a host of issues they would encounter in the role play. Our intent was to encourage students to make observations and raise questions, but not to answer them at this stage. Especially with respect to GMOs, they will return to this question later in the role play. (It's worth mentioning here that some background on GMOs is helpful to students as they engage in the role play. One film that we've found especially useful is *The Future of Food* (www.thefutureoffood.com). The entire film is worth viewing, but even showing the first 20 minutes or so offers ex-

cellent background on the issue of genetically altered food crops, and why farmers like those affiliated with La Vía Campesina would resist the introduction of GMOs.)

4. Project the names of the six organizations and introduce the role play to students. Tell them that each of them will portray a member of one of these organizations affiliated with La Vía Campesina. Tell students that they will learn about their organization and have a chance to meet with other members of La Vía Campesina from around the world. They have all traveled to a big meeting of the G7, a gathering of representatives of the world's seven wealthiest countries. This meeting will focus on the G7's plan to "feed the world," and La Vía Campesina has come to offer its counter-vision.

5. Divide students into six groups of roughly equal size. Distribute roles to each group—everyone in each group receives the same role. Ask students to read their roles carefully. Tell them that you'll be asking them to write an interior monologue from the perspective of someone in their organization, so they should take notes on whatever might help them to write—for example, what worries them, what hopes do they have, how have their lives changed in recent years, why is their organization important to them? Encourage students to read these out loud in the small groups, as both reading and hearing the role sheets will often help students internalize the life conditions of the individual they will be portraying. As students read their roles, distribute name placards and colored markers and ask students to write their group's name on them.

6. After students finish reading their roles, ask them to write an interior monologue. On occasion, students will simply re-copy much of what is in their role sheet. To avoid that possibility, encourage students to invent a family and to attempt to personalize the role with their hopes and fears, and with specific details, even stories, from an imagined life. It is often helpful to prompt students with an example. Here are a few excerpts you might

use from an interior monologue one of our students wrote for the role of Haiti's Group of 4:

> My town was devastated by the earthquake exactly five years ago. My mother and father died, along with the lady who sold us salt and sugar, the man who gave my sister and me candy when we went into town, and many others…After our houses had been reduced to dust, and after our crops had been damaged beyond repair, Monsanto, the company of all evils, offered us a lot of seeds. But they were all hybrid seeds and covered in disgusting chemicals. We were working to rebuild and restart farming, but Monsanto felt the need to step in. The seeds were enticing. I quickly learned from other farmers in the Group of 4, however, that this was not a gift. It was a trap to get us to purchase patented seeds from Monsanto year after year. My older sister joined other Group of 4 members and burned the bag of seeds without a second thought.

Another student wrote from the standpoint of a member of the Korean Women's Peasant Association. In hers, she coined the wonderful phrase the "hope group":

> By joining this association, or what my grandchildren refer to as the "hope group," I will fight for my family and the rice fields I grew up on. I refuse to see the smoke of factories settle over my home. I refuse to have the bitter smoke bite away at my life.

7. Ask students to read aloud their interior monologues in their small groups. This helps work them into their roles and helps flesh out some of the issues that they'll be exploring throughout the role play.

8. Distribute the "La Vía Campesina Mixer" questions to every student. Tell students that they will be doing a mixer activity to meet other activists from La Vía Campesina organizations around the world. They will share experiences and discover commonalities and differences with other groups. Ask students to read over the handout and to put a check mark next to each question that may apply to their group and for which they would be able to offer information to representatives from other groups. (Recommended: As students do this, distribute the nametags and have everyone write their group name in large print.) Some instructions that we offer students prior to beginning the mixer sharing:

- The aim is to have conversations with at least one representative from each of the La Vía Campesina organizations. Students must use a different individual to answer each of the nine questions.
- Tell students this is not a contest to see who can fill these out the fastest, so they should take their time, have a substantial conversation with each individual, but remember that they can use only one individual per question, so once they have finished talking with someone, they should move on to another representative.
- This is a one-on-one activity. Discourage students from grouping up, which can lead to some students passively listening and copying down information they hear from others. Nor may students pass their role sheets to each other, as this is a conversation-based activity. It is also meant to be a lively, get-up-and-move-around exercise, so discourage students from remaining at their seats and waiting for others to come to them.
- Role plays can invite stereotyping. One way to reduce this possibility is to tell people not to adopt accents as they attempt to represent individuals in their groups.
- Throughout the role play, an important aim is for students to identify with the La Vía Campesina activists they represent. Occasionally, we hear students distance themselves from their characters, saying things like, "If I were this person, I'd. . . ." Remind students to speak in the "I" voice, as if they are these individuals.

- Begin the mixer activity. In our recent experience teaching the role play, the mixer took at least 45 minutes.

9. Following the mixer activity, we asked students to pause and to write on three questions:

    1. What are some of the problems and fears that farmers from around the world are facing?

    2. What are some of the positive things that farmers in La Vía Campesina are doing around the world?

    3. As you talked with people from around the world, what did you notice about how their lives were similar to yours? In what ways were their lives different from yours?

    As we began our post-mixer discussion, one student commented on similarities he noticed: Everyone he met in La Vía Campesina was "worried about making food, not making people rich"—which seems a pretty astute summary of the movement's pro-community, anticorporate orientation. For the third question, one student addressed how all of them were similar: "We all want global help to stay local."

10. We followed the mixer activity by showing students a short (about 20 minutes) film, *La Vía Campesina in Movement . . . Food Sovereignty Now!*, which we found at the La Vía Campesina website (http://video.viacampesina. org). (The film is also available at the website in Spanish, French, and Portuguese.) It allows students to hear from a diverse array of La Vía Campesina members—including from some of the organizations that they represent in the role play—who offer their interpretations of "food sovereignty" and the aims of this mass movement. We asked students to listen carefully to these definitions and goals as they watched the film. The video introduces students to small farmers from around the world who are anything but the powerless victims they are often portrayed as. The individuals in the film are defiant, dignified, determined, eloquent, and organized. Students will not yet be familiar with every issue that is raised in the film, or every event or organization referred

to, but our aim in showing the film was to help students imagine the kind of people that they were representing in the activity and appreciate what motivates them. And it did that. As one of our students commented after the film, "I loved how people work with each other and support each other, even though they are from different places all over the world." (See other La Vía Campesina videos at http://vimeo.com/ viacampesina/videos.) Following the video, we asked students to write on the question "What does food sovereignty mean to these activists?"

11. Tell students that a G7 representative is coming to speak on the plan that the G7 has formulated to "Feed the World." In addition to offering a five-point proposal, the G7 will introduce its slogan at the meeting. In preparation for the meeting, tell students that in their group they should come up with a brief slogan that summarizes their perspective on food and agriculture. Afterward, they should write these on a large piece of paper and post them prominently on the wall, or attached to their desks. In our recent experience teaching this role play, students showed delightful imagination in coming up with slogans. These included: "Buy Local, Sell Local, Live Local"; "We Grow Our Own Power"; "Without Land There Is No Life"; "Mother's Earth Is for Us to Share"; "Fight for Fair Family Farms"; and the delightful "Food for the Many, Not for the Money." Prior to the arrival of the G7 representative, ask students from each group to share their slogans with the entire assembly. We told students that in a moment, they would hear from the G7 representative and that they should listen carefully to the presentation because afterward there would be time for questions and answers. We encouraged people to compare the analysis and proposals from the G7 with the specific information in their roles, information they had heard from fellow La Vía Campesina members, or material that they had encountered in the "Facts on Food, Farming, and Hunger" handout. We encouraged students to write their comments and questions on the "Proposal to Feed the World" as they listened to the G7 presentation.

12. Distribute a copy of the G7 "Proposal to Feed the World" to every student. The teacher plays the G7 representative. When we made our presentations to the classes that we worked with, we left the classroom for a moment to re-enter as the G7 representative and took our place at the podium and thanked the assembled members of La Vía Campesina for agreeing to hear our plan. We began by highlighting the G7 slogan—one we invented for them, but apt—"Science+Investment+Free Trade= Prosperity." With some pomp, we read the "Proposal to Feed the World" aloud, stopping occasionally to emphasize a point.

13. No doubt, the G7 proposal is a lot to throw at students all at once, but we wanted them to see that all these components are part of a "neoliberal" strategy that places its faith in the so-called free market. Following the presentation, we told La Vía Campesina members that there would be a brief period of time to discuss the G7 proposals but that they should first turn to other members in their group or to others sitting around them and talk about what they thought of the five points in the G7 proposal and to brainstorm questions or comments they might want to make. The back-and-forth with students following the G7 presentation was one of the most spirited parts of the role play. La Vía Campesina members peppered the G7 representatives—us!—with questions and criticisms. In several classes, students drew on information in the "Facts on Food, Farming, and Hunger" to challenge the entire premise of the G7 proposal, that the root of hunger is not enough food in the world. "The problem isn't that there is too little food," one student said. "It's that people are too poor to buy the food. Nothing in your plan changes that, so people will still go hungry." You should conclude this back-and-forth while students are still full of questions and challenges for the G7 representative, as you'll want for this enthusiasm to carry over into coming up with an alternative response.

14. After the G7 presentation, ask students to write for a few minutes on "How does your vision of food sovereignty differ from the G7 proposal?"

15. Following the meeting with the G7 representative, we distributed "La Vía Campesina's Response to the G7's 'Proposal to Feed the World.'" In the 12 classes we taught this to, from this point forward, we took different paths. One route is simply to have students use these questions as a guide and to have each organization articulate some of the main planks or demands they propose to counter the G7 proposals. Another is to convene a La Vía Campesina meeting and to have the full assembly tackle each of these five questions one at a time. Each separate organization group could go over these one by one prior to convening as a whole group. To simulate the democratic processes—and difficulties—La Vía Campesina encounters in real life, you might remove yourself as teacher from this discussion, and let students fashion a process to make the decisions on their own. Whichever method you select, the idea is simply that students should attempt to develop an alternative to the free market vision of the G7. In each of our classes, students anticipated many of the real-world positions of La Vía Campesina—as well as the underlying cooperative, anticapitalist ethos of food sovereignty. As one student began the discussion: "If you're trying to end world hunger, you don't do it by trying to make money."

16. After students finished their responses to the G7 proposals, we asked them to think about how their alternative plans might go beyond food. We asked them to complete these statements:

- Our plan addresses the climate crisis because…
- Our plan creates more jobs because…
- Our plan is better for people's health because…
- Our plan is more likely to prevent people from being forced to move away from their homes because…
- Our plan helps reduce the huge gaps between the world's rich and poor because…

Students were astute in recognizing how knitted together all these issues were.

One student noted, "Our plan addresses the climate crisis because if we buy more locally, the need to use gas and other fuel would be reduced because we wouldn't have to ship many things across the globe. Also, we want to use agroecology more, which does not need fertilizers and chemicals to grow things." Another student addressed the migration issue: "When corporations come in and seize land, they often need large tracts of land to support their huge moneymaking enterprise. This will force small farmers off their land, like it did the Korean Women's Peasant Association. La Vía Campesina wants to ensure that all people are guaranteed the land they currently own, and the right to work their land and uphold their livelihood."

17. We concluded our La Vía Campesina unit by asking students to create metaphorical drawings—drawings that symbolized some aspect of the food sovereignty struggles that we had studied, and to write short explanations of their drawings. This assignment elicited an explosion of creativity. We were struck by how many student drawings focused not only on global exploitation, hunger, and the poisoning of the planet for profit, but also on the hope they found in La Vía Campesina's organizing and resistance. One image depicts a globe with stick figure people holding hands. A flaming Monsanto stands at the top of the world. The student's explanation reads: "Monsanto (alone) standing, separating itself from the world. The yellow in its flames represents greed. The orange represents power above others, because orange doesn't rhyme with anything and works alone. Red represents force, with which they pirate farms. The people are small farmers, all holding hands because they are connected and connected to the Earth." Another student's drawing echoes this theme, and celebrates La Vía Campesina's accomplishment of uniting farmers across nationalities and languages. It shows a world map with people connected continent to continent by lines and listening devices. On the lines are written phrases like "organic," "independence," and "food sovereignty." The student's explanation: "This drawing represents the interconnectedness of La Vía Campesina. The way that people connect with

each other through the same cause, despite their language barriers, is something truly moving. The supporters of La Vía Campesina are able to build ideas and new ways of living from each other and even fight for the same change throughout the world to end hunger. People reach globally to keep things local." Another drawing features a single sprout, being watered from above. The student's caption: "The sprout in the center represents food sovereignty because it is growing (and a growing concept) that is good and will thrive. The water that is perpetrating this idea is La Vía Campesina because it is a huge group and is clean and natural, and is trying to purify their place. Also, the sprout is a plant that is growing, making food for the locals."

18. One adaptation that we have made in workshops, where we have less time, is to ask participants to follow the G7 presentation by preparing for a demonstration to protest the G7's GMO-friendly, free market approach to agriculture. We ask them to make placards in their La Vía Campesina groups featuring slogans and images that embody their critique of the G7 proposals and that express their alternative visions for "feeding the world." This is a playful activity that mirrors the in-the-streets activism that is so much a part of La Vía Campesina's organizing.

* * *

As we introduce students to the intertwined environmental and social crises that afflict the world, it's important that at the same time we introduce them to the social movements that address these crises. La Vía Campesina strikes us as one of the most significant of these social movements. We wrote the role play because we are encouraged by La Vía Campesina's vibrant, democratic challenge to the corporatization and homogenization of our global food systems. The movement makes it clear that a system based on maximizing profit will never produce sustainable or equitable solutions to food crises and hunger. But the groups affiliated with La Vía Campesina don't stop with critique; their brilliance shows in the hundreds of examples of farmer-run

collectives around the world, providing communities with everything from seed banks to education in the innovative adaptations that farmers are developing in response to a quickly changing climate.

La Vía Campesina's responses to the many global crises are dispersed and distinctly local—rooted in the knowledge of particular communities and the context of specific places. This contrasts with the solutions offered by multinational agribusiness corporations like Monsanto that are the same everywhere, like the genetically modified crops that "resist drought," whether in the Midwestern plains of the United States or the northern plains of India. These may be marketed as solutions that will "feed the hungry," but at the end of the day they are simply a means to make money.

As La Vía Campesina points out, the market regards food as a commodity and delivers it only to those with the means to pay. But like the atmosphere and clean water, food is a collective human right—something the market can never adequately provide. Social movements—not free markets—are what secure and expand our rights. And that's perhaps the most important lesson that students can take away from the La Vía Campesina role play: the necessity of collective action in response to social and environmental crises. ⊕

Bill Bigelow (bbpdx@aol.com) is curriculum editor of Rethinking Schools *magazine. Chris Buehler, Julie Treick O'Neill, and Tim Swinehart teach at Lincoln High School in Portland, Oregon.*

Facts on Food, Farming, and Hunger

RINI TEMPLETON

Assignment: *Read the facts below, and come up with at least three things that strike you as interesting or surprising—or that you had not known before. Come up with at least three questions that these facts leave you with.*

- The U.N. Food and Agriculture Organization estimates that 870 million of the world's 7.1 billion people suffer from chronic undernourishment—that's about one out of every eight people in the world.

- In the United States, 16.7 million children (about 22.4 percent) live in households that are considered "food insecure"—that is, families that don't always know where their next meal will come from.

- There is enough food produced to feed everyone on Earth. In fact, the world produces enough food to feed more than one and a half times the number of people on the planet.

- World agriculture produces 17 percent more calories per person today than it did 30 years ago, even though during that time the population has grown 70 percent.

- There is no relationship between population density and hunger. The Netherlands, with very little hunger, has 401 people per square kilometer. Brazil, with a relatively high rate of hunger, has 23 people per square kilometer. India, with a great deal of hunger, has 369 people per square kilometer. South Korea, with relatively little hunger, has 487 people per square kilometer.

- Every day, almost 16,000 children die from diseases caused by malnutrition. That's one child every six seconds.

- One-fifth of all the hungry people in the world live in India.

- India is the world's second largest grain producer.

- Of the arable land in Mexico, 88 percent is used for the cultivation of export crops and grazing of cattle. Every year, almost 2.5 million acres of land used to grow food crops for local consumption is transferred to plantation crops—almost all for export.

- The Mexican government admits that at least 7 million people in Mexico suffer from a combination of extreme poverty and malnutrition.

- Half of the world's corn, 90 percent of the world's soy, and about 25 percent of the fish caught each year is used to feed livestock and farmed fish.

- Of all the privately held land in the world, nearly three-quarters is controlled by just 2.5 percent of all landowners.

- More than half of the rural population in the Third World is landless.

- In Bangladesh, almost 80 percent of the people work in agriculture, but they own only 5 percent of the country's land. In the United States, just 4 percent of the landowners own 47 percent of U.S. farmland.

- Almost half of the world's population lives in rural areas and grows their own food.

- Worldwide, women make up more than 50 percent of the agricultural workforce. In rural Africa, women produce about 80 percent of the food grown for domestic consumption. Yet they receive less than 10 percent of the loans provided to farmers and own only 2 percent of all farmland.

- More than 1 billion pounds of pesticides are used on crops on farms in the United States every year. About 5.6 billion pounds are used annually throughout the world.

- In the 1940s, U.S. farmers lost 7 percent of their crops to pests. Since the 1980s, this crop loss has increased to 13 percent, despite the vast increase in the amount of pesticides used on farmland.

- Since 1945, between 500 and 1,000 pest and weed species have developed pesticide and herbicide resistance.

- Just four multinational corporations—Monsanto, Syngenta, Bayer, and DuPont—control half of the world's seeds.

- After the massive 2010 earthquake in Haiti, Monsanto announced what one company representative called "a fabulous Easter gift" to Haitian farmers: 60,000 seed sacks of genetically modified corn seeds and other vegetable seeds.

- On June 10, 2010, about 10,000 farmers in Haiti marched three and a half miles to a town in central Haiti where they burned 400 tons of Monsanto seeds.

- La Vía Campesina is a social movement of small farmers around the world. It has a membership of 164 organizations in 73 countries. It is considered by many to be the largest social movement in the world, representing 200 million people.

Sources: Why Hunger?, Food First/Institute for Food and Development Policy, UN Food and Agriculture Organization, La Vía Campesina, World Hunger Education Service, *Population in Perspective: A Curriculum Resource*, *The New Internationalist*.

RINI TEMPLETON

Basque Union of Small Farmers and Ranchers (EHNE)

You live and work in the Basque region of Spain, on the border between Spain and France. Although the Basque Country—called *Euskal Herría*—is technically a part of Spain, Basque people have always considered themselves independent of Spain. Your language is the oldest language in Europe. In 1937, during the Spanish Civil War, the Basque village of Gernika (Guernica in Spanish) was the first place in Europe to be bombed by the Nazis. Spain's dictator, Francisco Franco, outlawed your language; people were forbidden to speak, write, or sing in the Basque language. But your people resisted. Today, the Basque region is still being invaded, but it is a different kind of invasion.

There are about 6,000 farmers in EHNE. You are small farmers, farming an average of about 15 acres. It's difficult, but you love it. With mountains and streams and greenery everywhere, could anyplace be more beautiful, more peaceful? Cities are fine—to visit—but it's here in the countryside where you want to raise your children. On your farms, you have some kiwi trees, and you also grow peppers, tomatoes, corn, cabbage, lettuce, and onions. Some farmers raise cattle and sheep for meat, as well as for dairy products like milk, cheese, and yogurt.

But what farmers in EHNE really "grow" is health and community. EHNE was one of the first organizations in the world to use the term "food sovereignty"—that communities have the right to control decisions about the food they grow and eat. The campaign slogan you used with community members was "Produce local, eat local. It's good for you, it's good for us." Small farmers came together in EHNE to support one another and to provide an alternative to the big Walmart-like supermarkets that seem to be spreading everywhere. The food in these stores is grown by corporations on huge farms, with herbicides, pesticides, and chemical fertilizers. It's trucked in from hundreds or even thousands of miles away. It's not healthy food and the long distance the food travels contributes to climate change because it needlessly uses fossil fuels.

All over the world, huge corporations are trying to invade our lands and control our food. But EHNE—along with other farmer organizations in La Vía Campesina—are fighting back. In the Basque Country, small farmers have partnered with community members, who agree to buy a weekly basket of food from local farmers. This way, consumers know who grows their food, and they can trust that it was produced healthfully, without poisons or GMOs—genetically modified organisms. Your organization also offers help to young farmers, to try to keep young people on the land, so they don't have to leave the countryside to go to the cities. The big supermarkets try to push the little stores out of business. But EHNE refuses to sell to the big stores. EHNE sells only to local small markets, restaurants, and community food cooperatives.

EHNE is also fighting another kind of land grab in the Basque Country. The Spanish government is building "bullet" trains throughout the countryside. Small farmers won't benefit from trains that will speed through your land going as fast as 150 miles per hour. In order to build the railroad, the government has seized land from small farmers. In one local election, 98 percent of the people voted against the new high-speed trains. The government paid no attention, and took almost 40 acres of land away from six farm families. In another community, the government took 125 acres from 10 cattle farmers. You need this land for farming and to keep young people from leaving your communities. Who benefits from this so-called development? Only the rich corporations, which can get their goods to faraway places more quickly, and the corporations that make tens of millions of dollars building the railroad.

EHNE demands food sovereignty. Food ought to benefit farmers and community members, not profit-hungry corporations. Small is good. Local is good. Organic is good. Cooperation is good. Greed is bad. ⊕

Landless Workers Movement of Brazil (MST)

Seeds, soil, water, and sunlight—these basic ingredients of nature have allowed farmers to subsist from the Earth for thousands of years. Your Indigenous ancestors in Brazil used the land to grow crops like peanuts, sweet potatoes, and corn, but with the arrival of Portuguese colonists in 1500, the land was stolen from Indigenous people. Land became owned by a small group of wealthy elites, in massive estates.

Brazil still has one of the highest rates of land inequality in the world—the wealthiest 2 percent of the population owns about half of the nation's farmable land. To fight this unequal land distribution, you and thousands of other peasant farmers organized in 1985 to create the Landless Workers Movement (MST in Portuguese). The MST's strategy is to "take back the land" for poor farmers through land occupations and court battles, in order to win legal rights to unused farmland. After 30 years of struggle, MST farmers have won rights to millions of acres of land. Your organization is now the largest social movement in Latin America.

Land occupations focus on large, private land holdings that aren't being fully used. In fact, Brazilians have a constitutional right to challenge ownership of any land that is not "fulfilling its social function"; at least 80 percent of the land must be used effectively, and environmental and labor standards respected. The occupations are difficult and dangerous, with occupiers facing threats and violence. Court battles can take years to win the legal right to the land. But out of this struggle, MST occupations have led to successful, worker-run cooperative farms that provide living wages for workers and use environmentally sustainable farming practices.

The Brazilian government supports huge farms and multinational agriculture corporations that export products like soybeans, coffee, sugarcane, ethanol, and beef around the world. This "get big, or get out" model of farming hurts small farmers like you. Wealthy farm owners and foreign agriculture corporations get richer, but it's become almost impossible for small family farms to survive. As a result, more than 25 million Brazilians—14 percent of the population—suffer from hunger.

Brazil's changing climate is also making life more difficult for small farmers. The *Nordeste* region, where you are from, is a good example. Despite the dry, semi-desert, families like yours have lived and farmed in this region for centuries, planting crops that grow well with the yearly cycles of drought and rain. But in recent years, the droughts last longer and when the rains finally come, they result in massive flooding. Many farmers have given up and moved to the cities, usually landing in the slums.

But your community of MST farmers is showing that there are more positive ways of doing things. MST farmers are developing new farming techniques called "agroecology," which work with the droughts and floods in the region. By using ditches and underground dams, farmers have developed new ways of "harvesting" water during the rainy season, to help keep soils moist and fertile during droughts. MST farmers avoid chemical fertilizers and pesticides by using diverse crop groupings that include fruit trees, corn, ground cover crops, and climbing vine crops. This crop combination creates a lush forest-like environment that naturally builds the health of the soil—reducing the need for chemical fertilizers and pesticides.

The MST can help spread these sustainable farming practices around Brazil, but it will take government support—financial support for water storage technology and real land reform. In your own region, you can see the difference that land reform makes. Agroecology systems take three to five years to establish, and are usually practiced only when MST farmers have won the right to land they actually own. The same farmers use conventional methods (field burning and heavy use of chemical pesticides) on land that is merely occupied by the MST, because they can't count on having the land in the future. By fighting for land reform and government financing for agroecology, the MST can help millions of Brazilians return to the land and use sustainable, small-scale farming to preserve the land for future generations. ⊕

Group of 4 and the Dessalines Brigade, Haiti

You live in Haiti, the poorest country in the Western Hemisphere. You are members of the Group of Four (G4) and the Dessalines Brigade. In 2007, four of Haiti's largest peasant organizations came together to form an alliance to promote sustainable farming practices and advocate for the rights of peasant farmers. The Dessalines Brigade was created between Haitian and Brazilian farmers to show that a solidarity exchange is possible between peoples, not just between governments and corporations.

Farming is in your blood. Prior to the European conquest of Haiti, the Taínos, the Indigenous people, practiced a diversified agriculture that provided food for all and was ecologically responsible. Your agricultural practices are also influenced by West African farming traditions called *Konbit,* brought over by enslaved Africans that emphasizes cooperation over competition. Difficult farming tasks that require many people—such as planting seeds, clearing a field, or harvesting crops—are done by asking friends, family, and neighbors to lend a hand. This work is filled with singing, joking, feasting, and lots of hard work. Today, two-fifths of all Haitians still make their living from the land and agriculture accounts for roughly a quarter of Haiti's total economy.

In January of 2010, your country was devastated by a 7.0 magnitude earthquake that killed an estimated 320,000 and displaced more than 1.5 million Haitians. In the years following the earthquake, Haiti was hit by a cholera epidemic, and several tropical storms damaged infrastructure and made farming more difficult. Following this disaster, Monsanto, a multinational seed corporation, donated more than $4 million worth of non-native hybrid seeds. They said, "[I]t was clear a donation of our products—quality corn and vegetable seeds—could really make a difference in the lives of Haitians."

You were skeptical. La Vía Campesina, an international movement you belong to, has called Monsanto one of the "principal enemies of peasant sustainable agriculture and food sovereignty for all peoples." Monsanto sells hybrid crops that do not produce seeds that can be saved for next season. All over the Global South, Monsanto attempts to talk farmers into using the seeds, and traps farmers into purchasing them year after year at outrageous costs and under their conditions. Monsanto also owns almost 650 seed patents and has sued small farmers who have accidentally grown their seeds when they migrated from farm to farm. In addition, these hybrid seeds are covered in chemicals. The seeds donated to Haiti were treated with two fungicides called Maxim XO and thiram. The U.S. Environmental Protection Agency says that these chemicals are so dangerous to farmers that they need to wear special gear to protect themselves from being poisoned. The EPA also banned these products for home garden use because most homes do not have the safety equipment to properly use the seeds.

Although the government and wealthy Haitian elite accepted these seeds, you rejected them. Thousands of you marched to protest Monsanto's increasing presence in your country and you burned the donated seeds as a symbol of your rejection. You called this seed donation by Monsanto "a new earthquake" and "a very strong attack on small agriculture, on farmers, on biodiversity, on Creole seeds...and on what is left of our environment in Haiti." You know that the future of Haiti depends largely on local production with local food for local consumption, what you call food sovereignty. One farmer said, "People in the U.S. need to help us produce, not give us food and seeds. They're ruining our chance to support ourselves." Monsanto's donated seeds and arrival in your country is a direct threat to this. ⊕

National Peasants Union of Mozambique (UNAC)

You live in Mozambique, a country in southeastern Africa, and are members of UNAC, the National Peasants Union of Mozambique. You are farmers. In fact, four out of every five people in Mozambique are farmers—that's about 15 million. To know something about farming in Mozambique you have to know about how the Portuguese colonized your country for hundreds of years, stole your land, and enslaved your people. The Portuguese forced your people to grow non-food crops like cotton and sisal to export—which benefited the Portuguese but no one else. Because your land was stolen, many Mozambicans, especially men, left the country to work in the gold and diamond mines in South Africa.

After a long war, your people drove the Portuguese out and finally, in 1975, Mozambique became independent. Independent, but poor. This was the start of a difficult period, but for once, Mozambique was for Mozambicans. As one farmer put it, "It was the first opportunity when people were feeling that we have lives, we are independent…We can determine our own future without interference from outside." The new revolutionary government took all the land back from the Portuguese and declared that the land was now owned by the people, and that everyone in Mozambique had a right to land and to farm. To you, landownership is a sacred right of all Mozambicans. It's right there in Article 109 of your constitution: "As a universal means for the creation of wealth and of social well being, the use and enjoyment of land shall be the right of all the Mozambican people."

When the Portuguese ruled your country, they wanted cash crops to be grown only for export. But now, farmers like you try to be self-sufficient, and grow all the food your community needs to survive: fruits like papaya, oranges, and coconuts; everyday food crops like maize (corn), peanuts, and beans. You also raise cows, goats, chickens, and ducks. You practice what UNAC calls "agroecology." That's a fancy way of saying that you don't want to use pesticides and chemical fertilizers, which are expensive and can also poison your food, animals, and water, and hurt the soil. You value the traditional farming knowledge passed down through generations—including saving and sharing seeds among different communities, rather than importing seeds from other countries. As UNAC says, "Peasants are the guardians of life, nature, and the planet."

UNAC connects hundreds of "associations" of farmers in Mozambique—and around the world, through La Vía Campesina, the peasant movement you belong to. The UNAC associations are like cooperatives where people help each other out. Sometimes they work each other's land together, or share their resources. If farmers in the association produce more food than the community needs, then they sell these products and use the proceeds to buy a truck or improve the community's school. UNAC also brings farmers from Mozambique together to share their knowledge and learn the stories of other peasants from around the world. UNAC pressures the government to offer loans to farmers, or "infrastructure," like irrigation or better roads.

UNAC has also organized farmers to resist the attempts of big foreign companies—and your own government—who are coming in to steal peasants' land. For example, a huge program called ProSavanna—a partnership between the governments of Mozambique, Brazil, and Japan—that would seize 34 million acres from peasants and turn the land over to corporations to grow cash crops like soybeans, sugarcane, and cotton. These are not to feed your people but to export, to make others rich. Their idea is that the peasants who will be thrown off the land can get jobs working on the cash crop plantations. Not only will these enterprises steal farmers' land, but they will also poison your water with chemical pesticides and fertilizers. This whole scheme feels like colonialism, and UNAC and farmers all over Mozambique are fighting these unconstitutional land grabs.

The goal of UNAC is to create "dignified and lasting livelihoods" and to produce "high-quality foods in sufficient quantities for the entire Mozambican nation." Under colonialism, Mozambique's exports were high. But Mozambicans starved. You will not allow that to happen again. ●

Korean Women's Peasant Association

In Korean folklore, the mung bean, or *nokdu*, represents the resilient spirit of the Korean peasants. Under even harsh conditions, *nokdu* sprouts and grows, feeding you when you need it most. You are deeply concerned with Korea's future and the future of the Earth, but like *nokdu*, you are resilient. You are part of the Korean Women's Peasant Association, with 30,000 members, and you are fighting for women farmers of South Korea. You are also affiliated with La Vía Campesina, the movement of small farmers around the world.

Shintobuli—a Korean proverb—says, "Your body and Earth are not two different things." How you, as women peasant farmers, live, shows this. In addition to housework, women have traditionally been responsible for everything to do with food: managing and sowing seeds, harvesting, storing, processing, and cooking food for your family. But the way you have lived is changing.

After the Korean War ended in 1953, the United States pushed Korea to become more industrial. This rapid industrialization in Korea has had big effects on you. The number of farmers declined from 50 percent of the population in the 1970s to 7 percent in the 2010s. More than one-fourth of the farmland disappeared to make way for the booming technology industry. Many Korean farmers quit farming because they could not make money from products they grew—in part because of cheap imported food from the United States—and flocked to the cities looking for work. This resulted in slums and urban sprawl.

But not everyone sees this as a problem. Huge Korean corporations, or what you call *chaebols*, have benefited greatly. The top 30 *chaebols* account for 82 percent of your country's exports. They want to control all food in Korea for profit, while opening up opportunities for the tech industry in farming. If South Korean *chaebols* and the politicians who represent them had their way, you, and most other farmers, would all but disappear under the logic that your farming practices are "uncompetitive" in the global marketplace.

The *chaebols* want free trade agreements with the United States and other industrialized nations. They argue that these agreements will strengthen global demand for Korean high-tech products. But you can't eat technology. These agreements, like the World Trade Organization (WTO) and Trans-Pacific Partnership, benefit the *chaebols*, but not small farmers like you. You and other Korean farmers oppose these agreements, because they open Korea up to cheap imports and give more control of agriculture to rich corporations. In protest, you have taken to the streets. You feel that an industrial food system in Korea has hurt women farmers like you. Other small Korean farmers are also speaking out. Lee Kyung Hae even committed suicide by stabbing himself in the heart while protesting the WTO in Mexico because of its devastating impacts on small Korean farmers. Lee said that the "WTO kills farmers," because the "free trade" pushed by the WTO forced Korean farmers to compete with rice produced by huge agribusiness farms in the United States and apples grown with incredibly cheap labor on Chinese farms.

Fortunately, the Korean Women's Peasant Association has helped create a movement to eat local food. You have created an agricultural supply chain that connects women peasant farmers in the countryside with consumers in the city. Your food is grown without pesticides, herbicides, GMOs, or other chemicals. You farm *with*, rather than *against* nature, like using sea snails to get rid of weeds in your rice paddy fields. You have also created "common cafeterias" where women farmers can eat the food you all grow instead of buying food from *chaebol*-owned restaurants. This allows you to sell and share food from *your* farms to people in *your* community and not support the people who are wrecking your way of life.

You are unwilling to accept a food system where the *chaebols* control all the food—one that is industrialized, chemical-dependent, energy-intensive, and benefits only the rich. You want a food system where you consume what you produce in your local community. How you farm promotes sustainability, equality, and community. All this is under threat and you will not accept this fate. ⊕

Tamil Nadu Women's Collective

Drought has parched the land in Tamil Nadu, the southern state in India where you live. The seasonal monsoon rains have disappeared, yet another victim of climate change. That, combined with the many Coca-Cola plants that draw vital irrigation water from the Tambirabarani River, leaves village wells dry. Farm production has dramatically decreased. You cannot water your crops with Coca-Cola.

You also face pressure from another multinational corporation, Monsanto, which has purchased seed companies in India, and now uses the same historically known brand names to deceive farmers, getting them to purchase their genetically modified seed. The Indian government has approved Monsanto's Bt cotton for commercial use. It tempts men all over India. They hope to provide for their families with seeds designed to withstand drought and pests, increase yields. But these seeds are expensive.

Farmers can borrow from a bank to buy the seed, but many cannot, and turn to unscrupulous money lenders who lend at high rates with a farmer's land as collateral. However, these genetically modified crops often fail. Then the farmers are unable to pay back the loans. Some men leave their villages to find work in the cities; others, in acts of desperation, commit suicide. In India today, one farmer kills himself every half-hour.

Tamil Nadu is different; in part, because of you. The Tamil Nadu Women's Collective is an organization of more than 100,000 women in 18 different regions and 1,500 villages throughout Tamil Nadu, striving to empower the most vulnerable in India today: women. You especially organize those who are divorced or widowed, and the *dalit*—the so-called "untouchables."

Proudly, as part of a GM-free campaign, the Tamil Nadu Women's Collective has educated people about the impact of genetically modified crops on the land and people. Few men here fall prey to the empty promises offered by Monsanto for Bt cotton. More importantly, the state of Tamil Nadu has banned field trials of genetically modified food crops.

As a woman, you have always been the actual farmer; 80 percent of women in rural India are. On your farm, you and the other women in your family do most of the work: you plant, thin, water, harvest, and store the crops; without you, there would be no crops. Men deal only with cash crops, like cotton or soya.

Many men in Tamil Nadu have resisted genetically modified seeds. However, they continue to seek a profit in cash crops using hybrid seeds and chemical fertilizers and pesticides, advertised to increase the seeds' production—even in drought conditions—on land that could be used for food crops. Hybrid seeds cannot be saved; they are often sterile or produce plants unlike the parent plant. You worry. Once again, you see farmers becoming dependent on seed companies, spending money they do not have on seeds and chemicals, chasing a dream of abundant crops.

That is why the Tamil Nadu Women's Collective is returning to what leading seed activist and sister Indian, Dr. Vandana Shiva, calls "the gift of the seed from the Earth and our ancestors."

Your collective is focused on shifting land use from cash crop to food crop cultivation, and planting traditional, native food crops such as millet and pulses (beans and lentils), crops that need little water and are so well suited to regional conditions that they require no chemically based fertilizers or pesticides. Women farmers are now able to feed themselves more than 15 days a month from your own crops. One goal of the collective is to guarantee members two meals a day without having to borrow.

Companies like Monsanto, through GMO or hybrid seeds, prevented your self-sufficiency. Now, you can save your own seed, reserving part of the crop to plant the following year. You have even created seed banks, where women share seeds from different crops, ensuring the preservation and biodiversity of traditional food crops. You feel more independent, more powerful. By returning to traditional crops and farming practices, the collective is protecting the Earth and the seed she so generously provides. ⬤

La Vía Campesina Mixer

Find a different person to answer each of these questions. You must talk to at least one individual from each of the five other La Vía Campesina organizations.

1. Find someone whose organization has taken direct action against a multinational corporation. What group does the person belong to? What action did the group take?

2. Find someone who is concerned about the takeover of farmland in this individual's country. What organization does the person belong to? How is farmland being threatened and what is the organization doing about it?

3. Find someone whose organization is practicing "agroecology." What organization does the person belong to? What does agroecology mean to this person and what steps is the organization taking to bring it about?

4. Find someone whose organization has an opinion about seeds. What is the organization, and why does this organization focus on seeds? What actions is the group taking—or has it taken?

5. Find an individual from a group that organizes women farmers. What specific issues affect women in this group's country? What steps has the organization taken to help women farmers?

6. Find a person from a country that once was colonized by a European country. What is the country? How was the country affected by colonization and how does it still affect farmers in this country today?

7. Find a person from an organization that is working for more local solutions to feeding people in its country. What is the organization? How is this organization working for more local solutions?

8. Find a person from an organization that has demanded changes from a country's government. What is the organization and country? What demands has this organization made of the government?

A Proposal to Feed the World—From the G7

The world's developed countries hope to build a partnership with small farmers for a new alliance to feed the world. Our challenges are enormous. Experts predict that the world's population will reach 9 billion people by 2050. If we are to feed everyone, we will need to double the amount of food we produce today.

Together, we can do it.

Leaders of the G7 countries—the world's most developed and wealthy economies—want to work with you, the small farmers of the world. We propose a partnership among G7 governments, private corporations, and farmers. To reduce poverty and end hunger, we commit ourselves to give you financial aid and technical support, and to help generate greater private investment in agriculture.

All countries need to be full participants in the global economy. As the global economy expands and reaches more of the world's people, poverty will turn to prosperity. But it will require everyone's work.

Here are the major parts of our vision:

1. **Huge new investments from private corporations will help modernize agriculture, develop underused land, and help farmers get their crops to global markets.**

    It will be the investment from the private sector that will spread science and technology throughout the world and lift people out of poverty. Multinational corporations like Cargill, Monsanto, and Dreyfus can introduce advanced plant breeding, biotechnology, and improved farm management and marketing practices.

2. **All countries must eliminate any restrictions on private foreign investments.**

    Today, in some countries, only people from that country may buy land. In some countries, land may not be bought or sold at all! This is wrong. Private investors need to be able to launch large-scale agricultural projects, if they are to demonstrate the newest, most productive farming techniques, and produce vast quantities of agricultural products for the global market. Some of these projects may require unproductive farmers to lose their land so that the land may be put to more productive uses. This will be painful, but in the long run, it is in everyone's best interest. Examples of new projects include large-scale production of soybeans, sunflower oil, cotton, bananas, and maize (corn); pig barns for pork production; or crops for biofuels or animal feed. These operations create jobs for individuals who may lose their farms as the land is put to more productive uses. It's essential that countries eliminate all barriers to private foreign investment, if everyone is to benefit from the global marketplace.

3. **Tariffs on imported agricultural produce must be eliminated.**

    Some countries have high tariffs to keep out agricultural products from other countries. This must stop. We need free trade. Tariffs on agricultural goods must be eliminated so that every country can be brought into the global market. Farmers—especially small farmers—have the right to compete in a market of 7 *billion* consumers, rather than the few million in just their own country. Consumers have the right to buy the cheapest food available, even if it was not produced in their home country. If tariffs are eliminated, foreign investment in agriculture will increase and everyone benefits.

4. **Private seed companies will provide the world—at reasonable prices—the most productive, high-yielding quality seeds. Thus, countries must offer protection to private seed companies in order to gain access to these benefits.**

This would include recognizing and enforcing all patents on seeds, especially genetically modified seeds, which require huge private investments and need to be protected. There should be no restrictions on the use of scientifically tested seeds. Furthermore, governments must stop giving away seeds to farmers, and leave the seed business to the seed experts: private companies. Furthermore, governments must allow for the inspection of seeds by private seed companies to make sure that all farmers comply with patent regulations.

5. **Governments must end subsidies to farmers and let the free market work to make farmers more productive, to double agricultural output, and to reward those farmers who are most productive.**

Too often, governments subsidize unproductive farmers by offering irrigation assistance, marketing help, cheap loans, or other supports. These may help an individual farmer, but they hurt food security because they keep the least efficient farmers in business and discriminate against the most efficient farmers. Government has a role to play, but it cannot take sides in favor of smaller, unproductive farmers and against larger, more productive farmers.

Friends, we are at a crossroads in world history. As the population booms and as the climate changes, we will either feed the world, or we will see hunger and misery increase. And food insecurity will lead to more and more armed conflict, as people fight for scarce resources. But this is unnecessary. Science and technology can lead the way. We can have a world of abundance, a world of justice, a world of freedom. Let us seize the moment. ⊕

La Vía Campesina's Response to the G7's "Proposal to Feed the World"

As members of La Vía Campesina how do you respond to the five-point proposal from leaders of the wealthy countries? As the world's largest social movement of small farmers, La Vía Campesina needs to decide what you think of these proposals, and come up with an alternative "Plan to Feed the World," if you oppose any of these proposals.

1. The G7 countries propose that "huge new investments" from private corporations and help getting your crops to global markets will improve the lives of small farmers. Do you agree? If not, specifically, what things will help small farmers? Come up with at least three proposals.

2. The G7 countries propose eliminating all restrictions on private investment in your countries. Should these restrictions be eliminated? If not, what kind of restrictions or rules should be put on private investment?

3. The G7 countries propose eliminating agricultural tariffs on produce. Do you support this? If not, what kind of tariffs or other restrictions should be placed on imported food?

4. The G7 countries insist that countries must respect patents on seeds, not restrict the use of genetically modified seeds, must not give seeds away to farmers, and must allow private inspections of farms. Do you agree? If not, what should countries' "seed policy" be?

5. The G7 countries call for the elimination of government subsidies to farmers. Do you agree? If not, what kind of support should governments provide to small farmers?

Greening for All: The Right to Access Healthy Food—*page 345*

By Marcy Rein and Clifton Ross

In "Greening for All," Marcy Rein and Clifton Ross point out that "North Richmond, like many depressed communities across the nation, has more than its share of liquor stores, but no stores that sell decent, much less organic, produce." Have students do a "food access map" of their community. What is the availability of "decent" food within a one-mile radius of their school or home?

Have students come up with a community garden plan. What resources are available to help "green" their community?

Ask students to discuss or write on:

- What criteria determine decent food? Local, fresh, organic, reasonably priced?
- Who benefits in Richmond from Verde Garden?
- Would your community benefit from a Verde Garden?
- How did the Hmong and Mien women "call into question the meaning of 'public property'"?
- Saleem Bey says, "We consider these lots public lands. We walk by them every day. We think that if you have a need and the land is available, the greater good supercedes ownership." Do you agree?
- Why do you think it was immigrants from the hill country of Laos who led the movement for community gardening in Richmond? What attitudes did the Laotian women bring to the work of creating gardens in Richmond that helped strengthen this work?
- How is gentrification a possible danger of community gardens?
- What percentage of the food consumed in your community is grown there? How could you find out? What would it take to achieve a 5 percent goal like Richmond's?
- Why do the authors title this article "Greening for All"?
- Grace Lee Boggs says that the Urban Agricultural Movement developed not just in response to "belly hunger." What does she mean by that?

"We Have the Right..."
A Youth Food Bill of Rights—*page 354*

The full "Youth Food Bill of Rights" is available at http://www.youthfoodbillofrights.com. The most recent version (2013) lists 17 "rights" youth activists have identified. This is an activity that is probably best completed after students have studied contemporary food and hunger issues. Before students read the list, ask them to imagine they have come together with other young people from around the country to decide which rights they believe all young people deserve when it comes to food. Talk with students about what makes something a "right." What distinguishes a right from a privilege? Students could complete these questions in small groups or in pairs. Begin by brainstorming a few possible rights with the full class. Ask them to begin each entry: "We have the right to…" Then they should finish the sentence with their demand. Ask students to come up with at least seven youth food rights. Ask them to share what they believe is the most important "food right" they listed. Discuss what makes these rights so important.

Afterward, distribute "Youth Bill of Food Rights." Ask the groups to read and write on the similarities and differences they notice between this list and what they came up with. Discuss:

- Are there any rights in the "Youth Bill of Food Rights" that you did not include, but think should be included?

- Are there any rights on the "Youth Bill of Food Rights" that you don't consider as important as others that you included? Why not?
- Which of the food rights could you demand at your own school?
- How would you go about working to effect these?

Encourage students to look at the full list of food rights generated by the youth activists and included at the www.youthfoodbillofrights.com website.

Maya Salsedo, whom we quote to open this reading, is a winner of a Brower Youth Award. Winners of this award are extraordinary youth activists, and we encourage you to introduce your students to these individuals and this diverse work through the film *Forces of Nature: Stories from the Brower Youth Awards,* distributed by the Video Project: www.videoproject.com. We reviewed one volume of this film in *Rethinking Schools* magazine:

> In short four- or five-minute segments, *Forces of Nature* introduces us to Brower Youth Award winners. These inspiring young people are not organizing classroom recycling programs: They fight mountaintop removal coal mining in Appalachia, work for green corridors and against nature-wrecking hotel development in Puerto Rico, develop organic youth-run farms in north Philadelphia, work on the Menominee reservation in Wisconsin to revitalize the Menominee language and traditional foods, and seek to ban the use of rainforest-destroying palm oil in Girl Scout cookies. These mini-films can be used in class separately or as a whole to show students diverse ways that young people make a difference.

10 Myths About Hunger—*page 361*
By Frances Moore Lappé and Joseph Collins

Before distributing to students "Ten Myths About Hunger," ask the class: "What are all the reasons you've heard about why hunger exists in the world? Even if you think they're wrong, let's hear them." List these reasons for everyone to see.

Divide students into 10 groups. Assign a different hunger myth to each group. Ask students to research their respective "myth" and "reality." En-

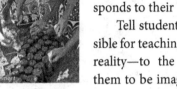

courage them to refer to the notes at the end of the reading, but also to use key search words and to explore the web. Ask them to note any information at odds with the "reality" that responds to their "myth."

Tell students they are responsible for teaching their myth—and reality—to the class. Encourage them to be imaginative: They can create a skit, a demonstration, or a simulation; design a poster or a metaphorical drawing; or create a game. Ask them to incorporate specific information from other historical study.

Ask students:

- Which of these myths seems most prevalent?
- How do myths like these develop?
- How are myths different from information that is simply wrong or even lies?
- Who is hurt by myths about hunger? Who benefits?
- Are some of these myths more harmful than others?
- If the world had a more accurate understanding about the roots of hunger in the world, how would that affect efforts to end it?

Here is the introduction to a widely used textbook, Holt McDougal's *Modern World History:* "As humanity moves further into the 21st century, another issue of growing concern is world hunger. Potential causes of famine are overpopulation, forces of nature, and war. Overpopulation occurs when there are too many people for the natural resources of an area to support. In some cases, it is war or natural catastrophes that push groups into starvation. Across the globe, nations are working to implement both temporary and more long-lasting measures aimed at reducing starvation and hunger." Use the information in the "10 Myths About Hunger"—and perhaps the La Vía Campesina role play—to critique this textbook excerpt.

Ask students to locate a textbook used at their school or another school and evaluate the accuracy and adequacy of this book's discussion of the causes of world hunger.

THE CYCLE OF COOPERATION

RICARDO LEVINS MORALES

AN AFTERWORD: RESOURCES FOR THE EARTH

As we indicate in this book's introduction, we see this collection more as an invitation to a curriculum conversation on teaching about the environmental crisis than as a definitive teaching statement. Here we offer some thoughts on resources we've found helpful in shaping these lessons and helping us identify key themes to pursue with students.

Publications

We've subscribed to **Orion magazine** during the entire time we worked on the book and found many key articles in its pages. These include essays

or excerpts from articles by Bill McKibben, Sandra Steingraber, Winona LaDuke, Derrick Jensen, Terry Tempest Williams, Gary Pace, and Erik Reece. And we could have included many more *Orion* pieces in the book—for example, from the always-provocative Rebecca Solnit. A number of quotes we feature in the book are borrowed from a poignant article by Sandra Steingraber, "The Fracking of Rachel Carson," based on a slide show of Steingraber's, available as a vimeo at the *Orion* website.

Another source for a number of articles we excerpt here is **Race, Poverty, and the Environment**. This journal features short, accessible articles that highlight the intersection of, well, race, poverty, and the environment; many would be useful for students.

We are surprised more teachers aren't aware of the indispensable magazine, the **New Internationalist**. The lively and attractive publication is published 10 times a year and includes many classroom-friendly articles. Each issue of the *New*

Internationalist addresses a different theme, like fracking, global land grabs, debt, climate change, and endangered languages. A free publication on critical social and environmental issues that is explicitly aimed at young people is **Indy-Kids**, published in print five times a year and on the web once a month. *IndyKids* "aims to inform children on current news and world events from a progressive perspective, and to inspire in children a passion for social justice and learning."

Our curriculum on environment and justice also benefits from articles in **The Progressive, In These Times, Monthly Review, The Nation, National Geographic, Green Teacher, NACLA's Report on the Americas, Utne Reader,** and **Rolling Stone,** which features articles by Jeff Goodell, whose work we also include in this book. **Yes!** magazine publishes articles that focus on individuals and movements making important strides for social and environmental justice. A one-year subscription to *Yes!* is available free to teachers. Fairness & Accuracy in Reporting (FAIR) publishes **Extra!**, a magazine with short classroom-friendly articles that examine how media covers—and fails to cover—critical social and environmental issues.

Of course, many of the articles in this volume first appeared in **Rethinking Schools** magazine, for which Tim writes regularly and where Bill is the curriculum editor. We urge you to subscribe to *Rethinking Schools* and to consider submitting articles on your experiences teaching about the environmental crisis. *Rethinking Schools* is the main in-print site for conversation on K–12 teaching for social and environmental justice, where the unfinished work of this volume will continue. In Canada, **Our Schools/Our Selves** is a fine publication that features articles on curriculum as well as critical issues in education.

Radio/TV

Where would we be without *Democracy Now!*, the most reliable source for news and analysis that we know of. *Democracy Now!*, with hosts Amy Goodman and Juan González, was covering the climate crisis, fracking, the Alberta Tar Sands and the Keystone XL Pipeline, and other environmental justice issues long before these concerns began pushing their way into most people's consciousness. Past shows can be heard online or downloaded in video or audio formats as podcasts. Short *Democracy Now!* segments are ideal for classroom use and can be the basis of compare-and-contrast assignments examining the treatment of key issues in the corporate-owned media. On the web, we especially rely on **Commondreams.org** for daily news and analysis.

Radio stations in the **Pacifica network** and other local community stations, like our own **KBOO** in Portland, carry many other excellent sources of news about the issues addressed in this book. David Barsamian's **Alternative Radio** features full-length talks and interviews with a number of this book's contributors, and others who address crucial issues, including Vandana Shiva, Sandra Steingraber, Naomi Klein, Arundhati Roy, Michelle Alexander, David Suzuki, and many more. Other radio shows we benefit from listening to include Dennis Bernstein's **Flashpoints**, **Making Contact**—recent shows were on "Scorched Earth: The Legacy of Agent Orange," and "Pesticides on the Playground"—and Dr. Helen Caldicott's **If You Love This Planet**. With passion and precision, Caldicott has been monitoring and speaking out on the impact of nuclear power for as long as we can remember. FAIR—see above—tracks media bias on its regular radio show **CounterSpin**.

Broadcast television is pretty much still a vast wasteland when it comes to honestly addressing the environmental crisis, but there are moments of hope, many provided by democracy's watchdog Bill Moyers—**Moyers & Company**, past shows of **Bill Moyers Journal**, and **NOW with Bill Moyers**. Aljazeera English's **Fault Lines** has also regularly featured shows that bring to life themes we address in this book—although as of our publication date, it appears, sadly, that not all Aljazeera English's content will be available on Aljazeera America. For its part, CNN shrugs that audiences are not interested in the climate crisis, so why bother? Here's CNN Worldwide President Jeff Zucker:

> Climate change is one of those stories that deserves more attention, but that we all talk about and we haven't figured out how to engage the audience in that story in a meaningful way, and that when we do do those stories, there does tend to be a tremendous amount of lack of interest on the audience's part.

Sources of DVDs

For excellent DVDs that can be used with students, **The Video Project** offers important educational media on the environment, science, health, and other global issues. The Video Project's *Forces of Nature, Volume One and Two*, is a wonderful resource that highlights winners of the Brower Youth Awards, which prompt students to think of imaginative ways they can take action on behalf of environmental justice. Another source for films we've used throughout our teaching careers is **California Newsreel**, which focuses on issues of racial justice. California Newsreel distributes *Maquilapolis*, a film we highlight in Chapter Five, and *Unnatural Causes: Is Inequality Making Us Sick?*, about race, class, and health. Still another source of fine films on environmental themes is **Bullfrog Films**, distributor of *A Fierce Green Fire*, about the history of the environmental movement, and *Come Hell or High Water*, described in Hardy Thames' article "Looking for Justice at Turkey Creek," in Chapter Two. The **New Day Films** cooperative distributes *Deep Down*, about the impact of mountaintop removal coal mining, described in "Coal at the Movies" in Chapter Four; and Jim Klein's *Taken for a Ride*, about the oil and auto industries' efforts to buy up and destroy streetcar lines throughout the country. And the **Zinn Education Project** has lists of film recommendations at each time period or theme listed at its website.

A film we first encountered at the International Forum on Globalization teach-in and that we mention in our introduction to this book is *The Story of Stuff*, narrated with biting wit by Annie Leonard. *Stuff* was followed by a series of other useful online films: *The Story of Bottled Water, The Story of Cosmetics, The Story of Electronics, The Story of Broke,* and *The Story of Change,* among others.

A film that deserves its own mention because of how essential it is in teaching about "development" and its impact on Indigenous cultures is Helena Norberg-Hodge's *Ancient Futures: Learning from Ladakh,* based on her fine book of the same name. For decades, Norberg-Hodge has lived on and off in Ladakh, in northern India, and with intimacy and insight she describes life in Ladakh and the impact of modernization. For teaching ideas, including a trial role play, see "Rethinking 'Primitive' Cultures" in *Rethinking Globalization.* Norberg-Hodge founded and directs **Local Futures** (formerly the **International Society for Ecology and Culture**).

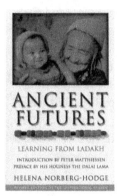

Organizations

It would be impossible to list all the organizations doing vital environmental justice work around the country, and we apologize to those we neglect here. Here, we'll just list some groups we've relied on for guidance in assembling the lessons and articles in the book. We might as well begin with the **International Forum on Globalization** (IFG), sponsor of the "Triple Crisis" teach-in that we describe in the introduction, and which emphasized the interconnections between the issues we seek to teach about. The IFG has monitored the impact of globalization and "neoliberal" reforms on the Global South since it was founded in 1994. They still sponsor teach-ins on key issues—the most recent of these on "Techno-Utopianism and the Fate of the Earth"—as well as reports, like one on the influence of the Koch brothers' money in politics.

The climate crisis is at the heart of this book, and **350.org** is at the heart of organizing for climate sanity—as its name suggests, advocating a return to 350 parts per million of carbon dioxide in the atmosphere, the number that could prevent the catastrophic planetary changes that we are already in the midst of. Launched by the indefatigable Bill McKibben, 350.org has organized direct actions around the country and the world, including the September 2014 "People's Climate March" in New York City. 350.org's website is a resource for teachers, with film suggestions, teaching guides, posters, and more.

A venerable organization, whose materials we've used since the early 1980s is **Food First**, known more formally as the **Institute for Food and Development Policy**. Food First has published books "exposing the root causes of hunger for four decades." Their periodic "Backgrounder" newsletters focus on key food and justice concerns, like global threats to the family farm, the Trans-Pacific Partnership, the Farm Bill, and land grabs around the world. Founded by Frances Moore Lappé and Joseph Collins, whose "10 Myths of Hunger" (p. 361) is included in this book and grows out of Food First work. Food First also sponsors Food Sovereignty Tours, taking travelers around the world to look at examples of food sovereignty struggles. One of these trips to the Basque Country was an inspiration for the role play on La Vía Campesina, included here on page 366.

And, of course, **La Vía Campesina** itself was a vital resource when we began researching and teaching about food sovereignty issues. Their website includes interviews and reports that can be excerpted for classroom use, as well as videos like the one we recommend in the La Vía Campesina role play, *La Vía Campesina in Movement...Food Sovereignty Now!*

The Council of Canadians was perhaps the first organization that alerted us to the dangers of global water privatization, especially through the work of Maude Barlow. Its resources offer educators a quick tutorial on key water issues. Similarly, **Food & Water Watch** addresses issues ranging from bottled water to fracking to genetically engineered foods to factory farms. Food and Water Watch also includes a film lending library, with important films like *Gasland* and *Gasland II, Flow, Blue Gold,* and related fact sheets—along with a link to the delightful short video *The Meatrix,* which could serve as a prompt for students to make their own short films or skits on key environmental justice issues.

The first special issue of *Rethinking Schools* magazine related to environmental justice themes was in the summer of 2006, when we published an issue titled "Feeding the Children: The Politics of Food in Our Schools and Classrooms." An organization that we relied on for several articles as well as for insights into these issues was the **Center for Ecoliteracy**. The center's website features a provocative essay series and resources on "Rethinking School Lunch."

The **Indigenous Environmental Network** is an excellent source for how the issues dealt with in this book ripple through Indigenous communities around the world, and how Indigenous communities are organizing in response. Another organization we have been following since we learned of their important work during the 1999 Seattle World Trade Organization protests is **Focus on the Global South**, a group that has offered crucial analysis and resources for understanding how communities in the Global South view the profound inequality and exploitation at the heart of the global economy. Too often, as teachers plan lessons on Asia, Africa, and Latin America, we fail to search out voices of scholars and activists from the Global South.

When Rethinking Schools discovered that the publisher Scholastic had been hired by the nonprofit arm of the coal industry, the American Coal Foundation, to produce a glossy 4th-grade pro-coal curriculum, we partnered with a number of organizations worth checking out and supporting. These include the **Campaign for a Commercial-Free Childhood**, which monitors how corporations insinuate themselves into the lives of children; the **Center for Biological Diversity**, which exposes the danger of diversity loss and fights against it; the environmental justice organization, **Friends of the Earth**; **Greenpeace**, which has done so much important direct-action work to defend the Earth; and the **Sierra Club**, whose Beyond Coal campaign has helped keep up the pressure against the use of the most climate-damaging fossil fuel.

Physicians for Social Responsibility offers a wealth of research on nuclear energy, climate change, and environmental toxins from medical and public health experts—an important example of the many connections between social and environmental issues.

In our book *Rethinking Globalization,* we include many other organizations that continue to inspire and inform our work. These include **Amazon Watch, Cultural Survival, the David Suzuki Foundation, ETC Group, the Institute for Policy Studies, International Rivers, the Institute for Ag-**riculture and Trade Policy, Oxfam, the Rainforest Action Network**, and the **Third World Network**.

Here in the Northwest, the **Sightline Institute** has done meticulous research on the fossil fuel industry's attempts to turn our region into a corridor for coal and liquefied natural gas exports, and for the trans-shipment of highly flammable oil from the Bakken fields in North Dakota. Other groups whose work here has helped inform articles included in *A People's Curriculum for the Earth* include **Columbia Riverkeeper** and **Rising Tide**, which is part of **Rising Tide North America**—an important network that links social and environmental justice work. All three groups are doing work that deserves to be known more widely.

Americans Against Fracking is a source for information, analysis, and activist resources on fracking. Its advisory board includes Josh Fox, who made the disturbing films *Gasland* and *Gasland II*, and writer and ecologist Sandra Steingraber. The organization maintains an email list monitoring developments around fracking. Sign up to receive these at gaslandthemovie.com.

Books

If it's impossible to list all the organizations doing outstanding work on behalf of a livable planet, it's no easier to recommend books. We'll begin with a few from contributors to this book.

We first heard Vandana Shiva speak in 1999 in Seattle at an International Forum on Globalization teach-in on the World Trade Organization. Her speaking and writing has helped shape our understanding of the problematic character of corporate-driven globalization. Her books include *Biopiracy: The Plunder of Nature and Knowledge; Stolen Harvest: The Hijacking of the Global Food Supply; Staying Alive: Women, Ecology, and Development;* and *Soil, Not Oil: Environmental Justice in an Age of Climate Crisis.* No education is complete without reading Vandana Shiva.

In 2011, we both attended a talk by Christian Parenti about his book *Tropic of Chaos*, which describes the climate change roots of contemporary social struggles—how conflicts that may at first seem like they have "ethnic" roots can be traced to their climate origins. Parenti's book helps puncture the myth that social issues and environmental issues are separate categories.

We open our first chapter with a talk by Van Jones about the link between the disposability of nature and people, and how "environmentalism" needs to address both. Jones' book *The Green Collar Economy* suggests that the environmental crisis can be an opportunity to address economic inequality. Also in our first chapter, we include an excerpt from a book by the eminent ecologist and prolific writer David Suzuki, *The Big Picture: Reflections on Science, Humanity, and a Quickly Changing Planet*. Suzuki has been doing brilliant work on behalf of a livable world for decades. This book is an accessible read for high school students. And in Chapter One we cite—and recommend—the fine book *Down to Earth: Nature's Role in American History*, by Ted Steinberg. Too often history boils down to "the things that people do," as if the Earth is merely humanity's inert stage. Better than any other book we know, *Down to Earth* probes the intersection of people and nature in U.S. history.

Ann Pelo's Chapter Two article, "A Pedagogy for Ecology," is excerpted from a remarkable collec-

tion she edited, *Rethinking Early Childhood Education*. As she suggests in her article, educators can nurture young children "who see themselves as part of a community of people anchored by fierce and determined love of place and who take responsibility for its well-being." Pelo's latest book, *The Goodness of Rain*, chronicles her "love of place" work with one child over the period of a year. C. A. (Chet) Bowers' work has been enormously influential in helping us recognize the way that our culture's root metaphors of progress and freedom of the autonomous individual find their way into the curriculum and help contribute to teaching that can unintentionally be hostile to the Earth. Chet's books are too numerous to list, but they include *Responsive Teaching: An Ecological Approach to Classroom Patterns of Language, Culture, and Thought*

(with David Flinders); *Education, Cultural Myths, and the Ecological Crisis: Toward Deep Changes;* and *Let Them Eat Data: How Computers Affect Education, Cultural Diversity, and the Prospects of Ecological Sustainability*.

Jeff Goodell wrote *Big Coal*, where we first learned of the work of mountaintop removal activist Maria Gunnoe, featured here in Chapter Four. *Big Coal* is the best introduction to the industry that digs out of the Earth the most carbon-intensive and climate-damaging fuel—and to some of the activists who are working to keep the stuff in the ground.

Sandra Steingraber's eloquent words appear throughout this book. We especially recommend two of her fine books, *Living Downstream: An Ecologist's Personal Investigation of Cancer and the Environment* and *Raising Elijah: Protecting Children in an Age of Environmental Crisis*. Steingraber approaches today's "toxic trespass" as a cancer survivor, a mom, a scientist, and an activist. Her books are required reading for those of us trying to figure out how to live in a world that doesn't poison us—they serve as a reminder that we can't remedy our toxic environment solely by making wise consumer choices.

Bill McKibben wrote the first book alerting the world to the dangers of global warming. He's been writing, speaking, and organizing ever since. See his many books, including *The End of Nature, Eaarth,* and *Fight Global Warming Now*. Perhaps begin with *The Bill McKibben Reader*.

Another individual who we've been learning from since his first book on the Vietnam War, *Endless War,* back in the early 1970s, is Michael Klare. Klare has been a keen observer of the link between militarism, empire, and resources. See, for example, his books *Resource Wars: The New Landscape of Global Conflict, Blood and Oil*, and most recently, *The Race for What's Left: The Global Scramble for the World's Last Resources*.

We first encountered Derrick Jensen's writing in his uncompromising *Orion* magazine columns. Jensen's books challenge us all to think more deeply about what we mean and whom we think about when we use the term "sustainability." His books include *Endgame, Deep Green Resistance, The Culture of Make Believe*, and *A Language Older Than Words*. Some of the images we include throughout the book come from Derrick Jensen and Stephanie

McMillan's illustrated book *As the World Burns*. And check out McMillan's hilarious and biting cartoons in books like *Capitalism Must Die!* and *Resistance to Ecocide.*

Elizabeth Royte's short article about the corrupt beginnings of the Keep America Beautiful campaign is excerpted from her book *Garbage Land,* which we first encountered in a book group here. *Garbage Land* offers abundant evidence that there is no ecological way to "dispose" of the items we no longer want.

Several articles in Chapter Five deal with toxic emissions on the Mexican side of the U.S.-Mexico border. A book that offers valuable background on the impact of "free trade" on workers and the broader community is *The Children of NAFTA,* by David Bacon. Also included in Chapter Five is an article by Winona LaDuke, whose writing we included in the Rethinking Schools book *Rethinking Columbus,* first published in 1991. LaDuke has been writing about the intersection of Indigenous rights, environmental racism, and feminism for decades. Her books include *All Our Relations: Native Struggles for Land and Life* and *Recovering the Sacred: The Power of Naming and Claiming.* Her most recent book is *The Militarization of Indian Country.*

It's hard to overstate the impact of Frances Moore Lappé's books, *Diet for a Small Planet* and, with Joseph Collins, *Food First: Beyond the Myth of Scarcity.* Lappé and Collins launched Food First, which has profoundly influenced our teaching, along with that of so many others. Both of us have used Food First materials since our early days in the classroom, 1978 and 2003, respectively. Lappé and Collins collaborated with Peter Rosset on *World Hunger: Twelve Myths,* which is the source for the article included in Chapter Six, trimmed to 10 myths. A new edition, *World Hunger: 10 Myths,* will be published in 2015.

And even though his writing makes only a small appearance in Chapter Six, we should acknowledge the impact of *The Omnivore's Dilemma* by Michael Pollan on our thinking about the politics of food. Pollan's tour through the U.S. food system demands that we rethink our industrialized way of farming and eating. Another insightful critique of the global food system can be found in Raj Patel's *Stuffed and Starved.* This sweeping analysis of the history and politics of our global system tells stories of farmers, consumers, and activists on five continents, arguing that free market capitalism will never equitably feed the world. Patel's sharp critique of capitalism continues in his next book, *The Value of Nothing,* an important source of inspiration for much of the writing on the commons that appears in *A People's Curriculum for the Earth.*

We especially relied on three books when doing background research for the La Vía Campesina role play in Chapter Six. The first, *La Vía Campesina: Globalization and the Power of Peasants,* by Annette Aurelie Desmarais, offers the history of La Vía Campesina in the broader context of corporate-led globalization. A second is the worthwhile reader *Food Sovereignty: Reconnecting Food, Nature, and Community,* edited by Hannah K. Wittman, Annette Aurelie Desmarais, and Nettie Wiebe. And finally, *Campesino a Campesino: Voices from Latin America's Farmer to Farmer Movement for Sustainable Agriculture,* by Eric Holt-Giménez, is filled with the grassroots wisdom of peasants themselves, talking about their vision of what has come to be called food sovereignty.

At the risk of sounding like a Rethinking Schools advertisement, we need to highlight the Rethinking Schools books that align with the vision of curriculum we have articulated in this book. *Rethinking Globalization: Teaching for Justice in an Unjust World* is a companion volume to this one. Although published in 2002, many of the articles and activities included in that book are as relevant as when it was first published, and it offers some of the essential "free trade" context that is essential to grasp—and to teach—when engaging students in learning about the environmental crisis. Other Rethinking Schools books that support the content and pedagogy of this book include *The Line Between Us: Teaching About the Border and Mexican Immigration, Rethinking Columbus, A People's History for the Classroom, Rethinking Mathematics: Teaching Social Justice by the Numbers, Rethinking Popular*

Culture and Media, Rethinking Multicultural Education, and *Teaching About the Wars*. Linda Christensen's two books *Reading, Writing, and Rising Up* and *Teaching for Joy and Justice* show how to blend poetry, story, and personal narrative with social analysis; she demonstrates how to ground this work in the power of students' own lives.

Finally, as we were just completing *A People's Curriculum for the Earth*, we received Naomi Klein's remarkable new book *This Changes Everything: Capitalism vs. the Climate.* Klein wrote the now-classic critique of "disaster capitalism," *The Shock Doctrine.* Her new book argues convincingly and passionately that the climate crisis has been super-heated by the globalized, deregulated capitalism that became nothing less than a religion beginning in the 1980s—and that at this late stage in the crisis, we need to radically change course. Klein poses the question: If there is a scientific consensus about the dire threat posed by continuing to pour greenhouse gases into the atmosphere, why do we continue along this path? She answers:

> We have not done the things to lower emissions because those things fundamentally conflict with deregulated capitalism, the reigning ideology for the entire period we have been struggling to find a way out of this crisis. We are stuck, because the actions that would give us the best chance of averting catastrophe—and would benefit the vast majority—are extremely threatening to an elite minority that has a stranglehold over our economy, our political process, and most of our major media outlets.

This analysis has profound implications for what we must do if we are to avoid nightmare scenarios in the not-so-distant future—a future that will fall most heavily on the very people who have done the least to create this crisis. Every school curriculum should be grounded in the analysis of Klein's book.

Continuing the Work

Rethinking Schools and the **Zinn Education Project** have partnered with **This Changes Everything**—which includes Naomi Klein's book, a forthcoming documentary film (2015) by her husband

Avi Lewis, and a public engagement project—to work with teachers to create more curriculum that explores the deep causes of the climate crisis and engages students in imagining the kind of social changes we need to effect immediately. This work will entail leading writing retreats for teachers doing exemplary work; publishing articles describing this work in *Rethinking Schools* magazine; and collecting and posting additional articles, lessons, and other resources at the Rethinking Schools, Zinn Education Project, and This Changes Everything websites.

And, as we indicate in our Introduction, this work will go on in the pages of *Rethinking Schools* magazine, as well as in the growing number of social justice education conferences around the country, our professional organizations, our unions, department meetings in our schools, PTA meetings, and everywhere people gather to talk about what young people should learn. But because of the power of the corporate-sponsored school reform agenda, we will need to fight for the right to teach about what matters, and to search out allies who share our concerns. A curriculum that prompts students to question the roots of the environmental crisis threatens the elites who have grown rich creat-

ing that crisis—and the school officials who, knowingly or not, follow their directives. We'll need lots of imagination, and we'll need to find new ways to share our work. But we'll also need to be willing to challenge—and perhaps to defy—the authorities who order us to stick to what's in the standards or the approved textbooks. There is too much at stake to be timid. ◉

INDEX

A

Abbey, Edward, 119, 120
Abe, Shinzo, 317, 318
Aboriginal people, uranium mining in lands of, 322–323
Abramovich, Roman, 81, 82, 95
activism. *See* grassroots activism
Adlington, Marianne, 225
African Americans
  cancer risks to, 312
  historic community (Turkey Creek), 61–66
agriculture. *See also* food
  biotech crops, 342–343
  carbon emissions in, 352
  community gardens, 345–347
  community supported (CSA), 334
  corn production, 349–351, 353
  enclosure of commons, 17–20, 22–24, 30
  environmentally sound, 362
  facts on, 374–375
  global warming impact on, 81, 96, 129, 136, 139, 187
  grain production, 14–16, 187, 362
  and land reform, 362
  La Vía Campesina small farmers movement role play, xii–xiii, 330, 331, 362, 365, 366–373, 376–385
  mechanization of, 13–14, 195
  policy shift from supply management, 350–351
  and seed exchange, 344
air pollution. *See also* carbon dioxide (CO2) emissions; greenhouse gas emissions
  and Asian dirty production, 5–6
  and breathing commons, 79–80
  from coal, 227–228
  dumping of toxins, 9, 12
  from fracking, 253, 259
  and particulate matter, 275–279
  and public health, 285, 286
air travel, greenhouse gases in, 101

Alaska
  Anchorage summit on climate change, 128, 131, 133, 142
  Gwich'in people, 73, 74–76, 96
  Indigenous peoples' land rights in, 323
  Yup'ik people, 128, 130, 135, 143–146
Alaskan Native Claims Settlement Act, 323
Alaskan Native Corporations, 323
Aljazeera, 390
Allen, Will, 347
Alternative Radio, 390
Amazon, Indigenous peoples of, 128, 131, 138
Amazon Watch, 392
Ambre Energy, 221, 223–225
America (Prentice Hall), 20, 21
Americans for Prosperity, 105
American Coal Foundation (ACF), 202, 203, 204, 392
American Geophysical Union conference, 277, 278
American Lung Association, 200, 284, 285
American Petroleum Institute, 242, 245
American Plastics Council, 272
Americans Against Fracking, 392
Americans, The (Holt McDougal), 355
America Pathways to the Present (Prentice Hall), 355
Anchorage summit on climate change, 128, 131, 133, 142
Ancient Futures: Learning from Ladakh, 391
Andean Regional Project on Adaptation to Climate Change (PRAA), 126
Anderson, Richard H., 101
anencephaly, 296
animals
  concentrated animal feeding

operations (CAFOs), 350
  and desertification, 225
  endangered, 158
  global warming impact on, 75, 96, 128, 157–160
  naming, 44
  and oil spills, 298–299
  and strip mining, 212–213
  toxins in, 9
  in urban wilderness, 58, 59–60
Arab Spring, 14
Aramayo, Garmán, 126
Archer Daniels Midland, 351
Arctic
  ice melting in, 113, 182, 188
  oil and gas development in, 75, 76, 81, 97, 135, 169, 188, 323
  permafrost melting in, 135
Arctic amplification, 113
Argott, Don, 327
armed forces, carbon dioxide production by, 154–156, 175
Arnowitt, Myron, 259
Arrhenius, Svante, 102, 108
arsenic, 9–10, 295
Asia, air pollution in, 5–6
"Ask Yourself These Questions" (Monbiot), 109
asthma
  and air quality, 275
  and nail salon toxins, 281, 282
  as public health issue, 283, 284–285, 286
Atomic States of America, The, 327
Australia, uranium mining in, 322–323
automobile ownership, and oil consumption, 195
Aymara people, 124–125, 137

B

Bacon, David, 394
Bacon, Francis, 25
Bady, Dianne, 213

Baikitea, Batee, 121
Bali Principles of Climate Justice, 171–173, 176
Balog, James, 174
Bambara people, 128, 129, 131, 136
Barlow, Maude, 391
Barsamian, David, 390
Bartolone, Pauline: "Combating Nail Salon Toxins," xi, 267, 280–282, 324–325
Basque Union of Small Farmers and Ranchers (EHNE), 367, 376
Bass, Rick, 43
bears
    Polar Bear/Global Warming Project, 157–160
    toxin consumption by, 9
"Before Today, I Was Afraid of Trees" (Larkin), 52–56
Beltrán, María Guadalupe, 97
Bennon, Brady
    "Environmental Crime on Trial," 231–237
    "Paradise Lost," 118–123
Bernstein, Dennis, 390
Berry, Wendell, 39, 331
Big Coal (Goodell), 206, 216, 393
Bigelow, Bill
    about, v
    "Climate Change Mixer," 92–101
    "Coal, Chocolate Chip Cookies, and Mountaintop Removal," 200–208
    "Coal at the Movies," 216–219
    "Exporting Coal and Climate Change," 220–225
    "Food, Farming, and Justice," 366–385
    "How My Schooling Taught Me Contempt for the Earth," 36–41
    "Hunger on Trial," 355–360
    "Mystery of the Three Scary Numbers, The," xii, 180–190
    "Teaching the Climate Crisis," 79–91
    "Thingamabob Game, The," 147–153
    "Who's to Blame for the Climate Crisis?" 163–170
Biggers, Jeff, 178
Big Picture, The: Reflections on Science, Humanity, and a Quickly Changing Planet (Suzuki), 393
"Big Talk, The" (Steingraber),

161–162, 176
Bill Moyers' Journal, 63, 390
biomimicry, 6, 29
biosphere, interconnections in, 8–12
birds
    naming, 44
    in urban wilderness, 59
birth defects, 228, 289, 295, 296, 320
Bittman, Mark: "Two Food Systems," 365
"Black Waters" (Ritchie), 178
Blankenship, Don, 210, 214, 215
Blue Skies Campaign, 223
Boggs, Grace Lee, 34, 346, 386
Bolivia, glacier melting in, 124–126, 137, 175
Bonds, Julia (Judy), 213, 219
book resources, 392–395
Book of Yaak, The (Bass), 43
Boston Associates, 22
Botany of Desire, The (Pollan), 356
bottle bills, 271, 272
bottled water vs. tap water, 304–307
Bovine Growth Hormone, in milk, 339–341
Bowen, Jaime, 229, 262
Bowers, C. A., 40, 393
Boyce, Greg, 222
Boyle, Robert, 25, 31
Bragg, Billy, 18–19
Brasch, Walter: "Life and Death in the Frack Zone," 258–259
Brazil, Landless Workers Movement of Brazil (MST), 367, 377
bread
    mechanized production of, 13–14
    price of, 14–16
Britain, and coal technology, 194–195
British Petroleum (BP), 193
    Gulf of Mexico oil spill, 231–237, 238–240
Brower, David, 7
Brower Youth Award, 354, 387
Buehler, Chris, xii
"Food, Farming, and Justice," 366–385
Bull, Peter, 216–217
Bullfrog Films, 390
Bunch, Darius, 69
Burke, Mike, 317–318
Burning the Future, 179, 204, 205, 217
Bush, George W., 213
"butterfly effect," 11–12
Butternut Hollow Pond (Heinz), 11
Butz, Earl, 350–351

C

Caldicott, Helen, 316–317, 325, 390
California
    anti-nuclear activism in, 314–316
    clean air gains in, 5–6
    hidden ecological curriculum in, 37–40
    incarceration rate in, 6
    nail salon regulation in, 281, 282
California Healthy Nail Salon Collaborative, 280, 281, 282
California Newsreel, 390
Campaign for a Commercial-Free Childhood, 392
Campaign for Safe Cosmetics, 281
Campesino a Campesino: Voices from Latin America's Farmer to Farmer Movement for Sustainable Agriculture (Holt-Giménez), 394
Canada
    Keystone XL Pipeline proposal of, 187, 221, 241–246
    uranium mining in, 101
Cancer Alley, ix, 4, 5
Cancer Prevention Institute of California (CPIC), 281
cancer risk
    disparities in, 309–313
    and fracking, 258, 259
    and growth hormone-treated milk, 339, 340
    and nail salon chemicals, 281, 282
    and nuclear contamination, 318
    and plastic toxicity, 4, 5
    and toxic waste, 289, 295
capitalist system
    Klein's critique of, 395
    Thingamabob Game simulation of, xii, 147–150
    trial role play on global warming, xii, 167, 170
    trial role play on oil spill, 234–235, 236, 237
carbon cap, 166
carbon cycle, 111, 114–116
carbon dioxide (CO_2) emissions
    in agriculture, 352
    from coal, 95, 115, 166, 168, 189, 191–192, 200–201, 221
    in developing countries, 168
    565-gigaton limit, 181, 185–186, 187
    historical eras of, 194–197

nonhuman sources of, 114
reduction rate for, 189
rise in levels of, 103–105, 107, 108, 109, 181, 186, 188
and sea level rise, 111
sequestration, 217
Thingamabob Game simulation, 85, 86, 87, 148, 149, 151
from unconventional fuels, 196
in U.S., 166
in warfare, 154–156
carbon footprint, 155, 175
"Carbon Matters" (Dean), 110–117
Carbon Tracker Initiative, 181, 186
Cargill, 16, 351
caribou, and Arctic climate change, 75, 96
Carson, Rachel, 162, 251, 266, 287
Caruso, David, 68
Cavanaugh, Francine, 218–219
Center for Biological Diversity, 392
Center for Ecoliteracy, 392
Centers for Disease Control (CDC), 295, 296
Cerda, Johnson, 132
Cerda, Magdalena, 292
César Chávez Day, 158, 160
Chabon, Sophie, 228–229, 262
Chandler, Lowell, 223
Charter of the Forest, 19
Chasing Ice, 174
Chávez, César, 158, 160
chemical toxins. *See* toxic trespass
chemical wastes, from Mexican border factories, 288–296
chemistry curriculum, maple-sugaring field trip, 52, 54–56
Cheney, Dick, 250
Cheney, Ian, 349, 350, 351
Chevron, 193
Children of NAFTA, The (Bacon), 394
China, carbon dioxide emissions of, 168
Chipko movement, 3, 26
Christensen, Linda, 122
    "Reading Chilpancingo," 267, 288–293
Citizens Energy, 323
civil disobedience, 90–91, 100
"Civilization's Last Chance" (McKibben), 88
Clean Air Act, 103, 250
Clean Water Act, 237, 250

climate change. *See also* carbon dioxide (CO2) emissions; glaciers, melting of; greenhouse gas emission; ice melting
activism, 73, 82, 89–91, 97, 100, 101
Anchorage summit on, 128, 131, 133, 142
children's books about, 162
Copenhagen conference on, 131–132, 185, 192
evidence for, 72, 188
explaining to children, 161–162, 176
and flooding, 110–111
and grain production, 15–16, 187
and Indigenous peoples, 73, 74–77, 96, 124–125, 127–129, 174
and migration, 82, 126, 134
science of, 106–116
and sea level rise, 77–78, 81, 96, 111–112, 181
skeptical attitude toward, 114
and social justice, 80, 82–83, 84, 171–173, 176
and temperature limit, 181, 185–186, 189, 192
and temperature rise, 107–108, 109, 114, 181, 182, 187
timeline for, 102–105, 174–175
climate change curriculum. *See also* fossil fuel curriculum
activist role play, 82, 89–91
carbon footprint calculation, 175
empathy building, 118–120, 129–130
first-person narratives, 143–146
glacier melting in, 174, 175
Indigenous Peoples' Climate Summit role play, 128, 129–142, 176
interdisciplinary collaboration in, xiii
mixer activity, 80–83, 92–101
parental questioning of, 111, 112–113, 114, 115
poetry writing activity, 122–123
Polar Bear Project, 157–160
political context for, xiii–xiv
scientific evidence in, 110–116
storytelling activities, 87–89, 120–121
and textbook bias, 83–84, 91
Thingamabob Game simulation,

84–87, 147–153
Three Scary Numbers activity, 182–190
trial role play, 163–170
"Climate Change in Kwigillingok" (McClanahan), 143–146
"Climate Change Mixer" (Bigelow), 92–101
"Climate Change Timeline," 102–105
Climate Hot Map, 113
Cloues, Rachel: "Polar Bears on Mission Street," 73, 157–160
CNN, 390
coal. *See also* mountaintop removal mining
    alternatives to, 216–217
    Asian export plan, 189, 220–225
    carbon emissions from, 95, 99, 115, 166, 168, 189, 192–193, 200–201, 221
    films about, 216–219, 224
    in global warming trial role play, 169
    and hidden curriculum, 200–201, 220–221
    and industrialization, 194–195, 218
    mercury emissions from, 227–228, 262
    methane emissions from, 109
    reserves, 186, 192, 193, 225
    teaching ideas, 201–209, 221–225, 262
"Coal, Chocolate Chip Cookies, and Mountaintop Removal" (Bigelow), 200–208
Coal Mountain Elementary, 202, 204
"Coal at the Movies" (Bigelow), 216–219
Coal River Mountain Watch, 198, 213, 217–218
colleges and universities, fossil fuel divestment campaign in, 184, 260
Collins, Joseph, 391, 394
    "10 Myths About Hunger," 361–364, 387
colonial America
    dispossession of Native Americans, 20–21, 22, 30
    fish runs in, 21–22
colonialism, and Irish Potato Famine, 355–360
Columbia Riverkeeper, 392
Columbia Slough, 57, 58–60
Columbus, Christopher, 20

Colundalur, Nash: "A Deadly Drought," 15
"Combating Nail Salon Toxins" (Bartolone), xi, 267, 280–282, 324–325
Combs, Ken, 64
Come Hell or High Water: The Battle for Turkey Creek, 63, 65
Commondreams.org, 390
commons
    atmospheric, 79–80
    enclosure of, 17–20, 22–24, 30, 31
"Commons, The" (Rovics), 24
community supported agriculture (CSA), 334
concentrated animal feeding operations (CAFOs), 350
ConocoPhillips, 193
consumer culture, in trial role play on global warming, 167
Cookie Mining game, 202–204, 207
Copenhagen Climate Accord, 181, 185
Copenhagen climate summit, 131–132, 185, 192
Corbett, Tom, 259
corn prices, 14–15
corn production, 349–351, 353
corporations. *See also* capitalist system
    Arctic oil and gas development by, 75, 76, 81, 97, 135, 169, 188
    food system control by, 331, 335–336, 338–344
    influence on FDA, 339
    and Keep America Beautiful campaign, 271–272
    toxic wastes from Mexican border factories of, 288–296
Corporation, The, 340
cosmetics industry, FDA regulation of, 281
CO2 emissions. *See* carbon dioxide (CO2) emissions
Council of Canadians, 391
CounterSpin radio show, 390
Crist, Eileen, 162
Crowder, Jill, 68
"Crying Indian" advertising campaign, 267, 271, 324
Cultural Survival, 392
Curole, Margaret and Kevin, 233, 238–240, 262
curriculum. *See* climate change curriculum; environmental education; films; food curriculum; fossil fuel curriculum; resources; science and math curriculum; water curriculum

D

"Deadly Drought, A" (Colundalur), 15
Dean, Jana: "Carbon Matters," 110–117
death-of-nature idea, 3, 25
Deep Down, 217
deforestation, 26, 57, 114, 115, 362
Democracy Now!, 90, 317, 326, 390
desertification, 3, 189, 225
Desert Solitaire (Abbey), 119
Desmarais, Annette Aurelie, 394
Dessalines Brigade, Haiti, 367, 378
"developing" countries, in trial role play on global warming, 168
Devi, Bachni, 26
diabetes, 283, 284, 286
dibutyl phthalate, 281
Diet for a Small Planet (Lappé), 394
Diggers, 17, 18–19, 23, 345
Diné (Navajo) people role play, 128, 129–130, 131, 139
Dirty Business, 216–217, 224
"Dirty Oil and Shovel-Ready Jobs" (Mac Phail), 241–246
disease. *See* health impacts; *specific diseases*
disposability, 6
"Divesting from Fossil Fuels," 260
divestment campaign, fossil fuel, 184, 260
"Don't Take Our Voices Away" (O'Neill and Swinehart), 127–142
Down to Earth: Nature's Role in American History (Steinberg), 21, 393
drinking water
    vs. bottled water, 304–307
    contamination of, 198, 205, 211, 214, 217
drought, 15–16, 331
Duran, Carmen, 293
DVDs, sources for, 390–391

E

Earth in Crisis curriculum workgroup, ix, xi, 80, 118, 128
earthquakes, 247
Earth Summit, 171
East Bay Academy for Young Scientists (EBAYS), 274
ecological identity, 35, 42, 43
ecological illiteracy, 20, 21
ecologically responsible curriculum, 40–41
economic choices, inequality in, 5
economic development, 37, 63–65
Egypt, food prices in, 14–15
Electricity Fairy, The, 218
Eli Lilly and Company, 339
Ellis, Curt, 349, 350, 351
Ellis-Lamkins, Phaedra, 282
Emanuel, Kerry, 185
empathy building, 118–120, 129–130
enclosure of commons, 17–20, 22–24, 30, 31
"Enclosure of the Commons, The" (Shiva), 20
Endangered Species Act, 158
End of Oil, The (Roberts), 195
Enduring Vision, The: A History of the American People (Houghton Mifflin), 355
energy consumption, 269
energy policy, U.S., 166
energy production. *See also* coal; natural gas; nuclear power; oil and air quality, in science curriculum, 273–279
green power, 323
Engler, Mark, 198
English language learners (ELLs), 48–49
Enomenga, Moi, 98
"Environmental Crime on Trial" (Bennon), 231–237
environmental education. *See also* climate change curriculum; films; food curriculum; fossil fuel curriculum; science and math curriculum; water curriculum
    commons concept in, 18–20, 23–24, 79–80
    community-building activities, 34
    ecologically responsible, 40–41
    for English language learners (ELLs), 48–49
    with FOSS (Full Option Science System), 49, 51
    goals of, 53–54

and hidden curriculum, 36–40
history curriculum related to, 61–66
and indoor bias, 38
natural history in, 20–23
and partisan teaching, x
place-based, 35, 42–47, 58, 146
public health problems in, 283–287
Rethinking Schools Earth in Crisis curriculum, ix, xi, 118, 128
science curriculum related to, 52, 54–56
teach-in, xii
toxic site cleanup campaign, 67–70
toxic waste, from Mexican border factories, 289–293
in urban environment, 57–60
in urban schools, 53, 56
Environmental Health Coalition, 295, 296
environmental justice
and air quality, 275–276
and climate change, 80, 82–83, 84, 171–173, 176
and low-income communities, 62–66
principles of, 27–28, 31
in science curriculum, 54, 274
Environmental Protection Agency (EPA), 228, 255–257, 277, 282, 294
Environmental Science & Technology, 9
Environmental Working Group, 304
Epstein, Samuel, 339, 340
Erin Brockovich, 54
erosion, 100
estrogen, in wastewater, 10–11
ETC Group, 392
ethanol, 166
European Union, cosmetics industry regulation by, 281
Evans, Derrick, 62, 64, 65
"Exploring Our Urban Wilderness" (Hansen), 35, 57–60
export crop production, 363
"Exporting Coal and Climate Change" (Bigelow), 220–225
Extra! magazine, 389
ExxonMobil, 169, 187, 192–193
Exxon Valdez oil spill, 232, 301–302

F

"Facing Cancer" (Lindahl), 309–313

famine. *See* hunger
"Famine" (O'Connor), 358
"Farewell, Sweet Ice" (Gilbert), 73, 74–76, 174
farming. *See* agriculture
Fault Lines radio show, 390
films
   Ancient Futures: Learning from Ladakh, 391
   Atomic States of America, The, 327
   Burning the Future, 179, 205, 217
   Chasing Ice, 174
   Come Hell or High Water: The Battle for Turkey Creek, 63, 65
   Corporation, The, 340
   Deep Down, 217
   Dirty Business, 216–217, 224
   Electricity Fairy, The, 218
   Forces of Nature: Stories from the Brower Youth Awards, 387
   Gasland and Gasland II, 202, 205, 249–251, 391, 392
   Into Eternity, 327
   King Corn, 348, 349–351
   Last Mountain, The, 217–218
   La Vía Campesina in Movement . . . Food Sovereignty Now!, 370, 391
   Mountaintop Removal Road Show, The, 218
   Nuclear Savage: The Islands of Secret Project 4.1, 327
   On Coal River, 218–219
   Paradise Lost, 89, 118, 121–122
   sources for, 390–391
   Story of Stuff, The, ix, 390–391
fires, 75, 97, 138
fish
   human hormones in, 10–11
   and mercury poisoning, 226–227, 228, 262
   and ocean turbulence, 12
   in oil spills, 233
   runs, colonial, 21–22
   toxins in, 9
   and water level decline, 76
565 gigatons of carbon limit, 185–186
Flashpoints radio show, 390
Flinders, David, 40
flooding, 16, 98, 100, 179
   and sea level rise, 77–78, 110, 111–112
   from strip mining, 198, 210–215
   from timbering, 212
Florida, sea level rise in, 77–78

Fluke 983 machine, 275
Focus on the Global South, 392
Food. *See also* agriculture
   facts on, 374–375
   global trade in, 363
   industrial food system, 331, 335–336, 352
   inequality and poverty, 356, 361–363
   locally sourced, 336
   organic, 334–335, 336, 341
   prices, 14–16, 331
   school lunch program, 332–333, 336, 351–352
   toxins in, 9–10
   Youth Food Bill of Rights, 354, 386–387
food curriculum
   corn production in, 349–351, 353
   fact sheet, 374–375
   food choices in, 336, 338–339, 349, 352–353
   food sourcing in, 334–336, 337
   gardens in, 333, 345, 346, 386
   genetically modified food role play, 341–344
   grocery store field trip, 333–334
   growth hormone-treated milk in, 339–341
   Irish Potato Famine, 355–360
   La Vía Campesina small farmer role play, 331, 366–373, 376–385, 391
   organic farm field trip, 334
   readings/discussions, 351–353
   teaching ideas, 386–387
   USDA classroom visit, 336–337
Food and Drug Administration (FDA)
   and bottled water, 306
   and cosmetics industry, 281
   and genetically engineered milk, 339
   ties to food industry, 339–340
"Food, Farming, and the Earth," 331
"Food, Farming, and Justice" (Bigelow, Buehler, O'Neill, and Swinehart), 366–385
Food First, 391, 394
Food First: Beyond the Myth of Food Scarcity (Lappé and Collins), 394
"Food Secrets" (Thacker), 331, 332–337
Food Sovereignty: Reconnecting Food, Nature, and Community, 394

Food Sovereignty Tours, 391
Food & Water Watch, 391
Forces of Nature: Stories from the Brower Youth Awards, 387
foreign investment, 383
forest fires, and climate change, 75, 138
forests. *See* trees
For Every Child: The U.N. Convention on Rights of the Child in Words and Pictures, 299
"Forget Shorter Showers" (Jensen), xi, 267, 268–270, 324
formaldehyde, 281
FOSS (Full Option Science System), 49, 51, 114
fossil fuel curriculum
    coal/coal mining activities, 202–207, 221–225
    Keystone XL Pipeline role play, 241–246
    mountaintop removal activities, 261–262
    oil spill trial role play, 231–237
    plastic bag recycling action project, 236–237
    teaching ideas, 261–263
fossil fuels. *See also* coal; natural gas; oil
    college and university divestment campaign, 184, 260
    emissions limits for, 185–186, 187
    energy alternatives to, 249
    and industry regulation, 192–193
    reserves, 181, 186, 192, 193, 225
    teaching ideas, 261
    threat from, 100, 107, 108, 179, 181–182
    three scary numbers about, 182–190
    unconventional, 196–197
Foster, John Bellamy, 40
Fourier, Jean-Baptiste Joseph, 83, 102, 108
Fox, Josh, 250, 392
fracking
    EPA meeting on, 255–257
    film about, 202, 249–251
    health impact of, 258–259
    pollutants from, 196, 250, 253, 255, 259
    resistance to, 252–254
    teaching ideas, 262–263
    wastewater in, 247, 250, 257, 259

"Fracking Democracy" (Steingraber), 255–257, 263
"Fracking . . . Firsthand," 252–254, 262–263
"Fracking of Rachel Carson, The" (Steingraber), 256, 389
Freaks, Geeks, and Cool Kids (Milner), 65
free trade, 363
Friends of the Earth, 392
Fukushima Daichi nuclear meltdown, 99, 316–317, 325–326
Fullerton, John, 186, 192

G

Gaia hypothesis, 12
Gallagher, Thomas, 356
Garbage Land (Royte), 394
gardens
    community, 345–347, 386
    school, 333, 345, 346
Gardner, Howard, 56
Gasland and *Gasland II*, 202, 205, 249–251, 391, 392
gas taxes, 166
Gelfond, Rich, 226–227, 262
genetically modified food, 341–344
Gibson, Larry, 95
Gilbert, Matthew, xi, 96
    "Farewell, Sweet Ice," 73, 74–76, 174
Gilomen, Jen, 217
glaciers, melting of
    Andean, 124–126, 137, 175
    Arctic, 113, 182, 189
    in India, 98, 189
    Mt. Hood, 99, 128–129
    teaching ideas, 174, 175
Glencore Xstrata, 16
Global Climate Coalition, 104
global warming. *See* climate change; climate change curriculum
"Global Warming's Terrifying New Math" (McKibben), 180–181, 184
Gochfeld, Michael, 227
González-Carillo, Selene, "Transparency of Water," 303–308
Gonzalez, Juan, 390
"Goodbye, Miami" (Goodell), 77–78, 174
Goodell, Jeff, ix, 206, 216, 224, 393
    "Goodbye, Miami," 77–78, 174
    "They Can Bury Me in These Hills,

but I Ain't Leavin'," 210–215, 261–262
Goodman, Amy, 90, 127, 390
Goodness of Rain, The (Pelo), 393
Gordon, Deborah, 196
Gore, Al, 268
"Got Milk, Got Patents, Got Profits?" (Swinehart), 338–344
Gould, Stephen Jay, 83
grain production, 14–16, 187, 362
grassroots activism
    climate activism, 73, 97, 100, 101
    climate activism role play, 82, 89–91
    against coal export plan, 223, 224–225
    community garden movement, 345–347
    divestment campaign, college and university, 184, 260
    against fracking, 252–254
    and food system, 331
    in Keystone XL Pipeline role play, 243, 245, 246
    La Vía Campesina small farmer role play, xii–xiii, 330, 331, 366–373, 376–385, 391
    against mountaintop removal mining, 213–215, 217–218, 219
    against nuclear power, 314–315, 317–318, 325
    simple living/individual action as, xi, 268–270, 324
    stories about, 47
    toxic site cleanup campaign, 67–70
    against toxic wastes from Mexican border factories, 288–289, 293, 296
    against uranium mining, 323
Gray, Nigel, 355
Green Collar Economy, The (Jones), 393
Green, Marley, 218
"Green Revolution," 362
Green Teacher, 389
green technology, 269
Greenbelt Movement, 95
greenhouse effect, 83, 159
greenhouse gas emissions. *See also* carbon dioxide (CO2) emissions; fossil fuels
    and consumer culture, 167
    Copenhagen Climate Accord, 181, 185

and global temperature rise, 138
lingering effects of, 8
methane, 109
in warfare, 156
"Greening for All" (Rein and Ross),
345–347, 386
Greenland, ice melting in, 189
Greenpeace, 82, 97, 392
Griles, Steven, 213
"Grounding Our Teaching," 35
Group of 4 (Haiti), 367, 378
growth hormones, in milk, 339–341
G7 proposals, in La Vía Campesina
role play, 370–371, 383–385
Gulf of Mexico oil spill, 231–237, 238,
241–246
Gulla, Ron: "Fracking . . . Firsthand,"
252–254, 262–263
Gunnoe, Maria, xi, 179, 210–215, 217,
218, 261, 393
Gwich'in people, 73, 74–76, 96, 174,
323

H

Haag, Scott, 69
Haiti, Dessalines Brigade, 367, 378
Haney, Bill, 217–218
Hansell, Tom, 218
Hansen, James, 83, 88, 100, 103, 185,
187, 221, 244
Hansen, Mark: "Exploring Our Urban
Wilderness," 35, 57–60
Hathaway Scholarship, 248
hazardous wastes, from Mexican
border factories, 288–296
health care access, 313
health impacts
of fracking, 258–259
of growth hormone-treated milk,
339, 340
of nail salon products, 281, 282
of particulate matter in air,
275–276
from plastic toxicity, 5
and public health problems,
283–287
of radioactive contamination, 318,
320, 326
from toxic wastes, 289, 295–296
from uranium mining, 321–322
Hecker, Jim, 217
Heinz, Brian, 11
Henderson, Amanda, 132

Hen's Teeth and Horse's Toes (Gould),
83
Hernandez, Rafael, 97
Hewes, Billy, 65
hidden ecological curriculum, 36–40
Hightower, Jane, 228–229
Himalayas
deforestation of, 26
glacier melting in, 189
Hiroshima nuclear explosion,
319–320, 326
history curriculum
environmental issues in, 61–66
Irish potato famine in, 355–360
nature's role in, 18–23
Holt-Giménez, Eric, 394
Holt McDougal, *Modern World
History,* 19, 83–84, 91
Homes-Hines, Ella, 62
Horowitz, Adam Jonas, 327
"How to Be a Climate Hero"
(Schulman), 86
"How My Schooling Taught Me
Contempt for the Earth" (Bigelow),
36–41
Huffington Post, 184
hunger
facts on, 374
Irish Potato Famine, 355–360
myths about, 361–363, 387
and patterns of inequality, 356
"Hunger on Trial" (Bigelow), 355–360
hurricanes, 77–78, 110, 111
hydraulic fracturing. *See* fracking
Hydro Resources Inc., 139

I

ice melting. *See also* glaciers, melting
of
Arctic, 113, 182, 188, 189
permafrost, 75–76, 135, 144, 189
If You Love This Planet radio show,
390
Ignatiev, Noel, 357
I Love Mountains, 261–262
incarceration rates, 6–7, 29
Inconvenient Truth, An, 268
India
carbon dioxide emissions of, 168
glacier melting in, 98, 189
Tamil Nadu Women's Collective,
367, 369, 381
Indigenous Environmental

Network, 392
Indigenous peoples
climate change global summit
(Anchorage), 128, 131, 133
climate change impact on, 73,
74–77, 96, 124–125, 127–129,
174
in climate summit role play, 128,
129–142, 176
coal exports opposed by, 223
"Crying Indian" advertising
campaign, 271, 324
dispossession of, 20–21, 22, 30, 101
green energy projects of, 323
in hidden ecological curriculum,
38–40
in Keystone XL Pipeline role play,
243, 245
and uranium mining, 321–323, 326
individual action, xi, 268–270, 324
industrialization
and coal technology, 102, 194–195,
218
and water resources, 21–23
IndyKids, 389
insects
as classroom guest, 48, 49–51
naming, 44
stories of, 46
tree-infecting beetles, 114
Institute for Agriculture and Trade
Policy, 392
Institute for Food and Development
Policy (Food First), 391, 394
Institute for Policy Studies, 392
"Insult to the Moon, An" (Reece), 209
interconnectedness
in biosphere, 8–12, 29–30
as central idea, 3
in ecologically responsible
curriculum, 40–41
of food system, 13–16, 331
and industrialization, 21–23
and inequality, 3, 5–7
and technology, 10
"Interconnectedness—the Food Web"
(Lyman), 11–12
Intergovernmental Panel on Climate
Change (IPCC), xii, 103–105 107,
108, 179
internal combustion engine (ICE),
195
International Climate Justice
Network, 171

International Energy Agency (IEA), 186, 196, 220
International Forum on Globalization (IFG), ix, 391
International Rivers, 392
In These Times magazine, 389
Iraq war, carbon dioxide production in, 155–156
Irish in America, The, 357
Irish Potato Famine, 355–360
Iskander, Mona, 121

J

James, Sarah, 76
Japan, nuclear meltdown in, 99, 316–317
Jensen, Derrick, 389, 393–394
  "Forget Shorter Showers," xi, 267, 268–270, 324
Johansen, Bruce E.: "Remember the Carbon Footprint of War," 154–156, 175
Johnson, Rose, 63
Johnson v. M'Intosh, 30
Jones, Van, ix, 236, 393
  "Plastics and Poverty," 3, 4–7
Journal of Marine Research, 12
Joyce, Sheena M., 327

K

Kahn, Jose, 289, 292–293, 294, 295
Kakadu National Park, uranium mining in, 322–323
Kameya, Yukiko, 317–318, 325
Katona, Jacqui, 323
"'Kazue, Alive!' Hiroshima's Nuclear Refugees" (Miura), 319–320, 326
Keeling, Charles David, 103, 106–107, 108, 183
Keeling curve, 103, 107–108, 115, 117, 183, 188
"Keep America Beautiful?" (Royte), 267, 271–272, 324
Kennedy, Robert F., Jr., 218
Kenya, droughts in, 15
Kerr-McGee Corporation, 321
Kerry, John, 189
Keystone XL Pipeline, 105, 187, 221, 241–246
King Corn (film), 348, 349–351, 353
"King Corn" (Swinehart), 348–353
King, Martin Luther, Jr., 2, 6

Kingsnorth, Paul, 339–340
Kingsolver, Barbara, 333
Kiribati
  in climate summit role play, 128, 134
  documentary film about, 89, 118, 121–122
  poetry writing about, 122–123
  and sea level rise, 89, 112, 134
Klare, Michael T., ix, 393
  "A Short History of the Three Ages of Carbon—and the Dangers Ahead," 194–197, 261
Kleeb, Jane, 245
Klein, Naomi, xiii, xiv, 225, 245, 395
Kopp, Kevin, 52, 54–55
Korean Women's Peasant Association, 367, 369, 372, 380
Kormann, Carolyn: "Retreat of Andean Glaciers Foretells Global Water Woes," 124–126, 175
Korten, David, 150
Kraft Foods, 335, 337
krill, and ocean turbulence, 12
Kunzig, Robert, "Proof Positive," 106–108
Kwigillingok (Alaska), first-person narratives from, 143–146
Kyoto Protocol, 104, 105, 166

L

LaDuke, Winona, 389, 394
  "Uranium Mining, Native Resistance, and the Greener Path," xi, 267, 321–323, 326
lakes
  permafrost melting in, 135
  water level of, 75–76
Lampkin, Morgan, 69, 70
landfills, 9
Landless Workers Movement of Brazil (MST), 367, 377
landownership, Native American concept of, 20–21
land reform, 362
Lanphear, Nancy, 228
Lappé, Frances Moore, ix, 391, 394
  "10 Myths About Hunger," 361–364, 387
Lara, José, 259
Larkin, Doug: "Before Today, I Was Afraid of Trees," 52–56
Last Child in the Woods (Louv), 47

Last Mountain, The, 217–218
La Vía Campesina: Globalization and the Power of Peasants (Desmarais), 394
La Vía Campesina movement role play, xii–xiii, 330, 331, 366–373, 376–385, 391
La Vía Campesina in Movement . . . Food Sovereignty Now!, 370, 391
lead poisoning, 283, 284, 285, 286, 287, 289, 295
leaves, color change in, 45–46
Le, Lam, 282, 325
Lee, Barbara, 282, 325
Leonard, Annie, ix, 304–305, 308, 325, 390–391
"Lessons from a Garden Spider" (Lyman), 48–51
Lewis, Avi, 395
"Life and Death in the Frack Zone" (Brasch), 258–259
Lindahl, Amy, 232
  "Facing Cancer," 309–313
Liu Guoyue, 222
livestock, corn-fed, 350
Local Futures, 391
Loken, Chris, 81, 96
"Looking for Justice at Turkey Creek" (Thames), 61–66
Louis-Rosenberg, Mat, 198
Louv, Richard, 47, 53, 56
Lovejoy, Thomas, 185
Lovelace, Robert, 101
Lovelock, James, 12
low-income people
  and air quality, 275–276
  and cancer risk, 309–313
  climate change impact on, 127–128
  and economic choice, 5
  and economic development, 62–66
  in high school social system, 65
  incarceration rates for, 6–7, 29
  and public health problems, 283–287
Ludlow Massacre of 1914, 200
Lujan, Lourdes, 267, 288–289, 291, 292, 293
Lyman, Kate
  "Interconnectedness—the Food Web," 11–12
  "Lessons from a Garden Spider," 48–51

M

Maathai, Wangari, 80–81, 95

Mackay, Betsy, 19, 30

Mackenzie, Ueantabo, 89, 118, 119, 120, 121

Mac Phail, Abby: "Dirty Oil and Shovel-Ready Jobs," 241–246

Madsen, Michael, 327

magazines, resources for, 389

Magna Carta, 19

Mahan, Ella, 63, 65

Major Economies Forum, 192

Making Contact radio show, 390

Manning, Richard, 333

maple-sugaring field trip, 52, 54–56

Maquilapolis (City of Factories), 293

Margarula, Yvonne, 323

Marks-Block, Tony: "Science for the People," 267, 273–279

Marshall, John, 30

martens, and Arctic climate change, 75

Martin, Jacqueline Briggs, 298

Martinez, Mariana, 292

Martinez, Wenceslao, 295–296

Massey Energy Co., 210, 217

Matewan, 200

"Matter of Degrees, A" (McKibben), xii, 191–193

Maumoon, Mohamed Axam, 127

May, Beverly, 217

May, William F., 272

McCarthy, James, 169

McClanahan, Lauren G.: "Climate Change in Kwigillingok," 143–146

McKibben, Bill, ix, 80, 88, 105, 123, 180–181, 182, 184, 244–245, 389, 393

"A Matter of Degrees," xii, 191–193

McMillan, Stephanie, 393–394

"Measuring Water with Justice" (Peterson), 297–302

Medina, Lydia, 160

Medina, Rita C., 160

Merchant, Carolyn, 25

mercury poisoning, 226–229, 262

Merkel, Angela, 192

Merson, Martha: "Transparency of Water," 303–308

Mesa, Bienvenida, 345

Metales y Derivados, toxic wastes from, 289–293, 294–295

methane emissions, 109, 111, 255

methylmercury, 228

Mexican border factories, toxic waste from, 288–293

Miami Beach (Florida), and sea level change, 77–78

migration, and climate change, 82, 126, 134

milk

genetically engineered growth hormone in, 339–341

healthy image of, 338

locally sourced, 341

Miller, Larry: "Students Blow the Whistle on Toxic Oil Contamination," 67–70

mill-power concept, 22

Milner, Murray, Jr., 65

Milwaukee Journal, 69

Miura, Kazue: "Kazue, Alive!," 319–320, 326

Modern World History (Holt McDougal), ix, 19, 83–84, 91, 387

"Molly Craig" (Christensen), 122

Monbiot, George: "Ask Yourself These Questions," 109

Monoculture, 25

Monsanto, 339–340

Monthly Review, 389

Moore, Kathleen Dean, 47

Moore Oil Company, 67, 69, 70

Morrow Pacific Project, 223

Morse, Reilly, 63

Moseley, Clark A., 223

mountaintop removal mining

and drinking water contamination, 198, 205, 211, 214, 217

films about, 217–219

and flooding, 210–212

impact on environment, 198, 201, 208, 209, 210–211, 216

resistance to, 213–215, 217–218, 219

and slurry impoundment ponds, 211

teaching ideas, 201–209, 261–262

Mountaintop Removal Road Show, The, 218

movies. *See* films

Moyers, Bill, 267, 390

Mozambique, National Peasants Union of Mozambique (UNAC), 367, 379

Mubarak, Hosni, 16

multicultural education, 53

municipal waste production, 269

"Mystery of the Three Scary Numbers, The" (Bigelow), xii, 180–190

N

nail salon products, toxicity of, xi, 280–282, 324–325

naming flora and fauna, 44

National Environmental Policy Act, 237

National Geographic magazine, 83, 189, 389

National Healthy Nail & Beauty Salon Alliance, 281

National Institute of Environmental Health Sciences, 285

National Institute for Occupational Health and Safety (NIOSH), 259

National Oceanic and Atmospheric Administration (NOAA), 187, 298

National Peasants Union of Mozambique (UNAC), 367, 379

Nation, The, 389

Native peoples. *See* Indigenous peoples

natural gas. *See also* fracking

Arctic development of, 75, 76, 95

Gasland films, 202, 205, 249–251, 391, 392

reserves, 186, 192, 225

unconventional sources of, 196–197

and workplace safety, 258, 259

in Wyoming, 248–249, 250

"Natural Gas and Fracking" (Steingraber), 247

naturalist intelligence, 56

Natural Resources Defense Council (NRDC), "Polar Bear SOS" campaign of, 158, 160

nature. *See also* interconnectedness

at creation, 7

death of (Terra Nullius), 3, 25

and ecological illiteracy, 20, 21

enclosure of, 17–20, 22–24, 30

and environmental justice, 27–28, 31

Gaia hypothesis of, 12

historical role of, 20

and hunger, 361

naive knowledge of, 37

teaching ideas, 29–31

nature deficit disorder, 53, 56

Navajo (Diné)
  climate summit role play, 128,
    129–130, 131, 139
  uranium miners, 321–324, 326
New Day Films, 390
New Internationalist, The, 389
Nguyen, Connie, 282
Norberg-Hodge, Helena, 391
Norman, Jack, 69
North American Development Bank,
  295
North American Free Trade
  Agreement (NAFTA), 295, 296, 363
Novack, David, 217
Nowak, Mark, 202
nuclear explosion, at Hiroshima,
  319–320, 325–326, 326
nuclear power
  anti-nuclear activism, 314–315,
    317–318
  as clean energy alternative, 99, 325,
    327
  films on, 327
  Fukushima Daiichi meltdown, 99,
    101, 316–317
  teaching ideas, 325–326
  and uranium mining, 101
Nuclear Regulatory Commission, 322
*Nuclear Savage: The Islands of Secret
  Project 4.1*, 327
nuclear testing, 327
nutrition education. *See* food
curriculum

O

Obama, Barack, 131, 192
obesity, 283–284, 286
Occupational Safety and Health
  Administration (OSHA), 259, 281
ocean turbulence, 12
O'Connor, Sinead, 357, 358
Ohio Valley Environmental Coalition,
  213
oil
  and agricultural mechanization,
    13–14, 195, 352
  Arctic exploration and drilling, 75,
    76, 81, 95, 97, 169, 182–183, 188
  foreign sources of, 195, 196
  in global warming trial role play,
    169, 195
  pipeline, 100, 187, 221, 241–246
  and plastics production, 4

prices, 16
reserves, 186, 187, 192–193, 225
spills, 4, 69, 98, 231–237,
  238–240, 298–299, 301–302
unconventional sources of,
  196–197
and warfare mechanization, 195
Oil Change International, 156
Oliver, Mary, 47
Omnivore's Dilemma, The (Pollan),
  394
On Coal River, 218–219
O'Neill, Julie Treick, xii, 201, 202, 203,
  204, 293, 348, 351
  "Don't Take Our Voices Away,"
    127–142
  "Food, Farming, and Justice,"
    366–385
  "Our Dignity Can Defeat Anyone,"
    205
  "Teaching About Fracking,"
    248–251
Opland-Dobs, Danah: "Students
  Blow the Whistle on Toxic Oil
  Contamination," 67–70
Oregon, and coal export plan,
  220–225
Oregon Writing Project, 128, 232
organic food, 334–335, 336, 341
organizations, resource, 391–392
Orion magazine, 389, 393
Orr, David, 35
"Our Dignity Can Defeat Anyone"
  (O'Neill), 205
Our Schools/Our Selves magazine,
  389
"Our Winters Are Getting Warmer"
  (Carson), 162
"Outrageous Hope" (Pace), 314–315,
  325
overpopulation, 361–362, 387
Oxfam, 392

P

Pace, Gary, 389
  "Outrageous Hope," 314–315, 325
Pacifica network, 390
Paddy's Lament (Gallagher), 356, 357
"Paradise Lost" (Bennon), 118–123
Paradise Lost (PBS NOW
  documentary), 89, 118, 121–122
Parenti, Christian, 184, 392–393
  "Reading the World in a Loaf of

Bread," 3, 13–16, 30, 331
particulate matter, and air quality,
  275–279
partisan teaching, x
Patel, Raj, 19, 23, 356, 394
Paz, Oscar, 175
PBS NOW documentary, *Paradise
  Lost*, 118, 121–122
Peabody Energy, 193, 201
peasant farmers movement (La Vía
  Campesina role play), 330, 331,
  365, 366–373, 376–385, 391
"Pedagogy for Ecology, A" (Pelo), 35,
  42–47
Pelo, Ann, 393
  "A Pedagogy for Ecology," 35,
    42–47, 393
Pennsylvania, fracking in, 252–254,
  256, 262–263
People's History of the United States, A
  (Zinn), 20–21
permafrost, melting of, 75–76, 135,
  144, 189
Perry, Rick, 245
pesticides, 10, 239, 362, 374
Peterson, Bob: "Measuring Water
  with Justice," 297–302
petrochemical industry, and plastics
  production, 4
Pew Research Center, 73
Philpott, Tom, 351
photosynthesis, 114, 115
Physical Science: Concepts in Action,
  91
Physicians for Social Responsibility,
  392
Piggee, Samantha, 69
*Pine Island Paradox: Making
  Connections in a Disconnected
  World* (Moore), 47
place-based education, 35, 42–47, 58,
  146
place, power of, 119–120
plants. *See also* trees
  mapping, 58
  names of, 44
plastics
  and recycling, 5, 7, 29, 236–237,
    272
  toxicity of, 4–5
  "Plastics and Poverty" (Jones), 3,
    4–7, 29
Poe, Ted, 245
poetry writing, about global warming,

122–123

"Polar Bears on Mission Street" (Cloues), 73, 157–160

Pollan, Michael, 352, 356, 394

population growth, and hunger, 361–362, 387

Porcupine River Caribou Herd, 75

Portland (Oregon)
climate change issues in, 99, 128–129
food activism in, 352
urban wilderness in, 57–60

Potato Famine, Irish, 355–360

poverty. *See* hunger; low-income people

Powder River Basin Resource Council, 251

Prince William (Rand and Rand), 298

"Principles of Environmental Justice, The," 27–28, 31

private property concept, 20–21, 22, 30

Proceedings of the National Academy of Sciences, 10

Progressive, The, 389

"Proof Positive" (Kunzig), 106–108

property rights, xi–xii

public health, 283–287

Q

Quach, Thu, 281

Quigley, John, 239

R

race, and cancer risk, 312

Race, Poverty, and the Environment journal, 389

radio show resources, 390

Rahman, Anisur, 82, 98

Rahman, Atiq, 83

Rainforest Action Network, 392

rainforests, tropical, 138

Ramirez, Edson, 126

Rand, Gloria and Ted, 298

Ratliff, Terry, 217

"Reading Chilpancingo," (Christensen), 267, 288–293

"Reading the World in a Loaf of Bread" (Parenti), 3, 13–16, 30, 331

Reckoning at Eagle Creek (Biggers), 178

Recombinant Bovine Growth Hormone (rBGH), 339–341

recycling, plastic, 5, 7, 29, 236–237, 272

Reece, Erik, 389
"An Insult to the Moon," 209

Reich, Robert, 237

Rein, Marcy: "Greening for All," 345–347, 386

"Remember the Carbon Footprint of War" (Johansen), 154–156, 175

Report on the Americas (NACLA), 389

Republican Party, in Keystone XL Pipeline role play, 242–243, 245

resources
books, 392–395
DVDs, 390–391
organizations, 391–392
publications, 389
radio/TV, 390

Responsive Teaching (Bowers and Flinders), 40

Rethinking Columbus, 394

Rethinking Early Childhood Education (Pelo), 393

Rethinking Globalization, x, 392, 394

Rethinking Schools books, 394–395

Rethinking Schools Earth in Crisis curriculum workgroup, ix, xi, 80, 118, 128

Rethinking Schools-Global Exchange tour, 288, 297

Rethinking Schools magazine, 389, 391–392, 395

"Retreat of Andean Glaciers Foretells Global Water Woes" (Kormann), 124–126, 175

rice, toxins in, 9–10

Rich, Adrienne, 39

Rising Tide North America, 392

Ritchie, Jean: "Black Waters," 178

rivers
and industrialization, 21–22
level of, 75

Roberts, Paul, 195

Rodriguez Lomeli, Moises, 296

Rogers, Andrea, 223

Rolling Stone, 180, 182, 389

Romney, Mitt, 245

Rooted in Community (RIC), 354

Rose, Matthew, 222

Ross, Clifton: "Greening for All," 345–347, 386

Rosselson, Leon: "The World Turned Upside Down," 17, 18–19

Rosset, Peter, 394

Rovics, David: "The Commons," 24, 31

Royal Dutch Shell, 88

Royte, Elizabeth, 394
"Keep America Beautiful?," 267, 271–272, 324

Rubin, Sally, 217

Russia, wheat production in, 15–16

S

Safe Cosmetics Act, 281–282

Safe Drinking Water Act, 250

Salas, Kelley Dawson: "Teaching About Toxins," 283–287

Salazar, Ken, 186

Sale, Kirkpatrick, 269, 270

Salsedo, Maya, 354, 387

Sanchez, Adam, 183, 221, 223, 224

Saño, Naderev "Yeb," 72

Savant, Kim, 63–64

Sayles, John, 200

Schild, Cindy, 245

Scholastic publishers, 392

school lunch program, 332–333, 336, 351–352

Schulman, Audrey, 86

science and math curriculum. *See also* climate change curriculum; fossil fuel curriculum; water curriculum
and cancer disparities, 309–313
carbon cycle, 111, 114–116
and climate change, 73
energy production/air quality study, 273–279
environmental issues in, 52, 54–56, 68
FOSS (Full Option Science System), 49, 51
"Science for the People" (Marks-Block), 267, 273–279

Scotland, clearances in, 19, 30

sea level rise, 77–78, 81, 96, 111–112, 134, 174, 181

secondhand smoke, 79–80

seed exchange, 344

seed patenting, 342

Seja, Roger, 126

Selvage, Kathy, 218

sensual awareness, in place-based education, 44–45

Shafer, Diane, 214

Shirley, Joe, Jr., 322

Shiva, Vandana, ix, 20, 392
    "Two Views of Nature," 3, 25–26,
    31

"Short History of the Three Ages of
    Carbon—and the Dangers Ahead"
    (Klare), 194–197, 261

shrimping industry, and Asian
    competition, 239

Sierra Club, 392

Sightline Institute, 189, 221, 392

Silberberger, Jamie, 281, 324

Silent Spring (Carson), 251

silicosis, 259

simple living activism, 268–270, 324

"Skibbereen" (O'Connor), 57

Skinner, Eva, 63

Slater, Dashka, "This Much
    Mercury. . . ," 226–230, 262

Slater, Samuel, 21–22

slurry impoundment ponds, 211

Small, Gail, 223

"Smarter Than Your Average Planet"
    (Suzuki), 3, 8–10, 29–30

Smog, 102

Smiskin, Henry, 223

snowfall, decline in, 75

Snow Leopard Conservancy, 80, 98

Solnit, Rebecca, 389

Sopoaga, Enele 81, 96

Spain, Basque Union of Small
    Farmers and Ranchers (EHNE), 367,
    376

spider, as classroom guest, 48, 49–51

Statistics for Action (SfA), 303–304

"Stealing and Selling Nature"
    (Swinehart), 18–23, 30

Steinberg, Ted, 21, 22, 393

Steingraber, Sandra, xi, 267, 389, 392,
    393
    "Big Talk, The," 161–162, 176
    "Fracking Democracy," 255–257,
    263
    "Natural Gas and Fracking," 247

Story of Stuff, The, ix, 390–391

storytelling
    about environmental activism, 46
    in climate change curriculum,
    87–89, 120–121

Stout, Ben, 214, 217, 261

strip mining. *See* mountaintop
    removal mining

"Students Blow the Whistle on Toxic

Oil Contamination" (Miller and
    Opland-Dobs), 67–70

Stuffed and Starved (Patel), 356, 394

Sullivan, Kevin; "A Toxic Legacy on
    the Mexican Border," 267, 294–296

Suzuki, David, 188, 393

"Smarter Than Your Average Planet,"
    3, 8–10, 29–30

Suzuki (David) Foundation, 392

Swinehart, Tim
    about, v
    and climate change curriculum, 80,
    81, 82, 83, 86, 87, 90, 91
    and coal mining simulation, 201,
    202, 203, 204
    "Don't Take Our Voices Away,"
    127–142
    "Food, Farming, and Justice,"
    366–385
    "Got Milk, Got Patents, Got
    Profits?," 338–344
    "King Corn," 348–353
    "Stealing and Selling Nature,"
    18–23, 30
    and three scary numbers activity,
    182–184

T

Tamil Nadu Women's Collective, 367,
    369, 381

Tanaka, Nancy, 99

tariffs, 383

tar sands oil, and Keystone XL
    Pipeline proposal, 221, 241

Taylor, Michael R., 339–340

"Teaching About Fracking" (O'Neill),
    248–251

"Teaching About Toxins" (Salas),
    283–287

"Teaching the Climate Crisis"
    (Bigelow), 79–91

Teaching for Joy and Justice, 122

"Teaching in a Toxic World," 267

Teh, Ian, 202

television resources, 390

temperature, global
    rise in, 107–108, 109, 114, 137, 181,
    182, 187
    two degrees Celsius limit, xii, 181,
    185–186, 189, 192, 225

"10 Myths About Hunger" (Lappé
    and Collins), 361–364, 387

Terra Madre (Mother Earth), 25, 31

Terra Nullius (empty land), 3, 25

test tube rockets, 274

textbook bias
    on climate change, ix, 83–84, 91
    on enclosure of commons, 19
    on Irish Potato Famine, 355–356
    on social movements, 366

textile mills, 21, 22, 194

Thacker, Michi: "Food Secrets," 331,
    332–337

Thames, Hardy: "Looking for Justice
    at Turkey Creek," 61–66

"They Can Bury Me in These Hills,
    but I Ain't Leavin'" (Goodell), 179,
    210–215, 261–262

Thingamabob Game, The, xii, 84–87,
    147–153

"Thingamabob Game, The" (Bigelow),
    147–153

Third World Network, 392

This Changes Everything, 395

*This Changes Everything: Capitalism
    vs. the Climate* (Klein), 395

"This Much Mercury. . ." (Slater),
    226–230, 262

Thomas-Muller, Clayton, 131–132

350.org, 105, 391

"Three Scary Numbers" activity,
    182–184

Tillerson, Rex, 181, 187

Time for Kids magazine, 157

toluene, 281

Tong, Anote, 121, 122, 134

"Toxic Legacy on the Mexican Border,
    A" (Sullivan), 267, 294–296

toxic site cleanup campaign, 67–70

toxic trespass
    and air quality, 275–279
    defined, 267
    and dumping, 9–10, 12, 239
    effectiveness of personal solutions,
    268–270
    and fracking, 258–259
    and government regulations, 281
    from Mexican border factories,
    288–296
    from nail salon products, xi,
    280–282, 324–325
    and nuclear contamination, 99,
    316–320, 325–326
    and public health, 283–287
    in rice production, 9–10
    teaching ideas, 324–326
    and uranium mining, 321–323, 326

Tran, Alisha, 282
TransCanada Keystone XL Pipeline, 187, 221, 241–246
"Transparency of Water" (Gonzalez-Carillo and Merson), 303–308
trees
    beetles infecting, 114
    carbon-absorbing, 116
    color change in leaves, 45–46
    and deforestation, 26, 57, 114, 115, 362
    forest fires, 75
    Greenbelt Movement, 95
    habitat destruction from gas drilling, 256
    maple sugaring, 52, 54–55
    in rainforest, 138
    in urban wilderness, 58, 59
trial role play
    on global warming, xii, 163–170
    on Irish Potato Famine, 358, 359–360
    on oil spill, 231–237, 238
Tritch, Steve, 99
Tropic of Chaos (Parenti), 392–393
Truax, Ann, 289–290
Tunmore, Stephanie, 82, 97
Turkana nomads, 15
Turkey Creek community, 61–66
Tuvalu, 81, 96, 112
Tuvalu: Sea Level Rise in the Pacific, Loss of Land and Culture, 112
2 degrees Celsius limit, xii, 181, 185, 186, 189, 192, 225
"Two Food Systems" (Bittman), 365
2,795 gigatons of carbon reserves, 181, 186
"Two Views of Nature" (Shiva), 3, 25–26, 31
Tyndall, John, 102, 108

U

Uddin, Manowara, 222
unconventional fuels, 196–197
Union of Concerned Scientists, 113
U.N. Framework Convention on Climate Change (UNFCCC), 104
United Nations Copenhagen climate summit, 131–132, 185
United States History, 220
uranium mining, 101, 139, 321–323, 326
"Uranium Mining, Native Resistance,

and the Greener Path" (LaDuke), xi, 267, 321–323
urban environment
    community gardens in, 345–347, 386
    historic African American community, 61–66
    natural world explored in, 57–60
    and "nature deficit disorder," 53, 56
U.S. Department of Agriculture (USDA), 336–337, 350–351
U.S. Environmental Corporation, 70
U.S. government. See also Food and Drug Administration (FDA)
    foreign aid, 363
    in global warming trial role play, 166
    in oil spill trial role play, 234–235, 236, 237
Utne Reader, 389

V

"Valuable Films on Our Toxic Nuclear Legacy," 327
Value of Nothing, The (Patel), 19, 394
Verde Partnership Garden, 345–347
Video Project, The, 390
Vulnerable Planet, The (Foster), 40

W

Waldman, Ayelet, 228–229
Wall Street Journal, 115
Wangchuk, Rinchen, 80, 98
warfare
    carbon dioxide production in, 154–156, 175
    impact of global warming on, 189
    mechanization of, 195
waste disposal
    bottle bills, 271
    in distant landfills, 9
    Keep America Beautiful campaign, 267, 271–272
    municipal, 269
    of toxic materials, from Mexican border factories, 288–296
wastewater, in fracking, 247, 250, 257, 259
water curriculum
    bottled vs. tap water, 304–307
    clean water access in, 299
    costs of water in, 299–300

oil spills in, 298–299, 301–302
    scientific data in, 297–298, 303–304, 307–308
water level
    rivers and lakes, 75–76
    sea, 77–78, 91, 96, 111, 134
water pollution
    from fracking, 196, 250, 253, 259
    mercury, 228
    from mountaintop removal mining, 198, 205, 211, 214, 217
    and oil spills, 4, 69, 98, 231–237, 238–240
    radioactive, 322
    and sea level rise, 134
    as social justice issue, 62–63
    toxic chemical dumping, 9, 98, 239
    from uranium dust, 139
    from wastewater, 10–11, 77
water resources
    depletion of, 15, 126, 137, 269
    in fracking, 247, 250, 257, 259
    glacial melting, 98, 99, 113, 124–126, 137, 175
    and industrialization, 21–22
    permafrost melting, 75–76
watershed, 57, 59–60
water temperature, rise in, 113
Watson, Trisha Kehaulani, 100
Webb, Bo, 218, 219
"We Have the Right. . . ," 354
"We Know What's Goin' On" (Williams), 238–240, 262
Western Washington University, climate change curriculum in, 143–146
West Virginia, mountaintop removal mining in, 210–215, 217
"'We Want to Stop It Now': Fukushima's Nuclear Refugees," 316–318, 325–326
"When the Hunger Was Upon Us" (Gray), 355
White, E. B., 51
White, Flowers, 61
White, Warren, 63
"Whole Thing Is Connected, The," 3
"Who's to Blame for the Climate Crisis" (Bigelow), 163–170
Wiebe, Nettie, 394
wildfires, 97
wildlife. See animals; birds; fish; insects
Wiley, Ed, 218

Williams, Terry Tempest, 233, 389
    "We Know What's Goin' On,"
        238–240, 262
Willow Tree Loon (Martin), 298
Wilson, E. O., 162
wind turbines, 323
Wittman, Hannah K., 394
Women's Voices for the Earth (WVE),
    281
Wood, Adams, 218–219
World Bank, 124, 137
World Forum of Fish Harvesters and
    Fish Workers, 238
World Meteorological Organization,
    109
World Trade Organization (WTO),
    363
    and genetically modified food role
        play, 342–344
"World Turned Upside Down, The"
    (Rosselson), 17, 18–19
World War I, 155
Wyoming
    education funding in, 248
    natural gas production in, 248–249,
        250

X

Xin Hao, 223

Y

Yes! magazine, 389
Youth Food Bill of Rights, 354,
    386–387
Yukon Flats, forest fires in, 75
Yup'ik people, 128, 130, 135, 143–146

Z

Zeyhle, Eberhard, 15
Zimmer-Stucky, Jasmine, 222
Zinn, Howard, x, 20–21, 358
Zinn Education Project, 390, 395
Zoellick, Robert, 15
Zucker, Jeff, 390

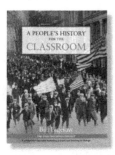
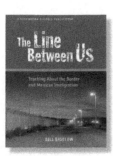

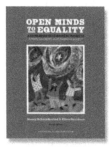

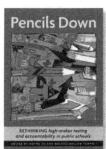